D1691630

Ed Campbell
R. Manter

BAE SYSTEMS
Advanced Technology Cen
Sowerby

CAIRS No.	10688
CLASS No.	P BLA
HELD BY	Hall R
Order No.	

shelf no
519.5

Causation, Prediction, and Search

Adaptive Computation and Machine Learning

Thomas Dietterich, Editor

Christopher Bishop, David Heckerman, Michael Jordan, and Michael Kearns, Associate Editors

Bioinformatics: The Machine Learning Approach
Pierre Baldi and Søren Brunak

Reinforcement Learning: An Introduction
Richard S. Sutton and Andrew G. Barto

Graphical Models for Machine Learning and Digital Communication
Brendan J. Frey

Learning in Graphical Models
Michael I. Jordan

Causation, Prediction, and Search, second edition
Peter Spirtes, Clark Glymour, and Richard Scheines

Causation, Prediction, and Search

Peter Spirtes, Clark Glymour, and Richard Scheines

with additional material by
David Heckerman, Christopher Meek,
Gregory F. Cooper, and Thomas Richardson

The MIT Press
Cambridge, Massachusetts
London, England

©2000 Massachusetts Institute of Technology

All rights reserved. No part of this book may be reproduced in any form by any electronic or mechanical means (including photocopying, recording, or information storage and retrieval) without permission in writing from the publisher.

Printed and bound in the United States of America.

Library of Congress Cataloging-in-Publication Data

Spirtes, Peter.
    Causation, prediction, and search.—2nd ed. / Peter Spirtes, Clark Glymour, and Richard Scheines ; with additional material by David Heckerman, Christopher Meek, Gregory F. Cooper, and Thomas Richardson.
       p. cm. — (Adaptive computation and machine learning)
    Includes bibliographical references and index.
    ISBN 0-262-19440-6 (hc : alk. paper)
    1. Mathematical statistics. I. Glymour, Clark N. II. Scheines, Richard. III. Title. IV. Series.

QA276 .S65 2000
519.—dc21
                                                                                                                                     00-026266

Contents

Preface to the Second Edition ix
Preface xi
Acknowledgments xv
Notational Conventions xvii

1 Introduction and Advertisement 1
2 Formal Preliminaries 5
3 Causation and Prediction: Axioms and Explications 19
4 Statistical Indistinguishability 59
5 Discovery Algorithms for Causally Sufficient Structures 73
6 Discovery Algorithms without Causal Sufficiency 123
7 Prediction 157
8 Regression, Causation, and Prediction 191
9 The Design of Empirical Studies 209
10 The Structure of the Unobserved 253
11 Elaborating Linear Theories with Unmeasured Variables 269
12 Prequels and Sequels 295
13 Proofs of Theorems 377

Notes 475
Glossary 481
References 495
Index 531

To my parents, Morris and Cecile Spirtes—P.S.

In memory of Lucille Lynch Schwartz Watkins Speede Tindall Preston—C. G.

To Martha, for her support and love—R.S.

It is with data affected by numerous causes that Statistics is mainly concerned. Experiment seeks to disentangle a complex of causes by removing all but one of them, or rather by concentrating on the study of one and reducing the others, as far as circumstances permit, to comparatively small residium. Statistics, denied this resource, must accept for analysis data subject to the influence of a host of causes, and must try to discover from the data themselves which causes are the important ones and how much of the observed effect is due to the operation of each.
—G. U. Yule and M. G. Kendall, 1950

The Theory of Estimation discusses the principles upon which observational data may be used to estimate, or to throw light upon the values of theoretical quantities, not known numerically, which enter into our specification of the causal system operating.
—Sir Ronald Fisher, 1956

George Box has [almost] said "The only way to find out what will happen when a complex system is disturbed is to disturb the system, not merely to observe it passively." These words of caution about "natural experiments" are uncomfortably strong. Yet in today's world we see no alternative to accepting them as, if anything, too weak.
—G. Mosteller and J. Tukey, 1977

Causal inference is one of the most important, most subtle, and most neglected of all the problems of Statistics.
—P. Dawid, 1979

Preface to the Second Edition

This second edition of *Causation, Prediction, and Search* is the culmination of almost twenty years of research on automation and causal inference, beginning in 1980 with a chapter of Glymour's *Theory and Evidence*, continuing with our book, *Discovering Causal Structure*, written with Kevin Kelly in 1987, and with an essay in 1990 (Spirtes et al. 1990) which laid out much of the research project we have followed in subsequent years. The thought—which one of us had—that the subject was more or less exhausted in 1993, when the first edition of this book appeared, has been proved entirely wrong.

For this edition we have substituted a new and briefer introduction, a discussion of d-separation that eliminates a misleading didacticism of the first edition, and an entirely new twelfth chapter, surveying and summarizing relevant results and applications since 1993. The original twelfth chapter was chiefly a series of conjectures, most of which have been proved correct, concerning cyclic graphs and feedback systems.

Our first debt for this edition is to our two former students, Chris Meek and Thomas Richardson. Much of the new work we describe is theirs. We are almost equally indebted to Gregory Cooper, David Heckerman, and Larry Wasserman, who have been wonderful, helpful colleagues and collaborators, and to Jaimie Robins, who, though often unhappy with the very idea of this book, helped with his insightfulness and fairness of mind. We have been encouraged by Judea Pearl's support, by his development of ideas presented here, particularly those on prediction first presented in chapter 7 of this book, and by his explorations of a multitude of new aspects of causal inference not considered here. We have been equally encouraged by the ingenious uses and modifications of our procedures provided by a number of scientists, including Bill Shipley, David Bessler and his collaborators, and Ludwig Litzka and his students. We owe a particular thanks to Cooper, Heckerman and Meek for permitting us to use in chapter 12 their survey of Bayesian search methods, and to Thomas Richardson for providing us with information about recent unpublished developments on chain graphs.

Preface

This book is intended for anyone, regardless of discipline, who is interested in the use of statistical methods to help obtain scientific explanations or to predict the outcomes of actions, experiments or policies.

Much of G. Udny Yule's work illustrates a vision of statistics whose goal is to investigate when and how causal influences may be reliably inferred, and their comparative strengths estimated, from statistical samples. Yule's enterprise has been largely replaced by Ronald Fisher's conception, in which there is a fundamental cleavage between experimental and non-experimental inquiry, and statistics is largely unable to aid in causal inference without randomized experimental trials. Every now and then members of the statistical community express misgivings about this turn of events, and, in our view, rightly so. Our work represents a return to something like Yule's conception of the enterprise of theoretical statistics and its potential practical benefits.

If intellectual history in the twentieth century had gone otherwise, there might have been a discipline to which our work belongs. As it happens, there is not. We develop material that belongs to statistics, to computer science, and to philosophy; the combination may not be entirely satisfactory for specialists in any of these subjects. We hope it is nonetheless satisfactory for its purpose. We are not statisticians by training or by association, and perhaps for that reason we tend to look at issues differently, and, from the perspective common in the discipline, no doubt oddly. We are struck by the fact that in the social and behavioral sciences, epidemiology, economics, market research, engineering, and even applied physics, statistical methods are routinely used to justify causal inferences from data not obtained from randomized experiments, and sample statistics are used to predict the effects of policies, manipulations, or experiments. Without these uses the profession of statistics would be a far smaller business. It may not strike many professional statisticians as particularly odd that the discipline thriving from such uses assures its audience that they are unwarranted, but it strikes us as very odd indeed. From our perspective outside the discipline, the most urgent questions about the application of statistics to such ends concern the conditions under which causal inferences and predictions of the effects of manipulations can and cannot reliably be made, and the most urgent need is a principled, rigorous theory with which to address these problems. To judge from the testimony of their books, a good many statisticians think any such theory is impossible. We think the common arguments against the possibility of inferring causes from statistics outside of experimental trials are unsound, and radical separations of the principles of experimental and observational study designs are unwise. Experimental and observational design may not always permit the same inferences, but they are subject to uniform principles.

The theory we develop follows necessarily from assumptions laid down in the statistical community over the last fifteen years. The underlying structure of the theory is essentially axiomatic. We will give two independent axioms on the relation between causal structures and probability distributions and deduce from them features of causal relationships and predictions that can and that cannot be reliably inferred from statistical constraints under a variety of background assumptions. Versions of all of the axioms can be found in papers by Lauritzen, Wermuth, Speed, Pearl, Rubin, Pratt, Schlaifer, and others. In most cases we will develop the theory in terms of probability distributions that can be thought of loosely as propensities that determine long run frequencies, but many of the probability distributions can alternatively be understood as (normative) subjective degrees of belief, and we will occasionally note Bayesian applications. From the axioms there follow a variety of theorems concerning estimation, sampling, latent variable existence and structure, regression, indistinguishability relations, experimental design, prediction, Simpson's paradox, and other topics. Foremost among the "other topics" are the discovery that statistical methods commonly used for causal inference are radically suboptimal, and that there exist asymptotically reliable, computationally efficient search procedures that conjecture causal relationships from the outcomes of statistical decisions made on the basis of sample data. (The procedures we will describe require statistical decisions about the independence of random variables; when we say such a procedure is "asymptotically reliable" we mean it provides correct information if the outcome of each of the requisite statistical decisions is true in the population under study.)

This much of the book is mathematics: where the axioms are accepted, so must the theorems be, including the existence of search procedures. The procedures we describe are applicable to both linear and discrete data and can be feasibly applied to a hundred or more variables so long as the causal relations between the variables are sufficiently sparse and the sample sufficiently large. These procedures have been implemented in a computer program, TETRAD II, which at the time of writing is publicly available.[1]

The theorems concerning the existence and properties of reliable discovery procedures of themselves tell us nothing about the reliabilities of the search procedures in the short run. The methods we describe require an unpredictable sequence of statistical decisions, which we have implemented as hypothesis tests. As is usual in such cases, in small samples the conventional p values of the individual tests may not provide good estimates of type 1 error probabilities for the search methods. We provide the results of extensive tests of various procedures on simulated data using Monte Carlo methods, and these tests give considerable evidence about reliability under the conditions of the simulations. The simulations illustrate an easy method for estimating the probabilities of error for any of the search methods we describe. The book also contains studies of one large pseudoempirical data set—a body of simulated data created by medical researchers to model emergency medicine diagnostic indicators and their causes—and a great many empirical data sets, most of which have been discussed by other authors in the context of specification searches.

A further aim of this work is to show that a proper understanding of the relationship between causality and probability can help to clarify diverse topics in the statistical literature, including the comparative power of experimentation versus observation,

Simpson's paradox, errors in regression models, retrospective versus prospective sampling, the perils of variable selection, and other topics. There are a number of relevant topics we do not consider. They include problems of estimation with discrete latent variables, optimizing statistical decisions, many details of sampling designs, time series, and a full theory of "nonrecursive" causal structures—that is, finite graphical representations of systems with feedback.

Causation, Prediction, and Search is not intended to be a textbook, and it is not fitted out with the associated paraphernalia. There are open problems but no exercises. In a textbook everything ought to be presented as if it were complete and tidy, even if it isn't. We make no such pretenses in this book, and the chapters are rich in unsolved problems and open questions. Textbooks don't usually pause much to argue points of view; we pause quite a lot.

The various theorems in this book often have a graph theoretic character; many of them are long, difficult case arguments of a kind quite unfamiliar in statistics. In order not to interrupt the flow of the discussion we have placed all proofs but one in a chapter at the end of the book. In the few cases where detailed proofs are available in the published literature, we have simply referred the reader to them. Where proofs of important results have not been published or are not readily available we have given the demonstrations in some detail.

The structure of the book is as follows. Chapter 1 concerns the motivation for the book in the context of current statistical practice and advertises some of the results. Chapter 2 introduces the mathematical ideas necessary to the investigation, and chapter 3 gives the formal framework a causal interpretation, lays down the axioms, notes circumstances in which they are likely to fail, and provides a few fundamental theorems. The next two chapters work out the consequences of two of the axioms for some fundamental issues in contexts in which it is known, or assumed, that there are no unmeasured common causes affecting measured variables. In chapter 4 we give graphical characterizations of necessary and sufficient conditions for causal hypotheses to be statistically indistinguishable from one another in each of several senses. In chapter 5 we criticize features of model specification procedures commonly recommended in statistics, and we describe feasible algorithms that from properties of population distributions extract correct information about causal structure, assuming the axioms apply and that no unmeasured common causes are at work. The algorithms are illustrated for a variety of empirical and simulated samples. Chapter 6 extends the analysis of chapter 5 to contexts in which it cannot be assumed that no unmeasured common causes act on measured variables. From both a theoretical and practical perspective, this chapter and the next form the center of the book, but they are especially difficult. Chapter 7 addresses the fundamental issue of predicting the effects of manipulations, policies, or experiments. As an easy corollary, the chapter unifies directed graphical models with Donald Rubin's "counterfactual" framework for analyzing prediction. Chapter 8 applies the results of the preceding chapters to the subject of regression. We argue that even when standard statistical assumptions are satisfied multiple regression is a defective and unreliable way to assess causal influence even in the large sample limit, and various automated regression model specification searches only make matters worse. We show that the algorithms of chapter 6 are more reliable in principle, and we compare the performances of these algorithms against various multiple regression procedures on a variety of simulated and empirical data sets. Chapter 9 considers the design of empirical studies in the light of the results of earlier chapters, including issues of retrospective and prospective sampling, the comparative power of experimental and observational designs, selection of variables, and the design of ethical clinical trials. The chapter concludes with a look back at some aspects of the dispute over smoking and lung cancer. Chapters 10 and 11 further consider the linear case, and analyze algorithms for discovering or elaborating causal relations among measured and unmeasured variables in linear systems. Chapter 12 is a brief consideration of a variety of open questions. Proofs are given in chapter 13.

We have tried to make this work self-contained, but it is admittedly and unavoidably difficult. The reader will be aided by a previous reading of Pearl 1988, Whittaker 1990, or Neopolitan 1990.

Acknowledgments

One source of the ideas in this book is in work we began ten years ago at the University of Pittsburgh. We drew many ideas about causality, statistics, and search from the psychometric, economic, and sociological literature, beginning with Charles Spearman's project at the turn of the century and including the work of Herbert Simon, Hubert Blalock, and Herbert Costner.

We obtained a new perspective on the enterprise from Judea Pearl's *Probabilistic Reasoning in Intelligent Systems*, which appeared the next year. Although not principally concerned with discovery, Pearl's book showed us how to connect conditional independence with causal structure quite generally, and that connection proved essential to establishing general, reliable discovery procedures. We have since profited from correspondence and conversation with Pearl and with Dan Geiger and Thomas Verma, and from several of their papers. Pearl's work drew on the papers of Wermuth (1980), Kiiveri and Speed (1982), Wermuth and Lauritzen (1983), and Kiiveri, Speed, and Carlin (1984), which in the early 1980s had already provided the foundations for a rigorous study of causal inference. Paul Holland introduced one of us to the Rubin framework some years ago, but we only recently realized it's logical connections with directed graphical models. We were further helped by J. Whittaker's (1990) excellent account of the properties of undirected graphical models.

We have learned a great deal from Gregory Cooper at the University of Pittsburgh who provided us with data, comments, Bayesian algorithms and the picture and description of the ALARM network which we consider in several places. Over the years we have learned useful things from Kenneth Bollen. Chris Meek provided essential help in obtaining an important theorem that derives various claims made by Rubin, Pratt, and Schlaifer from axioms on directed graphical models.

Steve Fienberg and several students from Carnegie Mellon's department of statistics joined with us in a seminar on graphical models from which we learned a great deal. We are indebted to him for his openness, intelligence, and helpfulness in our research, and to Elizabeth Slate for guiding us through several papers in the Rubin framework. We are obliged to Nancy Cartwright for her courteous but salient criticism of the approach taken in our previous book and continued here. Her comments prompted our work on parameters in chapter 4. We are indebted to Brian Skyrms for his interest and encouragement over many years, and to Marek Druzdzel for helpful comments and encouragement. We have also been helped by Linda Bouck, Ronald Christensen, Jan Callahan, David Papineau, John Earman, Dan Hausman, Joe Hill, Michael Meyer, Teddy Seidenfeld, Dana Scott, Jay Kadane, Steven Klepper, Herb Simon, Peter Slezak, Steve Sorensen, John Worrall, and Andrea Woody. We are indebted to Ernest Seneca for putting us in contact with Dr. Rick Linthurst, and we are especially grateful to Dr. Linthurst for making his doctoral thesis available to us.

Our work has been supported by many institutions. They, and those who made decisions on their behalf, deserve our thanks. They include Carnegie Mellon University, the National Science Foundation programs in History and Philosophy of Science, in Economics, and in Knowledge and Database Systems, the Office of Naval Research, the Navy Personnel Research and Development Center, the John Simon Guggenheim Memorial Foundation, Susan Chipman, Stanley Collyer, Helen Gigley, Peter Machamer, Steve Sorensen, Teddy Seidenfeld, and Ron Overmann. The Navy Personnel Research and Development Center provided us the benefit of access to a number of challenging data analysis problems from which we have learned a great deal.

Notational Conventions

Text
In the text, each technical term is written in boldface where it is defined.

Variables:	capitalized, and in italics, e.g., X
Values of variables:	lower case, and in italics, e.g., $X = x$
Sets:	capitalized, and in boldface, e.g., **V**
Values of sets of variables:	lower case, and in boldface, e.g., **V** = **v**
Members of **X** that are not members of **Y**:	**X\Y**
Error variables:	ε, δ, e
Independence of **X** and **Y**:	**X** ⊥⊥ **Y**
Independence of **X** and **Y** conditional on **Z**:	**X** ⊥⊥ **Y**|**Z**
X ∪ **Y**:	**XY**
Covariance of X and Y:	$COV(X,Y)$ or γ_{XY}
Correlation of X and Y:	ρ_{XY}
Sample correlation of X and Y:	r_{XY}
Partial Correlation of X and Y, controlling for all members of set **Z**:	$\rho_{XY \cdot \mathbf{Z}}$

In all of the graphs that we consider, the vertices are random variables. Hence we use the terms "variables in a graph" and "vertices in a graph" interchangeably.

Figures
Figure numbers occur just below a figure, starting at 1 within each chapter. Where necessary, we distinguish between measured and unmeasured variables by boxing measured variables and circling unmeasured variables (except for error terms). Variables beginning with e, ε, or δ are understood to be "error," or "disturbance," variables. For example, in the figure below, X and Y are measured, T is not, and ε is an error term.

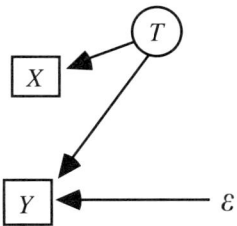

Figure n.1

We will neither box nor circle variables in graphs in which no distinction need be made between measured and unmeasured variables, for example, figure n.2.

Figure n.2

For simplicity, we state and prove our results for probability distributions over discrete random variables. However, under suitable integrability conditions, the results can be easily generalized to continuous distributions that have density functions by replacing the discrete variables by continuous variables, probability distributions by density functions, and summations by integrals.

If a description of a set of variables is a function of a graph G and variables in G, then we make G an optional argument to the function. For example, **Parents**(G,X) denotes the set of variables that are parents of X in graph G; if the context makes clear which graph is being referred to we will simply write **Parents**(X).

If a distribution is defined over a set of random variables **O** then we refer to the distribution as $P(\mathbf{O})$. An equation between distributions over random variables is understood to be true for all values of the random variables for which all of the distributions in the equation are defined. For example if X and Y each take the values 0 or 1 and $P(X = 0) \neq 0$ and $P(X = 1) \neq 0$ then $P(Y|X) = P(Y)$ means $P(Y = 0|X = 0) = P(Y = 0)$, $P(Y = 0|X = 1) = P(Y = 0)$, $P(Y = 1|X = 0) = P(Y = 1)$, and $P(Y = 1|X = 1) = P(Y = 1)$.

We sometimes use a special summation symbol, $\overrightarrow{\sum}$, which has the following properties:

(i) when sets of random variables are written beneath the special summation symbol, it is understood that the summation is to be taken over sets of values of the random variables, not the random variables themselves,

(ii) if a conditional probability distribution appears in the scope of such a summation symbol, the summation is to be taken only over values of the random variables for which the conditional probability distributions are defined,

(iii) if there are no values of the random variables under the special summation symbol for which the conditional probability distributions in the scope of the symbol are defined, then the summation is equal to zero.

Notational Conventions   xxi

For example, suppose that X, Y, and Z can each take on the values 0 or 1. Then if $P(Y=0, Z=0) \ne 0$

$$\sum_{X}^{\rightarrow} P(X|Y=0, Z=0) = P(X=0|Y=0, Z=0) + P(X=1|Y=0, Z=0)$$

However, if $P(Y=0,Z=0) = 0$, then $P(X=0|Y=0,Z=0)$ and $P(X=1|Y=0,Z=0)$ are not defined, so

$$\sum_{X}^{\rightarrow} P(X|Y=0, Z=0) = 0$$

We will adopt the following conventions for empty sets of variables. If $\mathbf{Y} = \varnothing$ then

(i) $P(\mathbf{X}|\mathbf{Y})$ means $P(\mathbf{X})$.
(ii) $\rho_{\mathbf{XZ.Y}}$ means $\rho_{\mathbf{XZ}}$.
(iii) $\mathbf{A} \perp\!\!\!\perp \mathbf{B}|\mathbf{Y}$ means $\mathbf{A} \perp\!\!\!\perp \mathbf{B}$.
(iv) $\mathbf{A} \perp\!\!\!\perp \mathbf{Y}$ is always true.

1 Introduction and Advertisement

1.1 The Issue

Adult judgments about which event is "the cause" of another event are loaded with topicality, interest, background knowledge about normal cases, and moral implications. Tell someone that Suzie was injured in an accident while John was driving her home, and then ask what further information is needed to decide whether John's actions caused her injury. People want to know John's condition, the detailed circumstances of the accident, including the condition of the roadway, of John's car, of the other driver if there was one, and so on (Ahn 1995, 1996). The responses show that in such contexts judgments about causation have a moral aspect, and an aspect that depends on an understanding of normal conditions and deviations from the normal. That sort of thing will vary with culture, background, and circumstance.

Causal claims have a subjunctive complexity—they are associated with claims about what did not happen, or has not happened yet, or what would have happened if some circumstance had been otherwise. If someone says their hair is brown because they dye it, we infer that if they had not dyed their hair it would have been some other color. That sort of counterfactual conditional is not always correct (someone with brown hair can dye their hair brown), and endless but indispensable complexities result. Our moral sense, our very notions of blame and regret, depend on subjunctive aspects of causal claims. In addition, the kinds of entities that are described as causes and effects are enormously varied, and the logical form of causal claims can vary from particular to general to universal. Events are causes—*the rise of the middle class caused the American Revolution; Constantine's conversion caused the triumph of Christianity in the Roman Empire; the discovery of penicillin saved millions of lives.* Features or properties, or their changes, are often cited as causes—*the pH of the liquid caused it to turn pink when phenolthalein was added; the heat caused the butter to melt.* Objects or persons are cited as causes—*my daughter gave me a cold.* Even relationships, or instances of them, can be described as causes—*her love for him caused her to leave the country.* Descriptions of effects can be equally varied. The salient effect of a preventive cause, for example, can be a circumstance or event that doesn't exist—*she prevented the catastrophe.*

The variation—the looseness—of causal claims has provided a reason for many people to dismiss the very idea of causation as prescientific. Bertrand Russell claimed as much, and Karl Pearson proposed to replace the idea of causation entirely by the idea of correlation. To this day, some writers try to avoid the issue by euphemism, as though a new word would clarify things, and at almost any conference of statisticians or social scientists (but not, anymore, of philosophers) there is someone—often not alone—who is eager to say he doesn't "believe in causality." But he acts as if he does; we all do, all the time: we ask people to do things, or do them ourselves, because we want what we think will result from the actions—*turn down the volume on the radio, its too loud*—and we blame people for the unhappy effects of their actions. The skeptic about causality pushes the brake pedal to make his car slow, flips a switch to make a lamp glow, puts his money in the bank to collect interest.

Francis Bacon claimed knowledge is power, and he was talking about the power of control supplied by causal knowledge. One of the greatest mysteries of the human condition is that in a few short years a newborn infant comes to control much of her environment, knows how to climb up to things, how to turn on the television with the remote control, how to make a balloon expand and a soap bubble form, how to summon others and how to avoid them. Developmental psychology has hardly begun to crack how all that causal knowledge, all that power to control, is acquired so quickly. And a great deal—perhaps most—of our scientific inquiry aims to find out something about causal relationships. Billions are spent each year to discover the effects of drug treatments alone, and similar sums to estimate the likely results of possible social and economic policies. Those who claim not to believe in causality may provide consulting services to clients who want to predict the effects of alternative business strategies, or who want to know how to judge the bearing of some body of data on a causal hypothesis. Loose as the notion may be, there is nothing serious to the claim that we can live and thrive without using the idea of causation, however we name it.

The baby and the scientist occupy two ends of the same question: how can observations be turned into causal knowledge, and how can causal knowledge, even if incomplete, be used to influence and control our environment? The theory of experimental design offers a route to causal knowledge, but while Fisher's discussions of the probabilistic and statistical aspects of experimental design were brilliant and rigorous, everything causal was left informal. Fisher did not provide as rigorous a theory regarding causal inference from non-experimental observations. Yet most of what we want to know about, and most of what we think we know, is not amenable to randomized clinical trials.

The question of prediction has been equally unsettled—the question is: if you know some causal relations, and you know some of the probability relations among some of the related variables, can you predict what will result if you intervene and alter the value of one or more of the variables. In many causal systems the probability of an event Y given an intervention to bring about an event X is different from the conditional probability of Y on X. In recent years some philosophers, economists, computer scientists, and statisticians have realized the importance to many different kinds of problems of the difference between predicting by conditioning and predicting by intervening. Philosophers have used the difference between conditioning and intervening to argue that the principle of maximum expected utility is not always rational. Whether they are right or wrong about that (and we think wrong: see Meek and Glymour 1994), the essential thing is to provide a general means of determining when the probability distribution of one variable can be calculated from an intervention that forces a probability distribution on the values of another variable, given partial causal and probabilistic knowledge of the undisturbed system, and, when the probability of an effect can be calculated, to provide a means of calculating it.

So we have three problems: first, the problem of clarifying the very idea of a causal system with sufficient precision for mathematical analysis and sufficient generality to capture a wide range of scientific practices; second, the problem of understanding the possibilities and limitations for discovering such causal structures from various kinds of data; and third, the problem of characterizing the probabilities predicted by a causal hypothesis given an intervention directly to force a value, or distribution of values, on one or more variables. This book attempts answers to all three of these questions.

Our answer to the problem of regimenting causal hypotheses uses a formalism developed by Terry Speed and his students and subsequently elaborated by Judea Pearl and his students, and gives it a causal interpretation previously suggested (Kiiveri and Speed 1982). Our answer to the problem of discovery turns on algorithms developed from the mathematics of the representation; that answer is supplemented in chapter 12 by a discussion of Bayesian algorithms lent us by David Heckerman, Christopher Meek, and Gregory Cooper, and by a discussion—based on joint work with Wasserman and Robins— of the convergence properties of any possible non-experimental discovery procedure. The assumptions of the theory of manipulation developed here has anticipations in the econometric literature (Strotz and Wold 1960), but by putting these assumptions in a graphical framework we are able to prove some novel theorems that follow from the assumptions.

One approach to clarifying the notion of causation—the philosophers' approach ever since Plato—is to try to define "causation" in other terms, to provide necessary and sufficient and noncircular conditions for one thing, or feature or event or circumstance, to cause another, the way one can define "bachelor" as "unmarried adult male human." Another approach to the same problem—the mathematician's approach ever since Euclid—is to provide axioms that use the notion of causation without defining it, and to investigate the necessary consequences of those assumptions. We have few fruitful examples of the first sort of clarification, but many of the second: Euclid's geometry, Newton's physics, Frege's logic and Hilbert's, Kolmogorov's probability. Some axiomatic theories—Newton's, for example—offer a substantive theory of nature, while others—Frege's and Hilbert's logics and Kolmogorov's probability—are a systematization and abstraction from practice and intuition—while still others—Euclid's Elements—are something of both. While we do not claim the success of these examples, they are the models of this book.

We use a formalism—directed graphical models—that is not in the least original with us; we claim some originality in explicitly stating the causal assumptions implicit in the causal interpretation of the graphs, and in extending the application of graphs to solving certain problems about manipulations. The representation invokes two ideas about causation that are fundamental and ancient. The first idea, which can be traced back at least to Bernoulli, is that the absence of causal relations is marked by independence in probability—in Bernoulli's examples, if the outcome of one trial has no influence on the outcome of another trial, then the probability of both outcomes equals the product of each outcome separately. The second idea, Bacon's again, is that probability is associated with control: if variation of one feature, X, causes variation of another feature Y, then Y can be changed by an appropriate intervention that alters X. It turns out that the representation captures what is common to a wide variety of statistical models of causal relations—for example: regression models, logistic regression models, structural equation models, latent factor models, and many models of categorical data—and captures how these models may be used in prediction and control. The general assumptions are given in chapter 3.

These axioms have implications for scientific discovery by experimental and non-experimental means. Our investigations require characterizing when two or alternative causal theories are, in various technical senses, indistinguishable by data, and characterizing when and which causal features are shared by all models indistinguishable from any particular model. These characterizations are given in chapter 4, and more recent work on equivalence is described in chapter 12. In chapters 5, 6, 10, and 11, and again in chapter 12, we describe algorithms for discovering causal structure from sample data. We evaluate the algorithms in terms of several different features. (1) Are they computationally feasible on realistic problems? (2) Are they reliable—do they in some sense converge to a description of features common to all models indistinguishable from the true model? (3) Are they as informative as possible for the features of the data they use? Our algorithmic results concern the discovery of causal structure in linear and nonlinear systems, systems with and without feedback, cases where there may, or may not, be unrecorded common causes of recorded variables, and cases in which membership in the observed sample is influenced by the variables under study. We investigate the reliabilities of the algorithms on simulated data, and illustrate their application with published data sets. In chapter 8, reliable procedures for causal inference are compared with regression in theory, on simulated data, and on real data sets. In chapter 9 we compare causal inference from experimental and non-experimental data. Besides reviews of work on equivalence, prediction and search algorithms, chapter 12 includes a consideration of the senses in which it is, and is not, possible to have a procedure that reliably converges to the truth about causal relations as they are represented here, and a discussion of procedures for learning feedback models, so far as such models can be represented by directed cyclic graphs, and a brief discussion of the combination of search methods with the Gibbs sampler and related procedures for estimation of posterior probabilities.

The discovery of causal relations is only half of the story. The other half concerns the use in prediction of causal knowledge, even partial and incomplete causal knowledge. The fundamentals of the theory of prediction are given in chapter 3, and their consequences are developed in detail in chapter 7. The theory of prediction has a number of limitations, some of which are considered in chapter 12. The final chapter, 13, provides detailed proofs of the theorems in the main text.

2 Formal Preliminaries

This chapter introduces some mathematical concepts used throughout the book. The chapter is meant to provide mathematically explicit definitions of the formal apparatus we use. It may be skipped in a first reading and referred to as needed, although the reader should be warned that for good reason we occasionally use nonstandard definitions of standard notions in graph theory. We assume the reader has some background in finite mathematics and statistics, including correlation analysis, but otherwise this chapter contains all of the mathematical concepts needed in this book. Some of the same mathematical objects defined here are given special interpretations in the next chapter, but here we treat everything entirely formally.

We consider a number of different kinds of graphs: directed graphs, undirected graphs, inducing path graphs, partially oriented inducing path graphs, and patterns. These different kinds of objects all contain a set of vertices and a set of edges. They differ in the kinds of edges they contain. Despite these differences, many graphical concepts such as undirected path, directed path, parent, etc., can be defined uniformly for all of these different kinds of objects. In order to provide this uniformity for the objects we need in our work, we modify the customary definitions in the theory of graphs.

2.1 Graphs

The undirected graph shown in figure 2.1 contains only undirected edges (e.g., $A - B$).

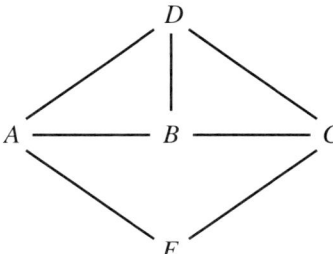

Figure 2.1

A directed graph, shown in figure 2.2, contains only directed edges (e.g., $A \rightarrow B$).

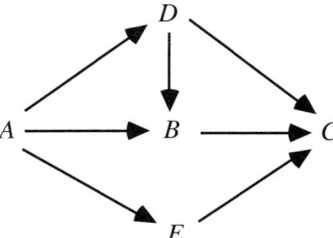

Figure 2.2

An inducing path graph, shown in figure 2.3, contains both directed edges (e.g., $A \to B$) and bi-directed edges (e.g., $B \leftrightarrow C$). (Inducing path graphs and their uses are explained in detail in chapter 6.)

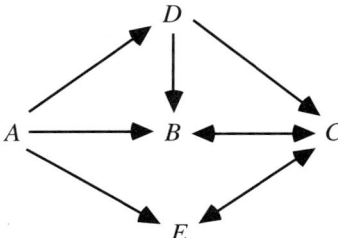

Figure 2.3

A partially oriented inducing path graph, shown in figure 2.4, contains directed edges (e.g., $B \to F$), bi-directed edges (e.g. $B \leftrightarrow C$), nondirected edges (e.g., E o-o D), and partially directed edges (e.g., A o$\to$ B.). (Partially oriented inducing path graphs and their uses are explained in detail in chapter 6.)

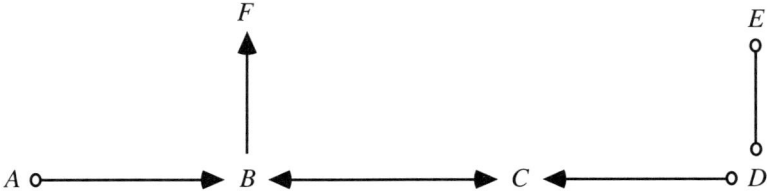

Figure 2.4

A pattern, shown in figure 2.5, contains undirected edges (e.g., $A - B$) and directed edges (e.g., $A \to E$). (Patterns and their uses are explained in detail in chapter 5.)

Formal Preliminaries

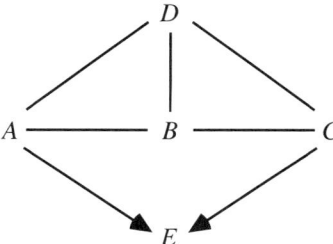

Figure 2.5

In the usual graph theoretic definition, a graph is an ordered pair <**V**,**E**> where **V** is a set of vertices, and **E** is a set of edges. The members of **E** are pairs of vertices (an ordered pair in a directed graph and an unordered pair in an undirected graph). For example, the edge $A \to B$ is represented by the ordered pair <A,B>. In directed graphs the ordering of the pair of vertices representing an edge in effect marks an arrowhead at one end of the edge. For our purposes we need to represent a larger variety of marks attached to the ends of undirected edges. In general, we allow that the end of an edge can be unmarked, or can be marked with an arrowhead, or can be marked with an "o."

In order to specify completely the type of an edge, therefore, we need to specify the variables and **marks** at each end. For example, the left end of "A o$\to B$" can be represented as the ordered pair $[A, o]$,[1] and the right end can be represented as the ordered pair $[B, >]$. The first member of the ordered pair is called an endpoint of an edge, for example, in $[A, o]$ the endpoint is A. The entire edge is a set of ordered pairs representing the endpoints, for example, $\{[A, o], [B, >]\}$. The edge $\{[B, >],[A, o]\}$ is the same as $\{[A, o],[B, >]\}$ since it doesn't matter which end of the edge is listed first.

Note that a directed edge such as $A \to B$ has no mark at the A endpoint; we consider the mark at the A endpoint to be empty, but when we write out the ordered pair we will use the notation EM to stand for the empty mark, for example, $[A,EM]$.

More formally, we say a **graph** is an ordered triple <**V**,**M**,**E**> where **V** is a non-empty set of vertices, **M** is a non-empty set of marks, and **E** is a set of sets of ordered pairs of the form $\{[V_1,M_1],[V_2,M_2]\}$, where V_1 and V_2 are in **V**, $V_1 \neq V_2$, and M_1 and M_2 are in **M**. Except in our discussion of systems with feedback we will always assume that in any graph, any pair of vertices V_1 and V_2 occur in at most one set in **E**, or, in other words, that there there is at most one edge between any two vertices. If G = <**V**,**M**,**E**> we say that G is **over V**.

For example, the directed graph of figure 2.2 can be represented as <$\{A,B,C,D,E\}$, $\{EM, >\}$, $\{\{[A,EM],[B, >]\}$, $\{[A,EM],[E, >]\}$, $\{[A,EM],[D, >]\}$, $\{[D,EM],[B, >]\}$, $\{[D,EM],[C, >]\}$, $\{[B,EM],[C, >]\}$, $\{[E,EM],[C, >]\}\}$>.

Each member $\{[V_1, M_1],[V_2,M_2]\}$ of **E** is called an **edge** (e.g., $\{[A,EM],[B, >]\}$ in figure 2.2.) Each ordered pair $[V_1, M_1]$ in an edge is called an **edge-end** (e.g., $[A,EM]$ is an edge-end of $\{[A,EM],[B, >]\}$.) Each vertex V_1 in an edge $\{[V_1, M_1],[V_2, M_2]\}$ is called an **endpoint** of the edge (e.g., A is an endpoint of $\{[A,EM],[B, >]\}$.) V_1 and V_2 are adjacent in G if and only if there is an edge in E with endpoints V_1 and V_2 (e.g., in figure 2.2, A and B are adjacent, but A and C are not.)

An **undirected graph** is a graph in which the set of marks $M = \{EM\}$. A **directed graph** is a graph in which the set of marks $M = \{EM, >\}$ and for each edge in **E**, one edge-end has mark EM and the other edge-end has mark ">."

An edge $\{<[A,EM],[B,>]\}$ is a **directed edge** from A to B. (Note that in an undirected graph there are no directed edges.) An edge $\{[A,M_1],[B,>]\}$ is **into** B. An edge $\{[A,EM],[B,M_2]\}$ is **out of** A. If there is a directed edge from A to B then A is a **parent** of B and B is a **child** (or **daughter**) of B. We denote the set of all parents of vertices in **V** as **Parents(V)** and the set of all children of vertices in **V** as **Children(V)**. The **indegree** of a vertex V is equal to the number of its parents; the **outdegree** is equal to the number of its children; and the **degree** is equal to the number of vertices adjacent to V. (In a directed graph, the degree of a vertex is equal to the sum of it's indegree and outdegree.) In figure 2.2, the parents of B are A and D, and the child of B is C. Hence, B is of indegree 2, outdegree 1, and degree 3.

We will treat an undirected path in a graph as a sequence of vertices that are adjacent in the graph. In other words for every pair X, Y adjacent on the path, there is an edge $\{[X,M_1],[Y,M_2]\}$ in the graph. For example, in figure 2.2, the sequence $<A,B,C,D>$ is an undirected path because each pair of variables adjacent in the sequence (A and B, B and C, and C and D) have corresponding edges in the graph. The set of edges in a path consists of those edges whose endpoints are adjacent in the sequence. In figure 2.2 the edges in path $<A,B,C,D>$ are $\{[A,EM],[B,>]\}$, $\{[B,EM],[C,>]\}$, and $\{[C,>],[D,EM]\}$.

More formally, an **undirected path** between A and B in a graph G is a sequence of vertices beginning with A and ending with B such that for every pair of vertices X and Y that are adjacent in the sequence there is an edge $\{[X,M_1],[Y,M_2]\}$ in G. An **edge** $\{[X,M_1],[Y,M_2]\}$ **is in path** U if and only if X and Y are adjacent to each other (in either order) in U. If an edge between X and Y is in path U we also say that X and Y are **adjacent** on U. If the edge containing X in an undirected path between X and Y is out of X then we say that the **path is out of** X; similarly, if the edge containing X in a path between X and Y is into X then we say that the **path is into** X. In order to simplify proofs we call a sequence that consists of a single vertex an **empty path.** A path that contains no vertex more than once is **acyclic**; otherwise it is **cyclic**. Two paths **intersect** iff they have a vertex in common; any such common vertex is a **point of intersection**. If path U is $<U_1,\ldots,U_n>$ and path V is $<U_n, V_1,\ldots,V_m>$, then the **concatenation** of U and V is $<U_1,\ldots U_n, V_1, \ldots ,V_m>$ denoted by U and$V$. The concatenation of U with an empty path is U, and the concatenation of an empty path with U is U. Ordinarily when we use the term "path" we will mean acyclic path; in referring to cyclic path we will always use the adjective.

A **directed path** from A to B in a graph G is a sequence of vertices beginning with A and ending with B such that for every pair of vertices X, Y, adjacent in the sequence and occurring in the sequence in that order, there is an edge $\{[X,EM],[Y,>]\}$ in G. A is the

source and *B* the **sink** of the path. For example, in figure 2.2 <*A,B,C*> is a directed path with source *A* and sink *C*. In contrast, in figure 2.2 <*A,B,D*> is an undirected path, but not a directed path because *B* and *D* occur in the sequence in that order, but the edge {[*B,EM*],[*D*, >]} is not in *G* (although {[*D,EM*],[*B*, >]} is in *G*.) Directed paths are therefore special cases of undirected paths. For a directed edge *e* from *U* to *V* ($U \rightarrow V$), **head**(*e*) = *V* and **tail**(*e*) = *U*. A **directed acyclic graph** is a directed graph that contains no directed cyclic paths.

A **semidirected path** between *A* and *B* in a partially oriented inducing path graph π is an undirected path *U* from *A* to *B* in which no edge contains an arrowhead pointing toward *A* (i.e., there is no arrowhead at *A* on *U*, and if *X* and *Y* are adjacent on the path, and *X* is between *A* and *Y* on the path, then there is no arrowhead at the *X* end of the edge between *X* and *Y*.) Of course every directed path is semidirected, but in graphs with "o" end marks there may be semidirected paths that are not directed.

A graph is **complete** if every pair of its vertices are adjacent. Figure 2.6 illustrates a complete undirected graph.

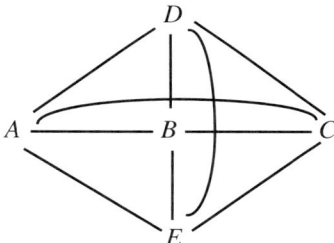

Figure 2.6

A graph is **connected** if there is an undirected path between any two vertices. Figures 2.1–2.6 are connected, but figure 2.7 is not.

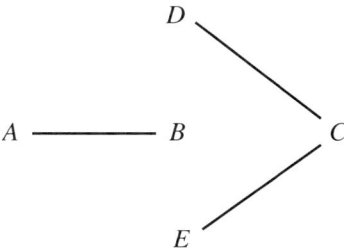

Figure 2.7

A **subgraph** of <**V,M,E**> is any graph <**V′,M′,E′**> such that **V′** is included in **V**, **M′** is included in **M**, and **E′** is included in **E**. Figure 2.7 is a subgraph of figure 2.1. The

subgraph of <V,M,E> over V', where **V'** is included in **V**, is the subgraph <V',M,E'> in which an edge is in **E'** if and only if it is in **E** and has both endpoints in **V'**.

A **clique** in graph G is any subgraph of G that is complete. In figure 2.1, for example, the subgraph $G' =$

<{A,B,D},{EM},{{[A,EM],[B,EM]},{[B,EM],[D,EM]},{[A,EM],[D,EM]}}>

is a clique with vertices A, B and D. A clique in G whose vertex set is not properly contained in any other clique in G is **maximal.** In figure 2.1, both G' and $G'' =$ <{A,B},{EM}, {{[A,EM],[B,EM]}}>, are cliques, but G'', unlike G', is not maximal because G'' is properly contained in G'.[2]

A **triangle** in a graph G is a complete subgraph of G with three vertices; in other words, vertices X, Y and Z form a triangle if and only if X and Y are adjacent, Y and Z are adjacent and X and Z are adjacent. In graph G a vertex V is a **collider on undirected path** U if and only if there are two distinct edges on U containing V as an endpoint and both are into V. Otherwise V is a **noncollider on** U. In graph G, vertex V is an **unshielded collider** on U if V is a collider on U, V is adjacent to distinct vertices V_1 and V_2 on U, and V_1 and V_2 are not adjacent in G. An **ancestor** of a vertex V is any vertex W such that there is a directed path from W to V. A **descendant** of a vertex V is any vertex W such that there is a directed path from V to W. In figure 2.2, A, B, C, D, and E are all ancestors of C, although neither A nor C is a parent of C. Similarly, C is a descendant of A, B, C, D, and E, although it is not a child of A or C. Since every vertex V is the source of a directed (empty) path from V to V, each vertex is its own descendant and its own ancestor, but not of course its own parent or its own child.

2.2 Probability

The vertices of the graphs we consider will always be random variables taking values in one of the following: a copy of the real line; a copy of the nonnegative reals; an interval of integers.

By a joint distribution on the vertices of a graph we mean a countably additive probability measure on the Cartesian product of these objects. We say that two random variables, X, Y are **independent** when the joint density of (X,Y) is the product of the density of X and the density of Y for all values of X and Y. We write this as $X \perp\!\!\!\perp Y$. We generalize in the obvious way when asserting that one set of variables is independent of another set of variables. When we say a set of random variables is **jointly independent** we mean that any two disjoint subsets of the set are independent of one another. We say that random variables X, Y are **independent conditional on Z** (or given **Z**), when the density of X, Y given **Z** equals the product of the density of X given **Z** and the density of Y given **Z**, for all values of X, Y, and for all values **z** of **Z** for which the density of **z** is not equal to 0. We generalize in the obvious way for sets of random variables, **X, Y, Z**. If **X**

Formal Preliminaries

is independent of **Y** given **Z** we write $X \perp\!\!\!\perp Y|Z$, and we say that the **order of the conditional independence** is equal to the number of variables in **Z**.

In the discrete case, we say that a distribution over **V** is positive if and only if for all values **v** of **V**, $P(\mathbf{v}) \neq 0$. (In general, a distribution over **V** is positive if the density function is nonzero for all **v**.) If **V** is included in **V'** and

$$P(\mathbf{V}) = \sum_{\mathbf{V'} \setminus \mathbf{V}}^{\rightarrow} P(\mathbf{V'})$$

we will say that $P(\mathbf{V})$ is the **marginal** of $P(\mathbf{V'})$ over **V**.

2.3 Graphs and Probability Distributions

We will examine several different graphical representations of conditional independence relations true in a distribution.

2.3.1 Directed Acyclic Graphs

A directed acyclic graph can be used to represent conditional independence relations in a probability distribution.

For a given graph G and vertex W let **Parents**(W) be the set of parents of W, and **Descendants**(W) be the set of descendants of W.

Markov Condition: A directed acyclic graph G over **V** and a probability distribution $P(\mathbf{V})$ satisfy the Markov condition if and only if for every W in **V**, W is independent of **V**\(**Descendants**(W) $\cup$ **Parents**(W)) given **Parents**(W).

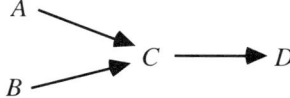

Figure 2.8

(Recall that W is its own descendant.) In the terminology of Pearl (1988) G is an **I-map** of P. In figure 2.8, the Markov Condition entails the following conditional independence relations:[3]

$A \perp\!\!\!\perp B$

$D \perp\!\!\!\perp \{A, B\} \mid C$

For all values of **v** of **V** for which $f(\mathbf{v}) \neq 0$, the joint density function $f(\mathbf{V})$ satisfying the Markov Condition is given by

$$f(\mathbf{V}) = \prod_{V \in \mathbf{V}} f(V|\mathbf{Parents}(V))$$

where $f(V|\mathbf{Parents}(V))$ denotes the density of V conditional on the (possibly empty) set of vertices that are parents of V. (See Kiiveri and Speed 1982. Recall our notation convention that if $\mathbf{Parents}(V) = \varnothing$, then $f(V|\mathbf{Parents}(V)) = f(V)$.)

If a joint distribution over discrete variables satisfies the Markov Condition for figure 2.8 it can be factored in the following way:

$P(A,B,C,D) = P(A)\ P(B)\ P(C|A,B)\ P(D\ |C)$

for all values of A, B, C, D such that $P(A,B,C,D) \neq 0$. In a directed acyclic graph G, vertices of zero indegree are said to be **exogenous**. If G satisfies the Markov Condition for a distribution P, then for every pair of exogenous variables V_1 and V_2, $V_1 \perp\!\!\!\perp V_2$ in P.

The Minimality Condition says, intuitively, that each edge in the graph prevents some conditional independence relation that would otherwise obtain.

Minimality Condition: If G is a directed acyclic graph over **V** and P a probability distribution over **V**, <G, P> satisfies the Minimality Condition if and only if for every proper subgraph H of G with vertex set **V**, <H,P> does not satisfy the Markov Condition.

Returning to the example of figure 2.8, a distribution P' which satisfies the Markov Condition, but in which A is independent of $\{B,C,D\}$ does not satisfy the Minimality Condition, because P' also satisfies the Markov Condition for the subgraph in which the edge between A and C is removed. In the terminology of Pearl (1988) if a distribution $P(\mathbf{V})$ satisifies the Markov and Minimality conditions for a directed acyclic graph G, then G is a **minimal I-map** of P.

If a distribution P satisfies the Markov and Minimality Conditions for directed acyclic graph G, we will say that G **represents** P. For any directed acyclic graph G and for any probability distribution P satisfying the Markov and Minimality Conditions, if variables A and B are statistically dependent, then either:

(i) there is a directed path in G from A to B; or
(ii) there is a directed path in G from B to A: or
(iii) there is a variable C and directed paths in G from C to B and from C to A.

A **trek** between distinct vertices A and B is an unordered pair of directed paths between A and B that have the same source, and intersect only at the source. The source of the pair

Formal Preliminaries 13

of paths is also called the source of the trek. Note that one of the paths in a trek may be an empty path.

2.3.2 Directed Independence Graphs

Directed independence graphs (Whittaker 1990) are another (almost equivalent) way of representing conditional independence relations true of a probability distribution. Say that directed acyclic graph G is a **directed independence graph** of $P(\mathbf{V})$ for an ordering $>$ of the vertices in G if and only if $A \rightarrow B$ occurs in G if and only if $\sim(A \perp\!\!\!\perp B \mid \mathbf{K}(B))$, where $\mathbf{K}(B)$ is the set of all vertices V such that $V \neq A$ and $V > B$.

THEOREM 2.1: If $P(\mathbf{V})$ is a positive distribution, then for any ordering of the variables in $\mathbf{V}$, P satisfies the Markov and Minimality conditions for the directed independence graph of $P(\mathbf{V})$ for that ordering.

If a distribution P is not positive, it is possible that the directed independence graph of P for a given ordering of variables is a subgraph of a directed acyclic graph for which P satisfies the Minimality and Markov conditions (Pearl 1988).

2.3.3 Faithfulness

Given any graph, the Markov condition determines a set of independence relations. These independence relations in turn may entail others, in the sense that every probability distribution having the independence relations given by the Markov condition will also have these further independence relations. In general, a probability distribution P on a graph G satisfying the Markov condition may include other independence relations besides those entailed by the Markov condition applied to the graph. For example, A and D might be independent in a distribution satisfying the Markov Condition for the graph in figure 2.9, even though the graph does not entail their independence.

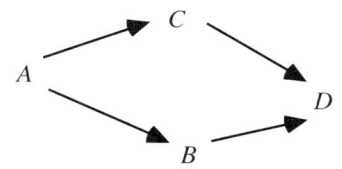

Figure 2.9

In linear models such an independence can arise if the product of the partial regression coefficients for D on C and C on A cancels the corresponding product of D on B and B on A.

If all and only the conditional independence relations true in P are entailed by the Markov condition applied to G, we will say that P and G are **faithful to one another.** We will, moreover, say that a distribution P is **faithful** provided there is some directed acyclic graph to which it is faithful. In the terminology of Pearl (1988) if P and G are faithful to one another then G is a **perfect map** of P and P is a **DAG-Isomorph** of G. If distribution *P* is faithful to directed acyclic graph *G*, *X* and *Y* are dependent if and only if there is a trek between *X* and *Y*.

2.3.4 d-separation

Following Pearl (1988), we say that for a graph *G*, if *X* and *Y* are vertices in *G*, $X \neq Y$, and **W** is a set of vertices in *G* not containing *X* or *Y*, then *X* and *Y* are **d-separated** given **W** in *G* if and only if there exists no undirected path *U* between *X* and *Y*, such that (i) every collider on *U* has a descendent in **W** and (ii) no other vertex on *U* is in **W**. We say that if $X \neq Y$, and *X* and *Y* are not in **W**, then *X* and *Y* are **d-connected** given set **W** if and only if they are not d-separated given **W**. If **U**, **V**, and **W** are disjoint sets of vertices in *G* and **U** and **V** are not empty then we say that **U** and **V** are d-separated given **W** if and only if every pair <*U*,*V*> in the cartesian product of **U** and **V** is d-separated given **W**. If **U**, **V**, and **W** are disjoint sets of vertices in *G* and **U** and **V** are not empty then we say that **U** and **V** are d-connected given **W** if and only if **U** and **V** are not d-separated given **W**. An illustration of d-connectedness is given in the directed acyclic graph in figure 2.10 (but note that the definition also applies to other sorts of graphs such as inducing path graphs, as explained in chapter 6).

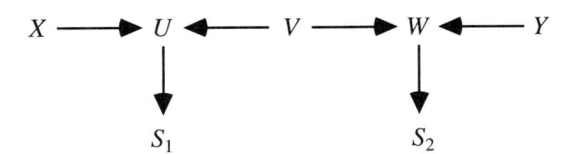

$$X \longrightarrow U \longleftarrow V \longrightarrow W \longleftarrow Y$$
$$ \downarrow \downarrow$$
$$ S_1 S_2$$

Figure 2.10

X and *Y* are d-separated given the empty set
X and *Y* are d-connected given set $\{S_1, S_2\}$
X and *Y* are d-separated given the set $\{S_1, S_2, V\}$

2.3.5 Linear Structures

A directed acyclic graph *G* over **V** **linearly represents** a distribution *P*(**V**) if and only if there exists a directed acyclic graph *G'* over **V'** and a distribution *P''*(**V'**) such that

(i) **V** is included in **V'**;
(ii) for each endogenous (that is, with positive indegree) variable *X* in **V**, there is a unique variable ε_X in **V'\V** with zero indegree, positive variance, outdegree equal to one, and a directed edge from ε_X to *X*;
(iii) *G* is the subgraph of *G'* over **V**;
(iv) each endogenous variable in *G* is a linear function of its parents in *G'*;
(v) in *P''*(**V'**) the correlation between any two exogenous variables in *G'* is zero;
(vi) *P*(**V**) is the marginal of *P''*(**V'**) over **V**.

Formal Preliminaries

The members of **V'\V** are called **error variables** and we call G' the **expanded graph**. Directed acyclic graph G **linearly implies** $\rho_{AB.\mathbf{H}} = 0$ if and only if $\rho_{AB.\mathbf{H}} = 0$ in all distributions linearly represented by G. (We assume all partial correlations exist for the distribution.) If G linearly represents $P(\mathbf{V})$ we say that the pair $<G,P(\mathbf{V})>$ is a linear model with directed acyclic graph G.

2.4 Undirected Independence Graphs

There is a well-known representation of statistical hypotheses about conditional independence by *undirected* graphs. The two representations, by directed and by undirected graphs, are closely related, but it is important not to confuse them.

An **undirected independence graph** G with a set of vertices **V represents** a probability distribution P if and only if there is no undirected edge between A and B just when A and B are conditionally independent given $\mathbf{V}\backslash\{A,B\}$ in P. If an undirected independence graph G represents a distribution P, A and B are independent conditional on some set **C** if and only if every undirected path between A and B contains a member of **C**.

Suppose we consider a particular directed acyclic graph G and faithful probability distribution P. Let U be the undirected graph of adjacencies **underlying** G; that is, U is the undirected graph with the same vertex set as G and the same adjacencies as G. Suppose that I is the undirected independence graph for the distribution P formed according to the definition just given. Then I and U are not in general the same, but U is always a subgraph of I. I and U will be the same if and only if G contains no unshielded colliders (Wermuth and Lauritzen 1983).

2.5 Deterministic and Pseudoindeterministic Systems

We will use the notion of a deterministic system in a technical sense: A joint probability distribution P on a set **V** of random variables represented by a directed acyclic graph G is **deterministic** if each of the vertices of G of nonzero indegree is a function of the vertices that are its immediate parents in G; we will also say that G is a **deterministic graph** of P. By "function" we mean that for each assignment of a unique value to each of the parent vertices, there is a unique value of the dependent vertex.

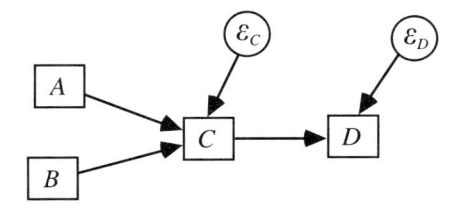

Figure 2.11

Suppose that the graph/distribution pair represented in figure 2.11 is deterministic, but that ε_C and ε_D are not measured. Were we to consider only the measured variables, that is,

A, B, C, and D, we would find that no variable has its value uniquely determined by the values of the others, although some of the variables are statistically dependent. The system looks indeterministic, although ε_C and ε_D are "hidden" variables which make it deterministic when they are added. Furthermore, it is not necessary to posit that two measured variables depend upon the same hidden variable, nor is it necessary to posit any dependence among the "hidden" variables in order to make the system deterministic. When a distribution represented by a directed acyclic graph among measured variables is not deterministic, but is embeddable in this way in a distribution represented by a directed acyclic graph that is, we say the distribution is pseudoindeterministic.

In contrast, consider figure 2.12. Again suppose that only A, B, C, and D, were measured. In this case we could not make the system deterministic by adding hidden variables unless either the hidden variables were associated or at least one hidden variable is adjacent to at least two of the measured variables.

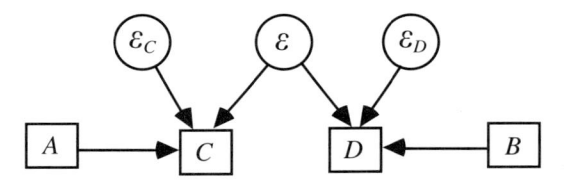

Figure 2.12

More formally, <G,P> is **pseudoindeterministic**, where P is a probability distribution over **V** and G is a directed acyclic graph over **V**, if and only if G is not a deterministic graph of P and there exists a distribution P' and a directed acyclic graph G' over a set of variables **V′** that properly includes **V** such that

(i) G' is a deterministic graph of P';
(ii) G is the subgraph of G' over **V**;
(iii) no vertex in **V** is an ancestor of a vertex in **V′\V**;
(iv) no vertex in **V′\V**, is the source of a trek connecting two vertices in **V**;
(v) P is the marginal of P';
(vi) G represents P.

If we say that <P,G> is **linear pseudoindeterministic** we mean that <P,G> is pseudoindeterministic and in addition in G', each vertex in **V′** is a linear function of its parents. A distribution linearly represented by a directed acyclic graph is pseudoindeterministic. (Analogous definitions apply to Boolean pseudoindeterministic pairs of graphs and distributions, etc.)

2.6 Background Notes

Drawing from purely graph-theoretical work of Lauritzen, Speed, and Vijayan (1978), and on statistical work in log-linear models (Bishop, Fienberg, and Holland 1975), in 1980 Darroch, Lauritzen, and Speed introduced undirected graphical representations of log-linear hypotheses of conditional independence. Based on Kiiveri's thesis work, Kiiveri and Speed (1982) introduced versions of the Markov Condition, defined the notion of recursive causal model, obtained maximum likelihood estimates for a multinomial distribution and provided a systematic survey of applications with both discrete and continuous variables. Shortly after, Kiiveri, Speed, and Carlin (1984) further developed the formal foundations. Wermuth and Lauritzen (1983) introduced the notion of a recursive diagram, or what we have called a directed independence graph. The definitions of minimality, d-separation, and faithfulness are due to Pearl (1988).

3 Causation and Prediction: Axioms and Explications

Views about the nature of causation divide very roughly into those that analyze causal influence as some sort of probabilistic relation, those that analyze causal influence as some sort of counterfactual relation (sometimes a counterfactual relation having to do with manipulations or interventions), and those that prefer not to talk of causation at all. We advocate no definition of causation, but in this chapter we try to make our usage systematic, and to make explicit our assumptions connecting causal structure with probability, counterfactuals and manipulations. With suitable metaphysical gyrations the assumptions could be endorsed from any of these points of view, perhaps including even the last.

3.1 Conditionals

Intelligent planning usually requires predicting the consequences of actions. Since actions change the states of affairs, assessing the consequences of actions not yet taken requires judging the truth or falsity of future conditional sentences—*If X were to be the case, then Y would be the case*. Judging the effects of past practice or policy requires judging the truth or falsity of counterfactual sentences—*If X had been the case, they Y would have been the case*.

Giving a detailed description of the conditions under which a future conditional or counterfactual conditional is true is a well known and difficult philosophical problem. Lewis (1973) notes that *If kangaroos had no tails, they would topple over* is true even though we can imagine circumstances in which kangaroos use crutches. We mean that if things were pretty much as they are—given the scarcity of crutches for kangaroos and the disinclination of kangaroos to use crutches—if kangaroos had no tails they would topple over. But making this intuition precise is not easy.

It is widely recognized that causal regularities entail counterfactual conditionals; indeed this is often taken to be the feature that distinguishes a causal law from generalizations that are true, as it were, by accident. *All of the coins in your pocket are made of silver* does not entail the counterfactual *If this penny were in your pocket then it would be made of silver*. But the causal law *All collisions of electrons and positrons release energy* does entail the counterfactual *If this electron were to collide with this positron then energy would be released*.

The connection between causal regularities and the truth of future conditional and counterfactual sentences makes the discovery of causal structure essential for intelligent planning in many contexts. A linear equation relating the fatality rate in automobile accidents to car weight may be true of a given population, but unless it describes a robust feature of the world it is useless for predicting what would happen to the fatality rate if car weight was manipulated through legislation. Even quite accurate parametric representations of the distribution of values in a population may be useless for planning unless they also reflect the causal structure among the variables.

3.2 Causation

We understand causation to be a relation between particular events: something happens and causes something else to happen. Each cause is a particular event and each effect is a particular event. An event A can have more than one cause, none of which alone suffice to produce A. An event A can also be overdetermined: it can have more than one set of causes that suffice for A to occur. We assume that causation is (usually) transitive, irreflexive, and antisymmetric. That is, i) if A is a cause of B and B is a cause of C, then A is also a cause of C, ii) an event A cannot cause itself, and iii) if A is a cause of B then B is not a cause of A.

3.2.1 Direct vs. Indirect Causation

The distinction between direct and indirect causes is relative to a set of events. If C is the event of striking a match, and A is the event of the match catching on fire, and no other events are considered, then C is a direct cause of A. If, however, we added B: the sulfur on the match tip achieved sufficient heat to combine with the oxygen, then we would no longer say that C directly caused A, but rather that C directly caused B and B directly caused A. Accordingly, we say that B is a **causal mediary** between C and A if C causes B and B causes A.

Having fixed a context and a set of events, what is it for one event to be a direct cause of another? The intuition is this: once the events that are direct causes of A occur, then whether A occurs or not no longer has anything to do with whether the events that are indirect causes of A occur. The direct causes *screen off* the indirect causes from the effect. If a child is exposed to chicken pox at her daycare center, becomes infected with the virus, and later breaks out in a rash, the infection screens off the event of exposure from the occurrence of the rash. Once she is infected, whether she gets the rash has nothing to do with whether she was exposed to the virus from her daycare or from her Saturday morning playgroup.

Suppose $\mathbf{V}$ is a set of events including C and A. C is a **direct cause** of A relative to $\mathbf{V}$ just in case C is a member of some set $\mathbf{C}$ included in $\mathbf{V}\backslash\{A\}$ such that (i) the events in $\mathbf{C}$ are causes of A, (ii) the events in $\mathbf{C}$, were they to occur, would cause A no matter whether the events in $\mathbf{V}\backslash(\{A\} \cup \mathbf{C})$ were or were not to occur, and (iii) no proper subset of $\mathbf{C}$ satisfies (i) and (ii).

3.2.2 Events and Variables

In order for causation to be connected with probabilities that can be estimated empirically, events must be sorted; some actual or possible events must be gathered together, declared to be of a type, and distinguished from other actual or possible events perhaps gathered into other types. The simplest classifications describe events as of a kind, for example, solar eclipses, or declines in the Dow Jones Industrial Average, and pair each event, A, of a kind with the event, $\neg A$, the nonoccurrence of A. Such classifications permit us to speak intelligibly of *variables* as causes. We do so through the introduction of Boolean variables that take events of a kind, or their absences, as values. We say that **Boolean variable C causes Boolean variable A** if and only if at least one member of a pair $(C, \neg C)$ causes at least one member of a pair $(A, \neg A)$. Ordinarily no one would bother with collecting events into a type and examining causal relations among such variables unless the causal relations among events of the two types had some generality—that is, lots of events of type A have events of type C as causes and lots of events of type C have effects of type A, or none do.

Events can be aggregated into variables X and Y, such that some events of kind X cause some events of kind Y and some events of kind Y cause some events of kind X. In such cases there will be no unambiguous direction to the causal relation between the variables.

Some events are of a quantity taking a certain value, such as bringing a particular pot of water to a temperature of 100 degrees centigrade. Scales of many kinds are associated with an array of possible events in which a particular system takes on a scale value or takes on a value within a set of scale values. We can also speak of the variables of such scales as causes and effects, at least for particular systems over particular time intervals. For any particular system S we say that **scaled variable Q causes scaled variable R** in S provided that there is a value (or set of values) q for Q and a value (or set of values) r for R and a possible event in which Q taking value q in S would cause an event in which R takes value r in S. In practice we usually form scales only when we think the causal relations among values of different measures are not confined to particular values or particular systems but are more general. We sometimes say that the value r for R is caused by the value q for Q if the system taking on the value q for Q caused it to take on the value r for R. If **K** is a collection of systems, we say that variable Q causes variable R in **K** provided that for every system S in **K**, Q causes R in S.

If our notion of causation between variables were strictly applied, almost every natural variable would count as a cause of almost every other natural variable, for no matter how remote two variables, A and B, may be, there is usually *some* physically possible—even if very unlikely—arrangement of systems such that variation in some values of A produces variation in some values of B. (A dictator could, we suppose, arrange circumstances so that the number of childbirths in Chicago is a function of the price of tea in China.) In practice, we always consider a restricted range of variation of other variables in judging whether A causes B. Strictly, therefore, our definitions of causal relations for variables should be relative to a set of possible values for other variables, but we will ignore this formality and trust to context. The notion of direct cause generalizes from events to variables in obvious parallel to the definition of causal dependence between variables: Variable C is a direct cause of variable A relative to **V** provided (i) C is a member of a set **C** of variables included in **V**, (ii) there exists a set of values **c** for variables in **C** and a value a for A such that were the variables in **C** to take on values **c**, they would cause A to take on value a no matter what the values of other variables in **V**, and (iii) no proper subset of **C** satisfies (i) and (ii). We say that a variable X is a **common cause** of variables Y and Z if and only if X is a direct cause of Y relative to $\{X,Y,Z\}$ and a direct cause of Z relative to $\{X,Y,Z\}$. If there is a sequence of variables in **V** beginning with A and ending with B such that for each pair of variables X and Y that are adjacent in the sequence in that order X is a direct cause of Y relative to **V**, then we say that there is a **causal chain** from A to B relative to **V**. A is an **indirect cause** of B relative to **V** if there is a causal chain from A to B relative to **V** of length greater than 2. We make the following two fundamental assumptions about causal relations: (i) if A is a cause of B then A is a direct cause or an indirect cause of B relative to **V**; (ii) if A, B, and C are in **V**, and there exists a causal chain from A to B relative to **V** that does not contain C, then for any set **V'** that contains A and B there is a causal chain from A to B relative to **V'** that does not contain C. When a cause is unmeasured it is sometimes called a **latent** variable. We say that two variables are **causally connected** in a system if one of them is the cause of the other or if they have a common cause. A **causal structure** for a population is an ordered pair <**V**,**E**> where **V** is a set of variables, and **E** is a set of ordered pairs of **V**, where <X,Y> is in **E** if and only X is a direct cause of Y relative to **V**. We assume that in the population A is a direct cause of B either for all units in the population or no units in the population, unless explicitly noted otherwise. If it is obvious which population is intended we do not explicitly mention it. If $P(\mathbf{V})$ is a distribution over **V** in a population with causal structure C = <**V**,**E**>, we say that C **generated** $P(\mathbf{V})$. Two causal structures <**V**,**E**> and <**V'**,**E'**> are **isomorphic** if and only if there is a one-to-one function f from **V** onto **V'** such that for any two members of A and B of **V**, <A,B> is in **E** if and only if <$f(A),f(B)$> is in **E'**. A set **V** of variables is **causally sufficient** for a population if and only if in the population every common cause of any two or more variables in **V** is in **V**, or has the same value for all units in the population.[1] We will often use the notion of causal sufficiency without explicitly mentioning the population.

3.2.3 Examples

Simple digital logic circuit elements present concrete examples of causal structures. They are not of much intrinsic interest to most people, but they have the virtue that given a description of such a circuit element almost everyone can agree about which events pertaining to the circuit cause which other events. In the element illustrated below, the variables X_1, X_2 and X_3 have two values, 1 and 0, accordingly as there is or is not a current through the corresponding line, and the semicircle represents an "and" gate. Current flows from top to bottom. The value of the variable X_3 is thus a simple Boolean function of the values of X_1 and X_2. If "•" represents Boolean multiplication, $X_3 = X_1 \bullet X_2$.

Causation and Prediction

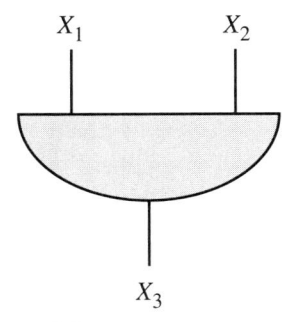

Figure 3.1

We understand the event of X_1 taking on value 1 and the event of X_2 taking on value 1 each to be a cause of the event in which X_3 takes on the value 1. We say that the Boolean variables X_1 and X_2 are each causes of the Boolean variable X_3.

The form of the causal structure does not depend on the sort of variables involved or the particular class of functions among them. Isomorphic causal structures might be realized by a system of linear dependencies of continuous variables. Thus consider three different variables X_1, X_2, and X_3, that represent the voltage in a given line and therefore range over the non-negative reals. Suppose we have a mechanism that outputs the sum of the voltage into it (figure 3.2).

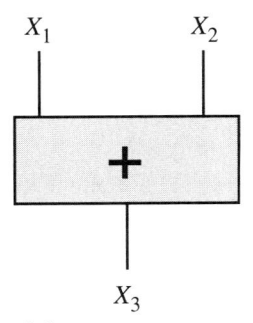

Figure 3.2

In this case $X_3 = X_1 + X_2$, but the causal structure is isomorphic to the causal structure in figure 1: X_1 and X_2 each are causes of X_3.

These examples suggest that the causal dependencies and the functional dependencies are related; X_3 is the effect of X_1 and X_2, and X_3 is a function of X_1 and X_2. In systems in which variables that are effects have their values uniquely determined by the values of all of the variables that are their direct causes, functional dependence can be inferred from causal dependence by expressing each variable or event as a function of its direct causes. The converse does not hold: from the fact that an equation correctly describes a system one cannot infer that the direct causal dependencies in the system are reflected in the functional dependencies in the equation. For example, if the equation $X_3 = X_1 + X_2$ is true of a system then the equation $X_2 = X_3 - X_1$ is equally true of that system, but if X_1 and X_2 cause X_3, then ordinarily X_3 and X_1 do not cause X_2.[2]

3.2.4 Representing Causal Relations with Directed Graphs

Using the notion of a direct cause, it is trivial to represent causal structures with directed graphs:

Causal Representation Convention: A directed graph $G = <\mathbf{V}, \mathbf{E}>$ represents a causally sufficient causal structure C for a population of units when the vertices of G denote the variables in C, and there is a directed edge from A to B in G if and only if A is a direct cause of B relative to $\mathbf{V}$.[3]

We call a directed acyclic graph that represents a causal structure a **causal graph**. Figure 3.3 is a causal graph for the circuit devices shown in figures 3.1 and 3.2.

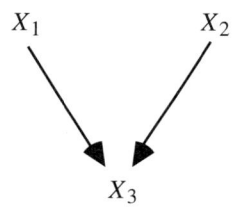

$X_1 \qquad X_2$
$\quad \searrow \swarrow$
$\quad X_3$

Figure 3.3

Consistently with our previous definition, if G is a causal graph and there is a vertex X in G and a directed path from X to Y that does not contain Z, and a directed path from X to Z that does not contain Y, we will say X is a **common cause** of Y and Z.

There are important limitations to the Causal Representation Convention. Suppose drugs A and B both reduce symptoms C, but the effect of A without B is quite trivial, while the effect of B alone is not. The directed graph representations we have considered in this chapter offer no means to represent this interaction and to distinguish it from other circumstances in which A and B alone each have an effect on C. The interaction is only represented through the probability distribution associated with the graph. Consider another example, a simple switch. Suppose as in figure 3.4 battery A has two states: charged and uncharged. Charge in battery A will cause bulb C to light up provided the switch B is on, but not otherwise. If A and B are independent random variables, then A and C are dependent conditional on B and on the empty set, and B and C are dependent conditional on A and the empty set, and A and B are dependent conditional on C. The directed acyclic graph representing the distribution over A, B, and C therefore looks like the directed graph shown above. There is nothing wrong with this conclusion except that it is not fully informative. The dependence of A and C arises entirely through the condition $B = 1$. When $B = 0$, A and C are independent.

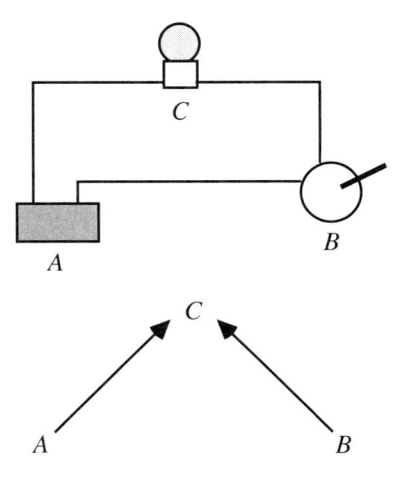

Figure 3.4

Since in discrete data the conditional independence facts, if known, identify the switch variables, a better representation would identify certain parents of a variable as switches. But a general representation of this sort would often not be very easy to grasp.[4] Work extending the directed acyclic graph representation to represent switches is described in Geiger and Heckerman (1991).

3.3 Causality and Probability
3.3.1 Deterministic Causal Structures

To good approximation the devices in figures 3.1 and 3.2 are **deterministic**, that is, the effects are deterministic functions of their direct causes. If each effect is a linear function of its direct causes in the population, we say the system is a **linear deterministic causal structure** in the population.

Variables in a causal graph that have zero indegree, that is, no causal input, are said to be **exogenous**. X_1 and X_2 are exogenous variables in the causal graph in figure 3.3. Variables that are not exogenous are **endogenous**. In a deterministic causal structure values for the exogenous variables determine unique values for the remaining variables.

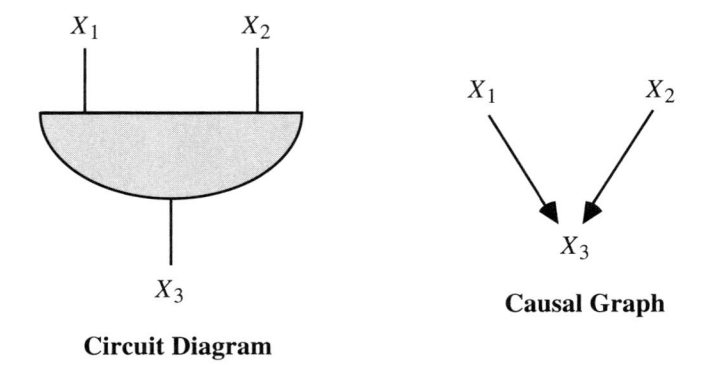

Figure 3.5

Circuit Diagram / Causal Graph

Consider the circuit element in figure 3.1 and its causal graph, both of which are shown in figure 3.5. Imagine an experiment to verify whether or not the device works as described. We would assign values to the exogenous variables, that is, decide whether to put current into X_1 and X_2, and then read whether or not X_3 has current. We can represent the experiment with the following table.

X_1	X_1	X_3
1	1	?
1	0	?
0	1	?
0	0	?

Suppose we were satisfied that the device usually worked as designed, but we wanted to know how often and in what way it fails. For each of a number of trials, we could randomly assign values to X_1 and X_2, and then read whether or not X_3 has current. That is, we could assign a probability to each state the set of exogenous variables could occupy. For example,

$P(X_1 = 1, X_2 = 1) = 0.2$
$P(X_1 = 1, X_2 = 0) = 0.3$
$P(X_1 = 0, X_2 = 1) = 0.2$
$P(X_1 = 0, X_2 = 0) = 0.3$

Because this causal structure is deterministic (even though the exogenous variables are random), a probability distribution over the exogenous variables determines a joint distribution for the entire set of variables in the system. For this example the joint distribution over (X_1, X_2, X_3) is:

$P(X_1 = 1, X_2 = 1, X_3 = 1) = 0.2$
$P(X_1 = 1, X_2 = 1, X_3 = 0) = 0.0$
$P(X_1 = 1, X_2 = 0, X_3 = 1) = 0.0$
$P(X_1 = 1, X_2 = 0, X_3 = 0) = 0.3$
$P(X_1 = 0, X_2 = 1, X_3 = 1) = 0.0$
$P(X_1 = 0, X_2 = 1, X_3 = 0) = 0.2$
$P(X_1 = 0, X_2 = 0, X_3 = 1) = 0.0$
$P(X_1 = 0, X_2 = 0, X_3 = 0) = 0.3$

We say that this distribution is generated by the causal structure of figure 3.5.

We use this example not to investigate sampling schemes for circuits but rather to illustrate how probability distributions are generated by deterministic causal devices. The only assumption we make about the connection between deterministic causal structures and the probability distributions they may generate involves the distributions we will allow over the exogenous variables. *We assume that the exogenous variables are jointly independent in a probability distribution over the variables in a causally sufficient structure.* This is in part a substantive assumption—that statistical dependence is produced by causal connection—and in part a convention about representation. If exogenous variables in a structure are not independent, we expect that the causal graph is incomplete and there is some further causal mechanism, not represented in the graph, responsible for the statistical dependence. Either some of the input variables are causes of others (in which case we have equivocated, and the causal graph is not actually the graph of the causal structure of the structure) or else some nonconstant common causes of observed variables have not been included in the description of the structure.

3.3.2 Pseudoindeterministic and Indeterministic Causal Structures

In practice, the variables people measure are seldom deterministic functions of one another. We call a causal structure over a set **V** of variables for a population in which some variable is not a determinate function of its immediate causes in **V** an **indeterministic causal structure** for the population. An indeterministic causal structure might be pseudoindeterministic. That is, a deterministic causal structure for which not all of the causes of variables in **V** are also members of **V** may appear to be indeterministic, even though there is no genuine indeterminism if the set of variables is enlarged by adding variables that are not common causes of variables in **V**. For example, suppose again that the device shown in figures 3.1 and 3.5 governs the current in line X_3. Suppose also that X_2 is hidden from us so that only X_1 and X_3 occur in the causal structure we investigate. We might still hypothesize that X_1 is a cause of X_3, thereby forming the causal graph on the right side of figure 3.6.

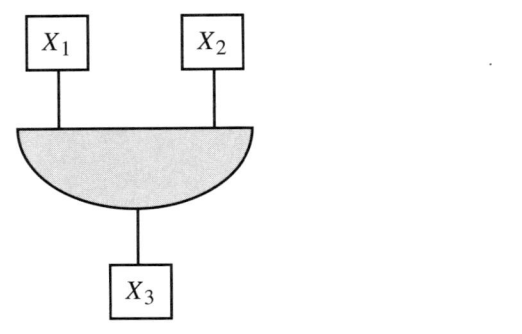

Actual Circuit Diagram   **Hypothesized Causal Graph**

Figure 3.6

Assuming that the joint distribution $P(X_1, X_2, X_3)$ generated by the actual circuit device is the same as the one given for figure 3.5 in section 3.3.1, the observed distribution $P(X_1, X_3)$ is just the marginal of $P(X_1, X_2, X_3)$, namely:

$P(X_1 = 1, X_3 = 1) = 0.2$
$P(X_1 = 1, X_3 = 0) = 0.3$
$P(X_1 = 0, X_3 = 1) = 0.0$
$P(X_1 = 0, X_3 = 0) = 0.5$

In the observed distribution X_3 is clearly not a function of its immediate parent X_1 and the causal structure appears to be indeterministic. We say the structure is pseudoindeterministic. More formally, causal structure $C = <V,E>$ is **pseudoindeterministic** for a population, if and only if C is not a deterministic causal structure for the population and there exists a causal structure C' for the population over a set of variables V' that properly includes V such that

(i) C' is a deterministic causal structure for the population;
(ii) If A and B are in V, then $<A,B>$ is in E if and only if $<A,B>$ is in E';
(iii) no variable in V is a cause of a variable in $V'\backslash V$;
(iv) no variable in $V'\backslash V$, is a common cause of two variable in V;

We say a structure is **linear pseudoindeterministic** if all functional dependencies in C' are linear. Structural equation models in the social sciences are usually assumed to be pseudoindeterministic causal structures. The error terms in such models are often interpreted as omitted causes.

It is at least conceivable that there are genuinely indeterministic structures, even genuinely indeterministic macroscopic structures, whose variables have a causal structure. We will assume that the same relations between conditional independence and causal structure that obtain for pseudoindeterministic structures hold as well for genuinely indeterministic causal relations, although as we will see later, there appear to be quantum mechanical systems for which that assumption must be carefully qualified. For a discussion of the case in which measured variables are exact functions of other measured variables see section 3.8.

3.4 The Axioms

We consider three conditions connecting probabilities with causal graphs: The Causal Markov Condition, the Causal Minimality Condition, and the Faithfulness Condition. These axioms are not independent. Consequences of various subsets of the conditions are investigated in the course of this book. We will consider justifications and objections to the conditions in the next section, but their importance—if not their truth—is evidenced by the fact that nearly every statistical model with a causal significance we have come upon in the social scientific literature satisfies all three: if the model were true, all three conditions would be met. While it is easy enough to construct models that violate the third of these conditions, Faithfulness, such models rarely occur in contemporary practice, and when they do, the fact that they have properties that are consequences of unfaithfulness is taken as an objection to them. In chapters 5 and 8 we will consider published log-linear models, regression models, and structural equation models satisfying the three conditions.

3.4.1 The Causal Markov Condition

The intuitions connecting causal graphs with the probability distributions they generate are unified and generalized in one fundamental condition:

Causal Markov Condition: Let G be a causal graph with vertex set **V** and P be a probability distribution over the vertices in **V** generated by the causal structure represented by G. G and P satisfy the Causal Markov Condition if and only if for every W in **V**, W is independent of **V**\(**Descendants**(W) $\cup$ **Parents**(W)) given **Parents**(W).

When G describes causal dependencies in a population with variables distributed as P satisfying the Causal Markov condition for G, we will sometimes say that P is **generated by** G. If **V** is not causally sufficient and **V** is a proper subset of the variables in a causal graph G generating a distribution P, we do *not* assume that the Causal Markov condition holds for the marginal over **V** of P.

The factorization results described in chapter 2 apply to the joint probability distribution for a set **V** of variables in a population of systems with a causal structure satisfying the Causal Markov Condition. If $P(V \mid \mathbf{Parents}\,(V))$ denotes the probability of V conditional on the (possibly empty) set of vertices that are direct causes of V, then

$$P(\mathbf{V}) = \prod_{V \in \mathbf{V}} P(V|\mathbf{Parents}(V))$$

for all values of **V** for which each $P(V|\mathbf{Parents}(V))$ is defined.

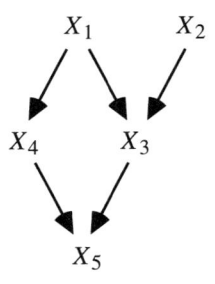

Figure 3.7

For the graph in figure 3.7, direct application of the Markov Condition yields a list of independence facts about the distribution generated by G.

$X_1 \perp\!\!\!\perp X_2$
$X_2 \perp\!\!\!\perp \{X_1, X_4\}$
$X_3 \perp\!\!\!\perp X_4 | \{X_1, X_2\}$
$X_4 \perp\!\!\!\perp \{X_2, X_3\} | X_1$
$X_5 \perp\!\!\!\perp \{X_1, X_2\} | \{X_3, X_4\}$

Other independence relations are entailed by these, for example

$\{X_4, X_5\} \perp\!\!\!\perp X_2 | \{X_1, X_3\}$

A discussion of axioms for conditional independence is found in Pearl 1988.

3.4.2 The Causal Minimality Condition

We will usually impose a further condition connecting probability with causality. The principle says that each direct causal connection prevents some independence or conditional independence relation that would otherwise obtain. For example, in the following causal graph G, C is a direct cause of A.

```
C ─────────► A
  ╲         ╱
   ╲       ╱
    ▼     ▼
      B
```

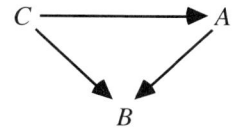

Figure 3.8

In a distribution P over $\{A,B,C\}$ for which $C \perp\!\!\!\perp A$, P satisfies the Markov condition even if the edge between C and A is removed from the graph.

Causal Minimality Condition: Let G be a causal graph with vertex set **V** and P a probability distribution on **V** generated by G. <G, P> satisfies the Causal Minimality condition if and only if for every proper subgraph H of G with vertex set **V**, the pair <H,P> does not satisfy the Causal Markov condition.

Since we will almost always give the graphs we consider a causal interpretation, we will in most cases hereafter simply describe these two conditions as the Markov and Minimality Conditions.

3.4.3 The Faithfulness Condition

Given a causal graph, the Markov condition determines a set of independence relations. These independence relations in turn may entail others, in the sense that every probability distribution having the independence relations given by the Markov condition will also have these further independence relations. In general a probability distribution P on a causal graph G satisfying the Markov condition may include other independence relations besides those entailed by the Markov condition applied to the graph. If, however, that does not occur, and all and only the independence relations of P are entailed by the Markov condition applied to G, we will say that P and G are **faithful to one another.** We will, moreover, say that a distribution P is faithful provided there is some directed acyclic graph to which it is faithful. So we consider a further axiom:

Faithfulness Condition: Let G be a causal graph and P a probability distribution generated by G. <G, P> satisfies the Faithfulness Condition if and only if every conditional independence relation true in P is entailed by the Causal Markov Condition applied to G.

Note that a distribution P is faithful to G if and only if it satisfies *both* the Markov and Faithfulness Conditions. The Faithfulness and Markov Conditions entail Minimality, but Minimality and Markov do not entail Faithfulness. We will sometimes use the weaker axiom or axioms and more often the stronger one. Faithfulness turns out to be important to discovering causal structure, and it also turns out to be the "normal" relation between probability distributions and causal structures.

3.5 Discussion of the Conditions

When and why should it be thought that probability and causality together satisfy these conditions, and when can we expect the conditions to be violated? When should the values of variables in a population be thought to be distributed in accordance with the conditions?

3.5.1 The Causal Markov and Minimality Conditions

If we consider probability distributions for the vertices of causal graphs of deterministic or pseudoindeterministic systems in which the exogenous variables are independently distributed, then the Markov Condition must be satisfied. We conjecture the Minimality Condition is true of all pseudoindeterministic systems. The warrant for the conditions lies in this fact, and in the history of human experience with systems that we can largely control or manipulate. Electrical devices, mechanical devices, chemical devices all satisfy the condition. Large areas of science and engineering—from auto mechanics to chemical kinetics to digital circuit design—would be impossible without using the principles to diagnose failures and infer mechanisms.

In an important class of cases the application of the Minimality and Markov Conditions may be unclear. In 1903 G. Udny Yule concluded his fundamental paper on the theory of association of attributes in statistics with a section "On the fallacies that may be caused by the mixing of distinct records." (Yule uses |AB | C| to denote "the association between A and B in the universe of C's" [p. 131]):

> It follows from the preceding work that we cannot infer independence of a pair of attributes within a sub-universe from the fact of independence within the universe at large. . . . The theorem is of considerable practical importance from its inverse application; that is, even if |AB| have a sensible positive or negative value we cannot be sure that nevertheless |AB | C | and |AB | γ | are not both zero. Some given attribute might, for instance, be inherited neither in the male line nor the female line; yet a mixed record might exhibit a considerable apparent inheritance. Suppose for instance that 50% of the fathers and of the sons exhibit the attribute, but only 10% of the mothers and daughters. Then if there be no inheritance in either line of descent the record must give (approximately)
>
> | | |
> |---|---|
> | fathers with attribute and sons with attribute: | 25% |
> | fathers with attribute and sons without attribute: | 25% |
> | fathers without attribute and sons with attribute: | 25% |
> | fathers without attribute and sons without attribute: | 25% |
> | mothers with attribute and daughters with attribute: | 1% |
> | mothers with attribute and daughters without attribute: | 9% |
> | mothers without attribute and daughters with attribute: | 9% |
> | mothers without attribute and daughters without attribute | 81% |

If these two records be mixed in equal proportions we get

parents with attribute and offspring with attribute	13%
parents with attribute and offspring without attribute	17%
parents without attribute and offspring with attribute	17%
parents without attribute and offspring without attribute	53%

Here 13/30 = 43 [and] 1/3% of the offspring of parents with the attribute possess the attribute themselves, but only 30% of offspring in general, that is, there is quite a large but illusory inheritance created simply by the mixture of the two distinct records. A similar illusory association, that is to say an association to which the most obvious physical meaning must not be assigned, may very probably occur in any other case in which different records are pooled together or in which only one record is made of a lot of heterogeneous material.

The fictitious association caused by mixing records finds its counterpart in the spurious correlation to which the same process may give rise in the case of continuous variables, a case to which attention was drawn and which was fully discussed by Professor Pearson in a recent memoir. If two separate records, for each of which the correlation is zero, be pooled together, a spurious correlation will necessarily be created unless the mean of one of the variables, at least, be the same in the two cases.

Yule's example seems to present a problem for the Causal Markov condition. Let a **mixture** over **V** be any population that consists of a combination of some finite number of subpopulations P_i each having different joint distributions over the variables in **V,** with each distribution satisfying the Causal Markov Condition for some graph. Consider a population that is a mixture of structures $<G,P_1>$ and $<G,P_2>$ where P_1 and P_2 are distinct and satisfy the Markov Condition for G. Let the proportions in the mixture be $n:m$.

Let $P(X,Y,Z) = nP_1(X,Y,Z) + mP_2(X,Y,Z)$, with $n + m = 1$. A little algebra shows that $P(XY|Z) = P(X|Z)P(Y|Z)$ if and only if

(1) $n^2 P_1(X,Y,Z)P_1(Z) + nmP_2(X,Y,Z)P_1(Z) + mnP_1(X,Y,Z)P_2(Z) + m^2 P_2(X,Y,Z)P_2(Z) =$

$n^2 P_1(X,Z)P_1(Y,Z) + nmP_1(X,Z)P_2(Y,Z) + mnP_2(X,Z)P_1(Y,Z) + m^2 P_2(X,Z)P_2(Y,Z).$

If $n, m > 0$ and in both distributions, X,Y are independent conditional on Z, that is, $P_1(X,Y|Z) = P_1(X|Z)P_1(Y|Z)$ and $P_2(X,Y|Z) = P_2(X|Z)P_2(Y|Z)$, then equation (1) reduces to

(2) $P_2(X|Z)P_2(Y|Z) + P_1(X|Z)P_1(Y|Z) = P_1(X|Z)P_2(Y|Z) + P_2(X|Z)P_1(Y|Z)$

The old but still rather surprising conclusion is that when we mix probability distributions we may find all possible conditional *dependence* relations. Thus, it seems, in many mixed populations conditional independence and dependence will not be a reliable guide to causal structure.

In the case of linear pseudoindeterministic systems, when populations with two different distributions each associated with a linear structure are mixed, vanishing correlations in each separate distribution will not produce vanishing correlations in the mixed distribution, and vanishing partial correlations in each separate distribution will not produce vanishing partial correlations in the mixed distribution. It is easy to verify that for any mixture of two distributions—based on linear structures or not—the covariance of two variables vanishes in the mixture if and only if

$$k_1 \text{COV}_1(XY) + k_2 \text{COV}_2(XY) = k_1 k_2 [\mu_1 X \mu_2 Y + \mu_1 Y \mu_2 X] + k_1(k_1 - 1)\mu_1 X \mu_1 Y + k_2(k_2 - 1) \mu_2 X \mu_2 Y$$

where the proportion of population 1 to population 2 is $n: m$ and $k_1 = n/(n+m)$, $k_2 = m/(n+m)$, and "μ_i" denotes the mean in population i.

So the situation is that we can have population 1 with causal graph G_1 and population 2 with causal graph G_2, and the joint population will have a distribution that does not satisfy the Markov Condition for either graph. The question is whether such a mixed population violates the Causal Markov Condition. *When a cause of membership in a subpopulation is rightly regarded as a common cause of the variables in V, the Causal Markov Condition is not violated in a mixed population*; instead, we have a population of systems satisfying the Causal Markov Condition but with a common cause (or causes) that may not have been measured. In some cases the cause of membership in a subpopulation may act like a latent switch variable of the kind considered in section 3.2.4; the distributions conditional on different values of the latent variable determine probability relations that are faithful to distinct causal graphs. In Yule's example, the missing common cause is gender. If, to take another example, we form a mixed sample of lead and copper pennies, within each subpopulation density and electrical conductivity will be independent, but in the mixed population they will be statistically dependent. We should say that is because chemical composition is a common cause of density and conductivity. In other cases the cause or causes of membership in relevant subpopulations may seem like *unnatural* kinds, or may at least not be the sort of causes a scientist seeks. Thus an important controversy (Caramazza 1986) in contemporary cognitive neuropsychology concerns the use of statistical results for samples of people selected by syndrome, for example subjects with Broca's aphasia. One aim of studying such groups may be to discover if two or more normal capacities have a common cause damaged in Broca's aphasics. Suppose in a sample of Broca's aphasics a correlation is observed in scores on tests of two cognitive skills. Should the psychologist conclude that the test performances have a common latent cause? Perhaps, but the common cause need not be any functional capacity—damaged or otherwise—that causes both skills. Instead, the sample of Broca's aphasics might be a mixture of people with different sorts of brain damage, and within each subgroup the skills in question might be independently distributed. The common cause is only a variable representing membership in a subpopulation.

There are contexts in which the statistics of mixtures do not reflect any variable for population membership. In linear models the correlations and partial correlations are determined by the linear coefficients and the variances of exogenous variables. These parameters themselves may be treated as random variables and the resulting population distribution is a (generally uncountable) mixture of distributions. Statistical, but not causal, inference has been extensively studied in such settings (Swamy 1971). If X is a random variable, we denote the expected value of X by $E(X)$.

THEOREM 3.1: Let M be a linear model with directed acyclic graph G and linear coefficients a_{ij}. Let M' be a linear model with directed acyclic graph G, such that the linear coefficients in M' are random variables a'_{ij} that are jointly independent of all other random variables in M', and $E(a'_{ij}) = a_{ij}$. Suppose the variances of the exogenous noncoefficient random variables are the same in M and M'. Then $\rho_{AB.C} = 0$ in M' if and only if $\rho_{AB.C} = 0$ in M.

Thus a population that is a mixture of linear pseudoindeterministic causally sufficient systems with the same causal graph and with parameters independently distributed will satisfy the Causal Markov Condition for that graph without any unmeasured common cause.

Professional philosophers have offered a spate of criticisms of consequences of the Causal Markov Condition. Most of them appear to depend on omitting relevant latent variables. Wesley Salmon (1984) claims that "[t]here is another, basically different, sort of common cause situation" that cannot appropriately be characterized in terms of the Causal Markov Condition. Salmon calls this other causal relation an "interactive fork."

One putative example of an "interactive fork" is from Davis (1988):

> Imagine a television set with a balky switch: it usually turns the set on, but not always. When the set is on, it produces both sound and picture. Then the probability of a picture given that the switch is on and given sound is greater than the probability of a picture given just that the switch is on. (Davis 1988, p. 156)

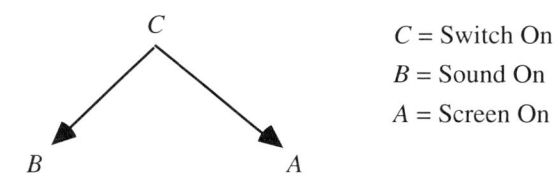

C = Switch On
B = Sound On
A = Screen On

Figure 3.9

So, $P(B|C) < P(B|A \text{ and } C)$.

Davis's example gives an inaccurate picture of the causal situation, which is better depicted in figure 3.10.

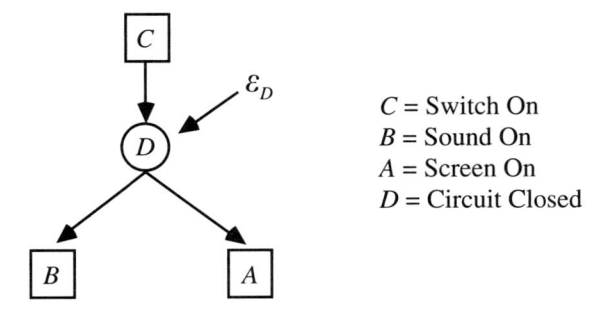

C = Switch On
B = Sound On
A = Screen On
D = Circuit Closed

Figure 3.10

The state of the circuit, or some variable downstream from the switch event, makes A and B independent.

Salmon's own illustration uses a slight variant of the following example from the game of pool (where we replace his events by Boolean variables).

C is the description of causal conditions relevant to both A and B, but A and B are not independent conditional on C.

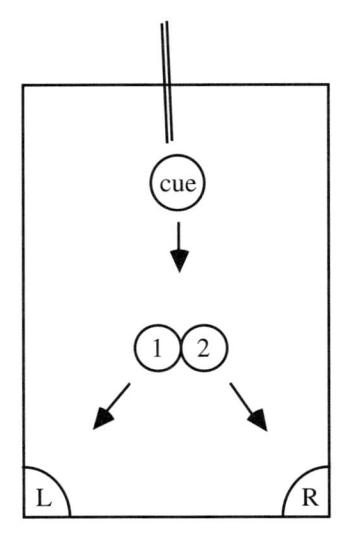

A = 1 ball in L.
B = 2 ball in R.
C = Collision of any sort between cue ball and either 1 or 2 ball.

Figure 3.11

Knowing C (that there was a collision) and A (that ball 1 dropped into its pocket) tells us more about whether B occurred (the 2 ball dropped into its pocket) than just knowing

C. *A* and *B* are not directly causally connected, and they are not independent conditional on *C*.

In Salmon's example, event *C* does not completely describe *all* of the common causes of *A* and *B*. *C* tells us that there was a collision of some sort between the cue ball and the 1 or 2 balls, but it does not tell us the nature of the collision. *A* informs us about the nature of the collision and therefore tells us more about *B*. Were the prior event more informative—for example, were it to specify the exact momentum of the cue ball on striking the two target balls—conditional independence would be regained. The example simply reflects a familiar problem in real data analysis that arises whenever some proxy variable is used in causal analysis or distinct values of a variable are collapsed. In our view these examples give no reason to doubt the Causal Markov Condition.

Elliott Sober (1987) argues that we routinely find correlations for which there are no common causes, or for which residual correlations remain after conditioning on known common causes. The correlation of bread prices in England and the sea level in Venice may have some common causes (perhaps the industrial revolution), but not enough to account for all of the dependency. His point seems to be Yule's: if we consider a series in which variable *A* increases with time and a series in which variable *B* increases with time, then *A* and *B* will be correlated in the population formed from all the units-at-times, even though *A* and *B* have no causal connection. Any such combined population is obviously a mixture of populations given by the time values.

There is a more fundamental objection to the Causal Markov Condition, namely that there exist nondeterministic causal systems for which, to the best of current knowledge, the condition is false. Consider pair production: a quantum mechanical event produces two particles which move off in different directions. Because of conservation laws, dynamical variables in the two particles must be correlated; if one has a component of spin up, for example, the other must have that spin component down. We can do experiments in which for pairs we measure either of two different components of spin at two spatially separated sensors and compute the correlations. Suppose there is some state *S* of the system at the moment the pair of particles is produced such that, conditional on *S*, the dynamical variables of the two particles are uncorrelated. J. S. Bell (1964) argued that on such an assumption there follows an inequality constraining the correlations of the measured dynamical variables. While the assumptions needed for the derivation are controversial, the empirical facts seem beyond doubt: Bell's inequality is violated in certain quantum mechanical experiments. In such experiments the correlated variables are associated with spatially remote subsystems, so unless principles constraining causal processes to act "locally" that is, not instantaneously over a distance, are abandoned, any statistical dependency is presumably *not* due to the effect of one sub-system on the other or to a common cause. Thus unless the locality principles are abandoned, the Causal Markov Condition appears to be false (Elby 1992).

In our view the apparent failure of the Causal Markov Condition in some quantum mechanical experiments is insufficient reason to abandon it in other contexts. We do not, for comparison, abandon the use of classical physics when computing orbits simply because classical dynamics is literally false. The Causal Markov Condition is used all the time in laboratory, medical and engineering settings, where an unwanted or unexpected statistical dependency is *prima facie* something to be accounted for. If we give up the Condition everywhere, then a statistical dependency between treatment assignment and the value of an outcome variable will never require a causal explanation and the central idea of experimental design will vanish. No weaker principle seems generally plausible; if, for example, we were to say only that the causal parents of Y make Y independent of more remote *causes*, then we would introduce a very odd discontinuity: So long as X has the least influence on Y, X and Y are independent conditional on the parents of X. But as soon as X has no influence on Y whatsoever, X and Y may be statistically dependent conditional on the parents of Y.

The basis for the Causal Markov Condition is, first, that it is necessarily true of populations of structurally alike pseudoindeterministic systems whose exogenous variables are distributed independently, and second, it is supported by almost all of our experience with systems that can be put through repetitive processes and whose fundamental propensities can be tested. Any persuasive case against the Condition would have to exhibit macroscopic systems for which it fails and give some powerful reason why we should think the macroscopic natural and social systems for which we wish causal explanations also fail to satisfy the condition. It seems to us that no such case has been made.

3.5.2 Faithfulness and Simpson's Paradox
Faithfulness can be violated in cases that realize variants of Simpson's "paradox" as Simpson originally presented it. We have already seen that both Yule and Pearson observed that two variables may be independent in subpopulations but dependent in a combined population. In 1948, M. G. Kendall used an example in his *Advanced Theory of Statistics* illustrating the reverse situation: two binary variables are independent but are dependent conditional on a third variable. Kendall's case was given a twist in a paper by Simpson (1951) a few years later, who thought his example introduced difficulties about the relation between causal dependencies and contingency tables. Subsequently the phenomenon the example exhibits has been referred to as "Simpson's paradox." Like examples have since become standard puzzlers in discussions of the connection between causality and probability.

Kendall's example[5] was as follows:

> Consider the case in which a number of patients are treated for a disease and there is noted the number of recoveries. Denoting A by recovery, ~A by nonrecovery, B by treatment, ~B by not-treatment,[6] suppose the frequencies are

	B	$\sim B$	Totals
A	100	200	300
$\sim A$	50	100	150
Totals	150	300	450

Here $(AB) = 100 = (A)(B)/N$, so that the attributes are independent. So far as can be seen, treatment exerts no effect on recovery. Denoting male sex by S_M and female sex by S_F, suppose the frequencies among males and females are

Males

	BS_M	$\sim B\ S_M$	Totals
AS_M	80	100	180
$\sim AS_M$	40	80	120
Totals	120	180	300

Females

	BS_F	$\sim B\ S_F$	Totals
AS_F	20	100	120
$\sim AS_F$	10	20	30
Totals	30	120	150

In the male group we now have

$Q_{AB.S_M} = 0.231$

and in the female group

$Q_{AB.S_F} = -0.429$

Thus among the males treatment is positively associated with recovery, and among the females negatively associated. The apparent independence in the two together is due to canceling of these associations in the sub-populations.

Kendall's example is thus of a mixture of two distributions, one for males and one for females, such that the positive association between two variables in one population is exactly canceled by the negative association in the other. There is nothing paradoxical in that, and one may find empirical examples for which the same structure is claimed. The mixed distribution will violate the Faithfulness Condition, because it will exhibit a statistical independence relation that does not follow from the Markov condition applied to the causal graph common to all units.

Kendall's explanation of his contingency table depends on the fact that in one population the association of two variables is positive, and in the other negative. But what can be going on if in *both sub-populations the association is positive*, and yet in the mixed population it vanishes? That is exactly the question Simpson posed in 1951.[7] Simpson gave the following table and commentary:

	Male		Female	
	Untreated	Treated	Untreated	Treated
Alive	4/52	8/52	2/52	12/52
Dead	3/52	5/52	3/52	15/52

> This time . . . there is a positive association between treatment and survival both among males and among females; but if we combine the tables we...find that there is no association between treatment and survival in the combined population. What is the "sensible" interpretation here? The treatment can hardly be rejected as valueless to the race when it is beneficial when applied to males and to females.

The question is what causal dependencies can produce such a table, and that question is properly known as "Simpson's paradox."[8]

In Simpson's example the variables G (male or female), T (treated or untreated) and S (survives or does not) are given an interpretation that imposes tacit restrictions on causal structure. When we read the example we naturally assume that gender G cannot be caused by treatment T or survival S, but may cause them. As with Kendall's example, the distribution in Simpson's table satisfies the Causal Markov Condition for a graph in which G causes T and S and T causes S. Simpson's distribution is not, however, faithful to such a graph, because T and S are independent in the distribution even though T is a parent of S in the graph.

Suppose for a moment that we ignore the interpretation that Simpson gave to the variables in his example, which was, after all, entirely imaginary, and let ourselves consider causal structures that would be excluded by that interpretation. To avoid substantive associations, we substitute A for T, B for G, and C for S and obtain graph (i) in figure 3.12. Distributions such as Simpson's and Kendall's can also be realized by a graph in which A and C are not adjacent but each causes B, as in graph (ii) in figure 3.12.[9]

With the substitution of variables just noted, Simpson's distribution is faithful to graph (ii) but not to graph (i); moreover (ii) is the only graph faithful to the distribution.

Causation and Prediction 41

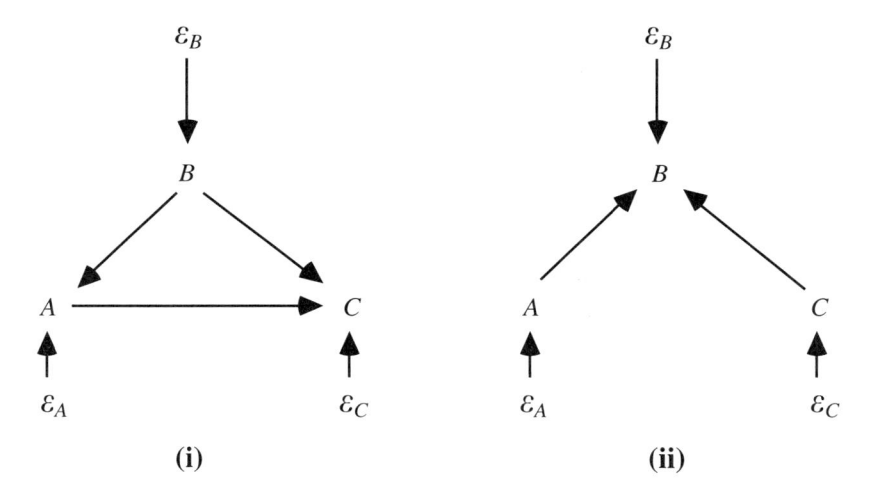

Figure 3.12

Judea Pearl (1988) offers a Bayesian example that illustrates why, when a causal structure like that in graph (ii) obtains, one should expect that A and C, though independent, are dependent conditional on B: Whether or not your car starts depends on whether or not the battery is charged and also on whether or not there is fuel in the tank, but these conditions are independent of one another. Suppose you find that your car won't start, and you hold in that case that there is some probability that the fuel tank is empty and some probability that the battery is dead. Suppose next you find that the battery is not dead. Doesn't the probability that the fuel tank is empty change when that information is added?

Were we to find that A and C are independent but dependent conditional on B, the Faithfulness Condition requires that if any causal structure obtains, it is structure (ii). Still, structure (i) is logically possible, and if the variables had the significance Simpson gives them we would of course prefer it. But if prior knowledge does not require structure (i), what do we lose by applying the Faithfulness Condition; what, in other words, do we lose by excluding causal structures that are not faithful to the distribution?

In the linear case, the parameter values—values of the linear coefficients and exogenous variances of a structure—form a real space, and the set of points in this space that create vanishing partial correlations not implied by the Markov Condition have Lebesgue measure zero.

THEOREM 3.2: Let M be a linear model with directed acyclic graph G and n linear coefficients $a_1,..., a_n$ and k positive variances of exogenous variables $v_1,..., v_k$. Let $M(<u_1,...,u_n, u_{n+1},...,u_{n+k}>)$ be the distributions consistent with specifying values $<u_1,...,u_n, u_{n+1},...,u_{n+k}>$ for $a_1,..., a_n$ and $v_1,...v_k$. Let Π be the set of probability measures P on the space $\Re^{n+k}$ of values of the parameters of M such that for every subset $\mathbf{V}$ of $\Re^{n+k}$ having

Lebesgue measure zero, $P(\mathbf{V}) = 0$. Let $\mathbf{Q}$ be the set of vectors of coefficient and variance values such that for all q in $\mathbf{Q}$ every probability distribution in with $M(q)$ has a vanishing partial correlation that is not linearly implied by G. Then for all P in Π, $P(\mathbf{Q}) = 0$.

The theorem can be strengthened a little; it is not really necessary that the set of exogenous and error variables be jointly independent—pairwise independence is sufficient. In the pseudoindeterministic case, faithfulness can be violated, if at all, only by very special choices of the functional dependencies between variables. Consider a population of linear, pseudoindeterministic systems in which the exogenous variables are independently and normally distributed. The conditional independence relations required by the Markov Condition will be automatically fulfilled for every possible value of the linear coefficients—they are guaranteed just by the way the device acts to compose linear functions. But conditional independence relations that are *not* required by the Markov Condition—the sorts of conditional independence relations that characterize distributions that are unfaithful to the causal structure of the devices—either cannot be produced at all or can only be produced if the linear coefficients satisfy very strong constraints.

The same moral applies to other classes of functions. While for discrete variables we have not attempted a formal proof of a theorem analogous to 3.2, such a result should be expected on intuitive grounds. The factorization formula for distributions satisfying the Markov Condition for a graph provides a natural parametrization of the distributions. If an exogenous variable has n values, it determines $n-1$ parametric dimensions consisting of a copy of the open interval $(0,1)$. If an endogenous variable X has n values, a conditional probability $P(X|\mathbf{Parents}(X))$ in the factorization determines another $n-1$ parametric dimensions consisting of a copy of $(0,1)$ for each vector of values of the parents of X. One expects that the set of probability values that generate conditional independence relations not entailed by the factorization itself will be measure zero in this parameter space. (Meek [1995] provides a proof of this conjecture.]

3.6 Bayesian Interpretations

We have interpreted the conditions as about frequencies in populations in which all units have the same causal structure. We wish to consider how the conditions can be given a Bayesian interpretation in which the probabilities are subjective. Current subjectivist interpretations hold that probability is an idealization of rational, subjective degree of belief. On a strict subjectivist view there can be finite frequencies, but there is no such thing as objective probability. One assumes the systems under study in the sciences are deterministic, and any appearance of indeterminacy is due simply to ignorance. The likelihood structures of Bayesian statistical models often look like ordinary un-Bayesian statistical models; Bayesians add a prior probability distribution over the free parameters. For example, Bayesian linear models specify a distribution over Θ, representing linear coefficients, variances, means, and so on. The Bayesian model is thus a mixture of ordinary linear models, and the joint distribution over the measured variables does not satisfy the conditions we have considered in this section.

Consider a study of systems with causal graph G. Suppose a Bayesian agent's degrees of belief are represented by a density, f, satisfying the condition $f(X|\textbf{Parents}(G,X)) = h(\textbf{Parents}(G,X); \Theta)$, where Θ is a parameter whose values determine a density for X conditional on its parents. Let the Bayesian agent also have a distribution over Θ. In such a case we understand the Causal Markov and Causal Minimality conditions to constrain the agent's degrees of belief conditional on Θ. The subjective joint distribution over the variables conditional on Θ will satisfy the conditions, but typically the unconditional joint distribution will not.

Suppose now that the agent entertains a set **G** of alternative possible causal structures, and holds that in each structure G in **G** $f(X|\textbf{Parents}(G,X)) = h(\textbf{Parents}(G,X); \Theta_G)$, as before. Then we understand the Causal Markov and Causal Minimality conditions to constrain the agent's degrees of belief conditional on Θ_G, G.

So understood, the conditions are normative principles about "reasonable" degrees of belief. In a later chapter we will consider in some detail a Bayesian proposal for clinical trials and argue that the assumptions the proposal makes about the degrees of belief of scientific experts accords with the Markov Condition.

3.7 Consequences of the Axioms

Consequences of the Causal Markov, Minimality, and Faithfulness Conditions are developed throughout this book, but some important connections between causal dependency and statistical dependency should be noted here.

3.7.1 d-Separation

Given a causal graph G, the Markov Condition axiomatizes the set of independence and conditional independence relations true of any distribution P faithful to G. But which conditional independence relations follow from the Markov Condition for a given graph may not be obvious. Suppose one wanted to know, for each pair of vertices X and Y and each set of vertices **Q** not containing X and Y, whether or not X and Y are independent conditional on **Q**, that is, all the atomic independence facts among sets of variables. Applying the Markov Condition directly to G, that is, applying the definition for each vertex, does not in general suffice.

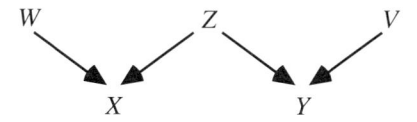

Figure 3.13

For example, in a distribution faithful to the graph in figure 13, suppose we wanted to know whether X and Y are independent conditional on the set **Q** = $\{Z\}$. Applying the Markov Condition directly to figure 3.13, we obtain:

$W \perp\!\!\!\perp \{Z,Y,V\}$

$X \perp\!\!\!\perp \{Y,V\} \mid \{W,Z\}$

$Z \perp\!\!\!\perp \{W,V\}$

$Y \perp\!\!\!\perp \{W,X\} \mid \{V,Z\}$

$V \perp\!\!\!\perp \{W,X,Z\}$

It is not obvious that these facts entail $X \perp\!\!\!\perp Y \mid \{Z\}$. Pearl proposed a purely graphical characterization—which he called **d-separation**—of conditional independence, and Geiger, Pearl, and Verma (Geiger and Pearl 1989a; Verma 1987) proved that d-separation in fact characterizes all and only the conditional independence relations that follow from satisfying the Markov condition for a directed acyclic graph.

We define d-separation twice - first concisely and then in terms that make the idea more accessible and, in our experience, much easier to remember and apply.

D-separation (Definition 1): If X and Y are distinct vertices in a directed graph G, and **W** a set of vertices in G not containing X or Y, then X and Y are **d-separated** given **W** in G just in case there exists no undirected path U between X and Y, such that

(i) every collider on U has a descendent in **W,** and
(ii) no other vertex on U is in **W**.

X and Y are **d-connected** given **W** if and only if they are not d-separated by **W**. If **U**, **V**, and **W** are disjoint sets of vertices in G and **U** and **V** are not empty then we say that **U** and **V** are **d-separated** given **W** if and only if every pair $<U,V>$ in the cartesian product of **U** and **V** is d-separated given **W**. Similarly for d-connection among sets.

The second definition of d-separation relies on the notions of an "active path" and an "active vertex" on a path, where active can be thought of as carrying association. Again, X and Y must be distinct, **W** cannot contain X or Y, and the concepts are all relative to a directed graph G.

D-separation (Definition 2): X and Y are **d-separated** given **W** just in case there is no undirected path between X and Y that is active relative to **W**.

An **undirected path** U is **active** relative to **W** just in case every vertex on U is active relative to **W**.

A **vertex** V is **active** on a path relative to **W** just in case either
i) V is a collider, and V or any of its descendents is in **W**, or
ii) V is a noncollider and is *not* in **W**.

Causation and Prediction

Consider first the situation when **W** = ∅, which we call the "unconditional" case. If **W** = ∅, then each vertex on an undirected path U is active just in case it is a noncollider. Since all the vertices on a path must be active for the whole path to be active, U is unconditionally active just in case all of its vertices are noncolliders. Unconditionally active paths are just treks.[10]

For example, in the causal graph in figure 3.14, X and V are d-separated by the empty set because Y is a collider (inactive) on the only path between them.

Conditioning on a node flip-flops its status. Whereas X and Y are unconditionally d-connected in figure 3.14, they are d-separated given $\{W\}$, $\{Z\}$, or $\{W,Z\}$. This is because the vertices on the path between X and Y are W and Z, which are noncolliders and thus unconditionally active. Conditioning flips their status from active to inactive, and if either is inactive the whole path between X and Y become inactive. That conditioning on a noncollider makes it inactive is similar to the intuition behind the Markov Condition. A noncollider is either a common cause (Z in figure 3.14) or part of a directed path, for example, W. Effects are made independent when we condition on their common causes, and effects are made independent of their remote causes when we condition on their more proximate ones.

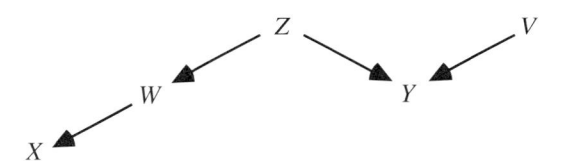

Figure 3.14

X and V are unconditionally d-separated in figure 3.14, but are d-connected given $\{Y\}$. This is because unconditionally, Y is the only inactive node on the path between X and V. Conditioning on Y makes it and thus the whole path active. That conditioning on a collider makes it active was discussed in section 3.5.2 above.

There is one additional twist, however. While conditioning on a collider activates it, so does conditioning on any of its descendants. In the graph in figure 3.15, for example, X and Y are d-connected given W. This is because although U is a collider on the only path between X and Y and thus unconditionally inactive, it is activated because one of its descendents W is in the conditioning set.

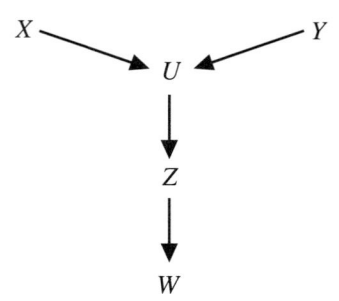

Figure 3.15

Checking whether two vertices X and Y are d-separated by a set $\mathbf{Q}$ in a graph G, and thus whether X and Y are independent conditional on $\mathbf{Q}$ in a distribution faithful to G should now be relatively straightforward. Check each undirected path between X and Y until one is found that is active relative to $\mathbf{Q}$, in which case X and Y are d-connected by $\mathbf{Q}$, or all paths have been checked and are inactive, in which case X and Y are d-separated by $\mathbf{Q}$. For each path, check each vertex on the path. If any are inactive the path is inactive. A vertex is inactive if it is a noncollider in $\mathbf{Q}$ or a collider with no descendents in $\mathbf{Q}$. A vertex is active if it is a noncollider not in $\mathbf{Q}$ or a collider with a descendant in $\mathbf{Q}$. In figure 16, for example, X and Y are d-connected given $\{U\}$, but they are d-separated given $\{V, Z\}$.

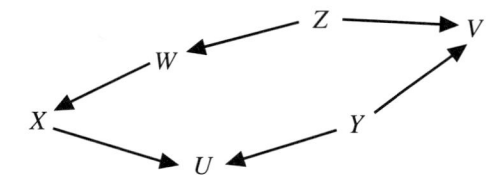

Figure 3.16

In the first case, conditioning on U activates the $X \rightarrow U \leftarrow Y$ path, and all it takes is one active path for $\{U\}$ to d-connect X and Y. In the second case, conditioning on $\{V,Z\}$ activates V on the $X \leftarrow W \leftarrow Z \rightarrow V \leftarrow Y$ path, but conditioning inactivates Z on this path and thus makes it inactive; the $X \rightarrow U \leftarrow Y$ path is also inactive given $\{V,Z\}$ because U is a collider on the path that is not in $\{V,Z\}$ and has no descendent in $\{V,Z\}$. Because all of the undirected paths between X and Y are inactive given $\{V,Z\}$, X and Y are d-separated given $\{V,Z\}$.

The essential results are the following:

THEOREM 3.3: $P(\mathbf{V})$ is faithful to directed acyclic graph G with vertex set $\mathbf{V}$ if and only if for all disjoint sets of vertices $\mathbf{X}$, $\mathbf{Y}$, and $\mathbf{Z}$, $\mathbf{X}$ and $\mathbf{Y}$ are independent conditional on $\mathbf{Z}$ if and only if $\mathbf{X}$ and $\mathbf{Y}$ are d-separated given $\mathbf{Z}$.

Theorem 3.4 provides a slightly more intuitive characterization of faithfulness, which motivates algorithms developed in chapter 5.

THEOREM 3.4: If $P(\mathbf{V})$ is faithful to some directed acyclic graph, then $P(\mathbf{V})$ is faithful to directed acyclic graph G with vertex set $\mathbf{V}$ if and only if

(i) for all vertices X, Y of G, X and Y are adjacent if and only if X and Y are dependent conditional on every set of vertices of G that does not include X or Y; and
(ii) for all vertices X, Y, Z such that X is adjacent to Y and Y is adjacent to Z and X and Z are not adjacent, $X \rightarrow Y \leftarrow Z$ is a subgraph of G if and only if X, Z are dependent conditional on every set containing Y but not X or Z.

The study of correlation is historically tied to the normal distribution, and for that distribution vanishing partial correlations and conditional independence are equivalent. But the Markov and Faithfulness Conditions tie vanishing correlation and partial correlation to graphical and causal structure for linear systems, without any normality assumption. Thus, for linear systems, correlational structure is a guide to causal structure. We will say that a distribution P is **linearly faithful** to a graph G if and only if for vertices A and B of G and all subset $\mathbf{C}$ of the vertices of G, A and B are d-separated given $\mathbf{C}$ if and only $\rho_{AB.\mathbf{C}} = 0$.

THEOREM 3.5: If G is a directed acyclic graph with vertex set $\mathbf{V}$, A and B are in $\mathbf{V}$, and $\mathbf{H}$ is included in $\mathbf{V}$, then G linearly implies $\rho_{AB.\mathbf{H}} = 0$ if and only A and B are d-separated given $\mathbf{H}$.

It follows that a distribution P is linearly faithful to a graph G if and only if for vertices A and B of G and all subsets $\mathbf{C}$ of the vertices of G, A and B are d-separated given $\mathbf{C}$ if and only $\rho_{AB.\mathbf{C}} = 0$. Theorem 3.5 is the general principle behind all of the path analysis examples (Wright 1934; Simon 1954; Blalock 1961; Heise 1975) connecting causal structure in "recursive" (i.e., acyclic) linear models with vanishing partial correlations.

In the chapters that follow we will frequently remark that some conditional independence or conditional dependence relation, or vanishing or nonvanishing partial correlation, follows from a causal structure, assuming the distribution is faithful. Conversely, we will often observe that given certain conditional independence and dependence relations, or partial correlation facts, the causal structure must have certain properties if the distribution is faithful. Whenever we make such claims, we are using tacit corollaries of theorems 3.3, 3.4, and 3.5.

3.7.2 The Manipulation Theorem

The fundamental aim of many empirical studies is to predict the effects of changes, whether the changes come about naturally or are imposed by deliberate policy. How can an observed distribution *P* be used to obtain reliable predictions of the effects of alternative policies that would *impose* a new marginal distribution on some set of variables? The very idea of imposing a policy that would directly change the distribution of some variable (e.g., drug use) necessitates that the resulting distribution P_{Man} will be different from *P*. *P* alone cannot be used to predict P_{Man}, but *P* and the causal structure can be.

Suppose that the Surgeon General is considering discouraging smoking, and he asks "What would the distribution of *Cancer* be if no one in the U.S were allowed to smoke?" Let **V** = {*Drinking*, *Smoking*, and *Cancer*}. For the purpose of illustration assume that in the actual population, the causal structure shown in figure 3.17 is correct.

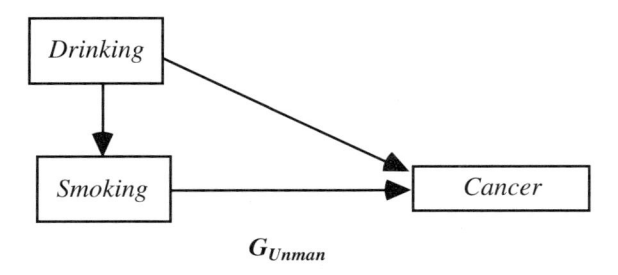

G_{Unman}

Figure 3.17

Let us call the population actually sampled (or produced by sampling and some experimental procedure) the **unmanipulated** population, and the hypothetical population for which smoking is banned the **manipulated** population. Suppose that if the policy of banning smoking were put into effect it would be completely effective, stopping everyone from smoking, but would not affect the value of *Drinking* in the population. Then the causal graph for the hypothetical manipulated population will be different than for the unmanipulated population, and the distribution of *Smoking* is different in the two populations. The manipulated causal graph is shown in figure 3.18.

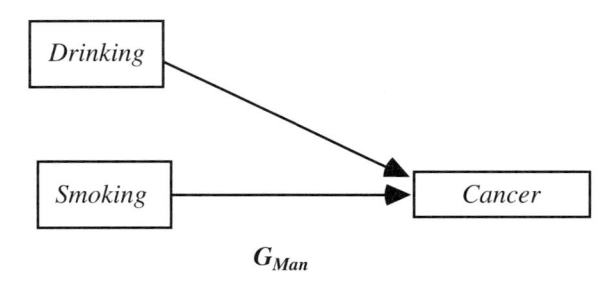

G_{Man}

Figure 3.18

Causation and Prediction

The difference between the unmanipulated graph and the manipulated graph is that some vertices that are parents of the manipulated variables in G_{Unman} may not (depending upon the precise form of the manipulation) be parents of manipulated variables in G_{Man} and vice-versa.

How can we describe the change in the distribution of *Smoking* that will result from banning smoking? One way is to note that the value of a variable that represents the policy of the federal government is different in the two populations. So we could introduce another variable into the causal graph, the *Ban Smoking* variable, which is a cause of *Smoking*. The full causal graph, including the new variable representing smoking policy, is then shown in figure 3.19. In the actual unmanipulated population the *Ban Smoking* variable is *off*, and in the hypothetical population the *Ban Smoking* variable is *on*. In the actual population we measure $P(Smoking|Ban\ Smoking = off)$; in the hypothetical population that would be produced if smoking were banned $P(Smoking = 0 | Ban\ Smoking = on) = 1$. For any subset $\mathbf{X}$ of $\mathbf{V} = \{Smoking, Drinking, Cancer\}$ in the causal graph, let $P_{Unman(Ban\ Smoking)}(\mathbf{X})$ be $P(\mathbf{X}|Ban\ Smoking = off)$ and let $P_{Man(Ban\ Smoking)}(\mathbf{X})$ be $P(\mathbf{X}|Ban\ Smoking = on)$.

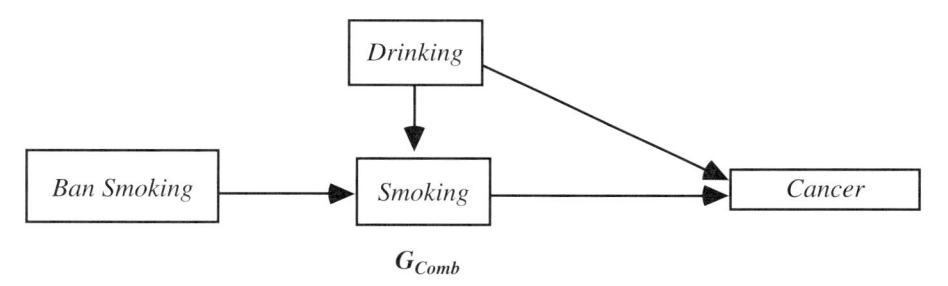

G_{Comb}

Figure 3.19

We can now ask if $P_{Unman(Ban\ Smoking)}(Cancer|Smoking) = P_{Man(Ban\ Smoking)}(Cancer|Smoking)$ (for those values of *Smoking* for which $P_{Man(Ban\ Smoking)}(Cancer|Smoking)$ is defined, namely *Smoking* = 0)? Clearly the answer is affirmative exactly when *Cancer* and *Ban Smoking* are independent given *Smoking*; but if the distribution is faithful this just reduces to the question of whether *Cancer* and *Ban Smoking* are d-separated given *Smoking*, which they are not in this causal graph. Further, $P_{Unman(Ban\ Smoking)}(Cancer) \neq P_{Man(Ban\ Smoking)}(Cancer)$ because *Cancer* is not d-separated from *Ban Smoking* given the empty set. But in contrast $P_{Unman(Ban\ Smoking)}(Cancer|Smoking,Drinking) = P_{Man(Ban\ Smoking)}(Cancer|Smoking,Drinking)$ (for those values of *Smoking* for which $P_{Man(Ban\ Smoking)}(Cancer|Smoking,Drinking)$ is defined, namely *Smoking* = 0), because *Ban Smoking* and *Cancer* are d-separated by {*Smoking, Drinking*}. The importance of this invariance is that we can predict the distribution of cancer if smoking is banned by considering the conditional distribution of cancer given drinking in the observed subpopulation of nonsmokers, and by considering the distribution of drinking in the unmanipulated population.

Note that one of the *inputs* to our conclusion about $P_{Man(Ban\ Smoking)}(Cancer)$ is that the ban on smoking is completely successful and that it does not affect *Drinking*; this knowledge does not come from the measurements that we have made on *Smoking*, *Drinking* and *Cancer*, but is assumed to come from some other source. Of course, if the assumption is incorrect, there is no guarantee that our calculation of $P_{Man(Ban\ Smoking)}(Cancer)$ will yield the correct result. If we had instead considered a policy that does not effectively ban smoking, but intervenes to make smoking less likely without affecting drinking, then the graph of the entire system including the manipulation variable *Ban Smoking*, would be the same as in figure 3.19, and the graph G_{Unman} would be as in figure 3.17, but the manipulated graph G_{Man} would look like figure 3.17 rather than 3.18. Intervention would alter but not remove the influence of drinking on smoking.

The analysis of prediction for a system involves three distinct graphs: a causal graph G_{Comb} which includes variables **W** representing manipulations, and a causal graph G_{Unman} which is the subgraph of G_{Comb} over a set of variables **V** not including the variables representing manipulations, and a graph G_{Man} over **V** which represents the causal relations among variables in **V** that result from a manipulation. G_{Man} may be a subgraph of G_{Unman} if the manipulation "breaks" causal dependencies in G_{Unman}; otherwise G_{Unman} and G_{Man} will be the same graph.

Here are the formal definitions: If G is a directed acyclic graph over a set of variables $\mathbf{V} \cup \mathbf{W}$, and $\mathbf{V} \cap \mathbf{W} = \emptyset$, then **W is exogenous with respect to V** in G if and only if there is no directed edge from any member of **V** to any member of **W**. If G_{Comb} is a directed acyclic graph over a set of variables $\mathbf{V} \cup \mathbf{W}$, and $P(\mathbf{V} \cup \mathbf{W})$ satisfies the Markov condition for G_{Comb}, then changing the value of **W** from w_1 to w_2 is a **manipulation** of G_{Comb} with respect to **V** if and only if **W** is exogenous with respect to **V**, and $P(\mathbf{V}|\mathbf{W} = w_1) \neq P(\mathbf{V}|\mathbf{W} = w_2)$.

We define $P_{Unman(\mathbf{W})}(\mathbf{V}) = P(\mathbf{V}|\mathbf{W} = w_1)$, and $P_{Man(\mathbf{W})}(\mathbf{V}) = P(\mathbf{V}|\mathbf{W} = w_2)$, and similarly for various marginal and conditional distributions formed from $P(\mathbf{V})$.

We refer to G_{Comb} as the **combined graph**, and the subgraph of G_{Comb} over **V** as the **unmanipulated graph** G_{Unman}. (Note that while $P_{Unman(\mathbf{W})}(\mathbf{V})$ satisfies the Markov Condition for G_{Unman}, it may also satisfy the Markov Condition for a subgraph of G_{Unman}. This is because G_{Comb}, and hence its subgraph G_{Unman}, may contain edges that are needed to represent the distribution of the manipulated subpopulation but not needed to represent the distribution of the unmanipulated subpopulation.)

V is in **Manipulated(W)** (that is, V is a variable directly influenced by one of the manipulation variables) if and only if V is in **Children(W)** $\cap$ **V**; we will also say that the variables in **Manipulated(W)** have been **directly manipulated**. We will refer to the variables in **W** as **policy variables**.

The **manipulated graph**, G_{Man} is a subgraph of G_{Unman} for which $P_{Man(\mathbf{W})}(\mathbf{V})$ satisfies the Markov Condition and which differs from G_{Unman} in at most the parents of members of **Manipulated(W)**. Exactly which subgraph G_{Man} is depends upon the details of the manipulation and what the causal graph of the subpopulation where $\mathbf{W} = \mathbf{w_2}$ is. For example, if smoking is banned, then G_{Man} contains no edge between income and smoking. On the other hand, if taxes are raised on cigarettes, G_{Man} does contain an edge between income and smoking. We will prove (in chapter 13) that given a manipulation as defined, there always exists a subgraph of G_{Unman} for which $P_{Man(\mathbf{w})}(\mathbf{V})$ satisfies the Markov Condition. All of our theorems about manipulations hold for any G_{Man} that is a subgraph of G_{Unman} for which $P_{Man(\mathbf{w})}(\mathbf{V})$ satisfies the Markov Condition, and which differs from G_{Unman} in at most the parents of members of **Manipulated(W)**.

These definitions entail the **Manipulation Theorem:**

THEOREM 3.6: (Manipulation Theorem): Given directed acyclic graph G_{Comb} over vertex set $\mathbf{V} \cup \mathbf{W}$ and distribution $P(\mathbf{V} \cup \mathbf{W})$ that satisfies the Markov condition for G_{Comb}, if changing the value of $\mathbf{W}$ from $\mathbf{w_1}$ to $\mathbf{w_2}$ is a manipulation of G_{Comb} with respect to $\mathbf{V}$, G_{Unman} is the unmanipulated graph, G_{Man} is the manipulated graph, and

$$P_{Unman(\mathbf{W})}(\mathbf{V}) = \prod_{X \in \mathbf{V}} P_{Unman(\mathbf{W})}(X | \mathbf{Parents}(G_{Unman}, X))$$

for all values of $\mathbf{V}$ for which the conditional distributions are defined, then

$$P_{Man(\mathbf{W})}(\mathbf{V}) =$$
$$\prod_{X \in \mathbf{Manipulated}(\mathbf{W})} P_{Man(\mathbf{W})}(X | \mathbf{Parents}(G_{Man}, X)) \times$$
$$\prod_{X \in \mathbf{V} \setminus \mathbf{Manipulated}(\mathbf{W})} P_{Unman(\mathbf{W})}(X | \mathbf{Parents}(G_{Unman}, X))$$

for all values of $\mathbf{V}$ for which each of the conditional distributions is defined.

The importance of this theorem is that if the causal structure and the direct effects of the manipulation (i.e., $P_{Man(\mathbf{W})}(X|\mathbf{Parents}(X))$ for each X in **Manipulated**(W) are known, then the joint distribution can be estimated from the unmanipulated population.

The Manipulation Theorem is not applicable when a causal mechanism between a pair of variables is reversible, in which case there can be two subpopulations in which the direction of the causal relationship between a pair of variables is reversed. For example, the movement of a motor of a car may cause the wheels to turn (as when the gas pedal is pressed), but also the turning of the wheels can cause the motor to move (as when the car rolls downhill).[11] An intervention in a causal system which reverses the direction of some causal relationship is not a manipulation in our technical sense because there is no one combined graph representing the causal relations in the combined population. We are not suggesting any non-experimental methods for determining whether a given mechanism is reversible. In some cases, such as smoking and yellow fingers, it is obvious from background knowledge that the mechanism is not reversible, because yellow fingers cannot cause smoking. In other cases, the relevant background knowledge may not be available, in which case it is not known whether the Manipulation Theorem is applicable.

Rubin (1977; 1978), and following him Pratt and Schlaifer (1988), have offered rules for when conditional probabilities in an observed population of systems will equal conditional probabilities for the same variables if the population is altered by a direct manipulation of some variables for all population units. We will show in chapter 7 that their various rules are direct consequences of the special case of the Manipulation Theorem, illustrated in the discussion of figures 3.17, 3.18, and 3.19, in which one variable is manipulated and the intervention makes that variable independent of its causes in the unmanipulated graph.

Because the Manipulation Theorem is a consequence of the Markov condition, it requires no separate justification. Although the Manipulation Theorem is abstract, it is just the general formulation of inferences that are routine, if not always correct. When, for example, a regression model is used to predict the effects of a policy that would force values on some of the regressors, we have an application of the Manipulation Theorem. Of course the prediction may be incorrect if the causal or statistical assumptions of the regression model are false, or if the changes actually carried out do not satisfy the conditions for a manipulation. There are striking examples of both sorts of failure. Application of the Manipulation Theorem may give misleading predictions if the values of variables for each unit depend on the values of other units and if that dependency is not represented in the causal graph. Some public policy debates illustrate absurd violations of this requirement. Recently a research institute funded by automobile insurers carried out a nonlinear regression of the rate of fatalities of occupants of various kinds of cars against car length, weight and other variables, finding unsurprisingly that the smaller the car the higher the fatality rate. This statistical analysis was then used by others to argue that proposed federal policies to downsize the American automobile fleet would increase highway fatalities. But of course the fatality rate in cars of a given size depends on the distribution of sizes of other cars in the fleet.

One can mistake which variables will be directly affected by a policy or intervention. Tacit applications of the Manipulation Theorem in such cases can lead to disappointment. As we will see in a later chapter, the literature on smoking, lung cancer and mortality provides vivid examples of predictions that went wrong, arguably because of misjudgments as to which variables would be directly manipulated by an intervention.

There is no reason why every intervention to deliberately alter the distribution of values of a set of variables **V** among units in a population (or sample) *must* satisfy the conditions for a direct manipulation of **V** and no others. But one of the chief aims in the design of experiments is to see to it that experimental manipulations are in fact direct manipulations of the intended variables and no others. The point of blind and double blind designs, for example, is exactly to obtain in experiment a direct manipulation of

Causation and Prediction

only the treatment variables. The concern with chronic wounding in drug trials with animals is essentially a worry that with respect to the outcome variables of interest, outcome variables as well as pharmacological variables have been directly manipulated. Typically, when we mistake the variables an intervention will directly manipulate, predictions of the outcomes of intervention will fail.

Our discussion in this section has assumed that the causal structure of the system is fully known. In chapters 6 and 7 we will consider when and how the effects of interventions can be predicted from an unmanipulated distribution, assuming the distribution is the marginal over the measured variables of a distribution faithful to an unknown causal graph, and assuming the intervention constitutes a direct manipulation in the sense we have defined here.

3.8 Determinism

Another way that the Faithfulness Condition can be violated is when there are deterministic relationships between variables. In this section, we will give some rules for determining what extra conditional independence relations are entailed by deterministic relationships among variables.

We will say that a set of variables **Z determines** the set of variables **A**, when every variable in **A** is a deterministic function of the variables in **Z**, and not every variable in **A** is a deterministic function of any proper subset of **Z**. When there are deterministic relationships among variables in a graph, there are conditional independencies that are entailed by the deterministic relationships and the Markov condition that are not entailed by the Markov condition alone. For example, if G is a directed acyclic graph over **V**, **V** contains **Z** and A, and **Z** determines A, then A is independent of **V**\(**Z** $\cup$ \{A\}) given **Z**. If **Z** is a proper subset of the parents of A then this entails that A is independent of its other parents given **Z**, and also independent of its descendants as well as its nondescendants given **Z**. But it could also be the case that the members of **Z** are children of A, in which case given its children, A is independent of all other variables including its parents. It is also possible that **Z** could contain nonparental ancestors of A. Each of these cases entails conditional independence relations not entailed by the Markov condition alone. For example, consider the graph in figure 3.20.

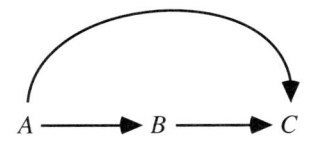

Figure 3.20

No conditional independence relations among A, B, or C are entailed by the Markov Condition alone. However, if the grandparent A determines the grandchild C, then $C \perp\!\!\!\perp B|A$. If the parent B determines the child C then $C \perp\!\!\!\perp A|B$. If the child C determines the parent B the $B \perp\!\!\!\perp A|C$.

Hence d-separability relations do not capture all of the conditional independencies entailed by the Markov condition and a set of deterministic relations. We will look for a graphical condition which entails the conditional independence of variables given the Markov condition and a set of deterministic relations among variables.

Geiger has proposed a simple, provably complete rule for graphically determining the conditional independencies entailed by the Markov and Minimality conditions and one kind of deterministic relationship among variables. Following Geiger (1990), in a directed acyclic graph G over **V** that includes A and **Z**, say that vertex A is a **deterministic variable** if it is a deterministic function of its parents in G. (Note that if a variable A has no parents in G, but has a constant value, then A is a deterministic variable.) A is **functionally determined** by **Z** if and only if A is in **Z**, or A is a deterministic variable and all of its parents are functionally determined by **Z**. If **X**, **Y**, and **Z** are three disjoint subsets of variables in **V**, **X** and **Y** are **D-separated** given **Z** if and only if there is no undirected path U between any member of **X** and any member of **Y** such that each collider has a descendant in **Z** and no other variable on U is functionally determined by **Z**. Geiger has shown that **X** and **Y** are D-separated given **Z** if and only if for every distribution that satisfies the Markov and Minimality Conditions for G, and the deterministic relations, **X** and **Y** are independent given **Z**. We will prove that Geiger's rule is correct for a much wider class of deterministic relations; we do not know if it is complete for this wider class of deterministic relationships.

Suppose G is a directed acyclic graph over **V**, and **Deterministic(V)** is a set of ordered tuples of variables in **V**, where for each tuple D in **Deterministic(V)**, if D is $<V_1,...,V_n>$ then V_n is a deterministic function of $V_1,...,V_{n-1}$ and is not a deterministic function of any subset of $V_1,...,V_{n-1}$; we also say $\{V_1,...,V_{n-1}\}$ **determines** V_n. Note that V_n could be an ancestor in G of members of $V_1,...,V_{n-1}$. Also, if A determines B and B determines A then **Deterministic(V)** contains both $<A,B>$ and $<B,A>$. We assume that **Deterministic(V)** is complete in the sense that if it entails some deterministic relationships among variables, those deterministic relations are in **Deterministic(V)**. (For example, if A determines B and B determines C, then A determines C.) **Det(Z)** is the set of variables determined by some subset of **Z**. If a variable A has a constant value, then we say that it is determined by the empty set, and is in **Det(Z)** for all **Z**.

Note that **Deterministic(V)** can entail dependencies between variables as well as independencies. If **Z** determines A, and Z is a member of **Z**, then A is dependent on **Z**\$\{Z\}$ given Z. (Other dependencies may be entailed by **Deterministic(V)** as well.) These dependencies may conflict with independencies entailed by satisfying the Markov Condition for a directed acyclic graph G, so not every **Deterministic(V)** is compatible

Causation and Prediction

with every directed acyclic graph with vertex set **V**. If **Deterministic(V)** and directed acyclic graph *G* are incompatible, theorem 3.7 stated below is vacuously true, but obviously it would be desirable to have a test for determining whether **Deterministic(V)** and *G* are compatible.

We will expand Geiger's concept of D-separability so that it is not limited to the kind of deterministic relations that he considers. If *G* is a directed acyclic graph with vertex set **V**, **Z** is a set of vertices not containing *X* or *Y*, $X \neq Y$, then *X* and *Y* are **D-separated** given **Z** and **Deterministic(V)** if and only if there is no undirected path *U* in *G* between *X* and *Y* such that each collider on *U* has a descendant in **Z**, and no other vertex on *U* is in **Det(Z)**; otherwise if $X \neq Y$ and *X* and *Y* are not in **Z**, then *X* and *Y* are **D-connected** given **Z** and **Deterministic(V)**. Similarly, if **X**, **Y**, and **Z** are disjoint sets of variables, and **X** and **Y** are non-empty, then **X** and **Y** are D-separated given **Z** and **Deterministic(V)** if and only if each pair <*X,Y*> in the Cartesian product of **X** and **Y** are **D-separated** given **Z** and **Deterministic(V)**; otherwise if **X**, **Y**, and **Z** are disjoint, and **X** and **Y** are non-empty, then **X** and **Y** are **D-connected** given **Z** and **Deterministic(V)**.

THEOREM 3.7: If *G* is a directed acyclic graph over **V**, **X**, **Y**, and **Z** are disjoint subsets of **V**, and *P*(**V**) satisfies the Markov condition for *G* and the deterministic relations in **Deterministic(V)** then if **X** and **Y** are D-separated given **Z** and **Deterministic(V)**, **X** and **Y** are independent given **Z** in *P*.

For example, suppose *G* is the graph in figure 3.21, and **Deterministic(V)** = {<*A,B*>, <*B,C*>, <*A,C*>}.

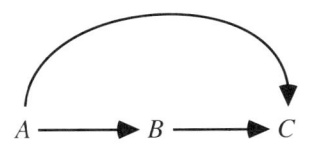

Figure 3.21

B and *C* are D-separated given *A* and **Deterministic(V)**, and *A* and *C* are D-separated given *B* and **Deterministic(V)**.

Suppose that *G* is still the graph in figure 3.21, but now **Deterministic(V)** = {<*A,B*>, <*B,A*>, <*B,C*>, <*C,B*>, <*A,C*>, <*C,A*>}. In addition to the previous D-separability relations, now *A* and *B* are D-separated given *C* and **Deterministic(V)** because *C* determines *A*.

In some cases, conditional independencies are entailed because a parent is determined by its child. Consider the graph in figure 3.22, where **Deterministic(V)** = {<*Y,W,Z*>,<*Z,Y*>,<*Z,W*>}. *X* and *T* are D-separated given *Z* and **Deterministic(V)** because Z determines Y and W, and Y and W are noncolliders on the only undirected path between X and T.

$$X \longrightarrow Y \longrightarrow Z \longleftarrow W \longleftarrow T$$

Figure 3.22

Finally, we note that it is possible that some nonparental ancestor X of Z determines Z, even though X does not determine any of the parents of Z. Let G be the graph in figure 3.23 and **Deterministic(V)** = {<X,Z>}. Suppose X, R, and Z each have two values, and Y has four values. Consider the following distribution (where we give the probability of each variable conditional on its parents):

$P(X = 0) = .2$
$P(R = 0) = .3$
$P(Y = 0 | X = 0, R = 0) = 1$
$P(Y = 1 | X = 0, R = 1) = 1$
$P(Y = 2 | X = 1, R = 0) = 1$
$P(Y = 3 | X = 1, R = 1) = 1$
$P(Z = 0 | Y = 0) = 1$
$P(Z = 0 | Y = 1) = 1$
$P(Z = 1 | Y = 2) = 1$
$P(Z = 1 | Y = 3) = 1$

In effect Y encodes the values of both R and X, and Z decodes Y to match the value of X.

$$X \longrightarrow Y \longrightarrow Z$$
$$\uparrow$$
$$R$$

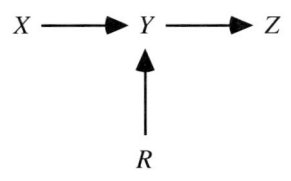

Figure 3.23

It follows that Y and Z are D-separated given X and **Deterministic(V)**, and X and Z are D-separated given Y and **Deterministic(V)**.

The following example points up an interesting difference between the set of distributions that satisfy the Markov condition for a given directed acyclic graph G, and the set of distributions that satisfy the Markov condition and a set of deterministic relationships among the variables in G. Suppose G is the graph shown in figure 3.24. For any directed acyclic graph, the set of probability distributions that satisfy the Markov condition for the graph includes some distributions that also satisfy the Minimality Condition for the graph. Suppose however, that **Deterministic**(**V**) = {<X,Y>}. In this case, among the distributions that satisfy the Markov Condition *and* the specified deterministic relations, there is no distribution that also satisfies the Minimality Condition. All distributions that satisfy the Markov Condition and the specified deterministic relation are faithful to the subgraph of figure 3.24 that does not contain the $Z \rightarrow Y$ edge. This suggests that to find all of the conditional independence relations entailed by satisfying the Markov Condition for a directed acyclic graph G and a set of deterministic relations, one would need to test for D-separability in various subgraphs G' of G with vertex set **V** in which for each Y in **V** no subset of **Parents**(G',Y) determines Y.

$X \longrightarrow Y \longleftarrow Z$

Figure 3.24

We will not consider algorithms for constructing causal graphs when such deterministic relations obtain, nor will we consider tests for deciding whether a set of variables X determines a variable Y.

3.9 Background Notes

The ambiguous use of hypotheses to represent both causal and statistical constraints is nearly as old as statistics. In modern form the use of the idea by Spearman (1904) early in the century might be taken to mark the origins of statistical psychometrics. Directed graphs at once representing both statistical hypotheses and causal claims were introduced by Sewell Wright (1934) and have been used ever since, especially in connection with linear models. For a number of particular graphs the connections between linear models and partial correlation constraints were described by Simon (1954) and by Blalock (1961) for theories without unmeasured common causes, and by Costner (1971) and Lazarsfeld and Henry (1968) for theories with latent variables, but no general characterization emerged. A distribution-free connection of graphical structure for linear models with partial correlation was developed in Glymour, Scheines, Spirtes and Kelly (1987), for first order partials only, but included cyclic graphs. Geiger and Pearl (1989a) showed that for any directed acyclic graph there exists a faithful distribution. The general characterization given here as theorem 3.5 is due to Spirtes (1989), but the connection between the Markov Condition, linearity and partial correlation seems to have been understood already by Simon and Blalock and is explicit in Kiiveri and Speed (1982). The Manipulation Theorem has been used tacitly innumerable times in experimental design and in the analysis of shocks in econometrics but seems never to have previously been explicitly formulated. A special case of it was first given in Spirtes, Glymour, Scheines, Meek, Fienberg, and Slate 1991. The Minimality Condition and the idea of d-separability are due to Pearl (1988), and the proof that d-separability determines the consequences of the Markov condition is due to Verma (1987), Pearl (1988), and Geiger (1989a). A result entailing theorem 3.4 was stated by Pearl, Geiger, and Verma (1990). Theorem 3.4 was used as the basis for a causal inference algorithm in Spirtes, Glymour, and Scheines (1990c). D-separability is described in Geiger (1990).

4 Statistical Indistinguishability

Without experimental manipulations, the resolving power of any possible method for inferring causal structure from statistical relationships is limited by statistical indistinguishability. If two causal structures can equally account for the same statistics, then no statistics can distinguish them. The notions of statistical indistinguishability for causal hypotheses vary with the restrictions one imposes on the connections between directed graphs representing causal structure and probabilities representing the associated joint distribution of the variables. If one requires only that the Markov and Minimality Conditions be satisfied, then two causal graphs will be indistinguishable if the same class of distributions satisfy those conditions for one of the graphs as for the other. A different statistical indistinguishability relation is obtained if one requires that distributions be faithful to graph structure; and still another is obtained if the distributions must be consistent with a linear structure, and so on. For each case of interest, the problem is to characterize the indistinguishability classes graph-theoretically, for only then will one have a general understanding of the causal structures that cannot be distinguished under the general assumptions connecting causal graphs and distributions.

There are a number of related considerations about the resolving power of any possible method of causal inference from statistical properties. Given axioms about the connections between graphs and distributions, what graph theoretic structure must two graphs share in order also to share *at least one* probability distribution satisfying the axioms? When, for example, do two distinct graphs admit one and the same distribution satisfying the Minimality and Markov Conditions? When do two distinct graphs admit one and the same distribution satisfying the Minimality and Markov Conditions for one and the Faithfulness and Markov Conditions for the other? Reversing the question, for any given probability distribution that satisfies the Markov and Minimality Conditions (or in addition the Faithfulness Condition) for *some* directed acyclic graph, what is the *set of all such graphs* consistent with the distribution and these conditions? Finally, there are relevant measure-theoretic questions. If procedures exist that will identify causal structure under a more restrictive assumption such as Faithfulness, but not always under weaker assumptions such as the Markov and Minimality Conditions, how likely are the cases in which the procedures fail? Under various natural measures on sets of distributions, for example, what is the measure of the set of distributions that satisfy the Minimality and Markov Conditions for a graph but are not faithful to the graph?

These are fundamental questions about the limits of any possible inference procedure—whether human or computerized—from non-experimental data to structure. We will provide answers for many of these questions when the system of measured variables is causally sufficient. Statistical indistinguishability is less well understood when graphs can contain variables representing unmeasured common causes.

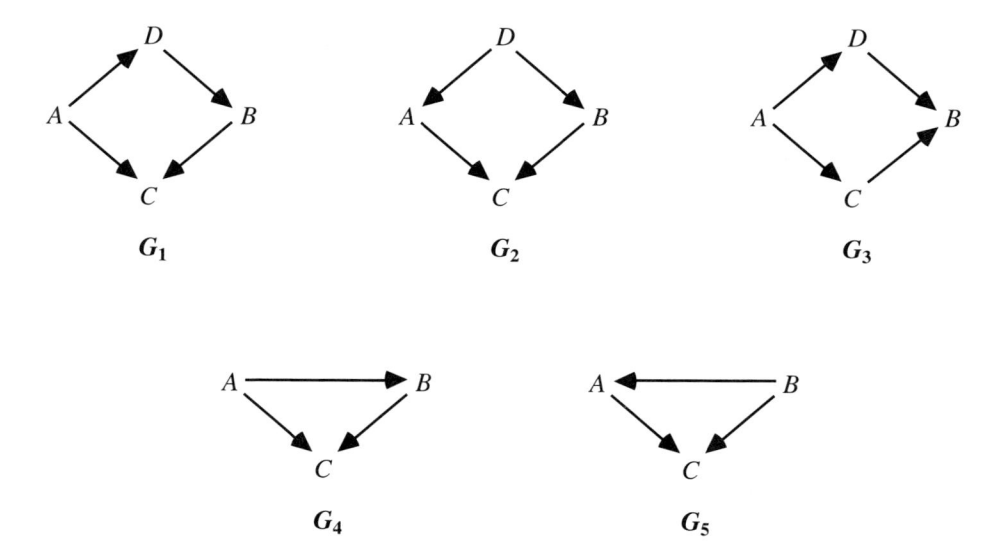

Figure 4.1

4.1 Strong Statistical Indistinguishability

We say that two directed acyclic graphs G, G' are **strongly statistically indistinguishable (s.s.i)** if and only if they have the same vertex set **V** and every distribution P on **V** satisfying the Minimality and Markov Conditions for G satisfies those conditions for G', and vice-versa.

That two structures are s.s.i. of course does not mean that the causal structures are one and the same, or that the difference between them is undetectable by any means whatsoever. From the correlation of two variables, X and Y, one cannot distinguish whether X causes Y, Y causes X or there is a third common cause, Z. But these alternatives may be distinguished by experiment or, as we will see, by other means.

Strong statistical indistinguishability is characterized by a simple relationship, namely that two graphs have the same underlying undirected graph and the same collisions:

THEOREM 4.1: Two directed acyclic graphs G_1, G_2, are strongly statistically indistinguishable if and only if (i) they have the same vertex set **V**, (ii) vertices V_1 and V_2 are adjacent in G_1 if and only if they are adjacent in G_2, and (iii) for every triple V_1, V_2, V_3 in **V**, the graph $V_1 \rightarrow V_2 \leftarrow V_3$ is a subgraph of G_1 if and only if it is a subgraph of G_2.

Given an arbitrary directed acyclic graph G, the graphs s.s.i. from G are exactly those that can be obtained by any set of reversals of the directions of edges in G that preserves all collisions in G. A decision as to whether or not two graphs are s.s.i. requires $O(n^3)$ computations, where n is the number of vertices.

In figure 4.1 graphs G_1 and G_2 are s.s.i., but G_1 and G_3, and G_2 and G_3 are not s.s.i.

Note, however, if a set of variables **V** is totally ordered, as for example by a known time order, and $P(\mathbf{V})$ is positive, then there is a *unique* graph for which $P(\mathbf{V})$ satisfies the Minimality and Markov conditions. (See corollary 3 in Pearl 1988.)

4.2 Faithful Indistinguishability

Suppose we assume that all pairs <G, P> are faithful: all and only the conditional independence relations true in *P* are a consequence of the Markov condition for *G*. We will say that two directed acyclic graphs, *G*, *G'* are **faithfully indistinguishable** (f.i.) if and only if every distribution faithful to *G* is faithful to *G'* and vice-versa. The problem is to characterize faithful indistinguishability graphically.

THEOREM 4.2: Two directed acyclic graphs *G* and *H* are faithfully indistinguishable if and only if (i) they have the same vertex set, (ii) any two vertices are adjacent in *G* if and only if they are adjacent in *H*, and (iii) any three vertices, *X*, *Y*, *Z*, such that *X* is adjacent to *Y* and *Y* is adjacent to *Z* but *X* is not adjacent to *Z* in *G* or *H*, are oriented as $X \rightarrow Y \leftarrow Z$ in *G* if and only if they are so oriented in *H*.

The question of faithful indistinguishability for two graphs can be decided in $O(n^3)$ where *n* is the number of vertices.

It is immediate from theorems 4.1 and 4.2 that if two graphs are strongly statistically indistinguishable they are faithfully indistinguishable, but not necessarily conversely. The graphs G_4 and G_5 in figure 4.1 are not s.s.i. but they are f.i.

A class of f.i. graphs may be represented by a **pattern**. A **pattern** Π is a mixed graph with directed and undirected edges. A graph *G* is in the set of graphs **represented by** Π if and only if:

(i) *G* has the same adjacency relations as Π;
(ii) if the edge between *A* and *B* is oriented $A \rightarrow B$ in Π, then it is oriented $A \rightarrow B$ in *G*;
(iii) if *Y* is an unshielded collider on the path <*X,Y,Z*> in *G* then *Y* is an unshielded collider on <*X,Y,Z*> in Π.

For example, the set of all complete, directed acyclic graphs on three vertices forms a faithful indistinguishability class that can be represented by a pattern consisting of the complete undirected graph on the same vertex set. When the pattern of the faithful indistinguishability class of a directed acyclic graph has no directed edges, and so is purely undirected, the statistical hypothesis represented by the directed graph is equivalent to the statistical hypothesis of the undirected independence graph corresponding to the pattern.

4.3 Weak Statistical Indistinguishability

The indistinguishability relations characterized in the two previous sections ask for the graphs that can accommodate the same class of probability distributions as a given graph. We can turn the tables, at least partly, by starting with a particular probability distribution on a set of variables and asking for the set of all directed acyclic graphs on those vertices that are consistent with the given distributions. The answers characterize how much the probabilities and our assumptions about the connection between probabilities and causes underdetermine the causal structure. Assuming Markov and Minimality only, the equivalence of these two conditions (under positivity) with the defining conditions for a directed independence graph provides an (impractical) algorithm for generating the set of all graphs that satisfy the two conditions for a given distribution P. For every ordering of the variables in P there is a directed acyclic graph G compatible with that ordering (i.e., A precedes B in the ordering only if A is not a descendant of B in G) satisfying the Minimality and Markov Conditions for P. It can be generated by assuming the ordering and the conditional independence relations in P and applying the definition of directed independence graph. An algorithm that does not assume positivity is given by Pearl (1988). According to that algorithm let Ord be a total ordering of the variables, and **Predecessors**(Ord,X) be the predecessors of X in the ordering Ord. For each variable X, let the parents of X in G be a smallest subset **R** of **Predecessors**(Ord,X) such that X is independent of **Predecessors**(Ord,X)**R** given **R** in P. It follows that P satisfies the Minimality and Markov Conditions for G.

The alternatives are more limited if we start with P and assume that any graph must be faithful to P. In that case all of the graphs faithful to P form a faithful indistinguishability class, that is, the set of all graphs f.i. from any one graph faithful to P. The next chapter presents a number of algorithms that generate the faithful indistinguishability class from properties of distributions.

Given axioms connecting causal graphs with probability distributions it makes sense to ask for which pairs G, G' of graphs there *exists* some probability distribution satisfying the axioms for both G and G'. Let us say that two graphs are **weakly faithfully indistinguishable** (w.f.i.) if and only if there exists a probability distribution faithful to both of them. We say that two graphs are **weakly statistically indistinguishable** (w.s.i.) if and only if there exists a probability distribution meeting the Minimality and Markov Conditions for both of them. Weak faithful indistinguishability proves to be equivalent to faithful indistinguishability:

THEOREM 4.3: Two directed acyclic graphs are faithfully indistinguishable if and only if some distribution faithful to one is faithful to the other and conversely; that is, they are f.i. if and only if they are w.f.i.

This theorem tells us that faithfulness divides the set of probability distributions over a vertex set into equivalence classes that exactly correspond to the equivalence classes of

Statistical Indistinguishability

graphs induced by faithful indistinguishability. It follows that if a distribution is faithful to some graph G then it is faithful to all and only the graphs faithfully indistinguishable from G.

There is no reason to expect so nice a match in general. Suppose we assume only the Minimality and Markov Conditions. Under what conditions will there exist a distribution P satisfying those axioms for two distinct graphs, G, and G'? The answer is *not*: exactly when G and G' are strongly statistically indistinguishable. The two graphs shown in figure 4.2 are not s.s.i., but there exist distributions that satisfy the Minimality and Markov Conditions for both.

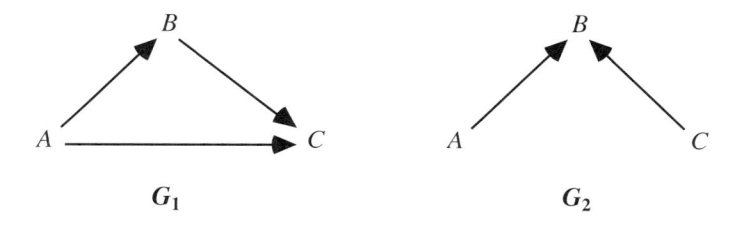

G_1 G_2

Figure 4.2

The distributions in Simpson's "paradox" provide another example, as we have already seen in chapter 3. We conjecture that if a distribution satisfies the Minimality and Markov Conditions for two graphs G and G', then G and G' have the same edges and the same colliders, save that triangles such as G_1 in one graph may be replaced by collisions such as G_2 in the other, provided appropriate conditions are met by other edges. We don't know how to characterize the "appropriate" conditions. There is, however, a related property of interest we can characterize.

No distribution that is faithful to graph G_1 in figure 4.2 can be faithful to graph G_2, but a distribution that satisfies the Minimality and Markov Conditions for G_1 can be faithful to graph G_2. Just when can this sort of thing happen? When, in other words, can the generalization of Simpson's "paradox" arise? If probability distribution P satisfies the Minimality and Markov Conditions for G, and P is faithful to graph H, what is the relation between G and H?

THEOREM 4.4: If probability distribution P satisfies the Markov and Minimality Conditions for directed acyclic graphs G and H, and P is faithful to H, then for all vertices X, Y, if X, Y are adjacent in H they are adjacent in G.

THEOREM 4.5: If probability distribution P satisfies the Markov and Minimality Conditions for directed acyclic graphs G and H, and P is faithful to graph H, then (i) for all X, Y, Z such that $X \rightarrow Y \leftarrow Z$ is in H and X is not adjacent to Z in H, either $X \rightarrow Y \leftarrow Z$

in G or X, Z are adjacent in G and (ii) for every triple X, Y, Z of vertices such that $X \to Y \leftarrow Z$ is in G and X is not adjacent to Z in G, if X is adjacent to Y in H and Y is adjacent to Z in H then $X \to Y \leftarrow Z$.

COROLLARY 4.1: If probability distribution P satisfies the Markov Condition for directed acyclic graph G, P is faithful to directed acyclic graph H, and G and H agree on an ordering of the variables (as, for example, by time) such that $X \to Y$ only if $X < Y$ in the order, then H is a subgraph of G.

4.4 Rigid Indistinguishability

In addition to the notions of strong, faithful and weak statistical indistinguishability, there is still another. Suppose two directed acyclic graphs, G and G', are statistically indistinguishable in some sense over a common set **O** of vertices. Then without experiment, no measurement of the variables in **O** will reliably determine which of the graphs correctly describes the causal structure that generated the data. It might be, however, that G and G' can be distinguished if other variables besides those in G or G' are measured and stand in appropriate causal relations to the variables in **O**. For example, the following simple graphs are both s.s.i., and f.i. (where A and B are assumed to be measured and in **O**).

$$A \longrightarrow B \qquad\qquad A \longleftarrow B$$
$$G_1 \qquad\qquad\qquad G_2$$

Figure 4.3

But if we also measure a variable C that is a cause of A or has a common cause with A and no connection with B save possibly through A, then the two structures can be distinguished.

For example, the graphs in figure 4.4 are not f.i. or s.s.i. It is equally easy to give examples of w.s.i. structures that can be embedded in graphs that are not w.s.i. Which causally sufficient structures can be distinguished by measuring extra variables? To answer the question we require some further definitions.

Let G_1, G_2 be two directed acyclic graphs with common vertex set **O**. Let H_1, H_2 be directed graphs having a common set **U** of vertices that includes **O** such that

Statistical Indistinguishability

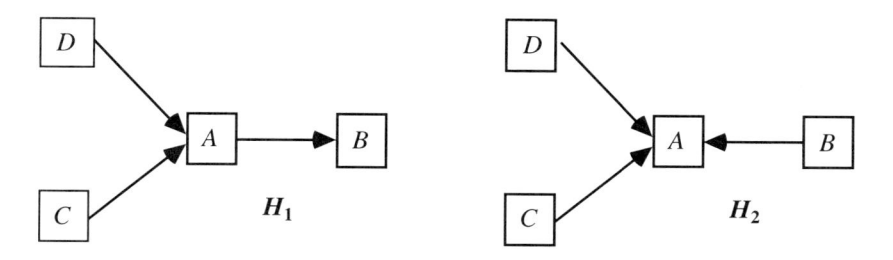

Figure 4.4

(i) the subgraph of H_1 over **O** is G_1 and the subgraph of H_2 over **O** is G_2;
(ii) every directed edge in H_1 but not in G_1 is in H_2 and every directed edge in H_2 but not in G_2 is in H_1.

We will say then that directed acyclic graphs G_1 and G_2 with common vertex set **O** have a **parallel embedding** in H_1 and H_2 over **O** and **U**. In figures 4.3 and 4.4, G_1 and G_2 have a parallel embedding in H_1 and H_2 over **O** = {A,B} and **U** = {A,B,C,D}. The question of whether two s.s.i. structures can be distinguished by measuring further variables then becomes the following: do the structures have parallel embeddings that are not s.s.i.? If no such embedding exists we will say the structures G_1 and G_2 are **rigidly statistically indistinguishable (r.s.i.)**.

THEOREM 4.6: No two distinct s.s.i. directed acyclic graphs with the same vertex set are rigidly statistically indistinguishable.

In other words, provided additional variables with the right causal structure exist and can be measured, the causal structure among a causally sufficient collection of measured variables can in principle be identified. The proof of theorem 4.6 also demonstrates a parallel result for faithfully indistinguishable structures. We conjecture that an analog of theorem 4.6 also holds for weak statistical indistinguishability assuming positivity.

4.5 The Linear Case

Parameter values can force conditional independencies or zero partial correlations that are not linearly implied by a graph. The graphs in figure 4.2 (reproduced in figure 4.5 with error variables explicitly included) illustrate the possibility: treat the vertices of the graphs as each attached to an "error" variable, and let the graphs plus error variables determine a set of linear equations. (We assume that any pair of exogenous variables, including the error terms, have zero covariance.) The result is, up to specification of the joint distribution of the exogenous variables, a structural equation model. A linear coefficient is attached to each directed edge. The correlation matrix, and hence all partial correlations, is determined by the linear coefficients and the variances of the exogenous variables.

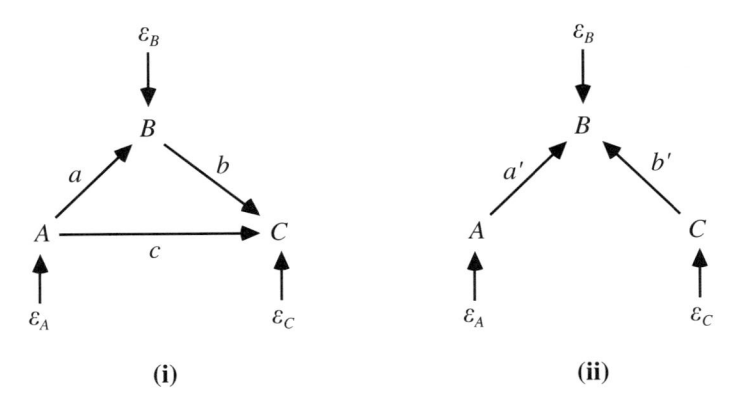

Figure 4.5

If, in the structure on the left, $ab = -c$, the A, C will be uncorrelated as in the model on the right. This sort of phenomenon—vanishing partial correlations produced by values of linear coefficients rather than by graphical structure—is bound to mislead any attempt to infer causal structure from correlations. When can it happen? We have already answered that question in the previous chapter, when we considered the conditions under which linear faithfulness might fail. In the linear case, the parameter values—values of the linear coefficients and exogenous variances of a structure with a directed acyclic graph G—form a real space, and the set of points in this space that create vanishing partial correlations not linearly implied by G have Lebesgue measure zero.

THEOREM 3.2: Let M be a linear model with directed acyclic graph G and n linear coefficients $a_1,..., a_n$ and k positive variances of exogenous variables $v_1,..., v_k$. Let $M(<u_1,...,u_n, u_{n+1},...,u_{n+k}>)$ be the distributions consistent with specifying values $<u_1,...,u_n, u_{n+1},...,u_{n+k}>$ for $a_1,..., a_n$ and $v_1,...v_k$. Let Π be the set of probability measures P on the space $\Re^{n+k}$ of values of the parameters of M such that for every subset $\mathbf{V}$ of $\Re^{n+k}$ having Lebesgue measure zero, $P(\mathbf{V}) = 0$. Let $\mathbf{Q}$ be the set of vectors of coefficient and variance values such that for all q in $\mathbf{Q}$ every probability distribution in with $M(q)$ has a vanishing partial correlation that is not linearly implied by G. Then for all P in Π, $P(\mathbf{Q}) = 0$.

Measure theoretic arguments of this sort are interesting but may not be entirely convincing. One could, after all, argue that in the general linear model absence of causal connection is marked by linear coefficients with the value zero, and thus form a set of measure zero, so by parity of reasoning everything is causally connected to everything else. In a recent book Nancy Cartwright (1989) objects that since in linear structures independence relations may be produced by special values of the linear coefficients and variances as well as by the causal structure, it is illegitimate to infer causal structure from such relations. In effect, she rejects any inference procedure that is unable to distinguish the true causal structure from w.s.i. alternatives. Such a position may be extreme, but it does serve to focus attention on two interesting questions: when is it impossible for two structures to be w.s.i. but not f.i. or s.s.i., and are there special marks or indicators that a distribution satisfies the Markov and Minimality conditions for two w.s.i. but not s.s.i. or f.i. causal structures? The answers to these questions are essentially just applications to the linear case of the theorems of the preceding sections.

We will assume, with Cartwright, that a time ordering of the variables is known. Pearl and Verma (Pearl 1988) have proved that for a positive distribution P and a given ordering of variables, there is only one directed acyclic graph for which P satisfies the Minimality and Markov Conditions. It follows that for a positive distribution with a given correlation matrix and a given ordering of a causally sufficient set of variables there is a unique directed acyclic graph that linearly represents the distribution and is consistent with the ordering.

In some cases at least, the positivity of a distribution can be tested for. (For example, in a bi-variate normal distribution the density function is everywhere nonzero if the correlation is not equal to one.) It follows for those cases that for a given ordering of variables either there is a unique directed acyclic graph for which P satisfies the Markov and Minimality Conditions, or it is detectable that more than one such directed acyclic graph exists. However, even if for a given ordering of variables there is a unique directed acyclic graph for which P satisfies the Markov and Minimality Conditions, algorithms for finding that graph are not feasible for large numbers of variables, because of the number and order of the conditional independence relations that they require be tested.

Suppose that we wrongly assume that a distribution is faithful to the causal graph that generated it. Then corollary 4.1 applies, which means, informally, that if faithfulness is assumed but not true, then conditional independence relations or vanishing partial correlations due to special parameter values can only produce erroneous causal inferences in which a true causal connection is *omitted*; no other sorts of error may arise. We will consider when this circumstance is revealed in the correlations.

Recall that a **trek** is an unordered pair of directed acyclic paths having a single common vertex that is the source of both paths (one of the paths in a pair may be the empty path defined in chapter 2). For standardized models, in which the mean of each variable is zero and non-error variables have unit variance, the correlation of two variables is given by the sum over all treks connecting X, Y of the product for each trek of the linear coefficients associated with the edges in that trek (we call this quantity the **trek sum**). For example, in directed acyclic graph (i) in figure 4.5, the trek sum between A and C is $ab + c$. We will use standardized systems throughout our examples in this section. The system of correlations determines all partial correlations of every order through the following formula.

$$\rho_{XY.\mathbf{Z}\cup\{R\}} = \frac{\rho_{XY.\mathbf{Z}} - \rho_{XR.\mathbf{Z}} \times \rho_{YR.\mathbf{Z}}}{\sqrt{1-\rho_{XR.\mathbf{Z}}^2} \times \sqrt{1-\rho_{YR.\mathbf{Z}}^2}}$$

Since the recursion relations give the same partial correlation between two variables on a set **U** no matter in what sequence the partials on the members of **U** are taken, a vanishing partial correlation corresponds to a system of equations in the coefficients of a standardized system.

Suppose now that special values of the linear parameters in a normal, standardized system *G* produce vanishing partial correlations that are exactly those linearly implied only by some false causal structure, say *H*. Then the parameter values must generate extra vanishing partial correlations not linearly implied by *G*. Any partial correlation is a function just of the trek sums connecting pairs of variables, and the trek sums in this case involve just the linear parameters in *G*. Hence each additional vanishing partial correlation not linearly implied by *G* determines a system of (nonlinear) equations in the parameters of *G* that must be satisfied in order to produce the coincidental vanishing partial correlation. (For example, in directed acyclic graph (i) of figure 4.5, the correlation between *A* and *C* is 0 only if the single equation *ab* = -*c* is satisfied). Now for some *G* and some *H* (a sub-graph of *G*), these systems of equations may have no simultaneous solution. In that case there are no values for the parameters of *G* that will produce partial correlations that are exactly those linearly implied by *H*. For other choices of *G* and a subgraph *H*, it may be that the system of equations has a solution, but only solutions that allow only a finite number of alternative values for one or more parameters and that require some error variance to vanish. Such a solution must "give itself away" by special correlation constraints that are not themselves vanishing partial correlation relations. Consider the following choices of *G* and *H*, where in each pair *G* is on the left hand side and *H* is on the right hand side.

Consider Figure 4.6. In (i) and (iii), coefficients and variances can be chosen for the graph on the left hand side so that it appears as though an edge does not occur, but only by making the coefficient labeled *b* equal to either 1 or –1. Since the variables are standardized, this requires that the error term for *Y* have zero variance and zero mean—that is, it vanishes. Thus in order for the true graph to be the one on the left hand side and the parameter values to produce vanishing partial correlations that are exactly those linearly implied by the graph on the right hand side, variable *Y* must be a linear function of variable *X* and only variable *X*. The same result obtains if the edges that are not eliminated in the first and last examples are replaced by directed paths of any length. Clearly in these cases special parameter values that create vanishing partial correlations not linearly implied by the true graph will be revealed by the correlations. In (ii) the edge between variables *X* and *Z* cannot be made to appear to be eliminated by any choice of parameter values for the true graph.

Statistical Indistinguishability

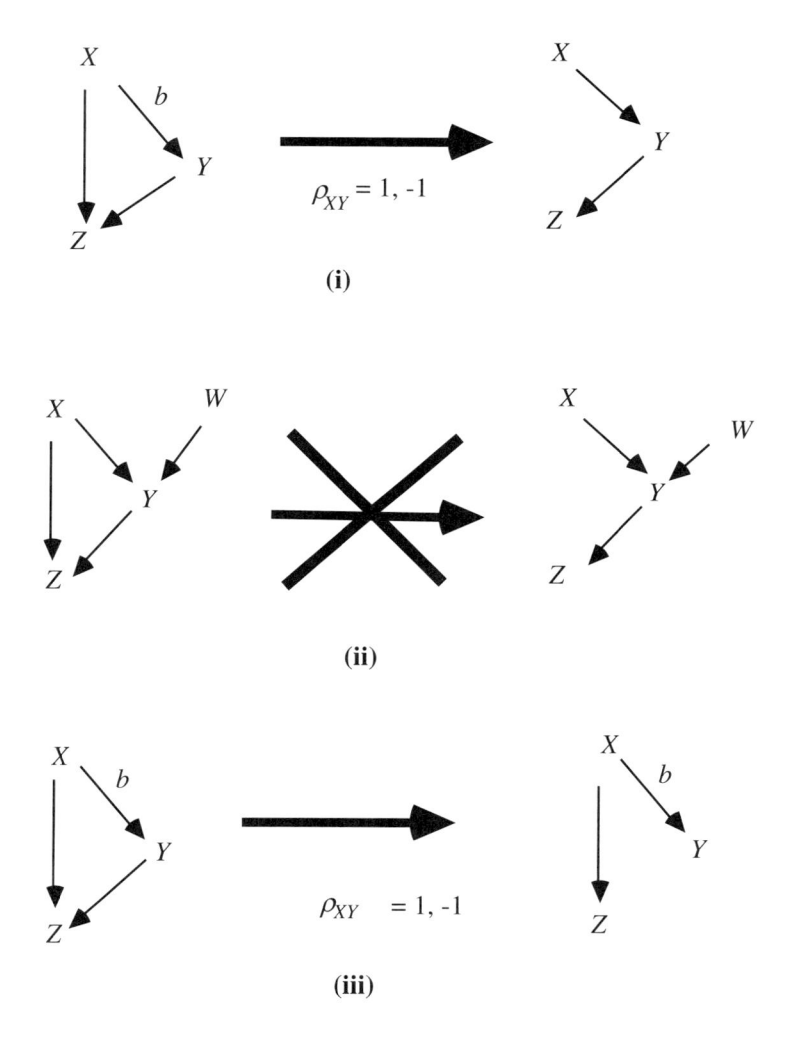

Figure 4.6

We conjecture that even without a prior time order, unless three edges form a triangle in G, if parameter values of G determine exactly the collection of vanishing partial correlations linearly implied by a graph H—whether or not H is a subgraph of G—then there are extra constraints on the correlations not entailed by the vanishing partial correlations.

4.6 Redefining Variables

The indistinguishability results so far considered relate alternative graphs over the same set of vertices. The vertices are interpreted as random variables whose values are subject to some system of measurement. New random variables can always be defined from a given set, for example by taking linear or Boolean combinations. For any specified apparatus of definitions, and any axioms connecting graphs with distributions, questions about indistinguishability classes arise parallel to those we have considered for fixed sets of variables. A distribution P over variable set **V** may correspond to a graph G, and a distribution P' over variable set **V**$'$ may correspond to a different graph G' (with P' and **V**$'$ obtained from P and **V** by defining new variables, ignoring old ones, and marginalizing). The differences between G and G' may in some cases be unimportant, and one may simply want to say that each graph correctly describes causal relations among its respective set of variables. That is not so, however, when the original variables are ordered by time, and redefinition of variables results in a distribution whose corresponding graphs have later events causing earlier events. Consider the following pair of graphs.

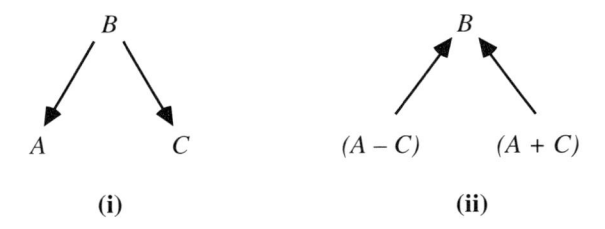

```
      B                          B
     / \                        / \
    ↙   ↘                      ↗   ↖
   A     C                  (A – C)  (A + C)

    (i)                          (ii)
```

Figure 4.7

In directed acyclic graph (i), A and C are effects of B; suppose that B occurs prior to A and C. By the procedure of definition and marginalization, a distribution faithful to graph (i) can be transformed into a distribution faithful to graph (ii). First, standardize A and C to form variables A' and C' with unit variance. Then consider the variables $(A' - C')$ and $(A' + C')$. Their covariance is equal to the expected value of $A'^2 - C'^2$ which is zero. Simple algebra shows that the partial correlation of $(A' - C')$ and $(A' + C')$ given B does not vanish. The marginal of the original distribution is therefore linearly faithful to (ii), and faithful to (ii) if the original distribution is normal.

Note that the transformation just illustrated is unstable; if the variances of A' and C' are unequal in the slightest, or if the transformation gives $(xA' + zC')$ and $(yA' + wC')$ for any values of x, y, z, and w such that $xy + wz + \rho_{A'C'}(zy + xw) \neq 0$ then the marginal on the transformed distribution will be faithful, not to (ii), but to all acyclic orientations of the complete graph on the three variables, a hypothesis that is not inconsistent with the time order.

Viewed from another perspective, a transformation of variables that produces a "coincidental" vanishing partial correlation is just another violation of the Faithfulness Condition. Consider the linear model in figure 4.8.

Statistical Indistinguishability 71

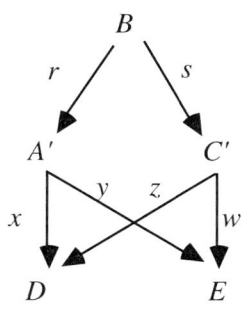

```
          B
       r/   \s
      /      \
     A'       C'
      \ y  z /
     x \ /\ / w
        X
       / \
      D   E
```

Figure 4.8

Let $A' = rB + \varepsilon_{A'}$, $C' = sB + \varepsilon_{C'}$, $D = xA' + zC' + \varepsilon_D$, and $E = yA' + wC' + \varepsilon_E$. If the variables are standardized, ρ_{DE} is equal to $xy + zw + rysz + rxsw = xy + zw + rs(yz + xw)$, which, since $rs = \rho_{A'C'}$, is the formula of the previous paragraph. If $\rho_{DE} = 0$, the Faithfulness Condition is violated. Hence the conditions under which we obtain a linear transformation of A and C that produces a "coincidental" zero correlation are identical to the conditions under which the treks between A and C exactly cancel each other in a violation of the Faithfulness Condition. We get the example of figure 4.7 when $D = A' + C'$ (i.e., $x = z = 1$), and $E = A' - C'$ (i.e., $y = -w = 1$) where the variances and means of the error terms have been set to zero. Since the set of parameter values that violate Faithfulness in this example has Lebesgue measure zero, so does the set of linear transformations of A and C that produce a "coincidental" zero correlation.

4.7 Background Notes

The underdetermination of linear statistical models by values of measured variables has been extensively discussed as the "identification problem," especially in econometrics (Fisher 1966) where the discussion has focused on the estimation of free parameters. The device of "instrumental variables," widely used for linear models, is in the spirit of Theorem 4.6 on rigid distinguishability, although instrumental variables are used to identify parameters in cyclic graphs or in systems with latent variables. The possibility of "rewriting" a pure linear regression model so that the outcome variable is treated as a cause seems to have been familiar for a long while, and we do not know the original source of the observation, which was brought to our attention by Judea Pearl.

Accounts of statistical indistinguishability in something like one or another of the senses investigated in this chapter have been proposed by Basmann (1965), Stetzl (1986) and Lee (1987). Basmann argued, in our terms, that for every simultaneous equation model with a cyclic graph (i.e., "nonrecursive") there exists a statistically indistinguishable model with an acyclic graph. The result is a weak indistinguishability theorem (see chapter 12). Stetzl and Lee focus exclusively on linear structural equation models with free parameters for linear coefficients and variances, and they define equivalence in terms of maximum likelihood estimates of the parameters and hence of the covariance matrix. No general graph theoretic characterizations are provided, although interesting attempts were made in Lee's thesis.

The notion of a pattern and theorem 4.2 are due to Verma and Pearl (1990b). We state some results about indistinguishability relations for causally insufficient graphs in chapter 6. A well-known result due to Suppes and Zanotti (1981) asserts that every joint distribution P on a set $\mathbf{X}$ of discrete variables is the marginal of some joint distribution P^* on $\mathbf{X} \cup \{T\}$ satisfying the Markov Condition for a graph G in which T is the common cause of all variables in $\mathbf{X}$ and there are no other directed edges. The result can be viewed as a weak indistinguishability theorem when causally insufficient structures are admitted. Except in special cases, P^* cannot be faithful to G.

5 Discovery Algorithms for Causally Sufficient Structures

5.1 Discovery Problems

A discovery problem is composed of a set of alternative structures, one of which is the source of data, but any of which, for all the investigator knows before the inquiry, could be the structure from which the data are obtained. There is something to be found out about the actual structure, whichever it is. It may be that we want to settle a particular hypothesis that is true in some of the possible structures and false in others, or it may be that we want to know the complete theory of a certain sort of phenomenon. In this book, and in much of the social sciences and epidemiology, the alternative structures in a discovery problem are typically directed acyclic graphs paired with joint probability distributions on their vertices. We usually want to know something about the structure of the graph that represents causal influences, and we may also want to know about the distribution of values of variables in the graph for a given population.

A discovery problem also includes a characterization of a kind of evidence; for example, data may be available for some of the variables but not others, and the data may include the actual probability or conditional independence relations or, more realistically, simply the values of the variables for random samples. Our theoretical discussions will usually consider discovery problems in which the data include the true conditional independence relations among the measured variables, but our examples and applications will always involve inferences from statistical samples.

A method solves a discovery problem in the limit if, as the sample size increases without bound, the method converges to the true answer to the question or to the true theory, *whatever (consistent with prior knowledge) the truth might be*. A procedure for inferring causes does not solve the problem posed if for some of the alternative possibilities it gives no answer or the wrong answer, although it may solve another, easier problem that arises when some of the alternative structures are excluded. Which causal discovery problems are solvable in the limit, and by what methods, are determinate, mathematical questions. The metaphysical wrangling lies entirely in motivating the problems, not in solving them. The remainder of this book is an introduction to the study of these formal questions and to the practical applications of particular answers.

5.2 Search Strategies in Statistics

The statistical literature is replete with procedures that use data to guide a search for some restricted parametrization of alternative distributions. When the representation of the statistical hypothesis is used to guide policy or practice, to predict what will happen if some of the variables are manipulated or to retrodict what would have happened if some of the variables had in the past been manipulated, then the statistical hypotheses are usually also causal hypotheses. In that case the first question is whether the search procedures are any good at finding causal structure.

Many of the search strategies proposed in the statistical literature are best-only beam searches, beginning either with an arbitrary model, or with a complete (or almost complete) structure that entails no constraints, or with a completely (or almost completely) constrained structure in which all variables are independent. Statisticians sometimes refer to the latter procedure as "forward" search, and the former procedures as "backward" search. Depending on which order is followed, the procedures iteratively apply a fit measure of some kind to determine which fixed parameter in the parametrization will most improve fit when freed—or which free parameter should be fixed. They then reestimate the modified structure to determine if a stopping criterion is satisfied. A "forward" procedure of this kind was proposed by Arthur Dempster (1972) for covariance structures, and a "backward" procedure was proposed by his student, Nanny Wermuth (1976), for both log-linear and linear systems whose distributions are "multiplicative"—in our terms, satisfy the Markov condition for some directed acyclic graph. Forward search algorithms using goodness of fit statistics were proposed for mutinormal linear systems by Byron (1972) and by Sorbom (1975) and versions of them have been automated in the LISREL (Joreskog and Sorbom 1984) and EQS (Bentler 1985) estimation packages. The latter program also contains a backward search procedure. Versions of the general strategy for log-linear parametrizations are described by Bishop, Fienberg and Holland (1975), by Fienberg (1977) by Aitkin (1979), by Christensen (1990) and many others. The same representations and search strategies have been used in the systems science literature by Klir and Parviz (1976) and others under the title of "reconstructability" analysis. Stepwise regression procedures in logistic regression can be viewed as versions of the same strategies. The same strategies have been applied to undirected and directed graph representations. They are illustrated for a variety of examples by Whittaker (1990).

In each of these cases the general statistical search strategy is unsatisfactory if the goal is not just to estimate the distribution but also to identify the causal structure or to predict the results of manipulations of some of the variables. When used to these ends, these searches are inefficient and unreliable for at least three reasons: (i) they often search a hypothesis space that excludes many causal hypotheses and includes many hypotheses of no causal significance; (ii) the specifications of distributions typically force the use of numerical procedures that for statistical or computational reasons unnecessarily limit search; (iii) restrictions requiring the search to output a single hypothesis entail that the search fails to output alternative hypotheses that may be indistinguishable given the evidence. We will consider each of these points in more detail.

5.2.1 The Wrong Hypothesis Space

In searching for the correct causal hypothesis the space of alternatives should, insofar as possible, include all causal hypotheses that have not been ruled out by background knowledge and no hypotheses that do not have a causal interpretation. The log-linear formalism, introduced by Birch in 1963, provides an important example of a search space poorly adapted to the goal of finding correct causal hypotheses. For discrete data a more appropriate search space turns out to be a sub-class of conjunctions of log-linear hypotheses.

The log-linear formalism provides a general framework for the analysis of contingency tables of any dimension. In the discrete case we are concerned with variables that take a finite number of values, whether ordered or not. For a system with four variables, for example, we will let i range over the values of the first variable, j the second, k the third and l the fourth. In a particular sample or population, x_{ijkl} will then denote the number of units that have value i for the first variable, value j for the second variable, k for the third and l for the fourth. We will refer to a particular vector of values for the four (or other number of) variables as a "cell." In the formalism the joint distribution over the cells is given by an equation for the logarithm of the expected value of each cell, expressed as the sum of a number of parameters. For example, in Birch's notation in which m_{ijk} denotes the expected number in cell i, j, k,

$$\ln(m_{ijk}) = u + u_{1i} + u_{2j} + u_{3k} + u_{12ij} + u_{13ik} + u_{23jk} + u_{123ijk}$$

The various u's are arbitrary parameters with an associated set of indices; only seven of the u terms can be independent for a system of three binary variables. The power of Birch's parametrization lies in at least two features. First, associations in multi-dimensional contingency tables that had long been studied in statistics can be represented as hypotheses that certain of the parameters are zero. For example Bartlett's representation of the hypothesis of no "three factor interaction" among three binary variables is given by the following relation among the cell probabilities:

$$p_{111}p_{122}p_{212}p_{221} = p_{222}p_{211}p_{121}p_{112}$$

Birch shows that a generalization of this condition to variables of any finite number of categories obtains if and only if various of the u terms are zero. Second, for each hypothesis obtained by setting some of the u terms to zero, there exist iterative methods for obtaining maximum likelihood estimates for a variety of sampling procedures.

Birch's results were extended by several researchers. A hypothesis in the log-linear parametrization has come to be treated as a specification that particular u terms vanish. There are direct maximum likelihood estimates of the expected cell counts for certain forms of such specifications, and for other specifications iterative algorithms have been developed that converge to the maximum likelihood estimates. Various formal motivations have been developed for focusing on particular classes of log-linear parametrizations. Using his information-based distance measure, for example, Kullback (1959) derived a class of log-linear relations that could be obtained in the same way from a slightly different perspective, the maximum entropy principle. Fienberg (1977) and others have urged restricting attention to "hierarchical models"—log linear parametrizations in which if a u term with a set of indices is put to zero so are all other u terms whose indices contain the first set. The motivation for the restriction is that these parametrizations bear a formal analogy to analysis of variance, so that the u_1 term, for example, may be thought of as the variation from the grand mean due to the action of the first variable.

To see the difficulties in representing causal structure in the log-linear formalism, consider the most fundamental causal relation of the preceding chapters, namely any collider $A \rightarrow B \leftarrow C$ in which A and C are not adjacent. Such a structure corresponds (assuming faithfulness) to two facts about conditional independence: first, A and C are independent conditional on some set of variables that does not contain B; second, A and C are dependent conditional on every set that does contain B but not A or C. In the very simplest case of this kind, in which A, B, and C are the only variables, A and B are independent, but dependent conditional on C. The hypothesis that these relations obtain cannot be expressed in the log-linear formalism by vanishing u terms. Birch himself observed that in a three variable system the hypothesis that in the marginal distribution two of the variables are independent cannot be expressed by the vanishing of any subset of parameters in the general log-linear expansion for the three variables. There are of course log-linear hypotheses that are *consistent* with marginal independence hypotheses, but do not entail them.

Another inappropriate search space is provided by the LISREL program. The LISREL formalism—at least as intended by its authors, Joreskog and Sorbom—allows search for structures corresponding to causal relations among measured variables when there are no unmeasured common causes, but when the search includes structures with unmeasured common causes, causal relations among measured variables are forbidden. Users have found ways around these restrictions (Glymour et al. 1987; Bollen 1989), rather to the dissatisfaction of the authors of the program (Joreskog and Sorbom 1990). LISREL owes these peculiarities to its ancestry in factor analysis, which provides still another example of an artificially contracted search space. Thurstone (1935) carefully and repeatedly emphasized that his "factors" were not to be taken as real causes but only as a mathematical simplifications of the measured correlations. Of course factors were immediately treated as hypothetical causes. But so applied, Thurstone's methods exclude *a priori* any causal relations among measured variables themselves, they exclude the possibility that measured variables are causes of unmeasured variables, and they cannot determine causal structure—only correlations—among the latent variables.

5.2.2 Computational and Statistical Limitations

Some searches examine only a small portion of the possible space of hypotheses because they require computationally intensive iterative algorithms in order to test each hypothesis. For example, the automatic model respecification procedure in LISREL re-estimates the entire model every time it examines a new hypothesis. One consequence is that the slowness of the search prohibits the procedure from examining large portions of the hypothesis space where the truth may be hiding.

Another common problem is that many searches require the determination of conditional independence relations that cannot be reliably tested. Many log-linear search procedures implicitly require the estimation of probabilities conditional on a set of variables whose size equals the total number of variables minus two, no matter what the true structure turns out to be. Estimates of higher order conditional probabilities and tests of higher order conditional independencies tend to be unreliable (especially with variables taking several discrete values) because at reasonable sample sizes most cells corresponding to an array of values of the variables will be empty or nearly empty. This disadvantage is not inherent in the log-linear formalism. An algorithm proposed by Fung and Crawford (1990) for searching the set of graphical models (the subset of the hierarchical log-linear models that can be represented by undirected independence graphs) reduces the need for testing high order conditional independencies. A version of the same problem arises for linear regression with a large number of regressors and small sample size, since in tests of the hypothesis that a regression coefficient vanishes, the sample size is effectively reduced by the number of other regressors, or the degrees of freedom are altered, so that the test may have little power against reasonable alternatives.

A related but equally fundamental difficulty is that searches for models of discrete data that use some measure of fit requiring model estimation at each (or any) stage are subject to an exponential increase in the number of cells that must be estimated as the number of variables increases. If, to take the simplest case, the variables are binary, then the number of cells for which an expected value must be computed is 2^n. When $n = 50$, say, the number of cells is astronomical.

One might think that these difficulties will beset any possible reliable search procedure. As we will see in this chapter and the next, that is not the case.

5.2.3 Generating a Single Hypothesis

If a kind of evidence is incapable of reliably distinguishing when one rather than another of several alternative hypotheses is correct, then an adequate search procedure should reflect this fact by outputting all of them. Producing only a single hypothesis in such circumstances misleads the user, and denies her information that may be vital in making decisions.

An example of this sort of flaw is illustrated by the LISREL and EQS programs. Beginning with a structure constructed from background knowledge, each of these programs searches for causal models among linear structures using a best-only beam search. At each stage they free the fixed parameter that is judged will most increase the fit of the model to the data. Since freeing a number of different fixed parameters may result in the very same improvement in fit, the programs employ an arbitrary tie-breaking procedure. The output of the search is a single linear model and any alternative statistically indistinguishable models are ignored.

In a later chapter we will describe a large simulation study of the reliabilities of the statistical search procedures implemented in the LISREL and EQS programs for linear models. Because of the computational problems and arbitrary choices from among indistinguishable models at various stages of search, we find that the procedures are of little value in discovering dependencies in the structures from which the data are generated, even when the programs are given most of the structure correctly to start with, including even correct linear coefficients and variances. The study involves systems with unmeasured variables, but we expect that similar results would be obtained in studies with causally sufficient systems.

5.2.4 Other Approaches

There are several exceptions to the generalization that statistical search strategies have been confined to generate-and-test-best-only procedures. Edwards and Havranek (1987) describe a form of procedure that tests models in sequence, under the assumption that if a model passes the test so will any more general model and if a model fails the test so will any more restricted model. Their proposal is to keep track of a bounding set of rejected hypotheses and a bounding set of accepted hypotheses until all possible hypotheses (in some parametrization) are classified. Apparently unknown to Edwards and Havranek, the same idea was earlier developed at length in the artificial intelligence literature under the name of "version spaces" (Mitchell 1977). For the applications they have in mind, no analysis of complexity or reliability is available.

5.2.5 Bayesian Methods

The best known discussion of search problems in statistics from a Bayesian perspective is Leamer's (1978). Leamer's book contains a number of interesting points, including a consideration of what a Bayesian should do upon meeting a novel hypothesis, but it does not contain a method for reliable search. Considering the use of regression methods in causal inference, for example, Leamer subsequntly recommended analyzing separately the sets of relevant regressors endorsed by any opinion, and giving separate Bayesian updates of distributions of parameters for each of these sets of regressors. The problem of deciding which variables actually influence an outcome of interest is effectively ignored.

A much more promising Bayesian approach to search has been developed by Cooper and Herskovits (1991, 1992). At present, their procedure is restricted to discrete variables and requires a total ordering such that no later variable can cause an earlier variable. Each directed graph compatible with the order is assigned a prior probability. The joint distribution of the variables assigns each vertex in the graph a distribution conditional on its parents, and these conditional probabilities parametrize the distributions for each graph. Using Dirchelet priors, a density function is imposed on the parameters for each graph. The data are used to update the density function by Bayes's rule. The probability of a graph is then just the integral of the density function over the distributions compatible with the graph. The probability of an edge is the sum of the probabilities of all graphs that contain it. Cooper and Herskovits use a greedy algorithm to construct the output graph in stages. For each vertex X in the graph, the algorithm considers the effect of adding to the parent set of X each individual predecessor of X that is not already a parent of X; it chooses the vertex whose addition to the parent set of X most increases

Discovery Algorithms for Causally Sufficient Structures   79

the posterior probability of the local structure consisting of X and its parents. Parents are added to X in this fashion until there is no single vertex that can be added to the parent set of X that will increase the posterior probability of the local structure. The program runs very well even on quite large sets of variables provided the true graph is sparse, and on discrete data with a prior ordering appears to determine adjacencies with remarkable accuracy. Its accuracy on dense graphs is not known at this time.

The Bayesian approach developed by Cooper and Herskovits has the advantages that appropriate prior degrees of belief can be used in search, that models are output with ratios of posterior distributions consistent with the specified prior distribution and the data, and that under appropriate assumptions[1] the method converges to the correct graph. Because the method can calculate the ratio of the posterior probabilities of any pair of graphs, it is possible to make inferences over multiple graphs weighted by the probability of the graph (although generally some heuristic to consider only the most probable graphs must be used because of the sheer number of possibilities.) The method works with Dirchelet priors because the relevant integrals are available analytically and posterior densities can therefore be rapidly evaluated without any numerical analysis. In view of the combinatorics of graphs, any other application of the search architecture must have the same feature. One problem is to extend the method to continuous variables, which depends on finding a family of conjugate priors that can be rapidly updated for parameters that describe graphs. Another, more fundamental, problem concerns whether the requirement of a prior ordering of the variables can be relaxed while preserving computational feasibility. Using a fixed ordering of the variables reduces the combinatorics enormously, but in many applied cases any such ordering may be uncertain. More recent work on Beyesian search algorithms is described in chapter 12.

5.3 The Wermuth-Lauritzen Algorithm

In 1983 Wermuth and Lauritzen defined what they called a *recursive diagram*. A recursive diagram is a directed acyclic graph G together with a total ordering of the vertices of the graph such that $V_1 \to V_2$ occurs only if $V_1 < V_2$ in the ordering. In addition there is a probability distribution P on the vertices such that V_i is a parent of V_k if and only if $V_i < V_k$ and V_i and V_k are dependent conditional on the set of all other variables previous to V_k in the ordering. Following Whittaker (1990), we call such systems *directed independence graphs*.

We can view this definition as an algorithm for constructing causal graphs from conditional independence relations and a time ordering of the variables. It has in fact been used in this way by some authors (Whittaker 1990). Given an ordering of the variables and a list of the conditional independence relations, proceed through the variables in their time order, and for each variable V_k to each variable V_i such that $V_i < V_k$ apply the dependence test in the definition, and add $V_i \to V_k$ if the test is passed. The procedure will correctly recover the directed graph from the order and the independence relations of a faithful distribution in which, for discrete variables, every combination of variable values has positive probability. In a sense, the discovery problem for causally sufficient faithful systems is solved. In practice, however, the Wermuth-Lauritzen algorithm is not feasible save for very small variable sets. The remaining issues are therefore these:

(i) how to remove the requirement that an ordering of the variables be known beforehand;
(ii) how to improve on the computational efficiency and statistical requirements of the Wermuth-Lauritzen procedure;
(iii) how to remove the tacit restriction to causally sufficient systems of variables.

In this chapter we will address the first two of these problems. The problem of causal inference when unmeasured common causes, or "latent variables," may be acting will be taken up in chapter 6.

5.4 New Algorithms

We will describe several algorithms for discovering causal structure (assuming causal sufficiency); they eliminate the need for a prior ordering of the variables and all but two of them improve computational efficiency and reduce the difficulty of statistical decisions in comparison with the Wermuth-Lauritzen algorithm. Some of the improvements are dramatic, others less so. Each of the search procedures described can also be used on discrete data to search for graphical log-linear models. (For each triple of variables, if $X \rightarrow Y \leftarrow Z$ occurs in the directed graph, and X is not adjacent to Z, add an undirected edge between X and Z; then remove all arrowheads from the graph. The result is an undirected independence graph.)

Under the following assumptions all of the algorithms presented in this section provably recover features of graphs faithful to the population distribution:

(i) The set of observed variables is causally sufficient.
(ii) Every unit in the population has the same causal relations among the variables.
(iii) The distribution of the observed variables is faithful to an acyclic directed graph of the causal structure (in the discrete case) or linearly faithful to such a graph (in the linear case).
(iv) The statistical decisions required by the algorithms are correct for the population.

The fourth requirement is unnecessarily strong, since the algorithms will in many cases succeed even if some statistical decisions are in error. Nonetheless, this is a strong set of assumptions that is often not met in practice, but it is no stronger than the assumptions that would be required to warrant most of the particular statistical models with a causal interpretation found in the medical, behavioral, and social scientific literature. In subsequent chapters we will examine the consequences of weakening some of these assumptions.

In practice, the algorithms take as input either a covariance matrix or cell counts. Where d-separation facts are needed by an algorithm, in the discrete case the procedure performs tests of conditional independence and in the linear continuous case tests for vanishing partial correlations. (Recall that if P is a discrete distribution faithful to a graph G, then A and B are d-separated given a set of variables **C** if and only A and B are conditionally independent given **C**, and if P is a distribution linearly faithful to a graph G, then A and B are d-separated given **C** if and only if $\rho_{AB.C} = 0$.) The algorithms construct the set of directed acyclic graphs that satisfy the given set of d-separability relations, if any such graph exists. Since the results of either kind of test are used only to determine the d-separation relations among the variables, we will speak as if the input to the algorithms is simply the d-separation relations themselves.[2]

Let us say that a graph G **faithfully represents a list of d-separations L** if and only if all and only the d-separations in L are true of G. A **list L of d-separations is faithful** if and only some acyclic directed graph faithfully represents L. In practice, even if a distribution is faithful to the causal structure that generates it, sampling error or minor violations of the assumptions of the statistical tests employed can lead to errors in judgment about the properties of the population. The robustness of the procedures to erroneous specification of the distribution family or to sampling variation can be investigated by Monte Carlo simulation methods.

Each of the following algorithms can have as output either a class of directed acyclic graphs, or else a single mixed object with both directed and undirected edges—the pattern that represents a class of graphs. Recall that pattern Π represents a set of directed acyclic graphs. A graph G is in the set of graphs represented by Π if and only if:

(i) G has the same adjacency relations as Π;
(ii) if the edge between A and B is oriented $A \rightarrow B$ in Π, then it is oriented $A \rightarrow B$ in G;
(iii) if Y is an unshielded collider on the path $<X,Y,Z>$ in G then Y is an unshielded collider on $<X,Y,Z>$ in Π.

If any of the algorithms use as input a covariance matrix from a distribution linearly faithful to G, or cell counts from a distribution faithful to G, we will say the input is **data faithful** to G. All of the algorithms we will discuss in this section have the following correctness property:

THEOREM 5.1: If the input to any of the algorithms is data faithful to G, the output of each of the algorithms is a pattern that represents the faithful indistinguishability class of G.

The algorithms do not, however, always provide a pattern that explicitly characterizes all of the orientation information implicit in the d-separation facts; a pattern may be produced that is consistent only with one orientation of an edge but does not explicitly contain the corresponding arrowhead.

5.4.1 The SGS Algorithm

The correctness of the SGS algorithm (Spirtes, Glymour, and Scheines 1990c) follows from theorem 3.4:

THEOREM 3.4: If P is faithful to some directed acyclic graph, then P is faithful to G if and only if

(i) for all vertices, X, Y of G, X and Y are adjacent if and only if X and Y are dependent conditional on every set of vertices of G that does not include X or Y; and
(ii) for all vertices X, Y, Z such that X is adjacent to Y and Y is adjacent to Z and X and Z are not adjacent, $X \rightarrow Y \leftarrow Z$ is a subgraph of G if and only if X, Z are dependent conditional on every set containing Y but not X or Z.

SGS Algorithm

A.) Form the complete undirected graph H on the vertex set $\mathbf{V}$.
B.) For each pair of vertices A and B, if there exists a subset $\mathbf{S}$ of $\mathbf{V}\backslash\{A,B\}$ such that A and B are d-separated given $\mathbf{S}$, remove the edge between A and B from H.
C.) Let K be the undirected graph resulting from step B). For each triple of vertices A, B, and C such that the pair A and B and the pair B and C are each adjacent in K (written as A - B - C) but the pair A and C are not adjacent in K, orient A - B - C as $A \rightarrow B \leftarrow C$ if and only if there is no subset $\mathbf{S}$ of $\{B\} \cup \mathbf{V}\backslash\{A,C\}$ that d-separates A and C.
D.) repeat
    If $A \rightarrow B$, B and C are adjacent, A and C are not adjacent, and there is no arrowhead at B, then orient B - C as $B \rightarrow C$.
    If there is a directed path from A to B, and an edge between A and B, then orient A - B as $A \rightarrow B$.
    until no more edges can be oriented.

5.4.1.1 Complexity

Reliability is one thing, efficiency another. Step B) of the SGS algorithm requires that for each pair of variables adjacent in G we look at all possible subsets of the remaining variables, and that, of course, is an exponential search. In the worst case that complexity is unavoidable if reliability is to be maintained. Two variables can be dependent conditional on a set $\mathbf{U}$ but independent on a superset or subset of $\mathbf{U}$. Any procedure that in the worst case does not examine the conditional independence relations of variables X, Y on all subsets of vertices not containing that pair will fail—there will be some structure the procedure does not get correctly.

5.4.1.2 Stability of SGS

We need to consider whether an algorithm remains reasonably reliable when the data are imperfect. We will use the notion of stability informally: If intuitively small errors of input produce intuitively large errors of output, the algorithm is not stable. For the SGS algorithm, an intuitively small error in input consists of a few d-separation relations that are falsely included or falsely excluded from the input. An intuitively small error for Step B is a few undirected edges erroneously included in or omitted from the output. An intuitively small error for Step C is a few edges misoriented.

Step B) of the SGS algorithm is stable. If, for example, a single correct d-separation relation is omitted from the input, the algorithm will nonetheless produce the correct undirected graph unless there is no other set besides **U** on which X, Y are d-separated. Even in that case Step B will make an error in postulating an X - Y connection, but no other errors. If X and Y are adjacent in the true graph, but it is incorrectly judged that X and Y are d-separated given **U**, the algorithm will fail to connect X and Y but no other error will be made.

Step C) of the SGS algorithm is less stable. A small error in either component of the input, either the undirected graph or the list of d-separation relations, can (and often will) produce large errors in the output. That is because the edges that occur in collisions determine the orientations of other edges in the graph, and if input errors lead the algorithm erroneously to include or exclude a collision, the error may affect the orientations of many other edges in the graph.

Suppose, for example, an edge connecting X, Z is erroneously omitted in the undirected graph input to step C), and X - Y - Z correctly occurs in the input. Then if X and Z are not d-separated by any subset of variables containing Y but not X, Z, the algorithm will mistakenly require a collision at Y, and this requirement will ramify through orientations of other edges. Or, if the true structure contains a collision at Y but X - Y is omitted in the input to step C), no unique orientation will be given to Y - Z, and this uncertainty may ramify through the orientations of other edges on paths including Z.

Instabilities may also arise in Step C) because of errors in the list of d-separation relations input, even when the underlying undirected graph is correct. If in the input to C), X is adjacent to Y and Y to Z but not X to Z and a d-separation relation between X and Z given **S** containing Y is omitted from the input, no orientation error will result unless no other set containing Y d-separates X and Z. But if in the true directed graph, the edges between X and Y and between Y and Z collide at Y, and a d-separation relation involving X and Z and some set **U** containing Y but not X or Z is erroneously included in the input, the algorithm will conclude that there is no collision at Y, and this error may be ramified to other edges.

A little reflection on Step C) reveals that its output may not be a collection of directed acyclic graphs if one of the four assumptions listed at the beginning of this section is violated. This is not necessarily a defect of the algorithm. If the algorithm finds that the edges X - Y - Z collide at Y, and Y - Z - W collide at Z, it will create a pattern with an edge $Y \leftrightarrow Z$. Double headed edges can occur when the causal structure is not causally sufficient, or when there is an error in input (as from sampling variation). They have a theoretical role in identifying the presence of unmeasured common causes, an issue discussed further in the next chapter.

5.4.2 The PC Algorithm

In the worst case, the SGS algorithm requires a number of d-separation tests that increases exponentially with the number of vertices, as must any algorithm based on conditional independence relations or vanishing partial correlations. But the SGS algorithm is very inefficient because for edges in the true graph the worst case is also the expected case. For any undirected edge that is in the graph G, the number of d-separation tests that must be conducted in stage B) of the algorithm is unaffected by the connectivity of the true graph, and therefore even for sparse graphs the algorithm rapidly becomes infeasible as the number of vertices increases. Besides problems of computational feasibility, the algorithm has problems of reliability when applied to sample data. The determination of higher order conditional independence relations from sample distributions is generally less reliable than is the determination of lower order independence relations. With, say, 37 variables taking three values each, to determine the conditional independence of two variables on the set of all remaining variables requires considering the relations among the frequencies of 3^{35} distinct states, only a fraction of which will be instantiated even in very large samples.

We should like an algorithm that has the same input/output relations as the SGS procedure for faithful distributions but which for sparse graphs does not require the testing of higher order independence relations in the discrete case, and in any case requires testing as few d-separation relations as possible. The following procedure (Spirtes, Glymour, and Scheines 1991) starts by forming the complete undirected graph, then "thins" that graph by removing edges with zero order conditional independence relations, thins again with first order conditional independence relations, and so on. The set of variables conditioned on need only be a subset of the set of variables adjacent to one or the other of the variables conditioned.

Let **Adjacencies**(C, A) be the set of vertices adjacent to A in directed acyclic graph C. (In the algorithm, the graph C is continually updated, so **Adjacencies**(C, A) is constantly changing as the algorithm progresses.)

PC Algorithm

A.) Form the complete undirected graph C on the vertex set **V**.
B.)
   $n = 0$.
   repeat
      repeat
         select an ordered pair of variables X and Y that are adjacent in C such that **Adjacencies**$(C,X)\setminus\{Y\}$ has cardinality greater than or equal to n, and a subset **S** of **Adjacencies**$(C,X)\setminus\{Y\}$ of cardinality n, and if X and Y are d-separated given **S** delete edge X - Y from C and record **S** in **Sepset**(X,Y) and **Sepset**(Y,X);

Discovery Algorithms for Causally Sufficient Structures

until all ordered pairs of adjacent variables X and Y such that **Adjacencies**$(C,X)\backslash\{Y\}$ has cardinality greater than or equal to n and all subsets **S** of **Adjacencies**$(C,X)\backslash\{Y\}$ of cardinality n have been tested for d-separation;
$n = n + 1$;

until for each ordered pair of adjacent vertices X, Y, **Adjacencies**$(C,X)\backslash\{Y\}$ is of cardinality less than n.

C.) For each triple of vertices X, Y, Z such that the pair X, Y and the pair Y, Z are each adjacent in C but the pair X, Z are not adjacent in C, orient $X - Y - Z$ as $X \rightarrow Y \leftarrow Z$ if and only if Y is not in **Sepset**(X,Z).

D.) repeat

If $A \rightarrow B$, B and C are adjacent, A and C are not adjacent, and there is no arrowhead at B, then orient $B - C$ as $B \rightarrow C$.

If there is a *directed* path from A to B, and an edge between A and B, then orient $A - B$ as $A \rightarrow B$.

until no more edges can be oriented.

Figure 5.1 traces the operation of the first two parts of the PC algorithm.

Although it does not in this case, stage B) of the algorithm may continue testing for some steps after the set of adjacencies in the true directed graph has been identified. The undirected graph at the bottom of figure 5.1 is now partially oriented in step C). The triples of variables with only two adjacencies among them are:

$A - B - C$; $A - B - D$;
$C - B - D$; $B - C - E$;
$B - D - E$; $C - E - D$

E is not in **Sepset**(C,D) so $C - E$ and $E - D$ collide at E. None of the other triples form colliders. The final pattern produced by the algorithm is shown in figure 5.2.

The pattern in figure 5.2 characterizes a faithful indistinguishability class. Every orientation of the undirected edges in figure 5.2 is permissible that does not include a collision at B.

5.4.2.1 Complexity

The complexity of the algorithm for a graph G is bounded by the largest degree in G. Let k be the maximal degree of any vertex and let n be the number of vertices. Then in the worst case the number of conditional independence tests required by the algorithm is bounded by

$$2\binom{n}{2}\sum_{i=0}^{k}\binom{n-1}{i}$$

which is bounded by

$$\frac{n^2(n-1)^{k-1}}{(k-1)!}$$

True Graph

Complete Undirected Graph

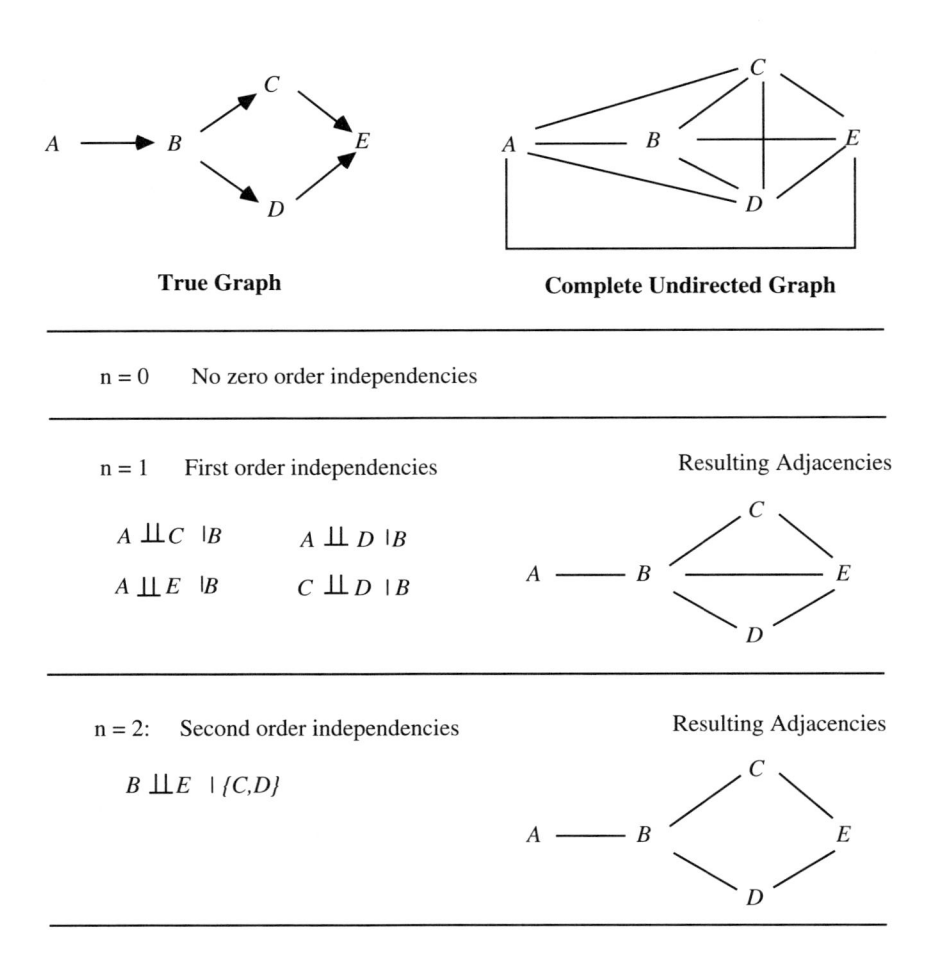

n = 0   No zero order independencies

n = 1   First order independencies   Resulting Adjacencies

$A \perp\!\!\!\perp C \mid B$ $A \perp\!\!\!\perp D \mid B$

$A \perp\!\!\!\perp E \mid B$ $C \perp\!\!\!\perp D \mid B$

n = 2:   Second order independencies   Resulting Adjacencies

$B \perp\!\!\!\perp E \mid \{C, D\}$

Figure 5.1

This is a loose upper bound even in the worst case; it assumes that in the worst case for *n* and *k*, no two variables are d-separated by a set of less than cardinality *k*, and for many values of *n* and *k* we have been unable to find graphs with that property. While we have no formal expected complexity analysis of the problem, the worst case is clearly rare, and

```
        C
      ↗   ↘
A ─── B      E
      ↘   ↗
        D
```

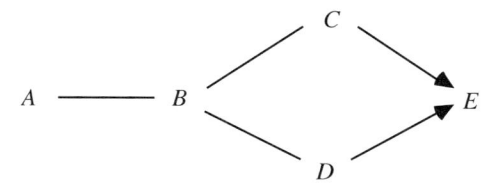

Figure 5.2

the average number of conditional independence tests required for graphs of maximal degree k is much smaller. In practice it is possible to recover sparse graphs with as many as a hundred variables. Of course the computational requirements increase exponentially with k.

The structure of the algorithm and the fact that it continues to test even after having found the correct graph suggest a natural heuristic for very large variable sets whose causal connections are expected to be sparse, namely to fix a bound on the order of conditional independence relations that will be tested.

5.4.2.2 Stability of PC

In theory, the PC Algorithm is unstable in both steps B) and C) although in practice step B) has proved to be much more reliable than step C).

If an edge is mistakenly removed from the true graph at an early stage of step B) of the algorithm, then other edges which are not in the true graph may be included in the output. Consider the following example.

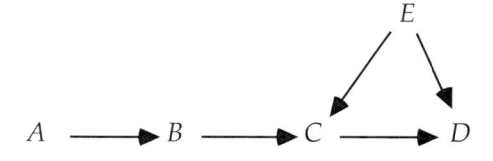

```
              E
            ↙   ↘
A ──→ B ──→ C ──→ D
```

Figure 5.3

If the edge E - D is mistakenly removed from the initial complete graph then at a subsequent stage of the search the edge B - D will not be removed, because E will no longer be in the adjacency set for D, and B and D are dependent on every subset of A and C. The omission of an edge can also lead to orientation errors. If an edge is mistakenly left in the graph and there are no additional errors in the list of d-separations in the input, the only further errors that result are that some edges which theoretically could be oriented, will not be oriented.

Step C) of the algorithm is unstable for the same reasons as in step C) of the SGS algorithm.

The PC algorithm is faster than the SGS algorithm because it tests fewer d-separation relations. Given a faithful list of d-separability relations, the two algorithms output the same set of pattern graphs. But if the list of d-separability relations is not faithful, due to sampling error for example, the two algorithms can output different pattern graphs. Consider the following example.

$$A \longrightarrow B \longrightarrow C \longrightarrow D \longrightarrow E$$

Figure 5.4

According to this graph, A and E are d-separated from each other given any non-empty subset of B, C, and D. If, after the A - C and E - C edges have been removed from the initial undirected graph, the procedure incorrectly judges that A and E are not d-separated given any non-empty subset of B and D, the PC algorithm will incorrectly include an edge between A and E, because it only tests whether A and E are d-separated given subsets of the adjacencies of A and E. On the other hand, because the SGS algorithm tests whether A and E are d-separated given any subset of $\mathbf{V}\backslash\{A,B\}$, it would properly recognize that there is not an edge between A and E because A and E are d-separated given C.

In contrast, if after the A - E and B - E edges are removed from the initial undirected graph, it is mistakenly judged that A and B are d-separated given E, the SGS algorithm will mistakenly remove the $A \rightarrow B$ edge. If the A - E and B - E edges are removed first, the PC algorithm, would correctly leave the $A \rightarrow B$ edge in, because it would not test whether A and B are d-separated given E.

Because the PC algorithm attempts to use "local" information to judge whether an edge exists or not, it is not guaranteed to produce a graph that is in some sense "closest" to an unfaithful distribution. Consider the example shown in figure 5.5.

In a distribution faithful to this graph, every variable is dependent on every other variable. Suppose a test determines that A and B are independent conditional on some other variable, either because of some coincidental parameter values, or because of sampling error. The PC algorithm would then remove the A - B edge in order to satisfy that constraint. In doing so, however, it would disconnect the graph. The resulting graph would entail that A and all of its descendants to the left are independent of B and all of its descendants. So, in order to satisfy one conditional independence constraint, the PC algorithm may produce a graph that violates a great many independence constraints. In a number of data sets the correlations between two variables do not vanish but the output pattern disconnects them. For greater reliability the procedure should be supplemented with a repair algorithm, for which the Cooper and Herskovits Bayesian procedure might suffice in the discrete case; alternatively, a variation of the procedure described in chapter 11 could be applied.

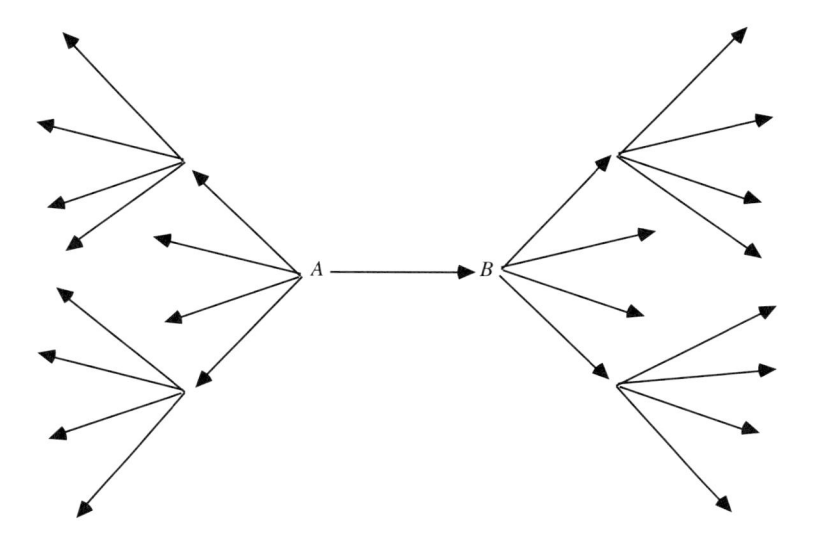

Figure 5.5

5.4.2.3 The PC* Algorithm

The PC algorithm is computationally efficient and asymptotically reliable, but on sample data the procedure takes unnecessary risks. In determining whether to eliminate an undirected edge between variables A and B, the procedure may test every subset of the adjacency set of A and of the adjacency set of B. But the independence or dependence of A and B on many of these subsets of variables may be entirely irrelevant to the causal relations between A and B. For a distribution faithful to a directed acyclic graph, if variables A and B are independent conditional given **Parents**(A) or given **Parents**(B) then they are independent given a subset of **Parents**(A) or given a subset of **Parents**(B) consisting only of vertices lying on undirected paths between A and B. It is sufficient, then, to test for the conditional independence of A and B given subsets of variables adjacent to A and subsets of variables adjacent to B that are on undirected paths between A and B. Call the modified algorithm PC*.

The PC and PC* algorithms yield the same output given a faithful list of conditional independence relations or correlations as input, but they may differ given conditional independence relations determined from tests on sample data. The PC* algorithm avoids one kind of error made by the PC algorithm. If, however, at an early stage the PC* algorithm mistakenly disconnects a path between X and Y it may then mistakenly leave the X - Y edge in the undirected graph, while the PC algorithm, given the same data, might avoid that error. Moreover, whatever increased reliability the PC* algorithm may have is bought at great cost, since the algorithm must at each stage of step B) keep track of all of the undirected paths in the graph it considers at that stage. The number of undirected paths is typically very large, and the memory requirements of the PC* algorithm are not feasible save for relatively small numbers of variables, in which case it may be the algorithm of choice. For large numbers of variables the PC algorithm must be used instead, although if the true graph is sparse, the PC algorithm can be used until the average degree of the undirected graph C is small, after which stage the PC* algorithm may be used. Later in this chapter we will describe the performance of the two algorithms on discrete data taken from Christensen 1990.

5.4.2.4 Speed-Up Heuristics for Ordering Tests

Step B of the PC algorithm selects some variable pair and some subset **S** of a given size to test for d-separation. The faster edges are eliminated from the complete graph, the smaller the search that has to be conducted at later stages of the algorithm, and the faster the algorithm runs. Hence, it is best to select first for testing those variable pairs A and B and subsets **S** for which A and B are most likely to be d-separated by **S**. We have considered three variants of the PC algorithm that use different methods of selecting the order of tests.

Heuristic 1 Test the variable pairs and subsets **S** in lexicographic order. (We will call this PC–1.)

Heuristic 2 First test those variables pairs that are least dependent[3] in probability. The conditioning subsets are selected by lexicographic order. (We will call this PC–2.)

Heuristic 3 For a given variable A, first test those variables B that are least probabilistically dependent on A, conditional on those subsets of variables that are most probabilistically dependent on A. (We will call this PC–3.)

The intuition behind heuristic 2 is that variables with the highest probabilistic dependence are most likely to be adjacent in the true graph, and hence not ever eliminated from the graph being constructed, while those with the smallest probabilistic dependence are least likely to be adjacent in the true graph. Of course, no such relation strictly holds.

The intuition behind heuristic 3 is similar. A variable B that is not genuinely adjacent to a variable A is d-separated from A given some subset of the variables that are adjacent to A or given some subset of the variables that are adjacent to B in the true graph. Assuming that variables with the highest probabilistic dependence upon A are most likely to be adjacent to A in the true graph, this suggests testing whether A is d-separated from variables with a low probabilistic dependence on A, conditional on variables with a high probabilistic dependence upon A.

5.4.3 The IG (Independence Graph) Algorithm

Verma and Pearl (1990) have suggested a variation of the SGS algorithm. In their alternative, the first step in searching for the directed acyclic graph is to construct the undirected independence graph N, that is, for each pair of variables A, B introduce an undirected edge between them if they are dependent conditional on the set of all other variables. In the undirected independence graph for a distribution faithful to a directed acyclic graph the parents of any variable form a maximal complete subgraph—a clique. Again for each pair of variables *A*, *B* adjacent in *N*, determine if *A* and *B* are d-separated given any subsets of variables in the cliques in *N* containing *A* or *B*. If so *A* is not adjacent to *B* in *G*. The complexity is thus a function of the size of the largest clique in *N*.

Determining the cliques in a graph would appear to require unnecessary computation, and in other than the worst case, testing for conditional independence of two variables conditional on all members of the maximal clique of one or the other will involve a test of unnecessarily high order. A better idea might be to blend the procedure with the PC algorithm: modify step A of the PC algorithm by setting the initial graph in the PC procedure to the *undirected independence graph*, rather than the complete undirected graph, and then proceed in the same way. We will call this algorithm IG (independence graph.)

The efficiency of these algorithms obviously depends upon how easily the independence graph can be constructed. The off-diagonal elements of the standardized inverse of the correlation matrix are the negatives of the partial correlation coefficients between the corresponding variables given the remaining variables (see e.g., Whittaker 1990). Hence in the linear case, the independence graph can be efficiently constructed by placing an edge between *A* and *B* if and only if the entry in the standardized inverse correlation matrix is nonzero. In the discrete case, Fung and Crawford (1990) have recently proposed a fast algorithm for constructing an independence graph from discrete data. We have not tested their procedure as a preprocessor for the PC algorithm.

5.4.4 Variable Selection

While prior knowledge of causal structure can sometimes make the results of the algorithms we have described more informative on real samples, correct selection of variables is essential for reliable inference, and for that algorithms (at least these algorithms) provide no help.

We can aggregate variables or we can aggregate distinct values of a variable. As in Salmon's imaginary example discussed in chapter 3, we sometimes measure a variable that is an imprecise version of a more precise natural variable; we fail, in other words, to distinguish values that have differing effects on other variables. Continuous variables are often deliberately collapsed into a few discrete categories, sometimes because contingency table methods offer the promise of statistical analysis free of the substantive assumptions that would otherwise be required about the form of the functional dependencies—e.g., linear or otherwise—and sometimes because some of the variables to be analyzed are necessarily discrete and there are few methods available for problems with mixtures of discrete and continuous variables. Sometimes, whether through ignorance or even deliberately, we may aggregate two or more distinct variables with distinct causal structures into a single scale. What effects can aggregation and collapse have on the reliability of causal inference?

We have already observed that if C is a cause of A and B and some proxy C' for C is used that is not so precise as C and not perfectly correlated with C, it may be that A and B are statistically dependent conditional on C'. Examples of this sort appear whenever a theory postulates a cause that is measured by proxies. Friedman (1957), for example, advocated a much discussed theory in which consumption is caused by "permanent" income which can only be measured by proxies; if Friedman's theory were true, regression of consumption on measured income would provide a biased estimate of the regression coefficient of consumption on permanent income and might leave unexplained correlations between consumption and other variables. Klepper (1988) has shown how, in the linear normal case, such errors may be bounded.

Suppose we are given variables A, B, C such that A and B are independent conditional on C. Let $C' = PROJ(C)$ where $PROJ(C)$ is a projection mapping the set of n values of C to a set of $m < n$ values. If there exist values c_1, c_2 for C such that $P(A,B|C=c_1) \neq P(A,B|C=c_2)$ and $PROJ(C=c_1) = PROJ(C=c_2)$, then A and B are not independent conditional on C'. Independence relations can be made to appear rather than disappear by collapsing values of a variable. Suppose that variable B, C are dependent. Let $C' = PROJ(C)$ where $PROJ(C)$ is a projection mapping the set of n values of C to a set of $m < n$ values. If there exists a value c_1 of C such that $P(C = c_1 | B) = P(C = c_1)$ and $PROJ(c_1)$ has a unique inverse and $PROJ(c_k) = PROJ(c_j)$ for all k, j not equal to 1, then B and C' are independent.

Pearl (personal communication) has pointed out that a very simple sort of aggregation can produce an unfaithful distribution. Suppose A causes C_1 and B causes C_2, and C_1 and C_2 are each binary, and there is no other causal connection among the variables. So $\{A, C_1\}$ is independent of the set $\{B, C_2\}$, but A and C_1 are dependent and so are B and C_2. Introduce variable C taking values 0, 1, 2, 3 coding the different value pairs for C_1 and C_2. Then the actual causal structure among A, B and C is shown in figure 5.6.

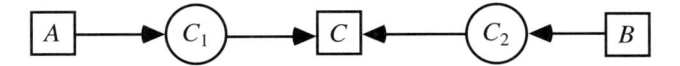

$$A \rightarrow C_1 \rightarrow C \leftarrow C_2 \leftarrow B$$

Figure 5.6

But in the joint distribution A and B are independent conditional on C, and so the joint distribution is not faithful to any causal structure whatsoever. In this case the unfaithfulness of the distribution is due to the fact that it is the marginal of a distribution that is unfaithful because of deterministic relationships among the variables: the independence of A and B given C follows directly from the application of D-separability (see chapter 3) to figure 5.6. This sort of thing may sometimes happen in practice, but it could always be tested for and in principle identified: Conditioning on A divides the values of C into two equivalence classes each containing values of C with the same conditional probability, and conditioning on B divides the values of C into a distinct pair of equivalence classes. Letting the equivalence classes induced by A be values of one variable and the equivalence classes induced by *B* be values of another variable recovers C_1 and C_2.

5.4.5 Incorporating Background Knowledge

A user of any of these algorithms may have a great deal of background knowledge—or at least belief—that could constrain the search. This knowledge might be about the existence or non-existence of certain edges in the graph, or it might be about the orientation of some of the edges, or it might be about the time order of the variables. How can this background knowledge be used by the algorithms?

The most common sort of reliable prior belief orders or partially orders the variables by time of occurrence: either measurements of *A* were taken before measurements of *B*, or *A* and *B* are believed to be exact measures of events that are so ordered. Any of the algorithms of this section can be easily modified to make two uses of such knowledge:

(i) In determining whether *A* and *B* are adjacent in the true graph by testing whether *B* is independent of *A* conditional on some subset of the current adjacencies of *A*, do not test for independence conditional on any set of variables that includes a variable that is later than *A*.

(ii) If *A* and *B* are adjacent and *B* is later than *A*, orient the edge as $A \rightarrow B$.

In the examples we give throughout this book the algorithms have been so modified, and we sometimes make use of common sense time order, always noting where such assumptions have been made.

Prior belief about whether one variable directly influences another can also be incorporated in these algorithms: if prior belief forbids an adjacency, for example, the algorithms need not bother to test for that adjacency; if prior belief requires than there be a direct influence of one variable on another, the corresponding directed edge is imposed and assumed in the orientation procedures for other edges. These procedures assume that prior belief should override the results of unconstrained search, a preference that may not always be judicious; they are nonetheless incorporated in versions of the TETRAD II program with the PC algorithm.

5.5 Statistical Decisions

The algorithms we have described are completely modular, and can be applied given any procedures for making the requisite statistical decisions about conditional independence or vanishing partial correlations. The better the decisions the better the performance to be expected from the algorithms. While tests of conditional independence relations form the most obvious class of such decisions, any statistical constraints that give d-separability relations for graphical structure will suffice. For example, in the linear normal case, vanishing partial correlation is equivalent to conditional independence, and the statistical decisions required by the algorithms could be provided by *t*-tests of the hypotheses that partial correlations vanish. But vanishing partial correlation marks d-separability whether or not the distribution is normal, so long as linearity and linear Faithfulness hold.[4] Hence under these assumptions the test of any statistic that vanishes when partial correlations vanish would suffice; one might, for example, use an *F* test for the square of the semipartial correlation coefficient, which equals the square of the *t*-test for a corresponding regression coefficient (Edwards 1976).

In the examples in this book we test whether $\rho_{XY.C} = 0$ using Fisher's *z*:

$$z(\rho_{XY.C}, n) = \frac{1}{2}\sqrt{n - |\mathbf{C}| - 3} \ln\left[\frac{(1 + \rho_{XY.C})}{(1 - \rho_{XY.C})}\right]$$

$\rho_{XY.C}$ = population partial correlation of *X* and *Y* given **C**, and |**C**| equals the number of variables in **C**. If *X*, *Y*, and **C** are normally distributed and $r_{XY.C}$ denotes the sample partial correlation of *X* and *Y* given **C**, the distribution of $z(\rho_{XY.C}, n) - z(r_{XY.C}, n)$ is standard normal (Anderson 1984).

In the discrete case, for simplicity consider two variables. Recall that we view the count in a particular cell, x_{ij}, as the value of a random variable obtained from sampling *N* units from a multinomial distribution. Let x_{i+} denote the sum of the counts in all cells in which the first variable has the value *i*, and similarly let x_{+j} denote the sum of the counts in all cells in which the second variable has the value *j*. On the hypothesis that the first and second variables are independent, the expected value of the random variable x_{ij} is:

$$E(x_{ij}) = \frac{x_{i+} x_{+j}}{N}$$

Analogously, we can compute the expected values of cells on any hypothesis of conditional independence from appropriate marginals. For example, on the hypothesis that the first variable is independent of the second conditional on the third, the expected value of the cell x_{ijk} is

$$E(x_{ijk}) = \frac{x_{i+k} x_{+jk}}{x_{++k}}$$

If there are more than three variables this formula applies to the expected value of the marginal count of the *i*, *j*, *k* values of the first three variables, obtained by summing over all other variables. The number of independent constraints that a conditional independence hypothesis places on a distribution is an exponential function of the order of the conditional independence relation and also depends on the number of distinct values each variable can assume.

Tests of such independence hypotheses have used—among others—two statistics:

$$X^2 = \sum \frac{(Observed - Expected)^2}{Expected}$$

$$G^2 = 2\sum (Observed) \ln\left(\frac{Observed}{Expected}\right)$$

each asymptotically distributed as χ^2 with appropriate degrees of freedom. In the examples in this book we calculate the degrees of freedom for a test of the independence of A and B conditional on $\mathbf{C}$ in the following way. Let $Cat(X)$ be a function which returns the number of categories of the variable X, and n be the number of variables in $\mathbf{C}$. Then the number of degrees of freedom (df) in the test is:

$$df = (Cat(A) - 1) \times (Cat(B) - 1) \times \prod_{i=1}^{n} Cat(C_i)$$

We assume that there are no structural zeroes. As a heuristic, for each cell of the distribution that has a zero entry, we reduce the number of degrees of freedom by one.[5]

Because the number of cells grows exponentially with the number of variables, it is easy to construct cases with far more cells than there are data points. In that event most cells in the full joint distribution will be empty, and even non-empty cells may have only small counts. Indeed, it can readily happen that some of the marginal totals are zero and in these cases the number of degrees of freedom must be reduced in the test. For reliable estimation and testing, Fienberg recommends that the sample size be at least five times the number of cells whose expected values are determined by the hypothesis under test.

For discrete data we fill out the PC algorithm with tests for independence using G^2 which in simulations we have found more often leads to the correct graph than does X^2. In testing the conditional independence of two variables given a set of other variables, if the sample size is less than ten times the number of cells to be fitted we assume the variables are conditionally dependent.

5.6 Reliability and Probabilities of Error

Most of the algorithms we have described require statistical decisions which, as we have just noted, can be implemented in the form of hypothesis tests. But the parameters of the tests cannot be given their ordinary significance. The usual comforts of a statistical test are the significance level, which offers assurance as to the limiting frequency with which a true null hypothesis would erroneously be failed by the test, and the power against an alternative, which is a function of the limiting frequency with which a false null hypothesis would not be rejected when a specified alternative hypothesis is true. Except in very large samples, neither the significance level nor the power of tests used within the search algorithms to decide statistical dependence measures the long run frequency of anything interesting about the search. What does?

The error probabilities one might naturally want to know for a search procedure include:

1. Given that model M is true, what is the probability that the procedure will return a conclusion inconsistent with M on sample size n?
2. Given that model M^* is true, what is the probability that the procedure will return a conclusion inconsistent with M^* but consistent with M on sample size n?
3. Given that model M is true, for samples of size n what is the probability that a search procedure will specify an adjacency not in M? What is the probability that a search procedure will omit an adjacency in M? What is the probability that a search procedure will add an arrowhead not in M to an edge that is in M? What is the probability that a search procedure will omit an arrowhead in M? What are these probabilities for any particular variable pair, A, B?

For large models, where we expect some errors of specification from most samples, questions of kind 3 are the most important.

There is little hope of obtaining analytic answers to these questions. In repeated tests of independence hypotheses in a sample, each using the same significance level, the probability that *some* true hypothesis will be rejected is not given by the significance level; depending on the number of hypotheses and the sample size, that probability may in fact be much higher than the significance level, but in any case the probability of some erroneous decision depends on which hypotheses are tested, and for all of the algorithms considered that in turn depends in a complex way on the actual structure. Further, each of the algorithms can produce correct output even though some required statistical decisions are made incorrectly. For example, suppose in graph G, vertices A and B are not adjacent. Suppose in fact A and B are independent conditional on C, on D, on C and D, and so on. If the hypothesis that A and B are independent conditional on C is rejected in the search procedure, and the algorithm goes on to test whether A and B are independent conditional on D, and decides in favor of the latter independence, then despite the earlier error the procedure will correctly conclude that A and B are not adjacent. Chapter 12 discusses various senses in which there do or do not exist confidence intervals for *any* search procedure for causal models.

For any particular M and M^* estimates of the answers to questions 1, 2, and 3 can be found empirically by Monte Carlo methods. Simulation packages for linear normal models are now common in commercial statistical packages, and the TETRAD II program contains a simulation package for linear and for discrete variable models with a variety of distributions. For small models it takes only a few minutes to generate a hundred or more samples and run the samples through the search procedures. Most of the time required is in counting the outcomes, a process that we have automated *ad hoc* for our simulations, and that can and should be automated in a general way for testing the reliabilities of particular search outcomes.

5.7 Estimation

There are wel- known methods for obtaining maximum likelihood estimates subject to a causal hypothesis under the assumption of normality, even with unmeasured variables (Joreskog 1981; Lohmoller 1989). A variety of computerized estimation methods, including ordinary and generalized least squares, are also available when the normality assumption is given up. In the discrete case, for a positive multinomial distribution, the maximum likelihood estimates (when they exist) for a cell subject to the independence constraints of the graph over a set of variables **V** can be obtained by substituting the marginal frequencies for probabilities in the factorization formula of chapter 3 (Kiiveri and Speed 1982).

$$P(\mathbf{V}) = \prod_{V \in \mathbf{V}} P(V | \mathbf{Parents}(V))$$

When there are unmeasured variables that act as common causes of measured variables, the pattern obtained from the procedures we have described can have edges with arrows at each end. In that rather common circumstance we do not know how to obtain a maximum likelihood estimate for the joint distribution of discrete measured variables.

5.8 Examples and Applications

We illustrate the algorithms for simulated and real data sets. With simulated data the examples illustrate the properties of the algorithms on samples of realistic sizes. In the empirical cases we often do not know whether an algorithm produces the truth. But it is at the very least interesting that in cases in which investigators have given some care to the treatment and explanation of their data, the algorithm reproduces or nearly reproduces the published accounts of causal relations. It is also interesting that in cases without these virtues the algorithm suggests quite different explanations from those advocated in published reports.

Studies of regression models and alternatives produced by the PC algorithm and by another procedure, the Fast Causal Inference (FCI) algorithm, are postponed until chapter 8, after latent variables and prediction have been considered in chapters 6 and 7, respectively.

5.8.1 The Causes of Publishing Productivity

In the social sciences there is a great deal of talk about the importance of "theory" in constructing causal explanations of bodies of data. Of course in explaining a data set one will always eliminate causal graphs that contradict common sense or that violate the time order of variables. But in addition, many practitioners require that every attempt to provide a causal explanation of observational data in the social sciences proceed through the particulars of principles in sociology, psychology, economics, political science, or whatever, and come accompanied with a denial of the possibility of determining a correct explanation from the statistical dependencies and common sense alone. In many of these cases the necessity of theory is badly exaggerated. Indeed, for every "recursive" structural equation model in the entire scientific literature, if the assumptions of the model are correct and no unmeasured common causes are postulated, then if the distribution is faithful the statistical dependencies in the population uniquely determine the undirected graph underlying the directed graph of causal relations. And in many cases the population statistics alone determine a direction of some, or even all, edges. When the variables are linearly ordered by time, so that variable A can be a cause of variable B only if A occurs later than B, the statistical dependencies and the time order determine a *unique* directed graph assuming only that the distribution is positive and the Markov and Minimality Conditions are satisfied. The efforts spent citing literature to justify specifications of causal dependencies are not misplaced, but in many cases effort would be better directed toward establishing the fundamental statistical assumptions, including the approximate homogeneity of the units, the correctness of the sampling assumptions, and sometimes the linearity of dependencies.

Here is a recent and rather vivid example. There is a considerable literature on causes of academic success, including publication and citation rates. A recent paper by Rodgers and Maranto (1989) considers hypotheses about the causes of academic productivity drawn from sociology, economics, and psychology, and produces a combined "theoretically based" model.

Their data were obtained in the following way: solicitations and questionnaires were sent to 932 members of the American Psychological Association who obtained doctoral degrees between 1966 and 1976 and were currently working academic psychologists. Equal numbers of male and female psychologists were sampled, and after deleting respondents who did not have degrees in psychology, did not take their first job in psychology, etc. a sample of 86 men and 76 women was obtained.

The response items were clustered into groups. For example, the *ABILITY* group consisted of measures of the mean *ACT*, *NMSQT,* and selectivity scores of the subject's undergraduate institution, together with membership in Phi Beta Kappa and undergraduate honors at graduation. Graduate Program Quality (*GPQ*) consisted of the scholarly quality of department faculty and program effectiveness using national rankings, the fraction of faculty with publications between 1978 and 1980, and whether an editor of a journal was on the department faculty. These response items were treated as indicators—that is, as effects—of the unmeasured variables *GPQ*, and *ABILITY*. Other measures were quality of first job (*QFJ*), *SEX*, citation rate (*CITES*) and publication rate (*PUBS*). In preliminary analyses they also used an aggregated measure of productivity (*PROD*). The various hypotheses Rodgers and Maranto considered were then treated as linear "structural equation models"[6] They report the following correlations among the cluster variables

ABILITY	GPQ	PREPROD	QFJ	SEX	CITES	PUBS
1.0						
.62	1.0					
.25	.09	1.0				
.16	.28	.07	1.0			
-.10	.00	.03	.10	1.0		
.29	.25	.34	.37	.13	1.0	
.18	.15	.19	.41	.43	.55	1.0

There follows a very elaborate explanation of causal theories suggested by the pieces of sociological, economic and psychological literature. Rodgers and Maranto estimate no fewer than six different sets of structural equations and corresponding causal theories. The six structures they consider are as shown in figure 5.7.

The labels on the graphs indicate simply the social scientific theory from which Rodgers and Maranto derived the causal graph. For example, the "Human Capital" and the "Screening" graphs were obtained from economic theory in the following way:

> In the human capital model (Becker 1964) education has a direct effect on productivity because it conveys relevant knowledge. People invest in education until its marginal cost (the extra expenses and foregone earnings for an additional year of education) is equal to its marginal benefit (the increase in lifetime earnings caused by another year of education). More able individuals are more productive in both work and the acquisition of skills than their less able counterparts. Thus, ability has a direct effect on productivity and an indirect effect through education, because more able individuals gain more from school. Work experience also increases productivity by providing on-the-job training. The quality as well as the quantity of education is relevant to the human capital framework.
>
> The screening hypothesis implicitly views ability as the primary determinant of productivity. Employers wish to hire the most productive applicants, but ability is not directly observable. Individuals invest in education as a means of signaling their ability to employers. The marginal cost of education is inversely related to ability, inducing a positive correlation between ability and the level of education. Therefore, by selecting applicants based on their education, employers hire by ability (Spence 1973). In this model, education does not affect productivity directly, but only through its association with ability. Variations in the quality of education are consistent with the screening model. (Wise 1975)

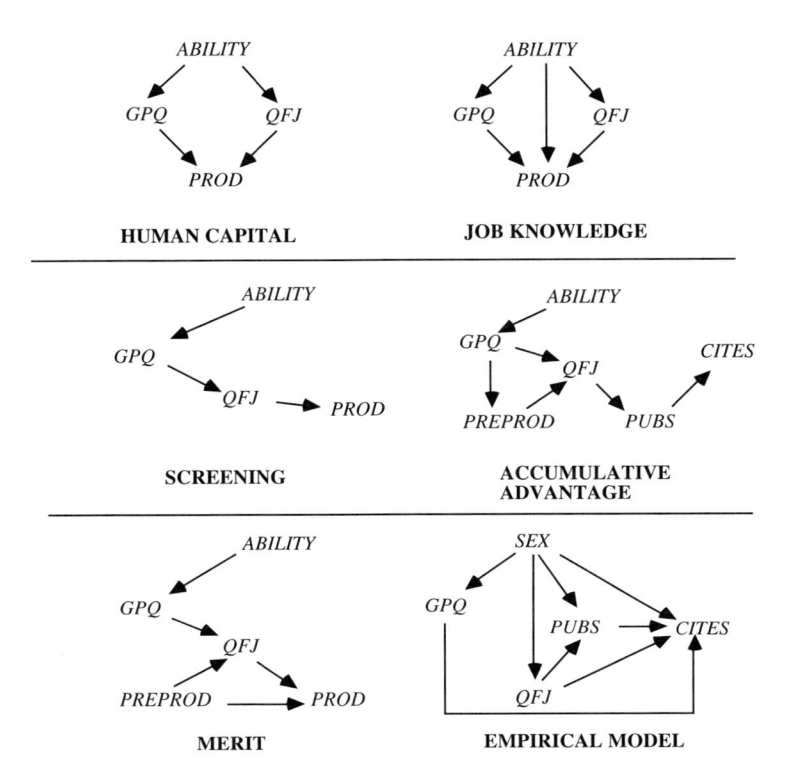

Figure 5.7

The "empirical model" was obtained from a previous study that did not appeal to social theory.

None of the structural equation systems based on these models save the phenomena. But combining all of the edges in the "theoretical" models, adding two more that seem plausible, and then throwing out statistically insignificant (at .05) dependencies, leads Rodgers and Maranto instead to propose a different causal structure that fits the data quite well.

It would appear that the tour through "theory" was nearly useless, but Rodgers and Maranto say otherwise:

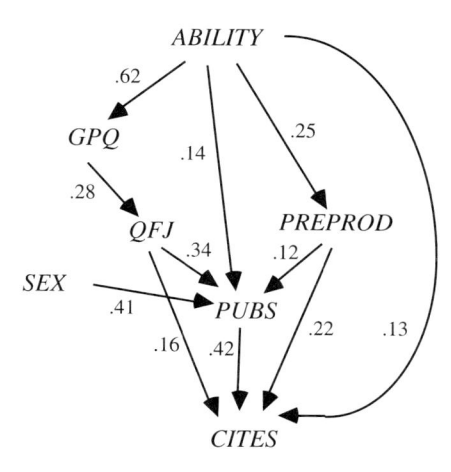

Discovery Algorithms for Causally Sufficient Structures

Figure 5.8

Causal models based solely on the pattern of observed correlations are highly suspect. Any data can be fitted by several alternative models. The construction of the best-fit model was thus guided by theory-based expectations. By using the two measures of productivity, *PUBS* and *CITES*, and the five causal antecedents, we initially estimated a composite model with all of the paths identified by the six theories. This model produced a large positive deviation between the observed and predicted correlation of *ABILITY* with *PREPROD*, suggesting that we omitted one or more important paths. Reexamination of our initial interpretation of the six theories led us to conclude that two paths had been overlooked. One such path is from *ABILITY* to *PREPROD*. . . . The other previously unspecified path is from *ABILITY* to *PUBS*. These two paths were added and all nonsignificant paths were deleted from the composite model to arrive at the best-fit model.

If the Rodgers and Maranto theory were completely correct, the undirected graph underlying their directed graph would be uniquely determined by the conditional independence relations, and the orientation would be almost uniquely determined; only the directions of the $GPQ \rightarrow QFJ$, $ABILITY \rightarrow GPQ$ and $ABILITY \rightarrow PREPROD$ edges could be changed, and only in a way that does not create a new collision.

When the PC algorithm is applied to their correlations with the common sense time order using a significance level of .1 for tests of zero partial correlations, the output is the graph on the left side of figure 5.9, which we show alongside the Rodgers and Maranto model.

ABILITY

GPQ

QFJ   PREPROD

SEX   PUBS

CITES

PC Output

ABILITY

GPQ

QFJ   PREPROD

SEX   PUBS

CITES

Rodgers and Maranto Graph

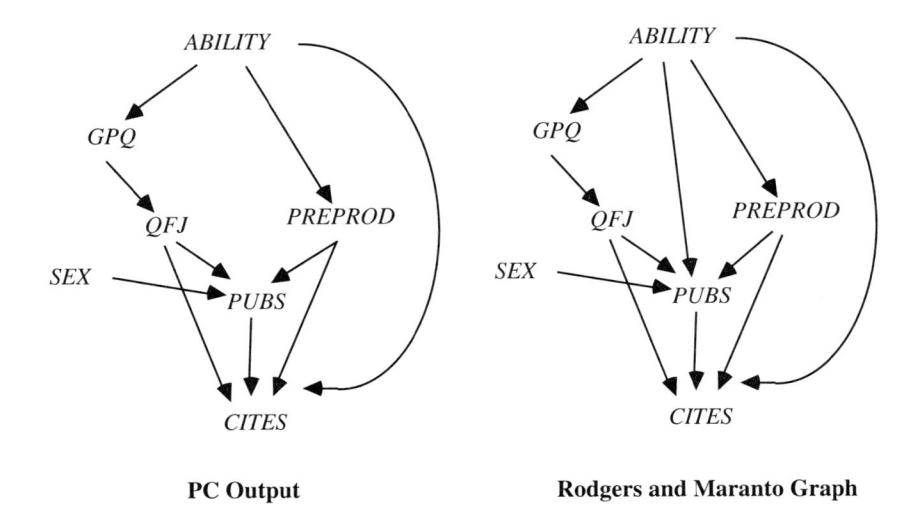

Figure 5.9

All but one of the edges in the Rodgers and Maranto model is produced instantaneously from the data and common sense knowledge of the domain—the time order of the variables. EQS gives this model a χ^2 of 13.58 with 11 degrees of freedom and $p = .257$. If the search procedure is repeated using .05 as the significance level, the program deletes the *PREPROD* → *PUBS* edge. When that model is estimated and tested with the EQS program we find that χ^2 is 19.2 with 12 degrees of freedom and a p value of .08, figures that should be taken as estimates of fit rather than of the probability of error.

Any claim that social scientific theory—other than common sense—is required to find the essentials of the Rodgers and Maranto model is clearly false. Nor do the preliminary results of Rodgers and Maranto's search afford any reason for confidence in social scientific theory. In contrast, we know a good deal about the reliability and limitations of the PC algorithm. The entire study with TETRAD and EQS takes a few minutes. A slight variant of the model is obtained using the SGS algorithm rather than the PC algorithm.

5.8.2 Education and Fertility

Rindfuss, Bumpass, and St. John (1980) were interested in the mutual influence in married women of education at time of marriage (*ED*) and age at which a first child is born (*AGE*). On theoretical grounds they argue at length for the model on the left in figure 5.10, where the regressors from top to bottom are as follows:

Discovery Algorithms for Causally Sufficient Structures

DADSO = father's occupation
RACE = race
NOSIB = absence of siblings
FARM = farm background
REGN = region of the United States
ADOLF = presence of two adults in the subject's childhood family
REL = religion
YCIG = cigarette smoking
FEC = whether the subject had a miscarriage.

Regressors are correlated. The sample size is 1766, and the covariances are given below.

DADSO	RACE	NOSIB	FARM	REGN	ADOLF	REL	YCIG	FEC	ED	AGE
456.676										
-.9201	.089									
-15.825	.1416	9.212								
-3.2442	.0124	.3908	.2209							
-1.3205	.0451	.2181	.0491	.2294						
-.4631	.0174	-.0458	-.0055	.0132	.1498					
.4768	-.0191	.0179	-.0295	-.0489	-.0085	.1772				
-0.3143	.0031	.0291	-.0096	-.0018	.0089	-.0014	.1170			
.2356	.0031	.0018	-.0045	-.0039	.0021	-.0003	.0009	.0888		
18.66	-.1567	-2.349	-.2052	-.2385	-.1434	-.0119	-.1380	.0267	5.5696	
16.213	-.2305	-1.4237	-.2262	-.3458	.1752	.1683	-.1702	.2626	3.6580	16.6832

Apparently to their surprise, the investigators found on estimating coefficients that the $AGE \rightarrow ED$ parameter is zero. Given the prior information that *ED* and *AGE* are not causes of the other variables, the PC algorithm (using .the 05 significance level for tests) directly finds the model on the right in figure 5.10, where connections among the regressors are not pictured. This case is discussed further in chapter 12, section 5.10.

5.8.3 The Female Orgasm

Bentler and Peeler (1979) obtained data from 281 female university undergraduates regarding personality and sexual response. They include the Eysenck Personality Inventory which measured neuroticism (*N*) and extraversion (*E*); a heterosexual behavior inventory (*HET*), a monosexual behavior inventory (*MONO*); a scale of negative attitudes toward masturbation (*ATM*) and an inventory of subjective assessments of coital and masturbatory experiences. Using factor analysis the investigators formed scales, thought to be unidimensional, from these responses, including two scales (*SCOR*) and (*SMOR*) from the subjective assessments of coital and masturbatory experiences.

The investigators were interested in two hypotheses: (1) subjective orgasm responses in masturbation and coitus are due to distinct internal processes; (2) extraversion,

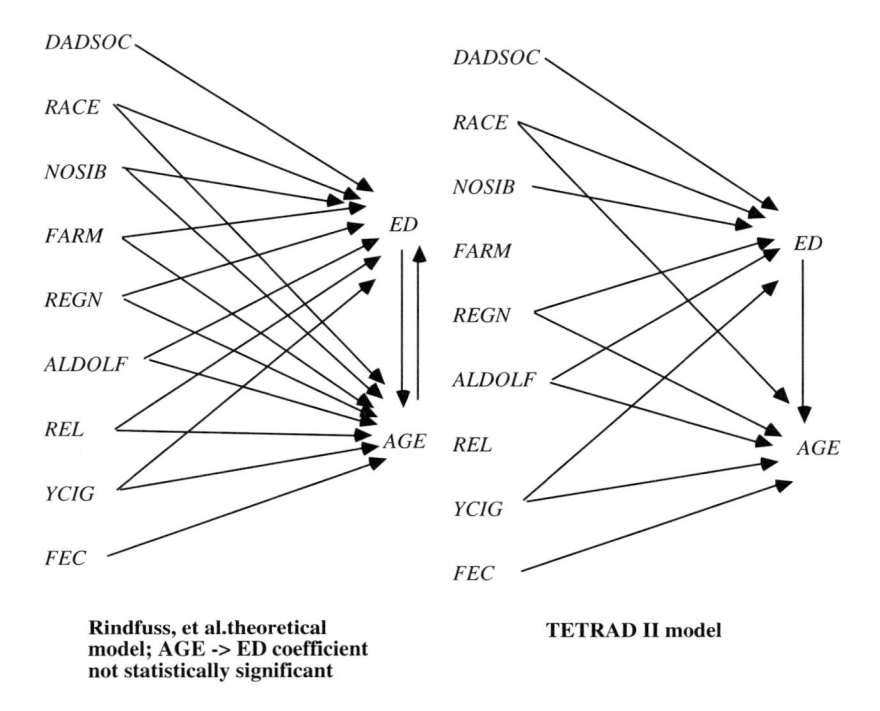

Rindfuss, et al. theoretical model; AGE -> ED coefficient not statistically significant

TETRAD II model

Figure 5.10

neuroticism and attitudes toward masturbation have no direct effect on orgasmic responsiveness, measured by *SCOR* and *SMOR*, but influence that phenomenon only through the history of the individual's sexual experience measured by *HET* and *MONO*.

We will not discuss the formation of the scales in this case, since the only data presented are the correlations of the scales and inventory scores, which are:

```
 E       N       ATM     HET     MONO    SCOR    SMOR
 1.0
-.132    1.0
 .009   -.136    1.0
 .22    -.166    .403    1.0
-.008    .008    .598    .282    1.0
 .119   -.076    .264    .514    .176    1.0
 .118   -.137    .368    .414    .336    .338    1.0
```

Bentler and Peeler offer two linear models to account for the correlations. The models and the probability values for the associated asymptotic χ^2 are shown in figure 5.11.

Discovery Algorithms for Causally Sufficient Structures 105

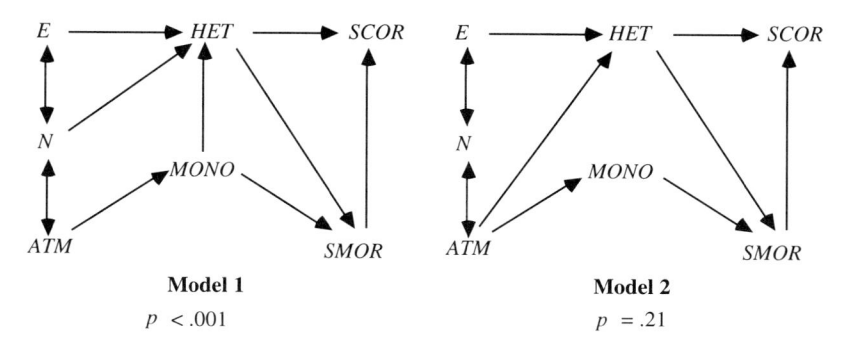

Figure 5.11 — Model 1 ($p < .001$) and Model 2 ($p = .21$)

Only the second model saves the phenomena. The authors write that

> it proved possible to develop a model of orgasmic responsiveness consistent with the hypothesis that extraversion (*e*), neuroticism (*n*), and attitudes toward masturbation (*atm*) influence orgasmic responsiveness only through the effect these variables have on heterosexual (*het*) and masturbatory (*mono*) experience. Consequently hypothesis 2 appears to be accepted. (p. 419)

The logic of the argument is not apparent. As the authors note "it must be remembered that other modes (sic) could conceivably also be developed that would equally well describe the data" (p. 419). But if the data could equally well be described, for example, by a model in which *ATM* has a direct effect on *SCOR* or on *SMOR*, there is no reason why hypothesis 2 should be accepted. Using the PC algorithm, one readily finds such a model.

The model on the right of figure 5.12 has an asymptotic χ^2 value of 17 with 12 degrees of freedom, with $p(\chi^2) = .148$.

The PC algorithm finds a model that cannot be rejected on the basis of the data and that postulates a direct effect of attitude toward masturbation on orgasmic experience during masturbation, contrary to Bentler and Peeler.

5.8.4 The American Occupational Structure

Blau and Duncan's (1967) study of the American occupational structure has been praised by the National Academy of Sciences as an exemplary piece of social research and criticized by one statistician (Freedman 1983a) as an abuse of science. Using a sample of 20,700 subjects, Blau and Duncan offered a preliminary theory of the role of education (*ED*), first job (J_1), father's education (*FE*), and father's occupation (*FO*) in determining one's occupation (*OCC*) in 1962. They present their theory in the graph in figure 5.13, in which the undirected edge represents an unexplained correlation.

**TETRAD pattern at .05
significance level**

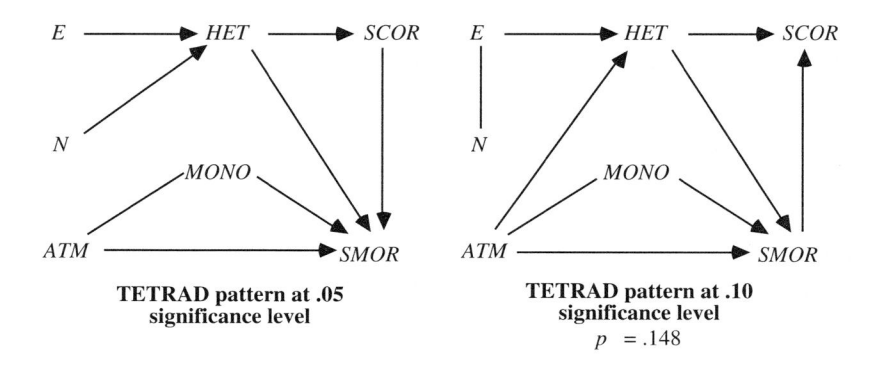

**TETRAD pattern at .10
significance level
$p = .148$**

Figure 5.12

Blau and Duncan argue that the dependencies are linear. Their salient conclusions are that father's education affects occupation and first job only through the father's occupation and the subject's education.

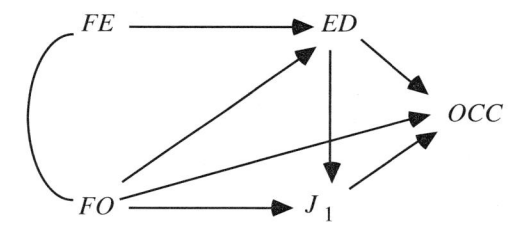

Figure 5.13

Blau and Duncan's theory was criticized by Freedman as arbitrary, unjustified, and statistically inadequate (Freedman 1983a). Indeed, if the theory is subjected to the asymptotic χ^2 likelihood ratio test of the EQS (Bentler 1985) or LISREL (Joreskog and Sorbom 1984) programs the model is decisively rejected ($p < .001$), and Freedman reports it is also rejected by a bootstrap test.

If the conventional .05 significance level is used to test for vanishing partial correlations, given a common sense ordering of the variables by time, from Blau and Duncan's covariances the PC algorithm produces the graph shown in figure 5.14.

```
    FE ─────────► ED
     ╲  ╲    ╱ │ ╲
      ╲  ╲  ╱  │  ╲
       ╲  ╳    │   ► OCC
        ╲╱ ╲   │  ╱
        ╱╲  ╲  │ ╱
       ╱  ╲  ╲ ▼╱
    FO ─────────► J₁
```

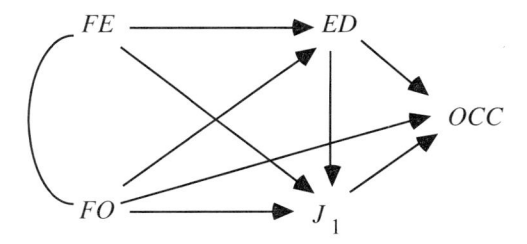

Figure 5.14

In this case every collider occurs in a triangle and there are no unshielded colliders. The data therefore do not determine the directions of the causal connections, but the time order of course determines the direction of each edge. We emphasize that the adjacencies are produced by the program entirely from the data, without any prior constraints. The model shown passes the same likelihood ratio test with $p > .3$.

The algorithm adds to Blau and Duncan's theory a direct connection between FE and J_1. The connection between FE and J_1 would only disappear if the significance level used to test for vanishing partial correlations were .0002. To determine a collection of vanishing partial correlations that are consistent with a directed edge from FE to OCC in 1962 one would have to reject hypotheses of vanishing partial correlations at a significance level greater than .3. The conditional independence relations found in the data at a significance level of .0001 are faithful to Blau and Duncan's directed graph.

Freedman argues that in the American population we should expect that the influences among these variables differ from family to family, and therefore that the assumption that all units in the population have the same structural coefficients is unwarranted. A similar conclusion can be reached in another way. We noted in chapter 3 that if a population consists of a mixture of subpopulations of linear systems with the same causal structure but different variances and linear coefficients, then unless the coefficients are independently distributed or the mixture is in special proportions, the population correlations will be different from those of any of the subpopulations, and variables independent in each subpopulation may be correlated in the whole. When subpopulations with distinct linear structures are mixed and these special conditions do not obtain, the directed graph found from the correlations will typically be complete. We see that in order to fit Blau and Duncan's data we need a graph that is only one edge short of being complete.

The same moral is if anything more vivid in another linear model built from the same empirical study by Duncan, Featherman, and Duncan (1972). They developed the following model of socioeconomic background and occupational achievement, where FE signifies father's education, FO father's occupational status, SIB the number of the respondent's siblings, ED the respondent's education, OCC the respondent's occupational status and INC the respondent's income.

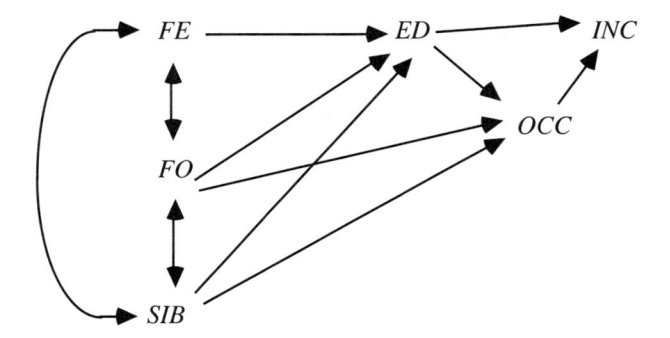

Figure 5.15

In this case the double headed arrows merely indicate a residual correlation. The model has four degrees of freedom, and entirely fails the EQS likelihood ratio test (χ^2 is 165). When the correlation matrix is given to the TETRAD II program along with an obvious time ordering of the variables, the PC algorithm produces a complete graph.

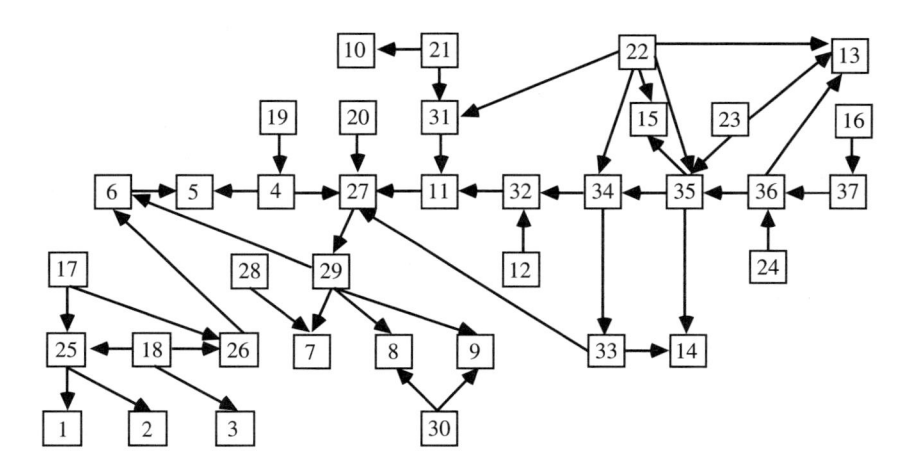

Figure 5.16

5.8.5 The ALARM Network

Recall the ALARM network developed to simulate causal relations in emergency medicine (figure 5.16).

The SGS and PC* algorithms will not run on a problem this large. We have applied the PC algorithm to a linear version of the ALARM network. Using the same directed graph, linear coefficients with values between .1 and .9 were randomly assigned to each directed edge in the graph. Using a joint normal distribution on the variables of zero indegree, three sets of simulated data were generated, each with a sample size of 2,000.

The covariance matrix and sample size were given to a version of the TETRAD II program with an implementation of the PC–1 algorithm. This implementation takes as input a covariance matrix, and it outputs a pattern. No information about the orientation of the variables was given to the program. In each trial the output pattern omitted two edges in the ALARM network; in one of the cases it also added one edge that was not present in the ALARM network.

In a related test, another ten samples were generated, each with 10,000 units. The results were scored as follows: We call the pattern the PC algorithm would generate given the population correlations the **true pattern**. We call the pattern the algorithm infers from the sample data the **output pattern**. An **edge existence error of commission (Co)** occurs when any pair of variables are adjacent in the output pattern but not in the true pattern. If an edge e between A and B occurs in both the true and output patterns, there is an **edge direction error of commission** when e has an arrowhead at A in the output pattern but not in the true pattern (and similarly for B.) **Errors of omission (Om)** are defined analogously in each case. The results are tabulated as the average over the trial distributions of the ratio of the number of actual errors to the number of possible errors of each kind. The results at sample size 10,000 are summarized below:

#trials	%Edge Existence Errors		%Edge Direction Errors	
	Commission	Omission	Commission	Omission
10	.06	4.1	17	3.5

For similar data from a similarly connected graph with 100 variables, for ten trials the PC–1 algorithm required an average of 134 seconds and the PC–3 algorithm required an average of 16 seconds.

Herskovits and Cooper (1990) generated discrete data for the ALARM network, using variables with two, three and four values. Given their data, the TETRAD II program with the PC algorithm reconstructs almost all of the undirected graph (it omitted two edges in one trial; and in another also added one edge) and orients most edges correctly. In most orientation errors an edge was oriented in both directions. Broken down by the same measures as were used for the linear data from the same network (with simulated data obtained from Herskovits and Cooper at sample size 10,000) the results are:

trial	%Edge Existence Errors		%Edge Direction Errors	
	Commission	Omission	Commission	Omission
1	0	4.3	27.1	10.0
2	0.2	4.3	5.0	10.4

5.8.6 Virginity

A retrospective study by Reiss, Banwart, and Foreman (1975) considered the relationship among a sample of undergraduate females between a number of attitudes, including attitude toward premarital intercourse, use of a university contraceptive clinic, and virginity. Two samples were obtained, one of women who had used the clinic and one of women who had not; the samples did not differ significantly in relevant background variables such as age, education, parental education, and so on. Fienberg gives the cross-classified data for three variables: Attitude toward extramarital intercourse (E) (always wrong; not always wrong); virginity (V) and use of the contraceptive clinic (C) (used; not used). All variables are binary. The PC and SGS procedures immediately produces the following pattern shown in figure 5.17, which is consistent with any of the orientations of the edges that do not produce a collision at V. One sensible interpretation is that attitude affects sexual behavior which causes clinic use. Fienberg (1977) obtains the same result with log linear methods.

$E \longrightarrow V \longrightarrow C$

Figure 5.17

5.8.7 The Leading Crowd

Coleman (1964) describes a study in which 3398 schoolboys were interviewed twice. At each interview each subject was asked to judge whether or not he was a member of the "leading crowd" and whether his attitude toward the leading crowd was favorable or unfavorable. The data have been reanalyzed by Goodman (1973a,b) and by Fienberg (1977). Using Fienberg's notation, let A and B stand for the questions at the first interview and C and D stand for the corresponding questions at the second interview. The data are given by Fienberg as follows:

Second Interview						
Membership Attitude			+	+	-	-
			+	-	+	-
Membership Attitude						
First Interview	+	+	458	140	110	49
	+	-	171	182	56	87
	-	+	184	75	531	281
	-	-	85	97	338	554

Fienberg summarizes his conclusions after a log-linear analysis in the path diagram in figure 5.18. He does not explain what interpretation is to be given to the double-headed arrow.

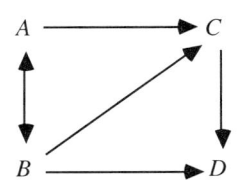

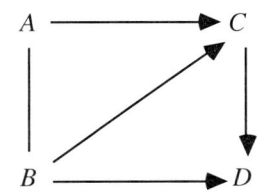

Discovery Algorithms for Causally Sufficient Structures 111

Figure 5.18   Figure 5.19

When the PC algorithm is told that *C* and *D* occur after *A* and *B*, with the usual .05 significance level for tests the program produces the pattern in figure 5.19.

Orienting the undirected edge in the PC model as a directed edge from *A* to *B* produces expected values for the various cell counts that are almost identical with Fienberg's (p. 127) expected counts.[7] Note, however, this is a nearly complete graph, which may indicate that the sample is a mixture of different causal structures.

5.8.8 Influences on College Plans

Sewell and Shah (1968) studied five variables from a sample of 10,318 Wisconsin high school seniors. The variables and their values are:

SEX	[male = 0, female = 1]
IQ = Intelligence Quotient,	[lowest = 0, highest = 2]
CP = college plans	[yes = 0, no = 1]
PE = parental encouragement	[low = 0, high = 1]
SES = socioeconomic status	[lowest = 0, highest = 3]

They offer the causal hypothesis shown in figure 5.20.

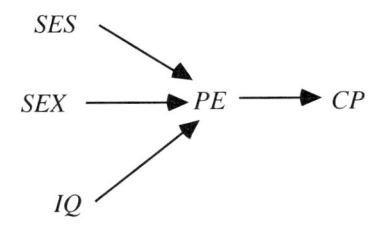

Figure 5.20

The data were reanalyzed by Fienberg (1977), who attempted to give a causal interpretation using log-linear models, but found a model that could not be given a graphical interpretation.

Given prior information that orders the variables by time as follows:

1 SEX
2 IQ   PE   SES
3 CP

so that later variables cannot be specified to be causes of earlier variables, the output with the PC algorithm is the structure shown in figure 5.21.

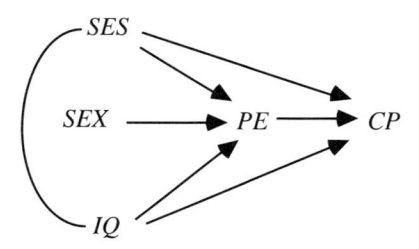

```
        SES
       ↙   ↘
SEX ──→ PE ──→ CP
       ↖   ↗
        IQ
```

Figure 5.21

The program cannot orient the edge between *IQ* and *SES*. It seems very unlikely that the child's intelligence causes the family socioeconomic status, and the only sensible interpretation is that *SES* causes *IQ*, or they have a common unmeasured cause. Choosing the former, we have a directed graph whose joint distribution can be estimated directly from the sample. We find, for example, that the maximum likelihood estimate of the probability that males have college plans (*CP*) is .35, while the probability for females is .31. Judged by this sample the probability a child with low *IQ*, no parental encouragement (*PE*) and low socioeconomic status (*SES*) plans to go to college is .011; more distressing, the probability that a child otherwise in the same conditions but with a high *IQ* plans to go to college is only .124.

5.8.9 Abortion Opinions

Christensen (1990) illustrates log-linear model selection and search procedures with a data set whose variables are Race (*R*) [white, nonwhite], Sex (*S*), Age (*A*) [six categories] and Opinion (*O*) on legalized abortion (supports, opposes, undecided). Forward selection procedures require fitting 43 log-linear models. A backward elimination method requires 22 fits; a method due to Aitkin requires 6 fits; another backward method due to Wermuth requires 23 fits. None of these methods would work at all on large variable sets.

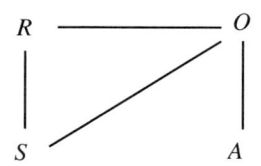

```
R ───────── O
|         ╱ |
|       ╱   |
S         A
```

Figure 5.22

Discovery Algorithms for Causally Sufficient Structures 113

Christensen suggests that the "best" log-linear model is an undirected conditional independence graphical model whose maximal cliques are [RSO] and [OA]. This is shown in figure 5.22.

Subsequently, Christensen proposes a recursive causal model (in the terminology of Kiiveri and Speed 1982) for the data. He suggests on substantive grounds a mixed graph and says "The undirected edge between R and S...represents an interaction between R and S." Figure 5.23 is not a causal model in the sense we have described. It can be interpreted as a *pattern* representing the equivalence class of causal graphs whose members are the two orientations of the R - S edge, but R and S in Christensen's data are very nearly independent.

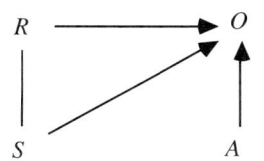

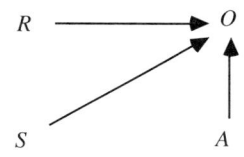

Figure 5.23

Figure 5.24

This example is small enough to use the PC* algorithm, which, with significance level .05 for independence tests, gives exactly figure 5.24. Assuming faithfulness, the statistical hypothesis of figure 5.24 is inconsistent with the independence of {R,S} and A conditional on O, required by the log-linear model of figure 5.22.

At a slightly lower significance level (.01) R and O are judged independent, and the same algorithm omits the $R \rightarrow O$ connection. On this data with significance level .05 the PC algorithm also produces the graph of figure 5.24 but with the $R \rightarrow O$ connection omitted. The difference in the outputs of the PC* and PC algorithms occur in the following way. Both algorithms produce at an intermediate stage the undirected graph underlying figure 5.24. In that undirected graph A does not lie on any undirected path between R and O. For that reason, the PC* algorithm never tests the conditional independence of R and O on A, and leaves the R - O edge in. In contrast, the PC algorithms does test the conditional independence of R and O on A, with a positive result, and removes the R - O edge.

5.8.10 Simulation Tests with Random Graphs

In order to test the speed and the reliability of the algorithms discussed in this chapter, we have tested the algorithms SGS, PC–1, PC–2, PC–3, and IG on a large number of simulated examples. The graphs themselves, the linear parameters, and the samples were all pseudorandomly generated. This section describes the sample generation procedures for both linear and discrete data and gives simulation results for the linear case. Simulation results with discrete data are considered in the chapter on regression.

The average degree of the vertices in the graphs considered are 2, 3, 4, or 5; the number of variables is 10 or 50; and the sample sizes are 100, 200, 500, 1000, 2000, and 5000. For each combination of these parameters, 10 graphs were generated, and a single distribution obtained faithful to each graph, and a single sample taken from each such distribution.

Because of its computational limitations, the SGS algorithm was tested only with graphs of 10 variables.

5.8.10.1 Sample Generation

All pseudorandom numbers were generated by the UNIX "random" utility. Each sample is generated in three stages:

(i) The graph is pseudorandomly generated.
(ii) The linear coefficients (in the linear case) or the conditional probabilities (in the discrete case) are pseudorandomly generated.
(iii) A sample for the model is pseudorandomly generated.

We will discuss each of these steps in more detail.

(i) The input to the random graph generator is an average degree and the number of variables. The variables are ordered so that an edge can only go from a variable lower in the order to a variable higher in the order, eliminating the possibility of cycles. Since some of the procedures use a lexicographic ordering, variable names were then randomly scrambled so that no systematic lexicographic relations obtained among variable pairs connected by edges. Each variable pair is assigned a probability p equal to

$$\frac{\text{average degree}}{\text{number of variables} - 1}$$

For each variable pair a number is drawn from a uniform distribution over the interval 0 to 1. The edge is placed in the graph if and only if the number drawn is less than or equal to p.[8]

(ii) For simulated continuous distributions, an "error" variable was introduced for each endogenous variable and values for the linear coefficients between .1 and .9 were generated randomly for each edge in the graph. For the discrete case, a range of values of variables is selected by hand, and for each variable taking n values, the unit interval is divided into n sub-intervals by random choice of cut-off points. A distribution (e.g., uniform) is then imposed on the unit interval.

(iii) In the discrete case for each such distribution produced, each sample unit is obtained by generating, for each exogenous variable, a random number between 0 and 1.0 according to the distribution and assigning the variable value according to the category into which the number falls. Values for the endogenous variables were obtained by choosing a value randomly with probability given by the conditional probabilities on the obtained values of the parents of the variable. In the linear case, the exogenous variables—including the error terms—were generated independently from a standard normal distribution, and values of endogenous variables were computed as linear functions of their parents.

5.8.10.2 Results

As before, reliability has several dimensions. A procedure may err by omitting undirected edges in the true graph or by including edges—directed or undirected—between vertices that are not adjacent in the true graph. For an edge that is not in the true graph, there is no fact of the matter about its orientation, but for edges that are in the true graph, a procedure may err by omitting an arrowhead in the true graph or by including an arrowhead not in the true graph. We count errors in the same way as in section 5.8.5.

Each of the procedures was run using a significance level of .05 on all trials. The five procedures tested are not equally reliable or equally fast. The SGS algorithm is much the slowest, but in several respects it proves reliable. The graphs on the following pages show the results. Each point on the graph is a number, which represents the average degree of the vertices in the directed graphs generating the data. We plot the run times and reliabilities of the PC–1 PC–2, PC–3, IG, and SGS algorithms against sample size for data from linear models based on randomly generated graphs with 10 variables, and similarly the reliabilities of the first four of these algorithms for linear models based on randomly generated graphs with 50 variables. In each case the results are plotted separately for graphs of degree 2, 3, 4, and 5.

The following qualitative conclusions can be drawn.

The rates of arrow and edge omission decrease dramatically with sample size up to about sample size 1000; after that the decreases are much more gradual. The rates of arrow and edge commission vary much less dramatically with sample size than do the rates of arrow and edge omission. As the average degree of the variables increases, the average error rates increase in a very roughly linear fashion, but the PC–2 algorithm tends to be less reliable than the other algorithms with respect to edge omissions when the average degree of the graph is high.

The PC–1, PC–3, IG, and SGS algorithms have compensating virtues and disadvantages. None of the procedures are reliable on all dimensions when the graphs are not sparse. One reliable dimension is the addition of edges: If two vertices are not adjacent in the true graph, there is very little chance they will be mistakenly output by any of these four procedures, no matter what the average degree of the graph and no matter what the sample size.

In contrast, at high average degree and low sample sizes the output of each of the procedures tends to omit over 50% of the edges in the true graph. At large sample sizes and low average degree only a few percent of the true edges are omitted, but with high average degree the percentage of edges omitted even at large sample sizes is significant. For example, at sample size 5000 and average degree 5, PC–1 omits over 30% of the edges in the true graph.

Arrow commission errors are much more common than edge commission errors. If an arrow does not occur in a graph, there is a considerable probability for any of the procedures that the arrow will be output, unless the sample size is large and the true graph is of low degree. For 10 variables with average degree about 2 and sample sizes of 1,000 or more, the SGS and IG algorithms are quite reliable, with errors of commission for arrows around 6%. Under the same conditions the error rates for the PC–1 and PC–2 algorithms run about 20%. In the case of the SGS algorithm, these relations are reversed for the question of arrow omission—if an arrow occurs in the true graph, what is the chance that the procedure will fail to include the arrow in its output? The answer is about 8% for PC–1 and PC–2 and about 20% for the SGS procedure. The IG algorithm, while much less reliable for arrow omissions at low sample sizes, is only slightly more unreliable at high sample sizes.

The return time of the PC–3 algorithm is dramatically smaller than the other algorithms. It's run time also does not increase as sharply with average degree, but the procedure does produce many more edge commission errors as the average degree increases.

The results suggest that the programs can reasonably be used in various ways according to the size of the problem, the questions one wants answered, and the character of the output. Roughly the same conclusions about reliability can be expected for the discrete case, but with lower absolute reliabilities. For larger number of variables the same patterns should hold, save that the SGS algorithm cannot be run at all.

More research needs to be done on local "repairs" to the graphs generated by these procedures, especially for edge omission errors and arrow commission errors. In order for the method to converge to the correct decisions with probability 1, the significance level used in making decisions should decrease as the sample sizes increase, and the use of higher significance levels (e.g., .2 at samples sizes less than 100, and .1 at sample sizes between 100 and 300) may improve performance at small sample sizes.

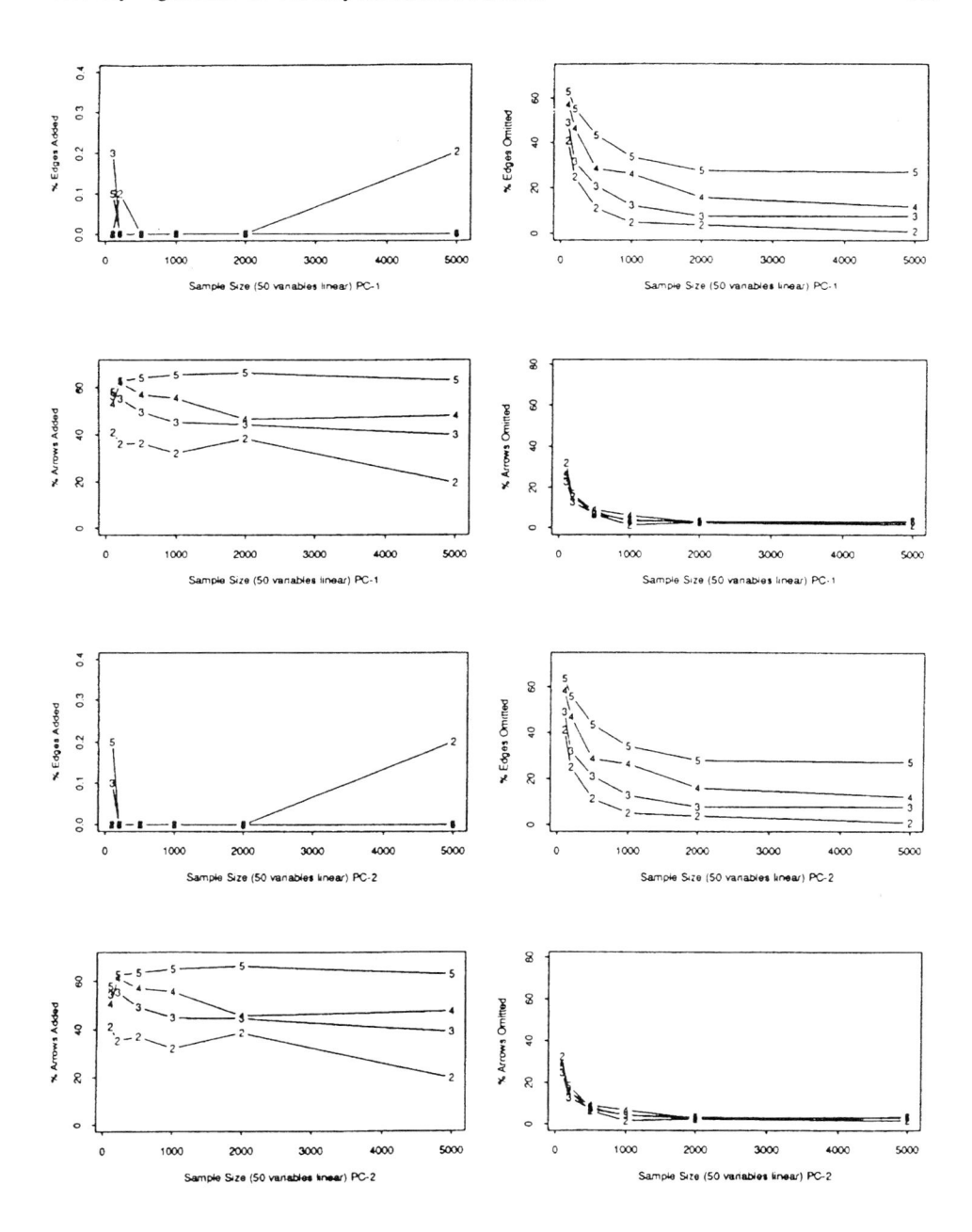

Discovery Algorithms for Causally Sufficient Structures

117

Figure 5.24a

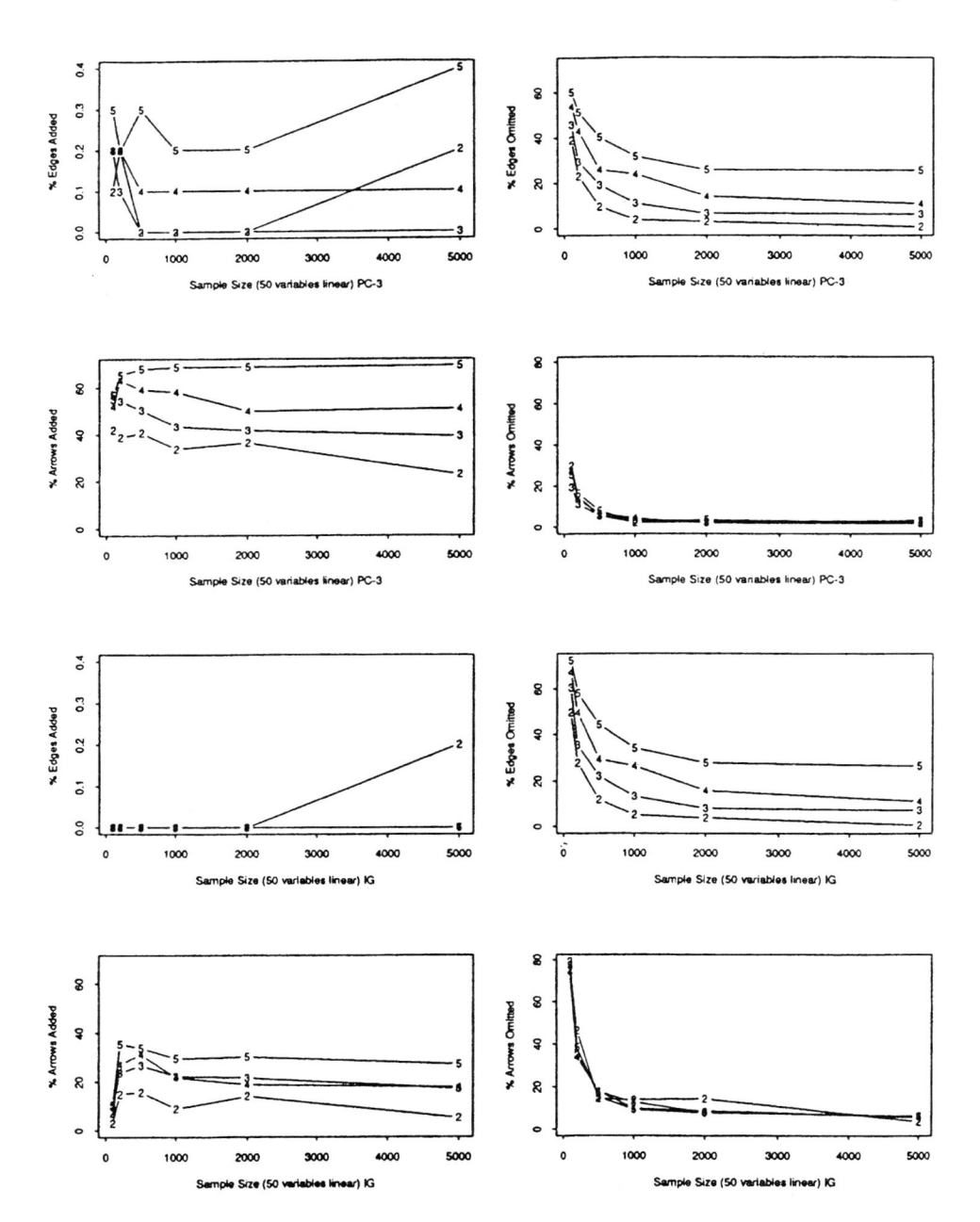

Figure 5.24b

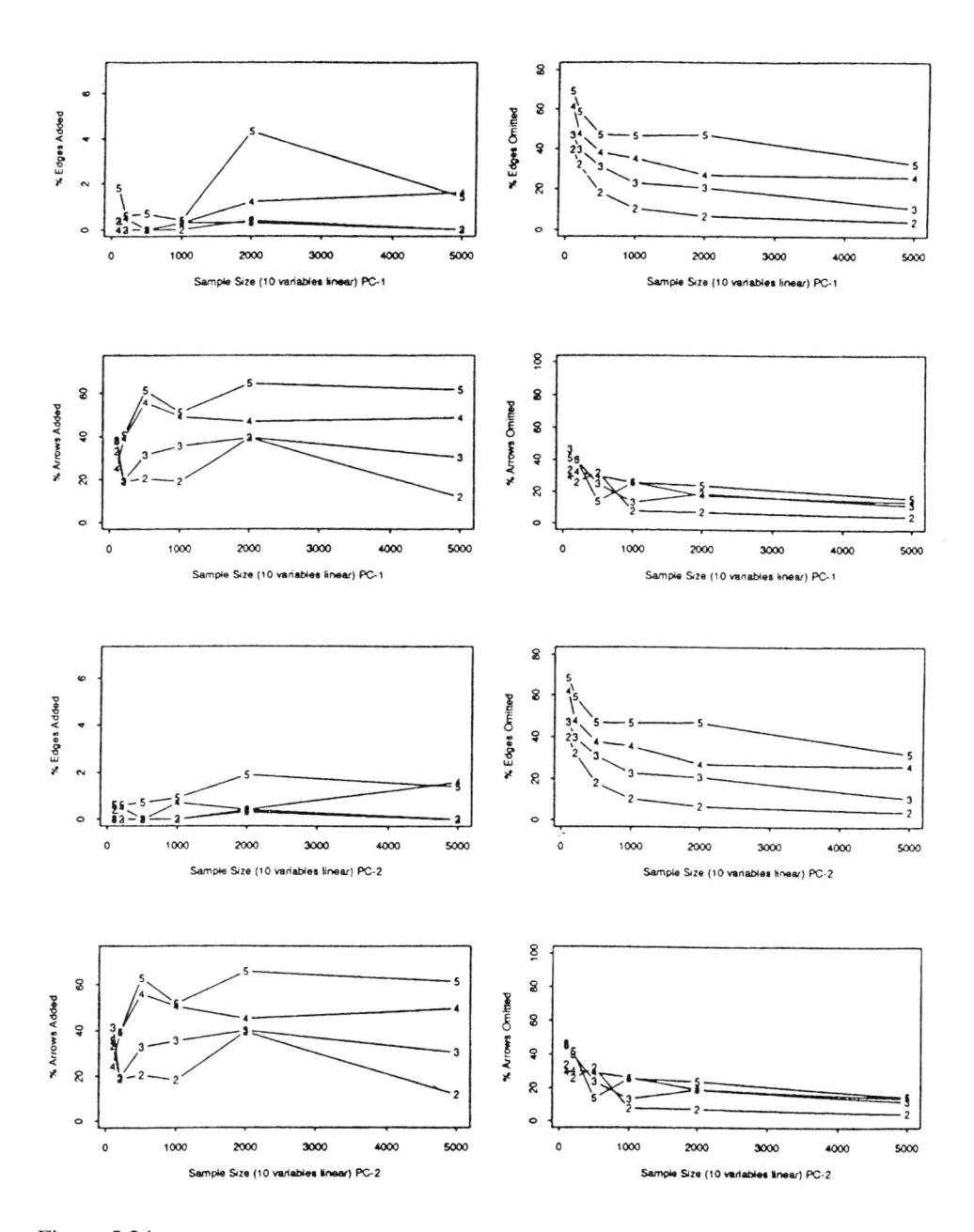

Discovery Algorithms for Causally Sufficient Structures 119

Figure 5.24c

120   Chapter 5

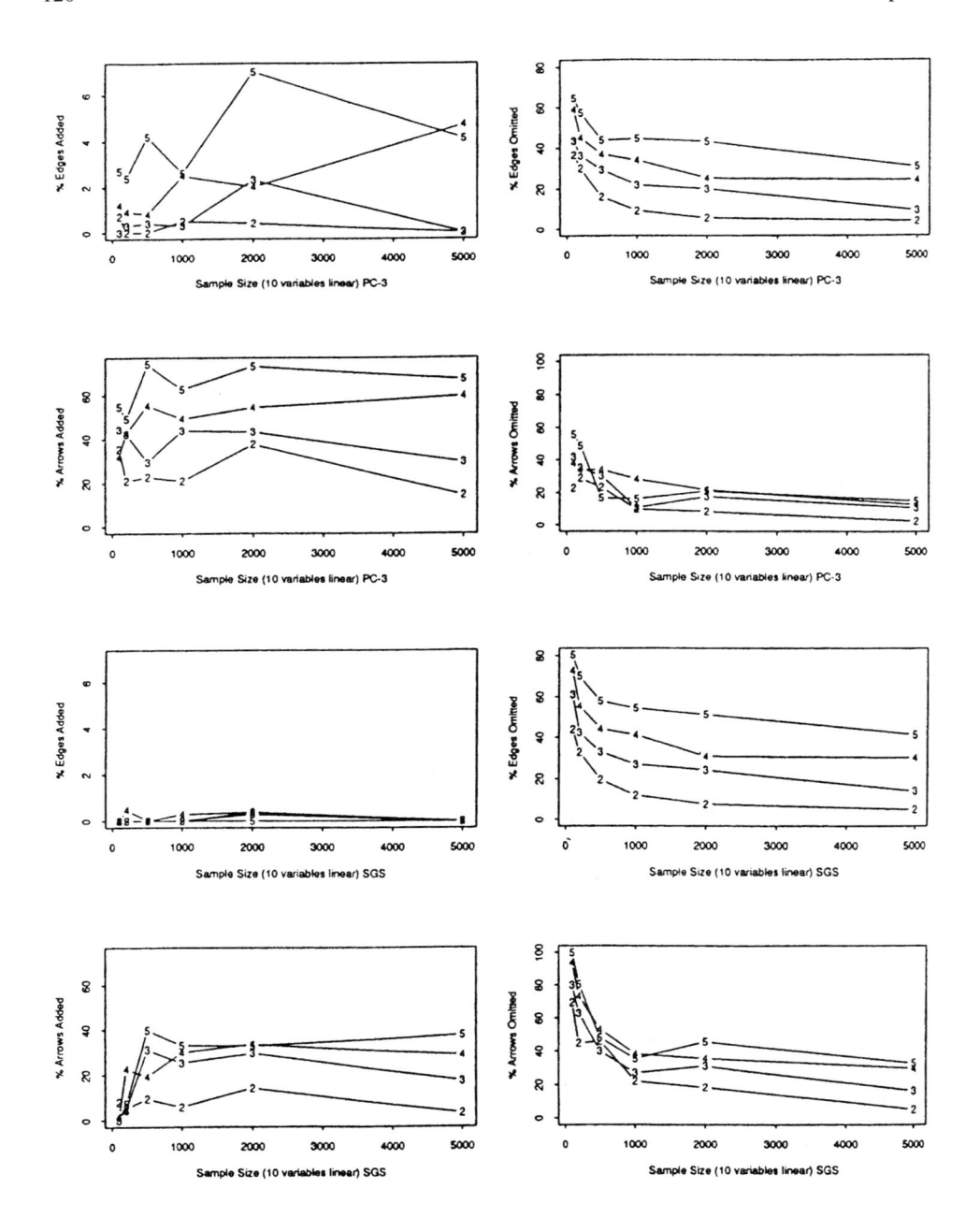

Figure 5.24d

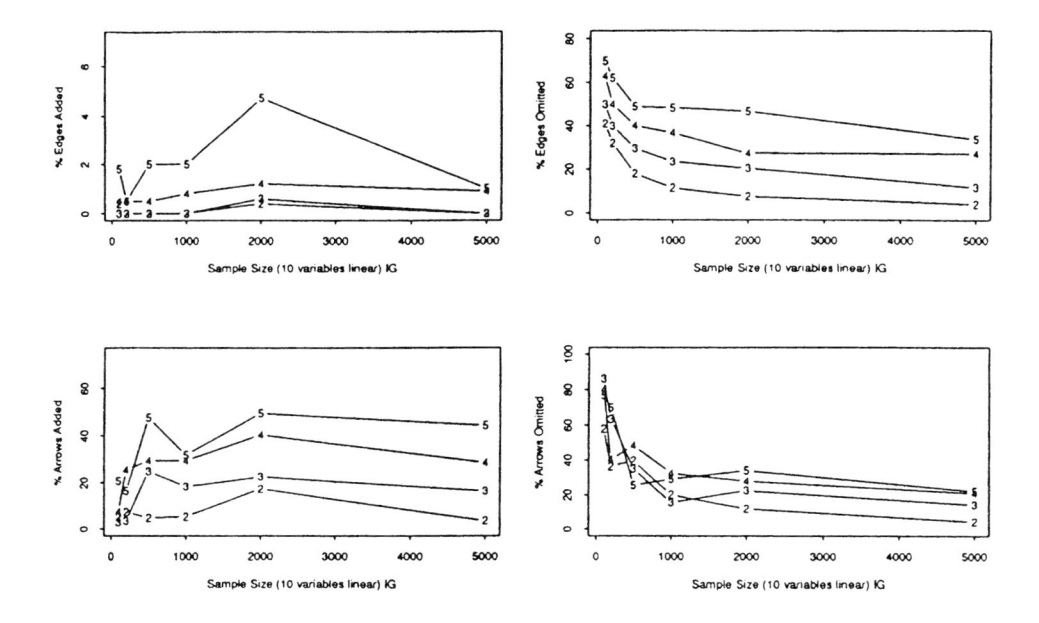

Figure 5.25e

5.9 Conclusion

This chapter describes several algorithms that can reliably recover sparse causal structures even for quite large numbers of variables, and illustrates their application. The algorithms have each been implemented using tests for conditional independence in the discrete case and for vanishing partial correlations in the linear case. We make no claim that these uses of tests to decide relevant probability relations is optimal, but any improvements in the statistical decision methods can be prefixed to the algorithms. With the exception of the PC* and SGS algorithms, the procedures described are feasible for large numbers of variables so long as the true causal graphs are sparse.

The algorithms we have described scarcely exhaust the possibilities, and a number of very simple alternative procedures should work reasonably well, at least for finding adjacency relations in the causal graph.

5.10 Background Notes

The idea of discovery problems is already contained in the notion of an estimation problem, and the requirement that an estimator be consistent is essentially a demand that it solve a particular kind of discovery problem. An extension of the idea to general nonstatistical settings was proposed by Putnam (1965) and independently by Gold (1965) and has subsequently been extensively developed in the literature of computer science, mathematical linguistics and logic (Osherson, Stob, and Weinstein 1986).

A more or less systematic search procedure for causal/statistical hypotheses can be found in the writings of Spearman (1904) and his students early in this century. A

Bayesian version of stepwise search was proposed by Harold Jeffreys (1957). Thurstone's (1935) factor analysis inaugurated a form of algorithmic search separated from any precise discovery problem: Thurstone did not view factor analysis as anything more than a device for finding simplifications of the data, and a similar view has been expressed in many subsequent proposals for statistical search. The vast statistical literature on search has focused almost exclusively on optimizing fitting functions.

The SGS algorithm was proposed by Glymour and Spirtes in 1989, and appeared in Spirtes, Glymour and Scheines (1990c). Verma and Pearl (1990b) subsequently proposed a more efficient version that examines cliques. A version of the PC algorithm was developed by Spirtes and Glymour (1990). The version presented here contains an improvement suggested by Pearl and Verma in the efficiency of step C) of the algorithm. Bayesian discovery procedures have been studied in Herskovits's thesis (1992).

The maximum likelihood estimation procedure for "recursive causal models" was developed in Kiiveri's (1982) doctoral thesis. The mathematical properties of the structures are further described in Kiiveri, Speed, and Carlin (1984).

6 Discovery Algorithms without Causal Sufficiency

6.1 Introduction

The preceding chapter complied with a common statistical fantasy, namely that in typical data sets it is known that no part of the statistical dependencies among measured variables are due to unmeasured common causes. We almost always fail to measure all of the causes of variables we do measure, and we often fail to measure variables that are causes of two or more measured variables. Any examination of collections of social science data gives the striking impression that variables in one study often seem relevant to those in other studies. Record keeping practices sometimes force econometricians to ignore variables in studies of one economy thought to have a causal role in studies of other economies (Klein 1961). In many studies in psychometrics, social psychology, and econometrics, the real variables of interest are unmeasured or measured only by proxies or "indicators." In epidemiological studies that claim to show that exposure to a risk factor causes disease, a burden of the argument is to show that the statistical association is not due to some common cause of risk factor and disease; since not everything imaginably relevant can be measured, the argument is radically incomplete unless a case can be made that unmeasured variables do not "confound" the association. If, as we believe, no reliable empirical study can proceed without considering whether relevant variables are unmeasured, then few published uncontrolled empirical studies are reliable.

In both experimental and non-experimental studies the unrecognized presence of unmeasured variables can lead to erroneous conclusions about the causal relations among the variables that are measured, and to erroneous predictions of the effects of policies that manipulate some of these variables. Until reliable, data-based methods are used to identify the presence or absence of unmeasured common causes, most causal inferences from observational data can be no more than guesswork at best and pseudoscience at worst. *Are such methods possible?* That question surely ought to be among the most important theoretical issues in statistics.

Statistical methods for detecting unmeasured common causes, or "confounding" in the terminology epidemiologists prefer, has been chiefly developed in psychometrics, where criteria for the existence and numbers of common causes have been sought since the turn of the century for special statistical models. The results include a literature on linear systems that contain criteria (e.g., Charles Spearman's [1904] vanishing tetrad differences) for latent variables that proved, however, to be neither necessary nor sufficient even assuming linearity. Criteria for two latent common causes were introduced by Kelley (1928), and related criteria are used in factor analysis, but they are not correct unless it is assumed that all statistical dependencies are due to unmeasured common causes. For problems in which the measured variables are discrete and their values a stochastic function of an unobserved continuous vector parameter θ, a number of criteria have been developed for the dimensionality of θ (Holland and Rosenbaum 1986). Suppes and Zanotti (1981) showed that for discrete variables there always exists a formal latent variable model in which all measured variables are effects of an unmeasured common cause and all pairs of measured variables are independent conditional on the latent variable. Their argument assumes the model must satisfy only the Markov Condition; the result does not hold if it is required that the distributions be faithful.

Among epidemiologists (Breslow and Day 1980; Kleinbaum, Kupper, and Morgenstern 1982) the criteria introduced by the Surgeon General's report on Smoking and Health (1964) are sometimes still advocated as a means for deciding whether a statistical dependency between exposure to risk factor A and disease B is "causal," apparently meaning that A causes B and A and B have no common causes. The criteria include (i) increase in response with dosage; (ii) that the statistical dependency between a risk factor and disease be specific to particular disease subgroups and to particular conditions of risk exposure; (iii) that the statistical association be strong; (iv) that exposure to a risk factor precede the period of increased risk; (v) lack of alternative explanations.

Even in causally sufficient systems, where all common causes of measured variables are themselves measured, such criteria do not separate causes from correlated variables. They fail even to come to grips with the problem of unmeasured "confounders." The problem with criterion (v) is exactly that there are too many alternative explanations of the data. Criterion (iv) is often of no use in deciding whether there are measured or unmeasured common causes at work. Criterion (iii) is defended on the grounds that "If an observed association is not causal, but simply the reflection of a causal association between some other factor and disease, then this latter factor must be more strongly related to disease (in terms of relative risk) than is the former factor," (Breslow and Day 1980). But the inference is incorrect: if there are two or more common causes, measured or not, none of them need be more strongly related to the disease than is the putative measured cause; and if A causes B and A and B *also* have a common cause, the latter need not be more strongly associated with B than is A. On behalf of Breslow and Day one might appeal to simplicity against all hypotheses of multiple common causes, but that would be an implausible claim in medical science, where multiple causal mechanisms abound. Nothing about the first two criteria separates the situation in which A and B have common causes from circumstances in which they do not.

In this chapter we present a more or less systematic account of how the presence of unmeasured common causes can mislead an investigator about causal relationships among measured variables, and of how the presence of unmeasured common causes can be detected. We deal with these questions separately for the general case and for the case in which all structures are linear. But the central aim of this chapter is to show how, assuming the Markov and Faithfulness conditions, reliable causal inferences can sometimes be made from appropriate sample data without any prior knowledge as to whether the system of measured variables is causally sufficient.

6.2 The PC Algorithm and Latent Variables

A natural idea is that a slight modification of the PC algorithm will give correct information about causal structure even when unmeasured variables may be present. Suppose that P' is a distribution over **V** that is faithful to a causal graph, and P is the marginal of P' over **O**, properly included in **V**. We will refer to the members of **O** as measured or observed variables. As we have already seen, if there are unmeasured common causes, the output of the PC algorithm can include bi-directed edges of the form $A \leftrightarrow B$. We could interpret a bi-directed edge between A and B to mean that there is an unmeasured cause C that directly causes A and B relative to **O**. We modify the algorithm by using a "o" on the end of an arrow to indicate that it is not known whether an arrowhead should occur in that place. We use a "*" as a metasymbol to stand for any of the three kinds of endmarks that an arrow can have: EM (empty mark), ">," or "o."

Modified PC Algorithm

A.) Form the complete undirected graph C on the vertex set **V**.

B.)
  $n = 0$.
  repeat
    repeat
      select an ordered pair of variables X and Y that are adjacent in C such that **Adjacencies**$(C,X)\setminus\{Y\}$ has cardinality greater than or equal to n, and a subset **S** of **Adjacencies**$(C,X)\setminus\{Y\}$ of cardinality n, and if X and Y are d-separated given **S** delete edge X - Y from C and record **S** in **Sepset**(X,Y) and **Sepset**(Y,X).
    until all ordered pairs of adjacent variables X and Y such that **Adjacencies**$(C,X)\setminus\{Y\}$ has cardinality greater than or equal to n and all subsets of **Adjacencies**$(C,X)\setminus\{Y\}$ of cardinality n have been tested for d-separation.
    $n = n + 1$.
  until for each ordered pair of adjacent vertices X, Y, **Adjacencies**$(C,X)\setminus\{Y\}$ is of cardinality less than n.

C.) Let F be the graph resulting from step B). If X and Y are adjacent in F, orient the edge between X and Y as X o-o Y.

D.) For each triple of vertices X, Y, Z such that the pair X, Y and the pair Y, Z are each adjacent in F but the pair X, Z are not adjacent in F, orient X *-* Y *-* Z as X *→ Y ←* Z if and only if Y is not in **Sepset**(X,Z).

E.) repeat
    If A *→ B, B *-* C, A and C are not adjacent, and there is no arrowhead at B on B *-* C, then orient B *-* C as $B \to C$.

If there is a *directed* path from A to B, and an edge between A and B, then orient the edge as A *→ B.

until no more edges can be oriented.

(When we say orient X *-* Y as X *→ Y we mean leave the same mark on the X end of the edge and put an arrowhead at the Y end of the edge.)

The result of this modification applied to the examples of the previous chapter is perfectly sensible. For example, in figure 6.1 we show both the model obtained from the Rodgers and Marantodata at significance level .1 by the PC algorithm and the model that would be obtained by the modified PC algorithm from a distribution faithful to the the graph in the PC output. (In each case with the known time order of the variables imposed as a constraint.)

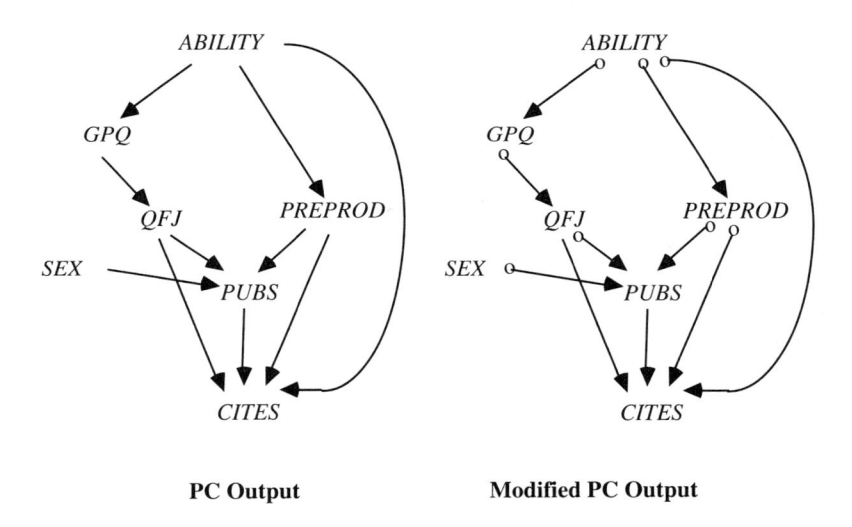

Figure 6.1

The output of the Modified PC Algorithm indicates that *GPQ* and *ABILITY*, for example, may be connected by an unmeasured common cause, but that *PUBS* is a direct cause of *CITES*, unconfounded by a common cause. Where a single vertex has "o" symbols for two or more edges connecting it with vertices that are not adjacent, a special restriction applies. *ABILITY*, for example has an edge to *GPQ* and to *PREPROD*, each with an "o" at the *ABILITY* end, and *GPQ* and *PREPROD* are not adjacent to one another. In that case the two "o" symbols cannot *both* be arrowheads. There cannot be an unmeasured cause of *ABILITY* and *GPQ* and an unmeasured cause of *ABILITY* and *PREPROD*, because if there were, *GPQ* and *PREPROD* would be dependent conditional on *ABILITY*, and the modified pattern entails instead that they are independent.

In many cases—perhaps most practical cases—in which the sampled distribution is the marginal of a distribution faithful to a graph with unmeasured variables, this simple modification of the PC algorithm gives a correct answer if the required statistical decisions are correctly made.

6.3 Mistakes

Unfortunately, this straightforward modification of the PC algorithm is not correct in general. An imaginary example will show why.

Everyone is familiar with a simple mistake occasioned by failing to recognize an unmeasured common cause of two variables, X, Y, where X is known to precede Y. The mistake is to think that X causes Y, and so to predict that a manipulation of X will change the distribution of Y. But there are more interesting cases that are seldom noticed, cases in which omitting a common cause of X and Y might lead one to think, erroneously, that some third variable Z directly causes Y. Consider an imaginary case:

A chemist has the following problem. According to received theory, which he very much doubts, chemicals A and B combine in a low yield mechanism to form chemical D through an intermediate C. Our chemist thinks there is another mechanism in which A and B combine to form D without the intermediate C. He wishes to do an experiment to establish the existence of the alternative mechanism. He can readily obtain reagents A and B, but available samples may be contaminated with varying quantities of D and other impurities. He can measure the concentration of the unstable alleged intermediate C photometrically, and he can measure the equilibrium concentration of D by standard methods. He can manipulate the concentrations of A and B, but he has no means to manipulate the concentration of C.

The chemist decides on the following experimental design. For each of ten different values of the concentration of A and B, a hundred trials will be run in which the reagents are mixed, the concentration of C is monitored, and the equilibrium concentration of D is measured. Then the chemist will calculate the partial correlation of A with D conditional on C, and likewise the partial correlation of B with D conditional on C. If there is an alternative mechanism by which A and B produce D without C, the chemist reasons, then there should be a positive correlation of A with D and of B with D in the samples in which the concentration of C is all the same; and if there is no such alternative mechanism, then when the concentration of C is controlled for, the concentrations of A, B on the one hand, and D on the other, should have zero correlation.

The chemist finds that the equilibrium concentrations of A, B on the one hand and of D on the other hand are correlated when C is controlled for—as they should be if A and B react to produce D directly—and he announces that he has established an alternative mechanism.

Alas, within the year his theory is disproved. Using the same reagents, another chemist performs a similar experiment in which, however, a masking agent reacts with the intermediate *C* preventing it from producing *D*. The second chemist finds no correlation in his experiment between the concentrations of *A* and *B* and the concentration of *D*. What went wrong with the first chemist's procedure?

By substituting a statistical control for the manipulation of *C* the chemist has run afoul of the fact that the marginal probability distribution with unmeasured variables can give the appearance of a spurious direct connection between two variables. The chemist's picture of the mechanism is given in graph G_1, and that is one way in which the observed results can be produced. Unfortunately, they can also be produced by the mechanisms in graph G_2, which is what happened in the chemist's case: impurities (*F*) in the reagents are causes of both *C* and *D*:

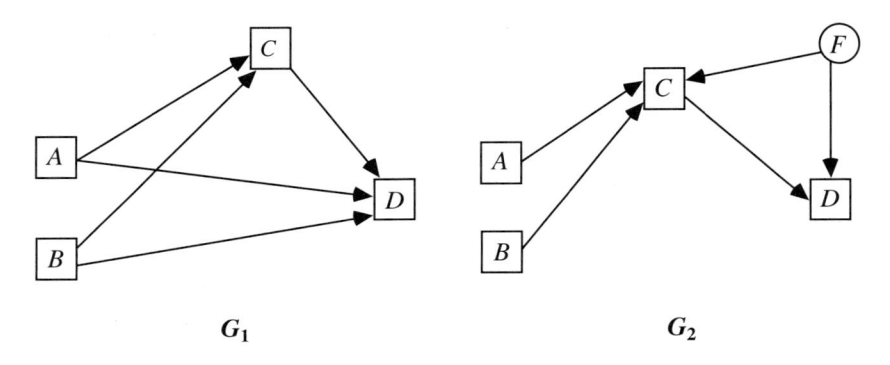

Figure 6.2

The general point is that a latent variable *F* acting on two measured variables *C* and *D* can produce statistical dependencies that suggest causal relations between *A* and *D* and between *B* and *D* that do not exist. For faithful distributions, if we use the SGS or PC algorithms, a structure such as G_2 will produce a directed edge from *A* to *D* in the output.

We can see the same point more analytically as follows: In a directed acyclic graph *G* over a set of variables **V**, if *A* and *D* are adjacent in *G*, then *A* and *D* are not d-separated given any subset of **V**\{*A,D*}. Hence under the assumption of causal sufficiency, either *A* is a direct cause of *D* or *D* is a direct cause of *A* relative to **V** if and only if *A* and *D* are independent conditional on no subset of **V**\{*A,D*}. However, if **O** is not causally sufficient, it is not the case that if *A* and *D* are independent conditional on every subset of **O**\{*A,D*} that either *A* is a direct cause of *D* relative to **O**, or *D* is a direct cause of *A* relative to **O**, or there is some latent variable *F* that is a common cause of both *A* and *D*.

This is illustrated by G_2 in figure 6.2, where **V** = {*A,B,C,D,F*} and **O** = {*A,C,D*}. **O** is not causally sufficient because *F* is a cause of both *C* and *D* which are in **O**, but *F* itself is not in **O**. *A* and *D* are not d-separated given any subset of **O**\{*A,D*}, so in any marginal of a distribution faithful to *G*, *A* and *D* are not independent conditional on any subset of

O\{A,D}, and the modified PC algorithm would leave an edge between A and D. Yet A is not a direct cause of D relative to **O**, D is not a direct cause of A relative to **O**, and there is no latent common cause of A and D. The directed acyclic graph G_1 shown in figure 6.2, in which A is a direct cause of D, and in which there is a path from A to D that does not go through C, has the same set of d-separation relations over {A,C,D} as does graph G_2. Hence, given faithful distributions, they cannot be distinguished by their conditional independence relations alone.

A further fundamental problem with the simple modification of the PC algorithm described above is that if we allow bi-directed edges in the graphs constructed by the PC algorithm, it is no longer the case that if A and B are d-separated given some subset of **O**, then they are d-separated given a subset of **Adjacencies**(A) or **Adjacencies**(B). Consider the graph in figure 6.3, where T_1 and T_2 are assumed to be unmeasured.

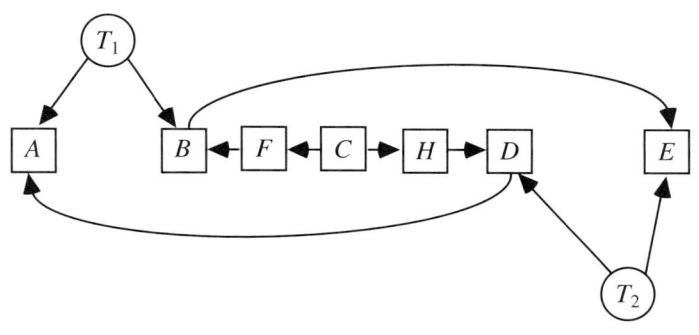

Figure 6.3

Among the measured variables, **Parents**(A) = {D} and **Parents**(E) = {B}, but A and E are not d-separated given any subset of {B} or any subset of {D}; the only sets that d-separate them are sets that contain F, C, or H. The Modified PC algorithm would correctly find that C, F, and H are not adjacent to A or E. It would then fail to test whether A and E are d-separated given any subset containing C. Hence it would fail to find that A and E are d-separated given {B,C,D} and would erroneously leave A and E adjacent. This means that it is not possible to determine which edges to remove from the graph by examining only local features (i.e., the adjacencies) of the graph constructed at a given stage of the algorithm. Similarly, once bi-directed edges are allowed in the output of the PC algorithm, it is not possible to extract all of the information about the orientation of edges by examining local features (i.e., pairs of edges sharing a common endpoint) of the graph constructed at a given stage of the algorithm.

Because of these problems, for full generality we must make major changes to the PC algorithm and in the interpretation of the output. We will show that there is a procedure, which we optimistically call the Fast Causal Inference (FCI) algorithm, that is feasible in large variable sets provided the true graph is sparse and there are not many bidirected edges chained together. The algorithm gives asymptotically correct information about causal structure when latent variables may be acting, assuming the measured distribution is the marginal of a distribution satisfying the Markov and Faithfulness conditions for the true graph. The FCI algorithm avoids the mistakes of the modified PC algorithm, and in some cases provides more information.

For example, with a marginal distribution over the boxed variables from the imaginary structure in figure 6.4, the modified PC algorithm gives the correct output shown in the first diagram in figure 6.5, whereas the FCI algorithm produces the correct and much more informative result in the second diagram in figure 6.5.

In figure 6.5, the double headed arrows indicate the presence of unmeasured common causes, and as in the modified PC algorithm the edges of the form o→ indicate that the algorithm cannot determine whether the circle at one end of the edge should be an arrowhead. Notice that the adjacencies among the set of variables {*Cilia damage*, *Heart disease*, *Lung capacity*, *Measured breathing dysfunction*} form a complete graph, but even so the edges can be completely oriented by the FCI algorithm.

The derivation of the FCI algorithm requires a variety of new graphical concepts and a rather intricate theory. We introduce Verma and Pearl's notions of an inducing path and an inducing path graph, and show that these objects provide information about causal structure. Then we consider algorithms that infer a class of inducing path graphs from the data.

6.4 Inducing Paths

Given a directed acyclic graph G over a set of variables **V**, and **O** a subset of **V**, Verma and Pearl (1991) have characterized the conditions under which two variables in **O** are not d-separated given any subset of **O**\{A,B}. If G is a directed acyclic graph over a set of variables **V**, **O** is a subset of **V** containing A and B, and $A \neq B$, then an undirected path U between A and B is an **inducing path relative to O** if and only if every member of **O** on U except for the endpoints is a collider on U, and every collider on U is an ancestor of either A or B. We will sometimes refer to members of **O** as **observed** variables.

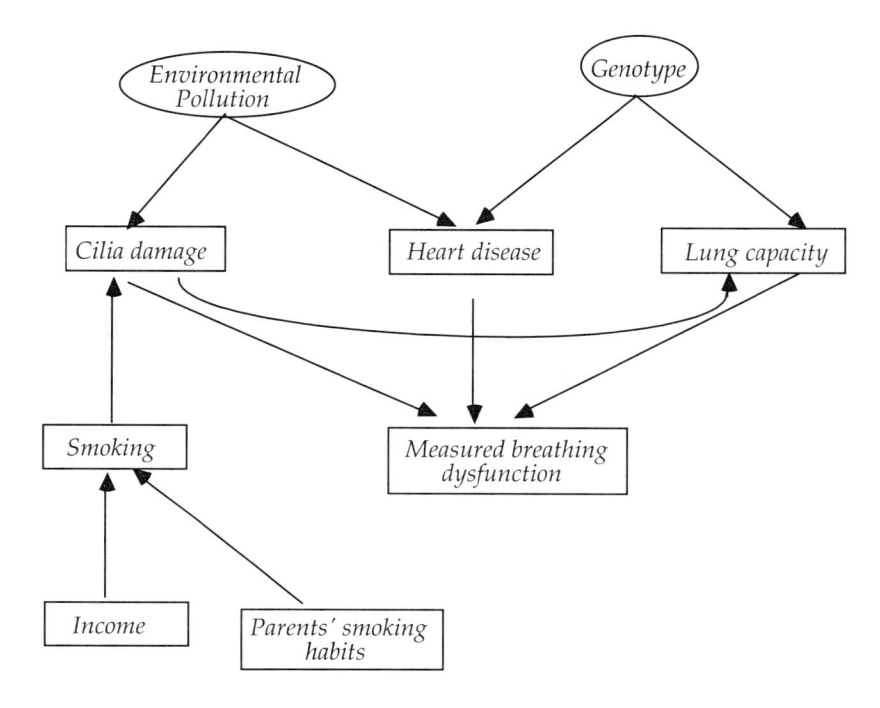

Discovery Algorithms without Causal Sufficiency

Figure 6.4

For example, in graph G_3, the path $U = <A,B,C,D,E,F>$ is an inducing path over $\mathbf{O} = \{A,B,D,F\}$ because each collider on U (B and D) is an ancestor of one of the endpoints, and each variable on U that is in $\mathbf{O}$ (except for the endpoints of U) is a collider on U. Similarly, U is an inducing path over $\mathbf{O} = \{A,B,F\}$. However, U is not an inducing path over $\mathbf{O} = \{A,B,C,D,F\}$ because C is in $\mathbf{O}$, but C is not a collider on U.

THEOREM 6.1: If G is a directed acyclic graph with vertex set $\mathbf{V}$, and $\mathbf{O}$ is a subset of $\mathbf{V}$ containing A and B, then A and B are not d-separated by any subset $\mathbf{Z}$ of $\mathbf{O}\backslash\{A,B\}$ if and only if there is an inducing path over the subset $\mathbf{O}$ between A and B.

It follows from theorem 6.1 and the fact that U is an inducing path over $\mathbf{O} = \{A,B,D,F\}$ that A and F are d-connected given every subset of $\{B,D\}$. Because in graph G_3 there is no inducing path between A and F over $\mathbf{O} = \{A,B,C,D,F\}$ it follows that A and F are d-separated given some subset of $\{B,C,D\}$ (in this case, $\{B,C,D\}$ itself.)

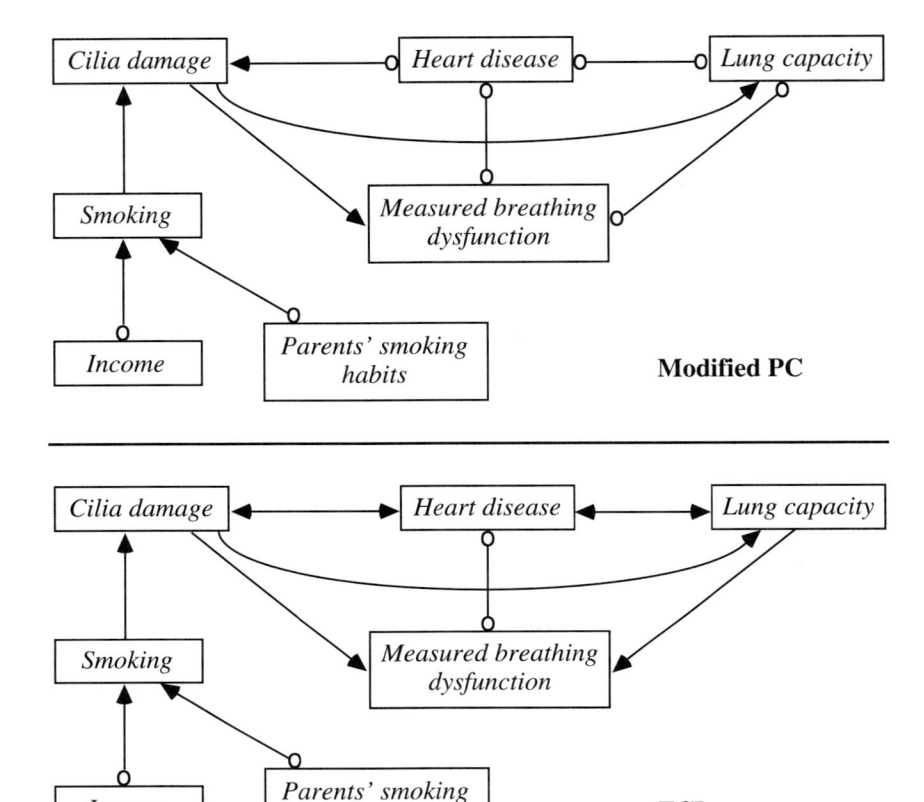

Figure 6.5

6.5 Inducing Path Graphs

The inducing paths relative to **O** in a graph G over **V** can be represented in the following structure described (but not named) in Verma and Pearl (1990b). G' is an **inducing path graph over O for directed acyclic graph** G if and only if **O** is a subset of the vertices in G, there is an edge between variables A and B with an arrowhead at A if and only if A and B are in **O** and there is an inducing path in G between A and B relative to **O** that is into

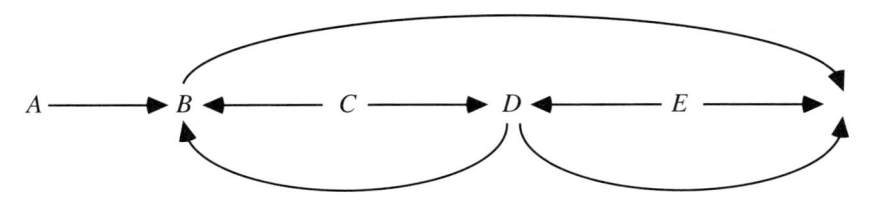

Figure 6.6: Graph G_3

Discovery Algorithms without Causal Sufficiency 133

A. (Using the notation of chapter 2, the set of marks in an inducing path graph is {>, EM}.) In an inducing path graph, there are two kinds of edges: $A \rightarrow B$ entails that every inducing path over **O** between *A* and *B* is out of *A* and into *B*, and $A \leftrightarrow B$ entails that there is an inducing path over **O** that is into *A* and into *B*. This latter kind of edge can only occur when there is a latent common cause of *A* and *B*.

Figures 6.7 through 6.9 depict the inducing path graphs of G_3 over **O** = {*A,B,D,E,F*}, **O** = {*A,B,D,F*} and **O** = {*A,B,F*} respectively. Note that in G_3 <*B,D*> is an inducing path between *B* and *D* over **O** = {*A,B,D,E,F*} that is out of *D*. However, in the inducing path graph the edge between *B* and *D* has an arrowhead at *D* because there is another inducing path <*B,C,D*> over **O** = {*A,B,D,E,F*} that is into *D*. There is no edge between *A* and *F* in the inducing path graph over **O** = {*A,B,D,E,F*}, but there is an edge between *A* and *F* in the inducing path graphs over **O** = {*A,B,D,F*} and **O** = {*A,B,F*}.

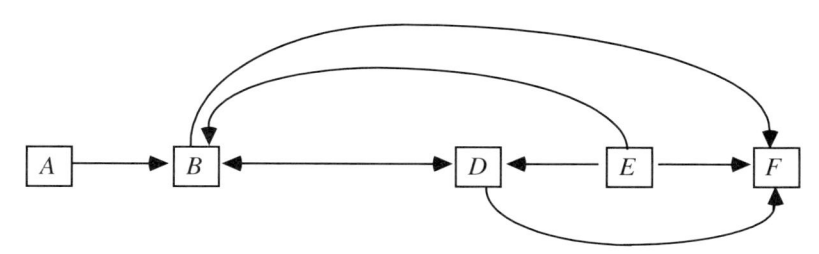

Figure 6.7

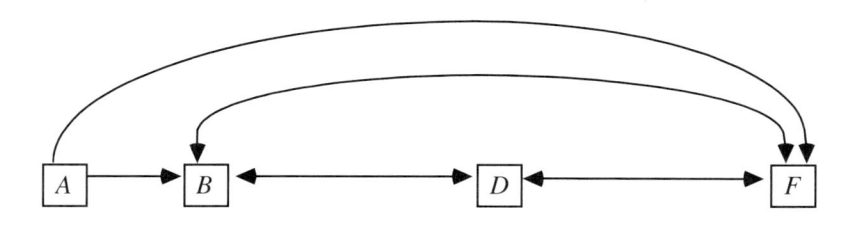

Figure 6.8

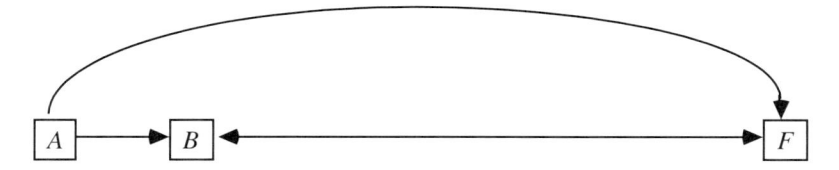

Figure 6.9. Inducing path graph of G_3 over {*A*, *B*, *F*}

We can extend without modification the concept of d-separability to inducing path graphs if the only kinds of edges that can occur on a directed path are edges with one arrowhead, and undirected paths may contain edges with either single or double arrowheads. If G is a directed acyclic graph, G' is the inducing path graph for G over **O**, and **X**, **Y**, and **S** are disjoint sets of variables included in **O**, then **X** and **Y** are **d-separated** given **S** in G' if and only they are d-separated given **S** in G.

Double-headed arrows make for a very important difference between d-separability relations in an inducing path graph and in a directed acyclic graph. In a directed acyclic graph over **O**, if A and B are d-separated given any subset of **O**\{A,B} then A and B are d-separated given either **Parents**(A) or **Parents**(B). This is not true in inducing path graphs. For example, in inducing path graph G_4, which is the inducing path graph of figure 6.3 over **O** = {A,B,C,D,E,F,H}, **Parents**(A) = {D} and **Parents**(E) = {B}, but A and E are not d-separated given any subset of {B} or any subset of {D}; all of the sets that d-separate A and E contain C, H, or F.

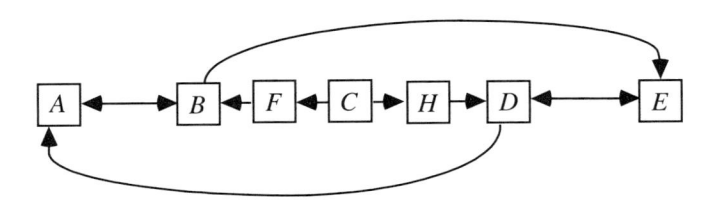

Figure 6.10. Inducing path graph G_4

There is, however, a kind of set of vertices in inducing path graphs that, so far as d-separability is concerned, behaves much like the parent sets in directed acyclic graphs.

If G' is an inducing path graph over **O** and $A \neq B$, let $V \in$ **D-SEP**(A,B) if and only if $A \neq V$ and there is an undirected path U between A and V such that every vertex on U is an ancestor of A or B and (except for the endpoints) is a collider on U.

THEOREM 6.2: In an inducing path graph G' over **O**, where A and B are in **O**, if A is not an ancestor of B, and A and B are not adjacent then A and B are d-separated given a subset of **D-SEP**(A,B).

In an inducing path graph either A is not an ancestor of B or B is not an ancestor of A. Thus we can determine whether A and B are adjacent in an inducing path graph without determining whether A and B are dependent conditional on *all* subsets of **O**.

If **O** is not a causally sufficient set of variables, then although we can infer the existence of an inducing path between A and B if A and B are dependent conditional on every subset of **O**\{A,B}, we cannot infer that either A is a direct cause of B relative to **O**, or that B is a direct cause of A relative to **O**, or that there is a latent common cause of A

and B. Nevertheless, the existence of an inducing path between A and B over $\mathbf{O}$ does contain information about the causal relationships between A and B, as the following lemma shows.

LEMMA 6.1.4: If G is a directed acyclic graph over $\mathbf{V}$, $\mathbf{O}$ is a subset of $\mathbf{V}$ that contains A and B, and G contains an inducing path over $\mathbf{O}$ between A and B that is out of A, and A and B are in $\mathbf{O}$, then there is a directed path from A to B in G.

It follows from lemma 6.1.4 that if $\mathbf{O}$ is a subset of $\mathbf{V}$ and we can determine that there is an inducing path between A and B over $\mathbf{O}$ that is out of A, then we can infer that A is a (possibly indirect) cause of B. Hence, if we can infer properties of the inducing path graph over $\mathbf{O}$ from the distribution over $\mathbf{O}$, we can draw inferences about the causal relationships among variables, regardless of what variables we have failed to measure. In the next section we describe algorithms for inferring properties of the inducing path graph over $\mathbf{O}$ from the distribution over $\mathbf{O}$.

6.6 Partially Oriented Inducing Path Graphs

A **partially oriented inducing path graph** can contain several sorts of edges: $A \to B$, $A \text{ o}\to B$, A o-o B, or $A \leftrightarrow B$. We use "*" as a metasymbol to represent any of the three kinds of ends (EM (the empty mark), ">," or "o"); the "*" symbol itself does not appear in a partially oriented inducing path graph. (We also use "*" as a metasymbol to represent the two kinds of ends (EM or ">") that can occur in an inducing path graph.)

A partially oriented inducing path graph π for directed acyclic graph G with inducing path graph G' over $\mathbf{O}$ is intended to represent the adjacencies in G', and some of the orientations of the edges in G' that are common to all inducing path graphs with the same d-connection relations as G'. If G' is an inducing path graph over $\mathbf{O}$, **Equiv**(G') is the set of inducing path graphs over the same vertices with the same d-connections as G'. Every inducing path graph in **Equiv**(G') shares the same set of adjacencies. We use the following definition:

π is a partially oriented inducing path graph of directed acyclic graph G with inducing path graph G' over O if and only if

(i). if there is any edge between A and B in π, it is one of the following kinds:
$A \to B$, $B \to A$, $A \text{ o}\to B$, $B \text{ o}\to A$, A o-o B, or $A \leftrightarrow B$;
(ii). π and G' have the same vertices;
(iii). π and G' have the same adjacencies;
(iv). if $A \text{ o}\to B$ is in π, then in every inducing path graph X in **Equiv**(G') either $A \to B$ or $A \leftrightarrow B$ is in X;
(v). if $A \to B$ is in π, then $A \to B$ is in every inducing path graph in **Equiv**(G');
(vi). if A *-* B *-* C is in π, then the edges between A and B, and B and C do not collide at B in any inducing path graph in **Equiv**(G');

(vii). if $A \leftrightarrow B$ is in π, then $A \leftrightarrow B$ is in every inducing path graph in **Equiv**(G');
(viii). if A o-o B is in π, then in every inducing path graph X in **Equiv**(G'), either $A \rightarrow B$, $A \leftrightarrow B$, or $A \leftarrow B$ is in X.

(Strictly speaking a partially oriented inducing path graph is not a graph as we have defined it because of the extra structure added by the underlining.) Note that an edge A *-o B does not constrain the edge between A and B either to be into or to be out of B in any subset of **Equiv**(G'). The adjacencies in a partially oriented inducing path graph π for G can be constructed by making A and B adjacent in π if and only if A and B are d-connected given every subset of **O**\{A,B}.

Once the adjacencies have been determined, it is trivial to construct an uninformative partially oriented inducing path graph π for G. Simply orient each edge A *-* B as A o-o B. Of course this particular partially oriented inducing path graph π for G is very uninformative about what features of the orientation of G' are common to all inducing path graphs in **Equiv**(G'). For example, figure 6.11 shows again the imaginary graph of causes of measured breathing dysfunction. Figure 6.12 shows an uninformative partially oriented inducing path graphs of graph G_5 over **O** = {*Cilia damage*, *Smoking*, *Heart disease*, *Lung capacity*, *Measured breathing dysfunction*, *Income*, *Parents' smoking habits*}.

Let us say that B is a **definite noncollider** on undirected path U if and only if either B is an endpoint of U, or there exist vertices A and C such that U contains one of the subpaths $A \leftarrow B$ *-* C, A *-* $B \rightarrow C$, or A *-* $\underline{B}$ *-* C. In a **maximally informative partially oriented inducing path graph** π for G with inducing path graph G',
(i) an edge A *-o B appears only if the edge between A and B is into B in some members of **Equiv**(G'), and out of B in other members of **Equiv**(G'.), and
(ii) for every pair of edges between A and B, and B and C, either the edges collide at B, or they are definite noncolliders at B, unless the edges collide in some members of **Equiv**(G) and not in others.

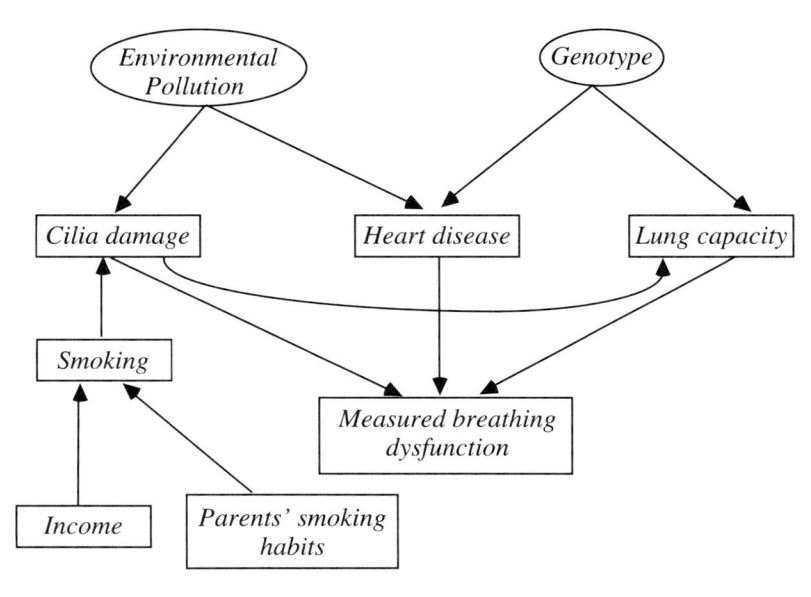

Figure 6.11. Graph G_5

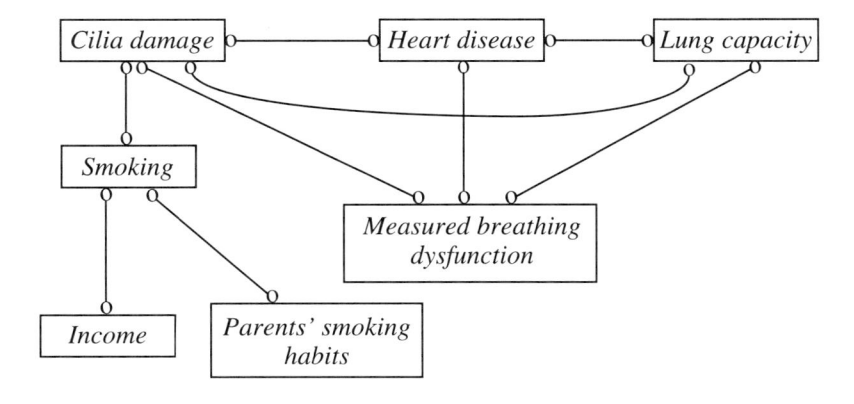

Figure 6.12. Uninformative partially oriented inducing graph of G_5 over O

Such a maximally informative partially oriented inducing path graph π for G could be oriented by the simple but inefficient algorithm of constructing every possible inducing path graph with the same adjacencies as G', throwing out the ones that do not have the same d-connection relations as G', and keeping track of which orientation features are common to all members of **Equiv**(G'). Of course, this is completely computationally infeasible. Figure 6.13 shows the maximally oriented partially oriented inducing path graph of graph G_5 over $\mathbf{O}$ = {*Cilia damage, Smoking, Heart disease, Lung capacity, Measured breathing dysfunction, Income, Parents' smoking habits*}.

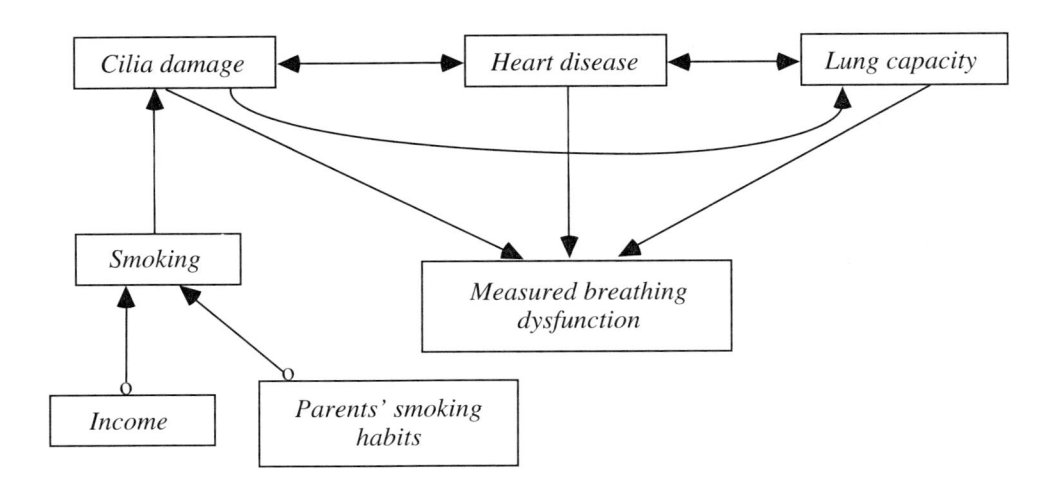

Figure 6.13. Maximally informative partially oriented inducing path graph of G_5 over O

Our goal is to state algorithms that construct a partially oriented inducing path graph for a directed acyclic graph G containing as much orientation information as is consistent with computational feasibility. The algorithm we propose is divided into two main parts. First, the adjacencies in the partially oriented inducing path graph are determined. Then the edges are oriented in so far as possible.

6.7 Algorithms for Causal Inference with Latent Common Causes

In order to state the algorithm, a few more definitions are needed. In a partially oriented inducing path graph π:

(i). A is a **parent** of B if and only if $A \rightarrow B$ in π.
(ii). B is a **collider** along path $\langle A,B,C \rangle$ if and only if $A \mathbin{*}\!\!\rightarrow B \leftarrow\!\!\mathbin{*} C$ in π.
(iii). An edge between B and A is **into** A if and only if $A \leftarrow\!\!\mathbin{*} B$ in π.
(iv). An edge between B and A is **out of** A if and only if $A \rightarrow B$ in π.
(v). In a partially oriented inducing path graph π', U is a **definite discriminating path** for B if and only if U is an undirected path between X and Y containing B, $B \neq X$, $B \neq Y$, every vertex on U except for B and the endpoints is a collider or a definite noncollider on U, and

(i) if V and V' are adjacent on U, and V' is between V and B on U, then $V \mathbin{*}\!\!\rightarrow V'$ on U,

(ii) if V is between X and B on U and V is a collider on U then $V \rightarrow Y$ in π, else $V \leftarrow^*$ Y in π,
(iii) if V is between Y and B on U and V is a collider on U then $V \rightarrow X$ in π, else $V \leftarrow^*$ X in π.
(iv) X and Y are not adjacent in π.

Figure 6.14 illustrates the concept of a definite discriminating path.

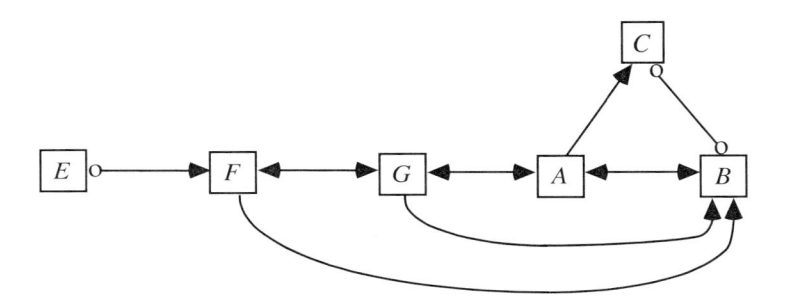

Figure 6.14. <E,F,G,A,C,B> is a definite discriminating path for C

In practice, the Causal Inference Algorithm and the Fast Causal Inference Algorithm (described later in this section) take as input either a covariance matrix or cell counts. Where d-separation facts are needed by the algorithms, the procedure performs tests of conditional independence (in the discrete case) or of vanishing partial correlations in the linear, continuous case. (Recall that if P is a discrete distribution faithful to a graph G, then A and B are d-separated given a set of variables **C** if and only A and B are conditionally independent given **C**, and if P is a distribution linearly faithful to a graph G, then A and B are d-separated given **C** if and only if $\rho_{AB.C} = 0$.) Both algorithms construct a partially oriented inducing path graph of some directed acyclic graph G, where G contains both measured and unmeasured variables.

Causal Inference Algorithm[1]

A.) Form the complete undirected graph Q on the vertex set **V**.
B.) If A and B are d-separated given any subset **S** of **V**, remove the edge between A and B, and record **S** in **Sepset**(A,B) and **Sepset**(B,A).
C.) Let F be the graph resulting from step B). Orient each edge as o-o. For each triple of vertices A, B, C such that the pair A, B and the pair B, C are each adjacent in F but the pair A, C are not adjacent in F, orient A *-* B *-* C as A *-→ B ←-* C if and only if B is not in **Sepset**(A,C), and orient A *-* B *-* C as A *-*<u> B </u>*-* C if and only if B is in **Sepset**(A,C).

D.) repeat
If there is a directed path from A to B, and an edge A *-* B, orient A *-* B as A *→ B,
else if B is a collider along <A,B,C> in π, B is adjacent to D, and D is in **Sepset**(A,C), then orient B *-* D as B ←* D,
else if U is a definite discriminating path between A and B for M in π, and P and R are adjacent to M on U, and P - M - R is a triangle, then
if M is in **Sepset**(A,B) then M is marked as a noncollider on subpath P *-* M *-* R
else P *-* M *-* R is oriented as P *→ M ← * R.
else if P *→ M *-* R then orient as P *→ M → R.²
until no more edges can be oriented.

If the CI or FCI algorithms use as input a covariance matrix from the marginal over **O** of a distribution linearly faithful to G, or cell counts from the marginal over **O** of a distribution faithful to G, we will say the input is data over **O** that is faithful to G.

THEOREM 6.3: If the input to the CI algorithm is data over **O** that is faithful to G, the output is a partially oriented inducing path graph of G over **O**.

If data over **O** = {Cilia damage, Smoking, Heart disease, Lung capacity, Measured breathing dysfunction, Income, Parents' smoking habits} that is faithful to the graph in figure 6.11 is input to the CI algorithm, the output is the maximally informative partially oriented inducing path graph over **O** shown in figure 6.13.

Unfortunately, the Causal Inference (CI) algorithm as stated is not practical for large numbers of variables because of the way the adjacencies are constructed. While it is theoretically correct to remove an edge between A and B from the complete graph if and only if A and B are d-separated given some subset of **O**\{A,B}, this is impractical for two reasons. First, there are too many subsets of **O**\{A,B} on which to test the conditional independence of A and B. Second, for discrete distributions, unless the sample sizes are enormous there are no reliable tests of independence of two variables conditional on a large set of other variables.

In order to determine that a given pair of vertices, such as X and Y are not adjacent in the inducing path graph, we have to find that X and Y are d-separated given some subset of **O**\{X, Y}. Of course, if X and Y are adjacent in the inducing path graph, they are d-connected given every subset of **O**\{X,Y}. We would like to be able to determine that X and Y are d-connected given every subset of **O**\{X,Y} without actually examining every subset of **O**\{X ,Y}.

In a directed acyclic graph over a causally sufficient set **V**, by using the PC algorithm we are able to reduce the order and number of d-separation tests performed because of the following fact: if X and Y are d-separated by any subset of **V**\{X,Y}, then they are d-separated either by **Parents**(X) or **Parents**(Y). While the PC algorithm is constructing the graph it does not know which variables are in **Parents**(X) or in **Parents**(Y), but as the

Discovery Algorithms without Causal Sufficiency 141

algorithm progresses it is able to determine that some variables are definitely not in **Parents**(X) or **Parents**(Y) because they are definitely not adjacent to X or Y. This reduces the number and the order of the d-separation tests that the PC algorithm performs (as compared to the SGS algorithm).

In contrast, an inducing path graph over **O** it is not the case that if X and Y are d-separated given some subset of **O**\{X,Y}, then X and Y are d-separated given either **Parents**(X) or given **Parents**(Y). However, if X and Y are d-separated given *some* subset of **O**\{X,Y}, then X and Y are d-separated given either **D-Sep**(X) or given **D-Sep**(Y). If we know that some variable V is not in **D-Sep**(X) and not in **D-Sep**(Y), we do not need to test whether X and Y are d-separated by any set containing V. Once again, we do not know which variables are in **D-Sep**(X) or **D-Sep**(Y) until we have constructed the graph. But there is an algorithm that can determine that some variables are *not* in **D-Sep**(X) or **D-Sep**(Y) as the algorithm progresses.

Let G be the directed acyclic graph of figure 6.3 (reproduced in figure 6.15). Let G' be the inducing path graph of G over **O** = {A,B,C,D,E,F,H}. A and E are not d-separated given any subset of the variables adjacent to A or adjacent to D (in both cases {B,D}). Because A and E are not adjacent in the inducing path graph of A and E, they are d-separated given some subset of **O**\{A,E}. Hence they are d-separated by either **D-Sep**(A,E) (equal to {B,D,F}) or by **D-Sep**(E,A)) (equal to {B,D,H}). (In this case A and E are d-separated by both **D-Sep**(A,E) and by **D-Sep**(E,A).) The problem is: how can we know to test whether A and E are d-separated given {B,D,H} or {B,D,F} without testing whether A and E are d-separated given every subset of **O**\{A,E}?

A variable V is in **D-Sep**(A,E) in G' if and only if $V \neq A$ and there is an undirected path between A and V on which every vertex except the endpoints is a collider, and each vertex is an ancestor of A or E. If we could find some method of determining that a variable V does not lie on such a path, then we would not have to test whether A and E were d-separated given any set containing V (unless of course V was in **D-Sep**(E,A).) We will illustrate the strategy on G. At any given stage of the algorithm we will call the graph constructed thus far π.

Graph G

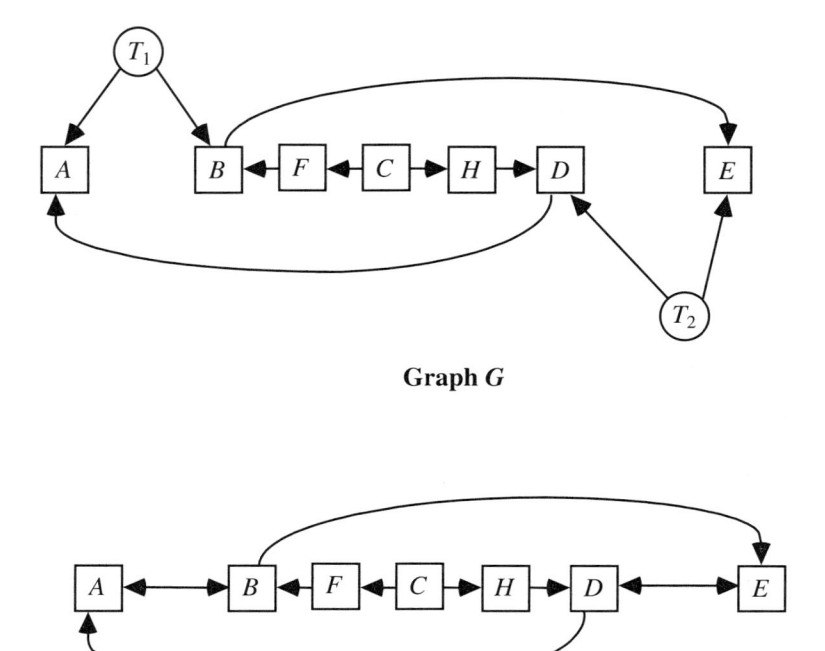

Inducing Path Graph G

Figure 6.15

The FCI algorithm determines which edges to remove from the complete graph in three stages. The first stage is just like the first stage of the PC Algorithm. We intialize π to the complete undirected graph, and then we remove an edge between X and Y if they are d-separated given subsets of vertices adjacent to X or Y in π. This will eliminate many, but perhaps not all of the edges that are not in the inducing path graph. When this operation is performed on data faithful to the graph in figure 6.15, the result is the graph in figure 6.16.

Note that A and E are still adjacent at this stage of the procedure because the algorithm, having correctly determined that A is not adjacent to F or H or C, and that E is not adjacent to F or H or C, never tested whether A and E are d-separated by any subset of variables containing F, H, or C.

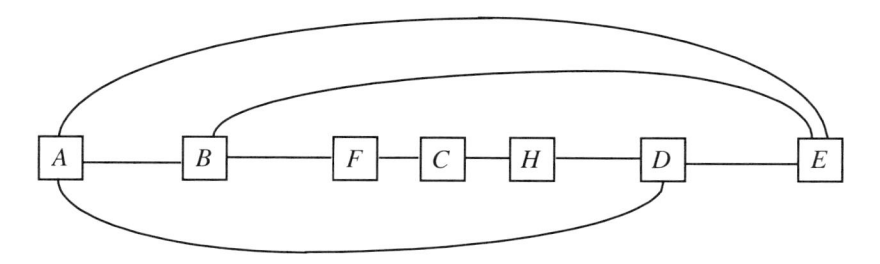

Figure 6.16

Second, we orient edges by determining whether they collide or not, just as in the PC algorithm. The graph at this stage of the algorithm is show in figure 6.17.

Figure 6.17 is essentially the graph constructed by the PC algorithm given data faithful to the graph in figure 6.15, after steps A), B), and C) have been performed.

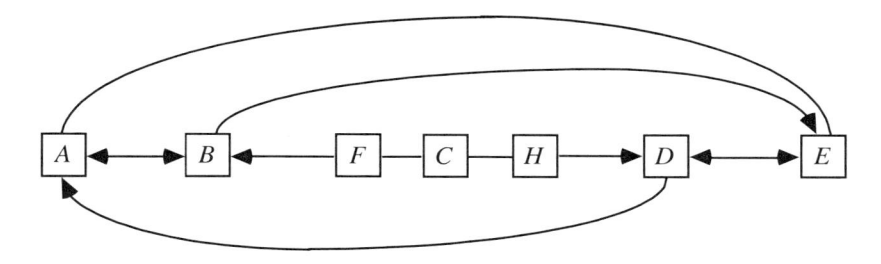

Figure 6.17

We can now determine that some vertices are definitely not in **D-Sep**(A,E) or in **D-Sep**(E,A); it is not necessary to test whether A and E are d-separated given any subset of $\mathbf{O}\backslash\{A,E\}$ that contains these vertices in order to find the correct adjacencies. At this stage of the algorithm, a necessary condition for a vertex V to be in **D-Sep**(A,E) in G' is that in π there is an undirected path U between A and V in which each vertex except for the endpoints is either a collider, or has its orientation hidden because it is in a triangle. Thus C and H are definitely not in **D-Sep**(A,E) and C and F are definitely not in **D-Sep**(E,A). All of the vertices that we have not definitely determined are not in **D-Sep**(A,E) in G' we place in **Possible-D-Sep**(A,E), and similarly for **Possible-D-Sep**(E,A). In this case, **Possible-D-Sep**(A,E) is $\{B,F,D\}$ and **Possible-D-Sep**(E,A) is $\{B,D,H\}$. We now know that if A and E are d-separated given any subset of $\mathbf{O}\backslash\{A,E\}$ then they are d-separated given some subset of **Possible-D-Sep**(A,E) or some subset of **Possible-D-Sep**(E,A). In this case we find that A and E are d-separated given a subset of **Possible-D-Sep**(A,E) (in this case the entire set) and hence remove the edge between A and E.

Once we have obtained the correct set of adjacencies, we unorient all of the edges, and then proceed to reorient them exactly as we did in the Causal Inference Algorithm. The resulting output is shown in figure 6.18.

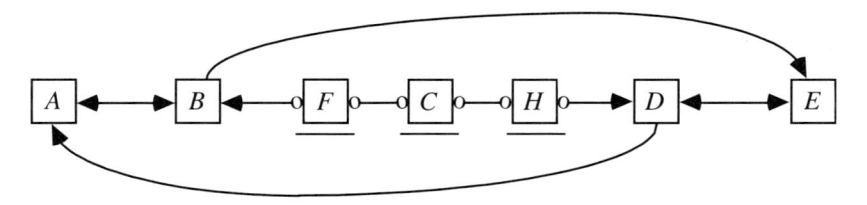

FCI Output

Figure 6.18

For a given partially constructed partially oriented inducing path graph π, **Possible-D-SEP**(A,B) is defined as follows: If $A \neq B$, V is in **Possible-D-Sep**(A,B) in π if and only if $V \neq A$, and there is an undirected path U between A and V in π such that for every subpath $<X,Y,Z>$ of U either Y is a collider on the subpath, or Y is not a definite noncollider and on U, and X, Y, and Z form a triangle in π.

Using this definition of **Possible-D-Sep**(A,E), we can prove that every vertex not in **Possible-D-Sep**(A,E) in π is not in **D-Sep**(A,E) in G'. However, it may be possible to determine from π that some members that we are including in **Possible-D-Sep**(A,E) are not in **D-Sep**(A,E) in G'. There is clearly a trade-off between reducing the size of **Possible-D-Sep**(A,E) (so that the number and order of tests of d-separability performed by the algorithm is reduced) and performing the extra work required to reduce the size of the set, while ensuring that it is still a superset of **D-Sep**(A,E) in G'. We do not know what the optimal balance is. If G is sparse (i.e., each vertex is not adjacent to a large number of other vertices in G), then the algorithm does not need to determine whether A and B are d-separated given **C** for any **C** containing a large number of variables.

Fast Causal Inference Algorithm

A). Form the complete undirected graph Q on the vertex set **V**.
B). $n = 0$.
repeat
  repeat
    select an ordered pair of variables X and Y that are adjacent in Q such that **Adjacencies**$(Q,X)\backslash\{Y\}$ has cardinality greater than or equal to n, and a subset **S** of **Adjacencies**$(Q,X)\backslash\{Y\}$ of cardinality n, and if X and Y are d-separated given **S** delete the edge between X and Y from Q, and record **S** in **Sepset**(X,Y) and **Sepset**(Y,X)

until all ordered variable pairs of adjacent variables *X* and *Y* such that **Adjacencies**(*Q,X*)\{*Y*} has cardinality greater than or equal to *n* and all subsets **S** of **Adjacencies**(*Q,X*)\{*Y*} of cardinality *n* have been tested for d-separation;
$n = n + 1$;
until for each ordered pair of adjacent vertices *X*, *Y*, **Adjacencies**(*Q,X*)\{*Y*} is of cardinality less than *n*.
C). Let *F'* be the undirected graph resulting from step B). Orient each edge as o-o. For each triple of vertices *A*, *B*, *C* such that the pair *A*, *B* and the pair *B*, *C* are each adjacent in *F'* but the pair *A*, *C* are not adjacent in *F'*, orient *A* *-* *B* *-* *C* as *A* *→ *B* ←* *C* if and only if *B* is not in **Sepset**(*A,C*).
D). For each pair of variables *A* and *B* adjacent in *F'*, if *A* and *B* are d-separated given any subset **S** of **Possible-D-SEP**(*A,B*)\{*A,B*} or any subset **S** of **Possible-D-SEP**(*B,A*)\{*A,B*} in *F* remove the edge between *A* and *B*, and record **S** in **Sepset**(*A,B*) and **Sepset**(*B,A*).

The algorithm then reorients an edge between any pair of variables *X* and *Y* as *X* o-o *Y*, and proceeds to reorient the edges in the same way as steps C) and D) of the Causal Inference algorithm.

THEOREM 6.4: If the input to the FCI algorithm is data over **O** that is faithful to *G*, the output is a partially oriented inducing path graph of *G* over **O**.

The Fast Causal Inference Algorithm (FCI) always produces a partially oriented inducing path graph for a graph *G* given correct statistical decisions from the marginal over the measured variables of a distribution faithful to *G*. We do not know whether the algorithm is complete, that is, whether it in every case produces a maximally informative partially oriented inducing path graph.

As with the CI algorithm, if the input to the FCI algorithm is data faithful to the graph of figure 6.11, the output is the maximally informative partially oriented inducing path graph of figure 6.13.

Two directed acyclic graphs *G* and *G'* that have the same FCI partially oriented inducing path graph over **O** have the same d-connection relations involving just members of **O**.

COROLLARY 6.4.1: If *G* is a directed acylic graph over **V**, *G'* is a directed acyclic graph over **V'**, and **O** is a subset of **V** and of **V'**, then *G* and *G'* have the same d-separation relations among only the variables in **O** if and only if they have the same FCI partially oriented inducing path graph over **O**.

Given a directed acyclic graph *G*, it is possible to determine what d-separation relations involving just members of **O** are true of *G* from the FCI partially oriented inducing path graph of *G* over **O**. In a partially oriented inducing path graph π, if $X \neq Y$, and X and Y are not in $\mathbf{Z}$, then an undirected path U between X and Y **definitely d-connects** X and Y given $\mathbf{Z}$ if and only if every collider on U has a descendant in $\mathbf{Z}$, every definite noncollider on U is not in $\mathbf{Z}$, and every other vertex on U is not in $\mathbf{Z}$ but has a descendant in $\mathbf{Z}$. In a partially oriented inducing path graph π, if $\mathbf{X}$, $\mathbf{Y}$, and $\mathbf{Z}$ are disjoint sets of variables, then $\mathbf{X}$ is definitely d-connected to $\mathbf{Y}$ given $\mathbf{Z}$ if and only if some member of $\mathbf{X}$ is d-connected to some member of $\mathbf{Y}$ given $\mathbf{Z}$.

COROLLARY 6.4.2: If G is a directed acyclic graph over $\mathbf{V}$, $\mathbf{O}$ is a subset of $\mathbf{V}$, π is the FCI partially oriented inducing path graph of G over $\mathbf{O}$, and $\mathbf{X}$, $\mathbf{Y}$, and $\mathbf{Z}$ are disjoint subsets of $\mathbf{O}$, then $\mathbf{X}$ is d-connected to $\mathbf{Y}$ given $\mathbf{Z}$ in G if and only if $\mathbf{X}$ is definitely d-connected to $\mathbf{Y}$ given $\mathbf{Z}$ in π.

These corollaries are proved in Spirtes and Verma 1992.

6.8 Theorems on Detectable Causal Influence

In this section we show that a number of different kinds of causal inferences can be drawn from a partially oriented inducing path graph.

THEOREM 6.5: If π is a partially oriented inducing path graph of directed acyclic graph G over $\mathbf{O}$, and there is a directed path U from A to B in π, then there is a directed path from A to B in G.

If G is a directed acyclic graph over $\mathbf{V}$, and $\mathbf{O}$ is included in $\mathbf{V}$, if the input to the CI algorithm is data faithful to G over $\mathbf{O}$, then we call the output of the CI algorithm the **CI partially oriented inducing path graph** of G over $\mathbf{O}$. We adopt a similar terminology for the FCI algorithm. A **semidirected path from** A to B in partially oriented inducing path graph π is an undirected path U from A to B in which no edge contains an arrowhead pointing toward A, that is, there is no arrowhead at A on U, and if X and Y are adjacent on the path, and X is between A and Y on the path, then there is no arrowhead at the X end of the edge between X and Y.

THEOREM 6.6: If π is the CI partially oriented inducing path graph of directed acyclic graph G over $\mathbf{O}$, and there is no semidirected path from A to B in π, then there is no directed path from A to B in G.

Recall that a trek between distinct variables A and B is either a directed path from A to B, a directed path from B to A, or a pair of directed paths from a vertex C to A and B respectively that intersect only at C. The following theorem states a sufficient condition for when the edges in a partially oriented inducing path graph indicate a trek in the graph that contains no measured vertices except for the endpoints.

THEOREM 6.7: If π is a partially oriented inducing path graph of directed acyclic graph G over $\mathbf{O}$, A and B are adjacent in π, and there is no undirected path between A and B in π

Discovery Algorithms without Causal Sufficiency 147

except for the edge between *A* and *B*, then in *G* there is a trek between *A* and *B* that contains no variables in **O** other than *A* or *B*.

THEOREM 6.8: If π is the CI partially oriented inducing path graph of directed acyclic graph *G* over **O**, and every semidirected path from *A* to *B* contains some member of **C** in π, then every directed path from *A* to *B* in *G* contains some member of **C**.

THEOREM 6.9: If π is a partially oriented inducing path graph of directed acyclic graph *G* over **O**, and $A \leftrightarrow B$ in π, then there is a latent common cause of *A* and *B* in *G*.

Parallel results holds for the FCI algorithm.

To illustrate the application of these theorems, condsider the maximally informative partially oriented inducing path graph in figure 6.13 of the causal structure of G_5. Applying theorem 6.5 we infer that *Smoking* causes *Cilia damage*, *Lung capacity,* and *Measured breathing dysfunction*. Applying theorem 6.6, we infer that *Smoking* does not cause *Heart disease* or *Income* or *Parents' Smoking Habits*. It is impossible to determine from the conditional independence relations among the measured variables whether *Income* causes *Smoking*, or there is a common cause of *Smoking* and *Income*. The statistics among the measured variables determine that *Cilia damage* and *Heart disease* have a latent common cause, *Cilia damage* does not cause *Heart disease*, and *Heart disease* does not cause *Cilia damage*.

We note here a topic that will be more fully explored in the next chapter. In the example from figure 6.11, in order to infer that smoking causes breathing dysfunction, it is necessary to measure two causes of *Smoking* (whose collision at *Smoking* orients the edge from *Smoking* to *Cilia damage*.) In general, this suggests that in the design of studies intended to determine if there is a causal path from variable *A* to variable *B*, it is useful to measure not only variables that might mediate the connection between *A* and *B*, but also to measure possible causes of *A*.

6.9 Nonindependence Constraints

The Markov and Faithfulness conditions applied to a causally insufficient graph may entail constraints on the marginal distribution of measured variables that are not conditional independence relations, and hence are not used in the FCI algorithm. Consider, the example in figure 6.19, due to Thomas Verma (Verma and Pearl 1991).

Assume *T* is unmeasured. Then a joint distribution faithful to the entire graph must satisfy the constraint that the quantity

```
        T
   ↙   ↓   ↘
A → B → C → D
```

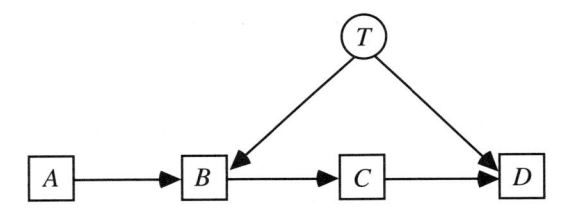

Figure 6.19

$$\sum_{B}^{\rightarrow} P(B|A)P(D|B,C,A)$$

is a function only of the values of C and D.

$$\sum_{B}^{\rightarrow} P(B|A)P(D|B,C,A) = \sum_{T}^{\rightarrow}\sum_{B}^{\rightarrow} P(B|A)P(D|B,C,A,T)P(T|B,C,A) =$$

$$\sum_{T}^{\rightarrow} P(D|C,T)\sum_{B}^{\rightarrow} P(B|A)P(T|B,A)$$

(because A, B are independent of D given $\{C, T\}$ and C is independent of T given $\{A, B\}$). Hence

$$\sum_{T}^{\rightarrow} P(D|C,T)\sum_{B}^{\rightarrow} P(B|A)P(T|B,A) = \sum_{T}^{\rightarrow} P(D|C,T)P(T|A) =$$

$$\sum_{T}^{\rightarrow} P(D|C,T)P(T) = g(C,D)$$

(because T and A are independent).

This constraint is not entailed if a directed edge from A to D is added to the graph. The moral is that there is further marginal structure not in the form of conditional independence relations that could in principle be used to help identify latent structure. We will see a similar point when we turn to linear models in the next section. A general theory of how Verma constraints arise is given by Desjardins (1999).

6.10 Generalized Statistical Indistinguishability and Linearity

Suppose that for whatever reasons an investigation were to be confined to linear structures and to probability distributions that are consistent with the assumption that each random variable is a linear function of its parents and of unmeasured factors. The effect of restrictions such as linearity is to make distinguishable causal structures that would otherwise be indistinguishable. That happens because the restriction, whatever it is, together with the conditional dependence and independence relations required by the Markov, Minimality or Faithfulness Conditions, entails additional constraints on the measured variables. These additional constraints may not be in the form of conditional independence relations. In the linear case they typically are not. Consider for example the two structures shown below, where the X variables are measured and the T variables are unmeasured.

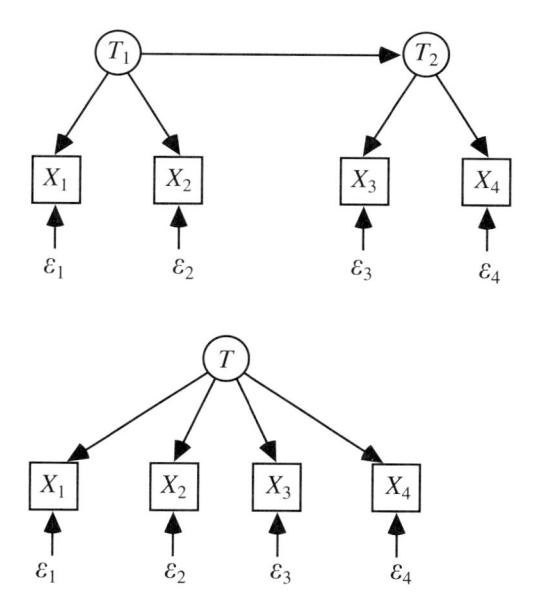

Figure 6.20

These structures each imply that in the marginal distribution over the measured variables every pair of variables is dependent conditional on every other set of measured variables. In each case the maximally informative partially oriented inducing path graph on the X variables is a complete undirected graph. By examining conditional independence relations among these variables, one could not tell which structure obtains. But if linearity is required, then it is easy to tell which structure obtains. For under the linearity assumption, the second structure entails all three of the following constraints on the correlations of the measured variables, while the first structure entails only the first of these constraints (where we denote the correlation between X_1 and X_2 as ρ_{12} in order to avoid subscripts with subscripts):

$\rho_{13}\rho_{24} - \rho_{14}\rho_{23} = 0$
$\rho_{12}\rho_{34} - \rho_{14}\rho_{23} = 0$
$\rho_{13}\rho_{24} - \rho_{12}\rho_{34} = 0$

Early in this century Charles Spearman (1928) called constraints of these sorts **vanishing tetrad differences**, and we will use his terminology.

Characterizing statistical indistinguishability under the linearity restriction thus presents an entirely new problem, and one for which we will offer no general solution. It is not true, for example, that conditional independence relations and vanishing tetrad differences jointly determine the faithful indistinguishability classes of linear structures with unmeasured variables. For example, each of the following linear structures entails that a single tetrad difference vanishes in the marginal distribution over A, B, C, and D, and has a partially oriented inducing path graph for these variables consisting of a complete undirected graph:

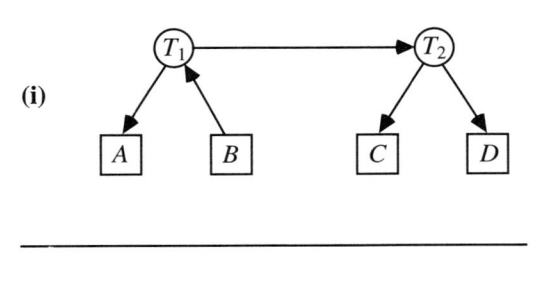

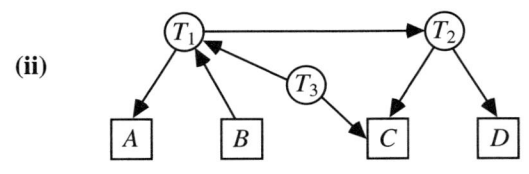

Figure 6.21

But the two graphs are not faithfully indistinguishable over the class of linear structures. Structure (ii) permits distributions consistent with linearity in which the correlation of A and B is positive, the correlation of B and C is positive and the correlation of A and C is negative. Structure (i) admits no distributions consistent with linearity whose marginals satisfy this condition.

Structures (i) and (ii) are not typical of the linear causal structures with unmeasured variables one finds in the social science literature. For practical purposes, the examination of vanishing tetrad constraints provides a powerful means to distinguish between alternative causal structures, even in structures that are only partially linear. Tests for hypotheses of vanishing tetrad differences were introduced by Wishart in the 1920s assuming normal variates, and asymptotically distribution free tests have been described by Bollen (1989).

Algorithms that take advantage of vanishing tetrad differences will be described and illustrated later in this book. In order to take that advantage, we need to be able to determine algorithmically when a structure with or without unmeasured common causes entails a particular vanishing tetrad difference among the measured variables. This question leads to an important theorem.

6.11 The Tetrad Representation Theorem

We wish to characterize entirely in graph theoretic terms a necessary and sufficient condition for a distribution on the vertices of an arbitrary directed acyclic graph G to **linearly imply** a vanishing tetrad difference, that is the tetrad difference vanishes in all of the distributions linearly represented by G. We will call a distribution linearly represented by some directed acyclic graph G a **linear model**. (A slightly more formal definition is given in chapter 13.) A linear model is uniquely determined by the directed acyclic graph G that represents it, and linear coefficients and the independent marginal distributions on the variables (including error terms) of zero indegree.

First some terminology: Given a trek $T(I,J)$ between vertices I and J, $I(T(I,J))$ denotes the directed path in $T(I,J)$ from the source of $T(I,J)$ to I and $J(T(I,J))$ denotes the directed path in $T(I,J)$ from the source of $T(I,J)$ to J. (Recall that one of the directed paths in a trek may be an empty path.) $\mathbf{T}(I,J)$ denotes the set of all treks between I and J.

In a directed acyclic graph G, if for all $T(K,L)$ in $\mathbf{T}(K,L)$ and all $T(I,J)$ in $\mathbf{T}(I,J)$, $L(T(K,L))$ and $J(T(I,J))$ intersect at a vertex Q, then Q is an $LJ(T(I,J),T(K,L))$ **choke point**. Similarly, if for all $T(K,L)$ in $\mathbf{T}(K,L)$ and all $T(I,J)$ in $\mathbf{T}(I,J)$, $L(T(K,L))$ and all $J(T(I,J))$ intersect at a vertex Q, and for all $T(I,L)$ in $\mathbf{T}(I,L)$ and all $T(J,K)$ in $\mathbf{T}(J,K)$, $L(T(I,L))$ and $J(T(J,K))$ also intersect at Q, then Q is an $LJ(T(I,J),T(K,L),T(I,L),T(J,K))$ **choke point**. Also see the definition of trek.

The fundamental theorem for vanishing tetrad differences in linear models is this:

TETRAD REPRESENTATION THEOREM 6.10: In a directed acyclic graph G, there exists an $LJ(T(I,J),T(K,L),T(I,L),T(J,K))$ or an $IK(T(I,J),T(K,L),T(I,L),T(J,K))$ choke point if and only if G linearly implies $\rho_{IJ}\rho_{KL} - \rho_{IL}\rho_{JK} = 0$.

A consequence of theorem 6.10 is

THEOREM 6.11: A directed acyclic graph G linearly implies $\rho_{IJ}\rho_{KL} - \rho_{IL}\rho_{JK} = 0$ only if either it linearly implies that ρ_{IJ} or $\rho_{KL} = 0$, and ρ_{IL} or $\rho_{JK} = 0$, or there is a (possibly empty) set **Q** of random variables in G that does not contain both I and K or both J and L such that G linearly implies that $\rho_{IJ.\mathbf{Q}} = \rho_{KL.\mathbf{Q}} = \rho_{IL.\mathbf{Q}} = \rho_{JK.\mathbf{Q}} = 0$.

Theorem 6.10 provides a fast algorithm for calculating the vanishing tetrad differences linearly implied by any directed acyclic graph. Theorem 6.11 provides a means to determine when unmeasured common causes are acting in linear structures. In later chapters we describe some of the implications of these facts for investigating the structure of causal relations among unmeasured variables.

6.12 An Example: Math Marks and Causal Interpretation

In several places in his recent text on graphical models in statistics, Whittaker (1990) discusses a data set from Mardia, Kent and Bibby (1979) concerning the grades of 88 students on examinations in five mathematical subjects: mechanics, vectors, algebra, analysis and statistics. The example illustrates one of the uses of the Tetrad Representation Theorem, and provides occasion to comment on some important differences of interpretation between our methods and those Whittaker describes. The variance/covariance matrix for the data is as follows:

Mechanics	*Vectors*	*Algebra*	*Analysis*	*Statistics*
302.29				
125.78	170.88			
100.43	84.19	111.60		
105.07	93.60	110.84	217.88	
116.07	97.89	120.49	153.77	294.37

When given these data, the PC algorithm immediately determines the pattern show in figure 6.22.

Whittaker obtains the same graph under a different interpretation. Recall that an undirected independence graph is any pair <G,P> where G is an undirected graph and P is a distribution such that vertices, X, Y in G are not adjacent if they are independent conditional on the set of all other vertices of G; or to state the contrapositive: if X, Y are dependent conditional on the set of all other vertices of G, then X, Y are adjacent in G. Undirected independence graphs hide much of the causal structure, and sometimes many of the independence relations. Thus if variables X and Z are causes of variable Y but X and Z are statistically independent and have no causal relations whatsoever, the undirected independence graph has an edge between X and Z. In effect, the independence graph fails to represent the conditional independence relations that hold among proper subsets of a set of variables.

Discovery Algorithms without Causal Sufficiency

```
    Mechanics              Analysis
           \              /
            \            /
             Algebra
            /            \
           /              \
    Vectors                Statistics
```

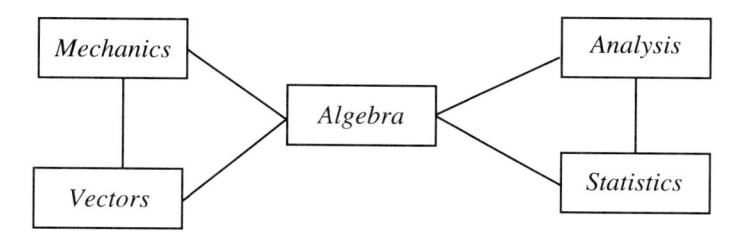

Figure 6.22

Every *undirected* pattern graph obtained from a faithful distribution (or sample) is a subgraph of the undirected independence graph obtained from that distribution. In the case at hand the two graphs are the same, but they need not be in general.

Whittaker claims that identifying the undirected independence graph is important for four reasons: (i) it reduces the complex five dimensional object into two simpler three dimensional objects—the two maximal cliques in the graph; (ii) it groups the variables into two sets; (iii) it highlights *Algebra* as the one crucial examination in analyzing the interrelationship between different subjects in exam performance; (iv) it asserts that *Algebra* and *Analysis* alone will be sufficient to predict *Statistics* and that *Algebra* and *Vectors* will be sufficient to predict *Mechanics*; but that all four marks are needed to predict *Algebra* (p. 6)

The second reason seems simply a consequence of the first, and the first seems of little consequence: the cognitive burden of noting that there are five variables is not very great. There is a long tradition in statistics of introducing representations on grounds that they simplify the data and in practice treating the objects of such reductions as causes. That is, for example, the history of factor analysis after Thurstone. But as with factor analysis, causal conclusions drawn from independence graphs would be unreliable. The third reason seems too vague to be worth much trouble. The assertion given in the fourth reason is sound, but only if "predict" is understood in all cases to have nothing to do with predicting the values of variables when they are deliberately altered, as by coaching. We suspect statistical analyses of such educational data are apt to be given a causal significance, and for such purposes directed graphical models better represent the hypotheses.

Applying theorem 6.11, the vanishing tetrad test for latent variables, we find that there are four vanishing tetrad differences that cannot be explained by vanishing partial correlations among the measured variables. This suggests that they are entailed by vanishing partial correlations involving latent variables, and thus suggests the introduction of latent variables. A natural idea in view of the mathematical structure of the subjects tested is that *Algebra* is an indicator of *Algebraic knowledge*, which is a factor in the *Knowledge of vector algebra* measured by *Vector* and *Mechanics* and is also a factor in *Knowledge of real analysis* that affects *Analysis* and *Statistics*. The explanation of the data then looks as shown in figure 6.23.

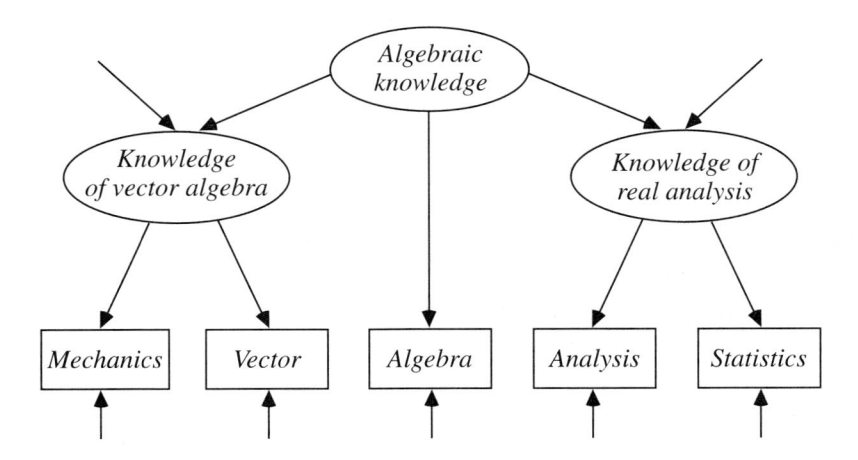

Figure 6.23

The arrows without notation attached to them indicate other sources of variation. Assuming a faithful distribution and linearity, this graph does not entail the vanishing first order partial correlations among the measured variables that the data suggest. But if the variance in *Algebra* due to factors other than algebraic knowledge is sufficiently small, a linear distribution faithful to this graph will to good approximation give exactly those vanishing partial correlations.

This structure (assuming linearity) entails eight vanishing tetrad differences, all of which the TETRAD II program identifies and tests and cannot reject ($p > .7$). The model itself, when treated as the null hypotheses in a likelihood ratio test, yields a p value of about .9, roughly the value Whittaker reports for the undirected graphical independence model.

6.13 Background Notes

In a series of papers (Pearl and Verma 1990, 1991, Verma and Pearl 1990a, 1990b, 1991) Verma and Pearl describe an "Inductive Causation" algorithm that outputs a structure that they call a pattern (or sometimes a "completed hybrid graph") of a directed acyclic graph G over a set of variables **O**. The most complete description of their theory appears in Verma and Pearl (1990b). The key ideas of an inducing path, an inducing path graph, and the proof of (what we call) theorem 6.1 all appear in this paper. Unfortunately, the two main claims about the output of the Inductive Causation Algorithm made in the paper, given in their lemma A2 and their theorem 2, are false (see Spirtes 1992).

Early versions of the Inductive Causation Algorithm did not distinguish between $A \rightarrow B$ and $A \; o\!\!\rightarrow B$, and hence could not be used to infer that A causes B as in theorem 6.5. This distinction was introduced (in a different notation) in order to prove a version of theorem 6.5 and theorem 6.6 in Spirtes and Glymour (1990); Verma and Pearl incorporated it in a subsequent version of the Inductive Causation Algorithm. The Inductive Causation Algorithm does not use definite discriminating paths to orient edges, and hence in some cases gives less orientation information than the FCI procedure. The output of the Inductive Causation Algorithm has no notation distinguishing between edges in triangles that definitely do not collide and merely unoriented edges. Like the CI algorithm, the Inductive Causation Algorithm cannot be applied to large numbers of variables because testing the independence of some pairs of variables conditional on every subset of $\mathbf{O}\backslash\{A,B\}$ is required.

The vanishing tetrad difference was used as the principle technique in model specification by Spearman and his followers. A brief account of their methods is given in Glymour, Scheines, Spirtes, and Kelly (1987). Spearman's inference to common causes from vanishing tetrad differences was challenged by Godfrey Thomson in a series of papers between 1916 and 1935. In our terms, Thomson's models all violated linear faithfulness.

7 Prediction

7.1 Introduction

The fundamental aim of many empirical studies is to predict the effects of changes, whether the changes come about naturally or are imposed by deliberate policy: Will the reduction of sources of environmental lead increase the intelligence of children in exposed regions? Will increased taxation of cigarettes decrease lung cancer? How large will these effects be? What will be the differential yield if a field is planted with one species of wheat rather than another; or the difference in number of polio cases per capita if all children are vaccinated against polio as against if none are; or the difference in recidivism rates if parolees are given $600 per month for six months as against if they are given nothing; or the reduction of lung cancer deaths in middle aged smokers if they are given help in quitting cigarette smoking; or the decline in gasoline consumption if an additional dollar tax per gallon is imposed?

One point of experimental designs of the sort found in randomized trials is to attempt to *create* samples that, from a statistical point of view, are from the very distributions that would result if the corresponding treatments were made general policy and applied everywhere. For such experiments under such assumptions, the problems of statistical inference are conventional, which is not to say they are easy, and the prediction of policy outcomes is not problematic in principle. But in empirical studies in the social sciences, in epidemiology, in economics, and in many other areas, we do not know or cannot reasonably assume that the observed sample is from the very distribution that would result if a policy were adopted. Implementing a policy may change relevant variables in ways not represented in the observed sample. The inference task is to move from a sample obtained from a distribution corresponding to passive observation or quasi-experimental manipulation, to conclusions about the distribution that would result if a policy were imposed. In our view one of the most fundamental questions of statistical inference is when, if ever, such inferences are possible, and, if ever they are possible, by what means. The answer, according to Mosteller and Tukey, is "never." We will see whether that answer withstands analysis.

7.2 Prediction Problems

The possibilities of prediction may be analyzed in a number of different sorts of circumstances, including at least the following:

Case 1: We know the causal graph, which variables will be directly manipulated, and what the direct manipulation will do to those variables. We want to predict the distribution of variables that will not be directly manipulated. More formally, we know the set **X** of variables being directly manipulated, $P(\mathbf{X} \mid \mathbf{Parents}(\mathbf{X}))$ in the manipulated distribution, and that **Parents(X)** in the manipulated population is a subset of **Parents(X)** in the unmanipulated population. That is essentially the circumstance that Rubin, Holland, Pratt, and Schlaifer address, and in that case the causal graph and the Manipulation Theorem specify a relevant formula for calculating the manipulated distribution in terms of marginal conditional probabilities from the unmanipulated distribution. The latter can be estimated from samples; we can find the distribution of **Y** (or of **Y** conditional on **Z**) under direct manipulation of **X** by taking the appropriate marginal of the calculated manipulated distribution.

Case 2: We know the set **X** of variables being directly manipulated, $P(\mathbf{X} \mid \mathbf{Parents(X)})$ in the manipulated distribution, that **Parents(X)** in the manipulated population is a subset of **Parents(X)** in the unmanipulated population, and that the measured variables are causally sufficient; unlike case 1, we do not know the causal graph. The causal graph must be conjectured from sample data. In this case the sample and the PC (or some other) algorithm determine a pattern representing a class of directed graphs, and properties of that class determine whether the distribution of **Y** following a direct manipulation of **X** can be predicted.

Case 3: The difficult, interesting and realistic case arises when we know the set **X** of variables being directly manipulated, we know $P(\mathbf{X} \mid \mathbf{Parents(X)})$ in the manipulated population, and that **Parents(X)** in the manipulated population is a subset of **Parents(X)** in the unmanipulated population, but prior knowledge and the sample leave open the possibility that there *may be* unmeasured common causes of the measured variables. If observational studies were treated without unsupported preconceptions, surely that would be the typical circumstance. It is chiefly because of this case that Mosteller and Tukey concluded that prediction from uncontrolled observations is not possible. One way of viewing the fundamental problem of predicting the distribution of *Y* or conditional distribution of *Y* on *Z* upon a direct manipulation of *X* can be formulated this way: *find conditions sufficient for prediction, and conditions necessary for prediction, given only a partially oriented inducing path graph and conditional independence facts true in the marginal (over the observed variables) of the unmanipulated distribution. Show how to calculate features of the predicted distribution from the observed distribution.* The ultimate aim of this chapter is to provide a partial solution to this problem.

We will take up these cases in turn. Case 1 is easy but we take time with it because of the connection with Rubin's theory. Case 2 is dealt with very briefly. In our view Case 3 describes the more typical and theoretically most interesting inference problems. The reader is warned that even when the proofs are postponed the issue is intricate and difficult.

7.3 Rubin-Holland-Pratt-Schlaifer Theory[1]

Rubin's framework has a simple and appealing intuition. In experimental or observational studies we sample from a population. Each unit in the population, whether a child or a national economy or a sample of a chemical, has a collection of properties. Among the properties of the units in the population, some are *dispositional*—they are propensities of a system to give a response to a treatment. A glass vase, for example, may be fragile, meaning that it has a disposition to break if struck sharply. A dispositional property isn't exhibited unless the appropriate treatment is applied—fragile vases don't break unless they are struck. Similarly, in a population of children, for each reading program each child has a disposition to produce a certain post-test score (or range of test scores) if exposed to that reading program. In experimental studies when we give different treatments to different units, we are attempting to estimate dispositional properties of units (or their averages, or the differences of their averages) from data in which only some of the units have been exposed to the circumstances in which that disposition is manifested. Rubin associates with each such dispositional quantity, Q, and each value x of relevant treatment variable X, a random variable, $Q_{Xf=x}$, whose value for each unit in the population is the value Q would have if that unit were to be given treatment x, or in other words if the system were forced to have X value equal to x. If unit i is actually given treatment $x1$ and a value of Q is measured for that unit, the measured value of Q equals the value of $Q_{Xf=x1}$.

Experimentation may give a set of paired values $<x, y_{Xf=x}>$, where $y_{Xf=x}$ is the value of the random variable $Y_{Xf=x}$. But for a unit i that is given treatment $x1$, we also want to know the value of $Y_{Xf=x2}$, $Y_{Xf=x3}$, and so on for each possible value of X, representing respectively the values for Y unit i is disposed to exhibit if unit i were exposed to treatment $x2$ or $x3$, that is, if the X value for these units were forced to be $x2$ or $x3$ rather than $x1$. These unobserved values depend on the causal structure of the system. For example, the value of Y that unit i is disposed to exhibit on treatment $x2$ might depend on the treatments given to other units. We will suppose that there is no dependence of this kind, but we will investigate in detail other sorts of connections between causal structure and Rubin's counterfactual random variables.

A typical inference problem in Rubin's framework is to estimate the distribution of $Y_{Xf=x}$ for some value x of X, over all units in the population, from a sample in which only some members have received the treatment x. A number of variations arise. Rather than forcing a unique value on X, we may contemplate forcing some specified distribution of values on X, or we may contemplate forcing different specified distributions on X depending on the (unforced) values of some other variables Z; our "experiment" may be purely observational so that an observed value q of variable Q for unit i when X is observed to have value x is not necessarily the same as $Q_{Xf=x}$. Answers to various problems such as these can be found in the papers cited. For example, in our paraphrasing, Pratt and Schlaifer claim the following:

When all units are systems in which Y is an effect of X and possibly of other variables, and no causes of Y other than X are measured, in order for the conditional distribution of Y on X = x to equal $Y_{Xf=x}$ for all values x of X, it is sufficient and "almost necessary" that X and each of the random variables $Y_{Xf=x}$ (where x ranges over all possible values of X) be statistically independent.

In our terminology, when the conditional distribution of Y on $X = x$ equals $Y_{Xf=x}$ for all values x of X we say that the conditional distribution of Y on X is "invariant"; in their terminology it is "observable." Pratt and Schlaifer's claim may be clarified with several examples, which will also serve to illustrate some tacit assumptions in the application of the framework. Suppose X and U, which is unobserved, are the only causes of Y, and they have no causal connection of any kind with one another, a circumstance that we will represent by the graph in figure 7.1.

X	Y	U	Xf	$U_{Xf=1}$	$Y_{Xf=1}$
1	1	0	1	0	1
1	2	1	1	1	2
1	3	2	1	2	3
2	2	0	1	0	1
2	3	1	1	1	2
2	4	2	1	2	3

Table 7.1

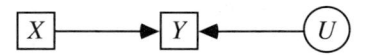

$X \longrightarrow Y \longleftarrow U$

Figure 7.1

For simplicity let's suppose the dependencies are all linear, and that for all possible values of X, Y and U, and all units, $Y = X + U$. Let Xf represent values of X that could possibly be *forced* on all units in the population. X is an observed variable; Xf is not. X is a random variable; Xf is not. Consider values in table 7.1.

Suppose for simplicity that each row (ignoring Xf, which is not a random variable) is equally probable. Here the X and Y columns give possible values of the measured variables. The U column gives possible values of the unmeasured variable U. Xf is a variable whose column indicates values of X that might be forced on a unit; we have not continued the table beyond Xf = 1. The $U_{Xf=1}$ column represents the range of values of U when X is forced to have the value 1; the $Y_{Xf=1}$ gives the range of values of Y when X is forced to have the value 1. Notice that in the table $Y_{Xf=1}$ is uniquely determined by the value of Xf and the value of $U_{Xf=1}$ and is independent of the value of X.

The table illustrates Pratt and Schlaifer's claim: $Y_{Xf=1}$ is independent of X and the distribution of Y conditional on X = 1 equals the distribution of $Y_{Xf=1}$. We constructed the table by letting $U = U_{Xf=1}$, and $Y_{Xf=1} = 1 + U_{Xf=1}$. In other words, we obtained the table by assuming that save for the distribution of X, the causal structure and probabilistic structure are completely unaltered if a value of X is forced on all units. By applying the same procedure with $Y_{Xf=2} = 2 + U_{Xf=2}$, the table can be extended to obtain values when Xf = 2 that satisfy Pratt and Schlaifer's claim.

Consider a different example in which, according to Pratt and Schlaifer's rule, the conditional probability of Y on X is *not* invariant under direct manipulation. In this case X causes Y and U causes Y, and there is no causal connection of any kind between X and U, as before, but in addition an unmeasured variable V is a common cause of both X and Y, a situation represented in figure 7.2.

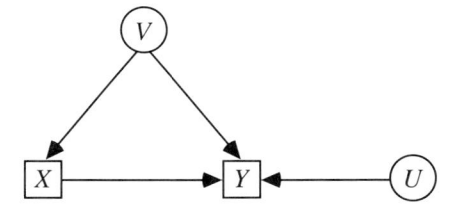

Figure 7.2

Consider the distribution shown in table 7.2, with the same conventions as in table 7.1. Again, assume all rows are equally probable, ignoring the value of Xf which is not a random variable. Notice that $Y_{xf=1}$ is now *dependent* on the value of X. And, just as Pratt and Schlaifer require, the conditional distribution of Y on X = 1 is *not* equal to the distribution of $Y_{Xf=1}$.

X	Y	U	Xf	$U_{Xf=1}$	$Y_{Xf=1}$
1	1	0	1	0	1
1	2	1	1	1	2
1	3	2	1	2	3
2	2	0	1	0	1
2	3	1	1	1	2
2	4	2	1	2	3

Table 7.2

The table was constructed so that when $X = 1$ is forced, and hence $Xf = 1$, the distributions of $U_{Xf=1}$, and $V_{Xf=1}$ are independent of Xf. In other words, while the system of equations

$Y = X + V + U$
$X = V$

was used to obtain the values of X, Y, and U, the assumptions $U_{Xf=1} = U$, $V_{Xf=1} = V$ and the equation

$Y_{Xf=1} = Xf + V_{Xf=1} + U_{Xf=1}$

were used to determine the values of $U_{Xf=1}$, $V_{Xf=1}$ and $Y_{Xf=1}$. The forced system was treated as if it were described by the diagram depicted in figure 7.3.

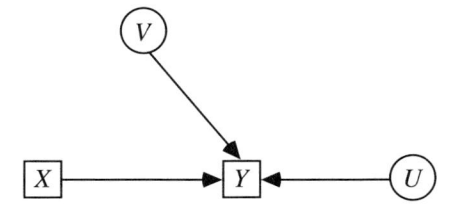

Figure 7.3

Prediction

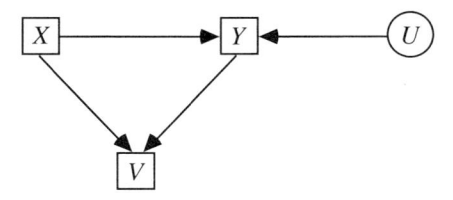

![Figure 7.4: Graph with X → Y, U → Y, X → V, Y → V]

Figure 7.4

For another example, suppose $Y = X + U$, but there is also a variable V that is dependent on both Y and X, so that the system can be depicted as in figure 7.4.

Table 7.3 is a table of values obtained by assuming $Y = X + U$ and $V = Y + X$ and these relations are unaltered by a direct manipulation of X.

X	Y	V	U	Xf	$V_{Xf=1}$	$U_{Xf=1}$	$Y_{Xf=1}$
0	0	0	0	1	2	0	1
0	1	1	1	1	3	1	2
0	2	2	2	1	4	2	3
1	1	2	0	1	2	0	1
1	2	3	1	1	3	1	2
1	3	4	2	1	4	2	3

Table 7.3

Again assume all rows are equally probable. Note that $Y_{Xf=1}$ is independent of X, and $Y_{Xf=1}$ has the same distribution as Y conditional on $X = 1$. So Pratt and Schlaifer's principle is again satisfied, and in addition the conditional probability of Y on X is invariant. The table was constructed by supposing the manipulated system satisfies the very same system of equations as the unmanipulated system, and in effect that the graph of dependencies in figure 7.4 is unaltered by forcing values on X.

Pratt and Schlaifer's rules, as we have reconstructed them, are consequences of the Markov Condition. So are other examples described by Rubin. To make the connection explicit we require some results. We will assume the technical definitions introduced in chapter 3, and we will need some further definitions.

If G is a directed acyclic graph over a set of variables $\mathbf{V} \cup \mathbf{W}$, $\mathbf{W}$ is exogenous with respect to $\mathbf{V}$ in G, $\mathbf{Y}$ and $\mathbf{Z}$ are disjoint subsets of $\mathbf{V}$, $P(\mathbf{V} \cup \mathbf{W})$ is a distribution that satisfies the Markov condition for G, and **Manipulated**($\mathbf{W}$) = $\mathbf{X}$, then $P(\mathbf{Y}|\mathbf{Z})$ is **invariant** under direct manipulation of $\mathbf{X}$ in G by changing $\mathbf{W}$ from $\mathbf{w}_1$ to $\mathbf{w}_2$ if and only if $P(\mathbf{Y}|\mathbf{Z},\mathbf{W}=\mathbf{w}_1) = P(\mathbf{Y}|\mathbf{Z},\mathbf{W}=\mathbf{w}_2)$ wherever they are both defined. Note that a sufficient condition for $P(\mathbf{Y}|\mathbf{Z})$ to be invariant under direct manipulation of $\mathbf{X}$ in G by changing $\mathbf{W}$ is that $\mathbf{W}$ be d-separated from $\mathbf{Y}$ given $\mathbf{Z}$ in G. In a directed acyclic graph G containing $\mathbf{Y}$ and $\mathbf{Z}$, $\mathbf{ND}(\mathbf{Y})$ is the set of all vertices that do not have a descendant in $\mathbf{Y}$. If $\mathbf{Y} \cap \mathbf{Z} = \varnothing$, then V is in $\mathbf{IV}(\mathbf{Y},\mathbf{Z})$ (informative variables for $\mathbf{Y}$ given $\mathbf{Z}$) if and only if V is d-connected to $\mathbf{Y}$ given $\mathbf{Z}$, and V is not in $\mathbf{ND}(\mathbf{YZ})$. (Note that this entails that V is not in $\mathbf{Y} \cup \mathbf{Z}$.) If $\mathbf{Y} \cap \mathbf{Z} = \varnothing$, W is in $\mathbf{IP}(\mathbf{Y},\mathbf{Z})$ (W has a parent who is an informative variable for $\mathbf{Y}$ given $\mathbf{Z}$) if and only if W is a member of $\mathbf{Z}$, and W has a parent in $\mathbf{IV}(\mathbf{Y},\mathbf{Z}) \cup \mathbf{Y}$. We will use the following result.

THEOREM 7.1: If G_{Comb} is a directed acyclic graph over $\mathbf{V} \cup \mathbf{W}$, $\mathbf{W}$ is exogenous with respect to $\mathbf{V}$ in G_{Comb}, $\mathbf{Y}$ and $\mathbf{Z}$ are disjoint subsets of $\mathbf{V}$, $P(\mathbf{V} \cup \mathbf{W})$ is a distribution that satisfies the Markov condition for G_{Comb}, no member of $\mathbf{X} \cap \mathbf{Z}$ is a member of $\mathbf{IP}(\mathbf{Y},\mathbf{Z})$ in G_{Unman}, and no member of $\mathbf{X}\backslash\mathbf{Z}$ is a member of $\mathbf{IV}(\mathbf{Y},\mathbf{Z})$ in G_{Unman}, then $P(\mathbf{Y}|\mathbf{Z})$ is invariant under a direct manipulation of $\mathbf{X}$ in G_{Comb} by changing $\mathbf{W}$ from $\mathbf{w}_1$ to $\mathbf{w}_2$.

The importance of theorem 7.1 is that whether $P(\mathbf{Y}|\mathbf{Z})$ is invariant under a direct manipulation of $\mathbf{X}$ in G_{Comb} by changing $\mathbf{W}$ from $\mathbf{w}_1$ to $\mathbf{w}_2$ is determined by properties of G_{Unman} alone. Therefore, we will sometimes speak of the invariance of $P(\mathbf{Y}|\mathbf{Z})$ under a direct manipulation of $\mathbf{X}$ in G_{Unman} without specifying $\mathbf{W}$ or G_{Comb}. (As the proofs show, a simpler but equivalent way of formulating theorem 7.1 is that $P(\mathbf{Y}|\mathbf{Z})$ is invariant under manipulation of $\mathbf{X}$ when $\mathbf{Y}$ is d-separated from the policy variables given $\mathbf{Z}$.)

Each of the preceding examples, and Pratt and Schlaifer's general rule, are consequences of a corollary to theorem 7.1:

COROLLARY 7.1: If G_{Comb} is a directed acyclic graph over $\mathbf{V} \cup \mathbf{W}$, $\mathbf{W}$ is exogenous with respect to $\mathbf{V}$ in G_{Comb}, X and Y are in $\mathbf{V}$, and $P(\mathbf{V} \cup \mathbf{W})$ is a distribution that satisfies the Markov condition for G_{Comb}, then $P(Y|X)$ is invariant under direct manipulation of X in G_{Comb} by changing $\mathbf{W}$ from $\mathbf{w}_1$ to $\mathbf{w}_2$ if in G_{Unman} no undirected path *into* X d-connects X and Y given the empty set of vertices. Equivalently, if (1) Y is not a (direct or indirect) cause of X, and (2) there is no common cause of X and Y in G_{Unman}.

In graphical terms, Pratt and Schlaifer's claim amounts to requiring that for "observability" (invariance) G and G'—the graph of a manipulated system obtained by removing from G all edges into X—and their associated probabilities must give the same conditional distribution of Y on X. Corollary 7.1 characterizes the sufficiency side of this claim. Pratt and Schlaifer say their condition is "almost necessary." What they mean, we take it, is that there are cases in which the antecedent of their condition fails to hold and the consequent does hold, and, furthermore, that when the antecedent fails to hold the consequent will not hold unless a special constraint is satisfied by the conditional probabilities. Parallel remarks apply to the graphical condition we have given. There exist cases in which there are d-connecting paths between X and Y given the empty set that are into X and the probability of Y when X is directly manipulated is equal to the original conditional probability of Y on X. Again the antecedent will fail and the consequent will hold only if a constraint is satisfied by the conditional probabilities, so the condition is "almost necessary."

It may happen that the distribution of Y when a value is forced on X cannot be predicted from the unforced conditional distribution of Y on X but, nonetheless, the conditional distribution of Y on $\mathbf{Z}$ when a value is forced on X can be predicted from the unforced conditional distribution of Y on X and $\mathbf{Z}$. Pratt and Schlaifer consider the general case in which, besides X and Y, some further variables $\mathbf{Z}$ are measured. Pratt and Schlaifer say that the law relating Y to X is "observable with concomitant $\mathbf{Z}$" when the unforced conditional distribution of Y on X and $\mathbf{Z}$ equals the conditional distribution of Y on $\mathbf{Z}$ in the population in which $\mathbf{X}$ is forced to have a particular value.

Pratt and Schlaifer claim sufficient and "almost necessary" conditions for observability with concomitants, namely that *for any value x of X the distribution of X be independent of the conditional distribution of $Y_{Xf=x}$ on the value of z of $Z_{Xf=x}$ when X is forced to have the value x*. This rule, too, is a special case of theorem 7.1.

Consider an example due to Rubin. (Rubin's X is Pratt and Schlaifer's Z; Rubin's T is Pratt and Schlaifer's X). In an educational experiment in which reading program assignments T are assigned on the basis of a randomly sampled value of some pretest variable X which shares one or more unmeasured common causes, V, with Y, the score on a post-test, we wish to predict the average difference τ in Y values if all students in the population were given treatment $T = 1$ as against if all students were given treatment $T = 2$. The situation in the experiment is represented in figure 7.5.

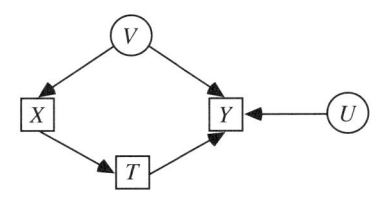

Figure 7.5

Provided the experimental sample is sufficiently representative, Rubin says that an unbiased estimate of τ can be obtained as follows: Let k range over values of X, from 1 to K, let $\overline{Y1k}$ be the average value of Y conditional on $T = 1$ and $X = k$, and analogously for $\overline{Y2k}$. Let $n1k$ be the number of units in the sample with $T = 1$ and $X = k$, and analogously for *n2k*. The numbers *n1* and *n2* represent the total number of units in the sample with $T = 1$ and $T = 2$ respectively.

Let $Y_{Tf=1}$ = expected value of Y if treatment 1 is forced on all units. According to Rubin, estimate $Y_{Tf=1}$ by:

$$\sum_{k=1}^{K} \frac{n1k + n2k}{n1 + n2} \overline{Y1k}$$

and estimate τ by:

$$\sum_{k=1}^{K} \frac{n1k + n2k}{n1 + n2} \left(\overline{Y1k} - \overline{Y2k} \right)$$

The basis for this choice may not be apparent. If we look at the hypothetical population in which every unit is forced to have $T = 1$, then it is clear from Rubin's tacit independence assumptions that he treats the manipulated population as if it had the causal structure shown in figure 7.6, as the following derivation shows.

$$\overline{Y}_{Tf=1} = \sum_{Y} \vec{Y} \times P(Y_{Tf=1}) =$$

$$\sum_{Y} \vec{Y} \times \sum_{k=1}^{K} P(Y_{Tf=1} | X_{Tf=1} = k, T_{Tf=1} = 1) P(X_{Tf=1} = k | T_{Tf=1} = 1) P(T_{Tf=1} = 1) =$$

$$\sum_{Y} \vec{Y} \times \sum_{k=1}^{K} P(Y_{Tf=1} | X_{Tf=1} = k, T_{Tf=1} = 1) P(X_{Tf=1} = k)$$

The second equality in the above equations hold because $P(T_{Tf=1} = 1) = 1$, and $X_{Tf=1}$ and $T_{Tf=1}$ are independent according to the causal graph shown in figure 7.6. By theorem 7.1, both $P(Y_{Tf=1}|X_{Tf=1},T_{Tf=1})$ and $P(X_{Tf=1})$ are invariant under direct manipulation of T in the graph of figure 7.5. This entails the following equation.

$$\overline{Y}_{Tf=1} = \sum_{Y} \vec{Y} \times \sum_{k=1}^{K} P(Y_{Tf=1} | X_{Tf=1} = k, T_{Tf=1} = 1) P(X_{Tf=1} = k) =$$

$$\sum_{k=1}^{K} P(X = k) \times \sum_{Y} \vec{Y} \times P(Y | X = k, T = 1) = \frac{n1k + n2k}{n1 + n2} \times \overline{Y1k}$$

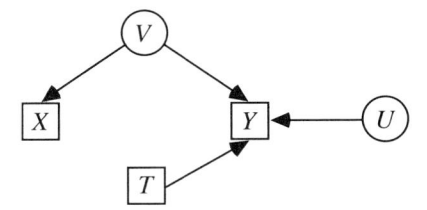

Figure 7.6

Note that X and T, unlike $X_{Tf=1}$ and $T_{Tf=1}$ are *not* independent. Rubin's tacit assumption that $X_{Tf=1}$ and $T_{Tf=1}$ are independent indicates that he is implicitly assuming that the causal graph of the manipulated population is the graph of figure 7.6, not the graph of figure 7.5, which is the causal structure of the unmanipulated population. $\overline{Y}_{Tf=2}$ can be derived in an analogous fashion.

The reconstruction we have given to Rubin's theory assumes that all units in the population have the same causal structure for the relevant variables, but not, of course, that the units are otherwise homogenous. It is conceivable that someone might know the counterfactuals required for prediction according to the Pratt and Schlaifer rules even though the relevant causal structure in the population (and in the sample from which inferences are to be made) differs from unit to unit. For example, it might somehow be known that A and B have no unmeasured common cause and that B does not cause A, and the population might in fact be a mixture of systems in which A causes B and systems in which A and B are independent. In that case the distribution of B if A is forced to have the value $A = a$ can be predicted from the conditional probability of B given $A = a$, indeed the probabilities are the same. For this, and analogously for other cases of prediction for populations with a mixture of causal structures, the predictions obtained by applying Pratt and Schlaifer's rule can be derived from the Markov Condition by considering whether the relevant conditional probabilities are invariant in each of the causally homogenous subpopulations. Thus if A and B have no causal connection, $P(B|A = a)$ equals the probability of B when A is forced to have value a, and if A causes B, $P(B|A = a)$ equals the probability of B when A is forced to have value a, and so the probability is also the same in any mixture of systems with these two causal structures.

7.4 Prediction with Causal Sufficiency

The Rubin framework is specialized in two dimensions. It assumes known various counterfactual (or causal) properties, and it addresses *invariance* of conditional probability. But we very often don't know the causal structure or the counterfactuals before considering the data, and we are interested not in invariance *per se* but only as an instrument in prediction. We need to be clearer about the goal. We suppose that the investigator knows (or estimates) a distribution $P_{Unman}(\mathbf{O})$ which is the marginal over $\mathbf{O}$ of a distribution faithful to an unknown causal graph G_{Unman}, with unknown vertex set $\mathbf{V}$ that includes $\mathbf{O}$. She also knows the variable, X, that is the member of $\mathbf{O}$ that will be directly manipulated, and the variables **Parents**(G_{Man},X) that will be direct causes of X in G_{Man}. She knows that X is the only variable directly manipulated. Finally she knows what the manipulation will do to X, that is, she knows $P_{Man}(X \mid \textbf{Parents}(G_{Man},X))$. The distribution of **Y** conditional on **Z** is **predictable** if in these circumstances $P_{Man}(\textbf{Y}|\textbf{Z})$ is uniquely determined no matter what the unknown causal graph, no matter what the manipulated and unmanipulated distributions over unobserved variables, and no matter how the manipulation is brought about consistent with the assumptions just specified. The goal is to discover when the distribution of **Y** conditional on **Z** is predictable, and how to obtain a prediction.

The assumption that $P_{Unman}(\textbf{O})$ is the marginal over **O** of a distribution faithful to the unmanipulated graph G_{Unman} may fail for several reasons. First, it may fail because of the particular parameters values of the distribution. If **W** is a set of policy variables, it also may fail because the $\textbf{w}_2$ (manipulated) subpopulation contains dependencies that are not in the $\textbf{w}_1$ (unmanipulated) subpopulation. For example, suppose that a battery is connected to a light bulb by a circuit that contains a switch. Let W be the state of the switch, w_1 be the unmanipulated subpopulation where the switch is off and w_2 be the manipulated subpopulation where the switch is on. In the w_1 subpopulation the state of the light bulb (on or off) is independent of the state of the battery (charged or not) because the bulb is always off. On the other hand in the w_2 subpopulation the state of the light bulb is dependent on the state of the battery. Hence in G_{Comb} there is an edge from the state of the battery to the state of the light bulb; it follows that there is also an edge from the state of the battery to the state of the light bulb in G_{Unman} (which is the subgraph of G_{Comb} that leaves out W.) This implies that the joint distribution over the state of the battery and the state of the light bulb in the w_1 subpopulation is not faithful to G_{Unman}. The results of the Prediction Algorithm are reliable only in circumstances where a manipulation does not introduce additional dependencies (which may or may not be part of one's background knowledge.)

Suppose we wish to make a prediction of the effect of an intervention or policy from observations of variables correctly believed to be causally sufficient for systems with a common but unknown causal structure. In that case the sample and the PC (or some other) algorithm determine a pattern representing a class of directed graphs, and properties of that class determine whether the distribution of Y following a direct manipulation of X can be predicted. Suppose for example that the pattern is X - Y - Z which represents the set of graphs in figure 7.7.

Prediction

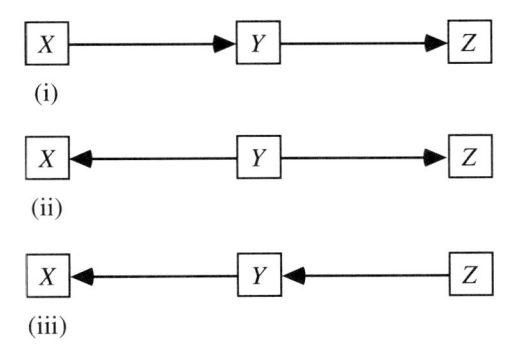

```
(i)   X ──────► Y ──────► Z

(ii)  X ◄────── Y ──────► Z

(iii) X ◄────── Y ◄────── Z
```

Figure 7.7

For each of these causal graphs, the distribution of Y after a direct manipulation of X can be calculated, but the result is different for the first graph than for the two others. $P_{Man}(Y)$ for each of the graphs can be calculated from the Manipulation Theorem and taking the appropriate marginal; the results for each graph are given below:

(i) $P_{Man}(Y) = \sum_{X} P_{Unman}(Y|X) \overrightarrow{P_{Man}(X)} \neq P_{Unman}(Y)$

(ii) $P_{Man}(Y) = P_{Unman}(Y)$

(iii) $P_{Man}(Y) = \sum_{Z} P_{Unman}(Y|Z) \overrightarrow{P_{Unman}(Z)} = P_{Unman}(Y)$

If every unit in the population is forced to have the same value of X, then for (i) the manipulated distribution of Y does not equal the unmanipulated distribution of Y. For (ii) and (iii) the manipulated distribution of Y equals the unmanipulated distribution. Since the pattern does not tell us which of these structures is correct, the distribution of Y on a manipulation of X cannot be predicted.

If a different pattern had been obtained a prediction would have been possible; for example the pattern $U - X \rightarrow Y \leftarrow Z$ can represent either of the graphs in figure 7.8.

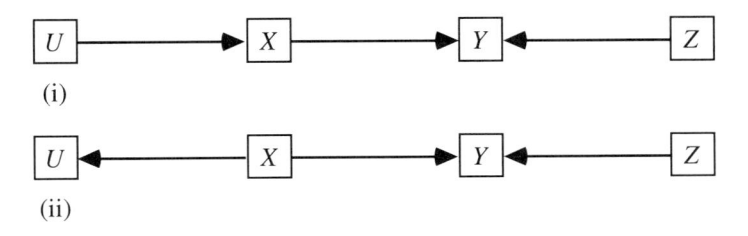

```
(i)  U ──────► X ──────► Y ◄────── Z

(ii) U ◄────── X ──────► Y ◄────── Z
```

Figure 7.8

$P_{Man}(Y)$ for each of the graphs can be calculated from the Manipulation Theorem and taking the appropriate marginal; the results for each graph are given below:

(i) $P_{Man}(Y) = \sum_{X} P_{Unman}(Y|\vec{X})P_{Man}(X)$

(ii) $P_{Man}(Y) = \sum_{X} P_{Unman}(Y|\vec{X})P_{Man}(X)$

(Note, however, that while $P_{Man}(Y)$ is the same for (i) and (ii), $P_{Man}(U,X,Y,Z)$ is not the same for (i) and (ii), so $P_{Man}(U,X,Y,Z)$ is not predictable.)

When it is known that the structure is causally sufficient, we can decide the predictability of the distribution of a variable (or conditional distribution of one set of variables on another set) by finding the pattern and applying the Manipulation Theorem and taking the appropriate marginal for every graph represented by the pattern. If all graphs give the same result, that is the prediction. Various computational shortcuts are possible, some of which are described in the Prediction Algorithm stated in the next section.

7.5 Prediction without Causal Sufficiency

We come finally to the most serious case, in which for all we know the causal structure of the manipulated systems will be different from the causal structure of the observed systems, the causal structure of the observed systems is unknown, and for all we know the observed statistical dependencies may be due to unobserved common causes. This is the situation that Mosteller and Tukey seem to think typical in non-experimental studies, and we agree. The question is whether, nonetheless, prediction is sometimes possible, and if so when and how.

Consider the following trivial example. If we have measured only smoking and lung cancer, we will find that they are correlated. The correlation could be produced by any of the three causal graphs depicted in figure 7.9.

Prediction 171

```
        (Genotype)
         /      \
        v        v
   [Smoking]  [Lung cancer]
           (i)

   [Smoking] ———→ [Lung cancer]
           (ii)

   [Smoking] ←——— [Lung cancer]
           (iii)
```

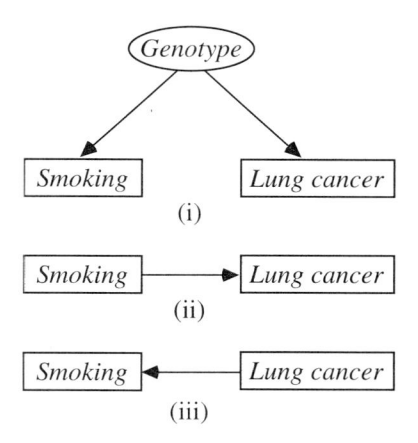

Figure 7.9

All three graphs yield the same maximally informative partially oriented inducing path graph, shown in figure 7.10.

```
   [Smoking] o————————o [Lung cancer]
```

Figure 7.10

If *Smoking* is directly manipulated in graphs (i) or (iii), then *P(Lung cancer)* will not change; but if *Smoking* is directly manipulated in graph (ii) then *P(Lung cancer)* will change. So it is not possible to predict the effects of the direct manipulation of *Smoking* from the marginal distribution of the measured variables.

In the causally sufficient case each complete orientation of the pattern yields a directed acyclic graph G. According to the Manipulation Theorem, for each directed acyclic graph G_{Unman} when we factor the distribution into a product of terms of the form $P_{Unman(\mathbf{W})}(V \mid \mathbf{Parents}(G_{Unman}, V))$ we can calculate the effect of manipulating a variable X simply by replacing $P_{Unman(\mathbf{W})}(X \mid \mathbf{Parents}(G_{Unman}, X))$ with $P_{Man(\mathbf{W})}(X \mid \mathbf{Parents}(G_{Man}, X))$ (where G_{Man} is the manipulated graph). This simple substitution works because each of the terms in the factorization other than $P_{Unman(\mathbf{W})}(X \mid \mathbf{Parents}(G_{Unman}, X))$ is guaranteed to be invariant under any direct manipulation of X in G_{Unman}, and hence can be estimated from frequencies in the unmanipulated population.

Let us now try and generalize this strategy to the causally nonsufficient case, where $P(\mathbf{O})$ is the marginal of a distribution $P(\mathbf{V})$ that is faithful to a directed acyclic graph G_{Unman}, and π is the partially oriented inducing path graph of G_{Unman}. We could search for a factorization of the distribution of $P(\mathbf{O})$ that is a product of terms of the form $P_{Unman}(V \mid \mathbf{M}(V))$ (where membership in the set $\mathbf{M}(V)$ is a function of V) in which each of the terms except $P_{Unman}(X \mid \mathbf{M}(X))$ is invariant under all direct manipulations of X in all directed acyclic graphs for which π is a partially oriented inducing path graph over $\mathbf{O}$. If we find such a factorization, then we can predict the effect of the manipulation by substituting the term $P_{Man}(X \mid \mathbf{Parents}(G_{Man},X))$ for $P_{Unman}(V \mid \mathbf{M}(X))$ where G_{Man} is the manipulated graph), just as we did in the causally sufficient case. We will not know which of the many directed acyclic graphs for which π is a partially oriented inducing path graph over $\mathbf{O}$ actually generated the distribution; however, it will not matter, because $P_{Man}(\mathbf{Y}|\mathbf{Z})$ will be the same for each of them. This is essentially the strategy that we adopt. However, the task of finding such a factorization is considerably more difficult in the causally nonsufficient case: unlike the causally sufficient case where we can simply construct a factorization in which each term except $P(X \mid \mathbf{Parents}(G_{Unman},X))$ is invariant under direct manipulation of $\mathbf{X}$ in G_{Unman}, in the causally nonsufficient case we have to *search* among different factorizations in order to find a factorization in which each term except $P_{Unman}(X \mid \mathbf{M}(X))$ is invariant under all direct manipulations of X for all directed acyclic graphs G that have partially oriented inducing path graph over $\mathbf{O}$ equal to π. Fortunately, as we will see, we do not have to search though every possible factorization of $P(\mathbf{O})$.

We will flesh out the details of this strategy and provide examples. We will use the FCI algorithm to construct a partially oriented inducing path graph π over $\mathbf{O}$ of G_{Unman}. Note that in view of Verma and Pearl's example described in chapter 6, it may be that some graphs for which π is a partially oriented inducing path graph over $\mathbf{O}$ may not represent any distribution with marginal $P_{Unman}(\mathbf{O})$ because of nonindependence constraints. From the theory developed in this book, we cannot hope to provide a computational procedure that decides predictability and obtains predictions whenever they are possible in principle, because we have no understanding of all constraints that graphs may entail for marginal distributions. But by considering only conditional independence constraints we can provide a sufficient condition for predictability.

Here is an example that provides a more detailed illustration of the strategy: Suppose we measure *Genotype* (*G*), *Smoking* (*S*), *Income* (*I*), *Parents' smoking habits* (*PSH*) and *Lung cancer* (*L*). Suppose the unmanipulated distribution is faithful to the unmanipulated graph that has the partially oriented inducing path graph shown in figure 7.11.

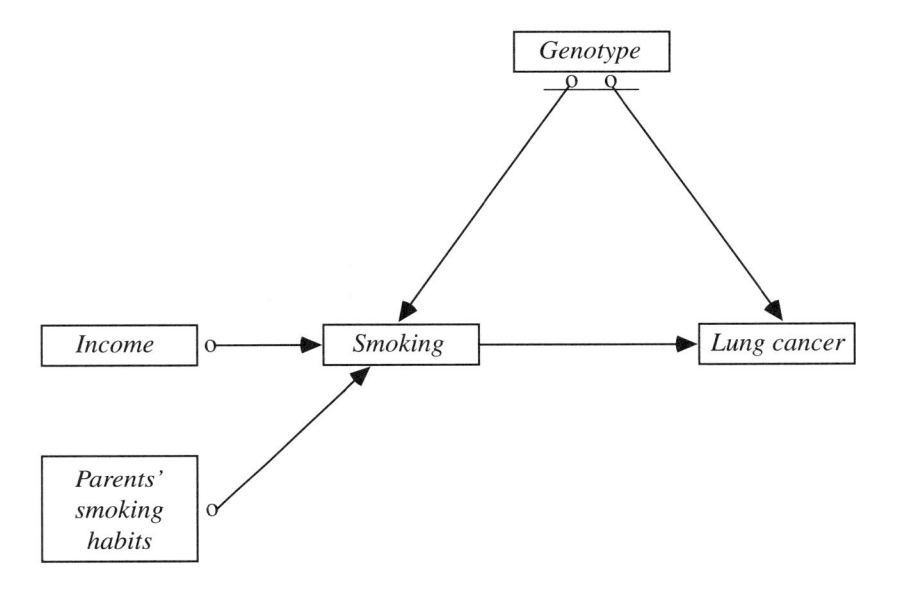

Figure 7.11

The partially oriented inducing path graph does not tell us whether *Income* and *Smoking* have a common unmeasured cause, or *Parents' smoking habits* and *Smoking* have a common unmeasured cause, and so on. The measured distribution might be produced by any of several structures, including, for example those in figure 7.12, where T_1 and T_2 are unmeasured.

If we directly manipulate *Smoking* so that *Income* and *Parents' smoking habits* are not parents of *Smoking* in the manipulated graph, then no matter which graph produced the marginal distribution, the partially oriented inducing path graph and the Manipulation Theorem tell us that if *Smoking* is directly manipulated then in the manipulated population the resulting causal graph will look like the graph shown in figure 7.13.

In this case, we can determine the distribution of *Lung cancer* given a direct manipulation of *Smoking*. Three steps are involved. Here, we simply give the results of carrying out each step. How each step is carried out is explained in more detail in the next section.

First, from the partially oriented inducing path graph we find a way to factor the joint distribution in the manipulated graph. Let P_{Unman} be the distribution on the measured

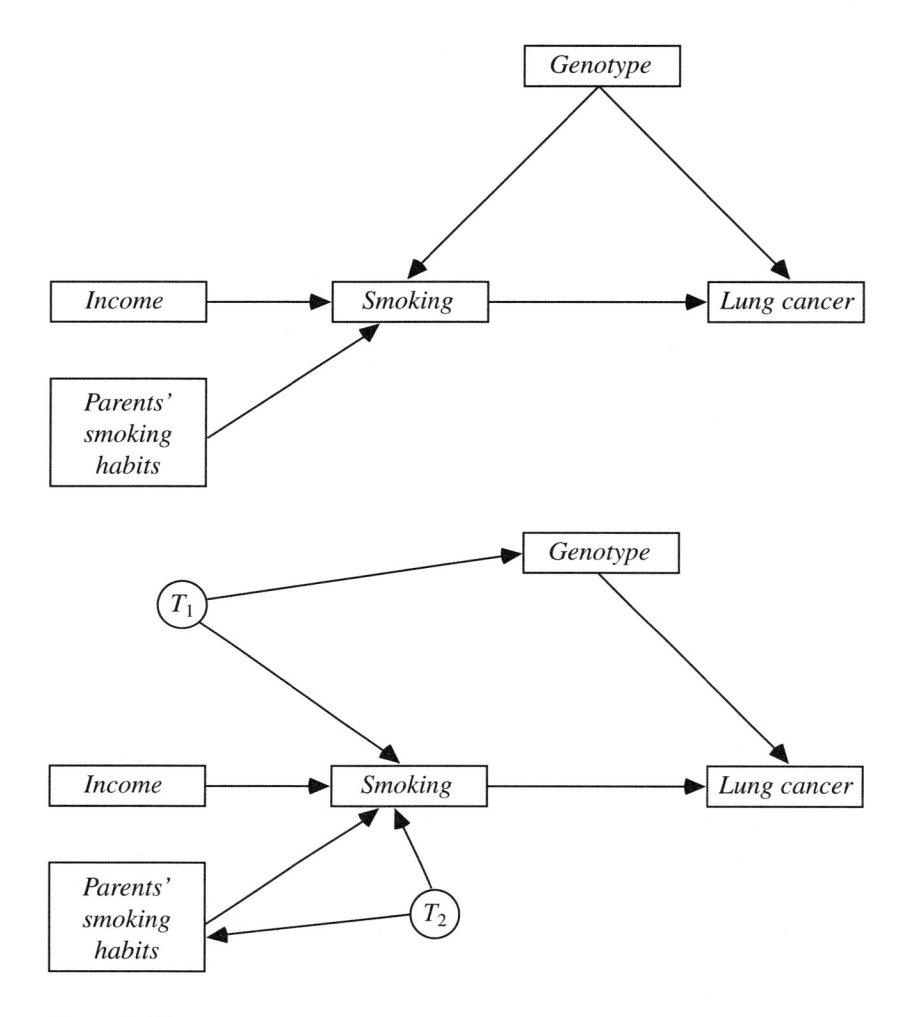

Figure 7.12

variables and let P_{Man} be the distribution that results from a direct manipulation of *Smoking*. It can be determined from the partially oriented inducing path graph that

$$P_{Man}(I, PSH, S, G, L) = P_{Man}(I) \times P_{Man}(PSH) \times P_{Man}(S) \times P_{Man}(G) \times P_{Man}(L \mid G, S)$$

Prediction

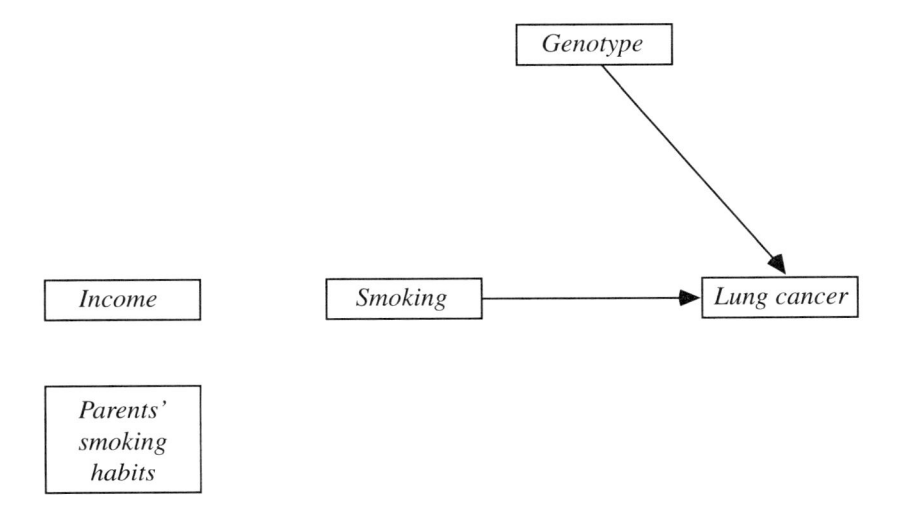

```
                    ┌──────────┐
                    │ Genotype │
                    └────┬─────┘
                         │
                         ▼
┌────────┐   ┌─────────┐   ┌─────────────┐
│ Income │   │ Smoking │──▶│ Lung cancer │
└────────┘   └─────────┘   └─────────────┘

┌──────────┐
│ Parents' │
│ smoking  │
│ habits   │
└──────────┘
```

Figure 7.13

where I = Income, PSH = Parents' smoking habits, S = Smoking, G = Genotype, and L = Lung cancer. This is the factorization of P_{Man} corresponding to the immediately preceding graph that represents the result of a direct manipulation of Smoking.

Second, we can determine from the partially oriented inducing path graph which factors in the expression just given for the joint distribution are needed to calculate $P_{Man}(L)$. In this case $P_{Man}(I)$ and $P_{Man}(PSH)$ prove irrelevant and we have:

$$P_{Man}(L) = \sum_{G,S}^{\rightarrow} P_{Man}(S) \times P_{Man}(G) \times P_{Man}(L|G,S)$$

Third, we can determine from the partially oriented inducing path graph that $P_{Man}(G)$ and $P_{Man}(L|G,S)$ are equal respectively to the corresponding unmanipulated probabilities, $P_{Unman}(G)$ and $P_{Unman}(L|G,S)$. Furthermore, $P_{Man}(S)$ is assumed to be known, since it is the quantity being manipulated. Hence, all three factors in the expression for $P_{Man}(L)$ are known, and $P_{Man}(L)$ can be calculated.

Note that $P_{Man}(L)$ can be predicted even though $P(L)$ is most definitely not invariant under a direct manipulation of S. The example should be enough to show that while Mosteller and Tukey's pessimism about prediction from observation may have been justified when they wrote, it was not well-founded.

The algorithm sketched in the example is described more formally below, where we have labeled each step by a letter for easy reference. Suppose $P_{Unman}(\mathbf{V})$ is the distribution before the manipulation, $P_{Man}(\mathbf{V})$ the manipulation after the distribution, and a single variable X in $\mathbf{X}$ is manipulated to have distribution

$P_{Man}(X \mid \mathbf{Parents}(G_{Man},X))$, where G_{Man} is the manipulated graph. We assume that $P_{Unman}(\mathbf{V})$ is faithful to the unmanipulaated graph G_{Unman}, that $\mathbf{Parents}(G_{Man},X)$ is known, that $P_{Man}(X \mid \mathbf{Parents}(G_{Man},X))$ is known, and that we are interested in predicting $P_{Man}(\mathbf{Y}|\mathbf{Z})$. The Prediction Algorithm is simplified by the fact that if $P_{Unman}(\mathbf{O})$ satisfies the Markov Condition for a graph G_{Unman}, then so does $P_{Man}(\mathbf{O})$, and hence any factorized expression for $P_{Unman}(\mathbf{Y}|\mathbf{Z})$ is also an expression for $P_{Man}(\mathbf{Y}|\mathbf{Z})$. Recall that a total order Ord of variables in a graph G' is **acceptable** for G' if and only if whenever $A \neq B$ and there is a directed path from A to B in G', A precedes B in Ord. If π is the FCI partially oriented inducing path graph of G over $\mathbf{O}$, then X is in **Definite-Nondescendants**$(\mathbf{Y})$ if and only if there is no semidirected path from any Y in $\mathbf{Y}$ to X in π. Recall that a directed acyclic graph G is a minimal I-map of distribution P if and only if P satisfies the Markov and Minimality Conditions for G.

Prediction Algorithm

A.) $P_{Man}(\mathbf{Y}|\mathbf{Z})$ = unknown.
B.) Generate partially oriented inducing path graph π from $P_{Unman}(\mathbf{O})$.
C.) For each ordering of variables acceptable for π in which the predecessors of X in Ord equals $\mathbf{Parents}(G_{Man},X) \cup \mathbf{Definite\text{-}Nondescendants}(X)$
 C1.) Form the minimal I-map F of $P_{Unman}(\mathbf{O})$ for that ordering;
 C2.) Extract an expression for $P_{Unman}(\mathbf{Y}|\mathbf{Z})$ from F; call it E;
 C3.) If for each $V \neq X$, the term $P_{Unman}(V|\mathbf{Parents}(F,V))$ in E is invariant in G_{Man} when X is directly manipulated then
  C3a). return $P_{Man}(\mathbf{Y}|\mathbf{Z}) = E'$, where E' is equal to E except that $P_{Unman}(X \mid \mathbf{Parents}(F,X))$ is replaced by $P_{Man}(X \mid \mathbf{Parents}(G_{Man},X))$
  C3b). exit

(The algorithm can also be applied to the case where a set $\mathbf{X}$ of variables is manipulated, as long as it is possible to find an ordering of variables such that for each X in $\mathbf{X}$ all of the predecessors of X are in **Definite-Nondescendants**(X) or $\mathbf{Parents}(G_{Man},X)$, there are no causal connections among the variables in $\mathbf{X}$, and if some X in $\mathbf{X}$ is a parent of some variable V not in $\mathbf{X}$, then every member of $\mathbf{X}$ is a predecessor of V.) The description leaves out important details. How can we find the partially oriented inducing path graph (step B), the graph for which $P_{Unman}(\mathbf{V})$ satisfies the Minimality and Markov conditions for a given ordering of variables (step C1), the expression E for $P_{Man}(\mathbf{Y}|\mathbf{Z})$ (step C2); how do we determine if a given conditional probability term that appears in the expression for $P_{Unman}(\mathbf{Y}|\mathbf{Z})$ is invariant under a direct manipulation of X in G_{Unman} when we do not know what G_{Unman} is (step C3)? The details are described below.
Step B: We carry out step B) with the FCI Algorithm.

Step C: Say steps C1) and C2) are *successful* if they produce an expression for $P_{Unman}(\mathbf{Y}|\mathbf{Z})$ in which for every V in $\mathbf{O}\backslash\{X\}$, $P_{Unman}(V \mid \textbf{Parents}(F,V))$ is invariant under direct manipulation of X in G_{Unman}. We conjecture that if there is an ordering of variables for which some directed acyclic graph makes C1) and C2) successful, then there is such an ordering that is acceptable for π. (Notice that the correctness of the algorithm does not depend upon the correctness of this conjecture, although if it is wrong the algorithm will be less informative than some other algorithm that searches a larger set of variable orderings.)

Step C1: For a given ordering Ord, let **Predecessors**(Ord,V) be the predecessors of V in Ord. For each V in F over $\mathbf{O}$, let **Parents**(V) be the smallest subset of **Predecessors**(V) such that V is independent of **Predecessors**$(Ord,V)\backslash$**Parents**(V) given **Parents**(V). Then F is a minimal I-map of $P(\mathbf{O})$. See Pearl 1988. Under the assumption that $P(\mathbf{O})$ is the marginal of a faithful distribution $P(\mathbf{V})$ we can test whether V is independent of **Predecessors**$(Ord,V)\backslash$**Parents**(V) given **Parents**(V) by testing whether each member of **Predecessors**$(Ord,V)\backslash$**Parents**(V) is independent of V given **Parents**(V). This clearly suggests testing whether small sets of variables are equal to **Parents**(V) first.

For inducing path graph G' and acceptable total ordering Ord, W is in **SP**(Ord,G',V) (separating predecessors of V in G' for ordering Ord) if and only if W precedes V in Ord and there is an undirected path U between W and V such that each vertex on U except for the endpoints precedes V in Ord and is a collider on U. If G is a directed acyclic graph over $\mathbf{V}$, G_{IP} is the inducing path graph of G over $\mathbf{O}$, Ord is an ordering acceptable for G_{IP}, and $P(\mathbf{V})$ is faithful to G, then the directed acyclic graph G_{Min} in which for each X in $\mathbf{O}$ **Parents**$(X) = $ **SP**(Ord,X) is a minimal I-map of $P(\mathbf{O})$. Of course we are not generally given G_{IP}. However, we can construct a partially oriented inducing path graph and identify sets of variables that narrow down the search for **SP**(Ord,X). For a partially oriented inducing path graph π and ordering Ord acceptable for π, let V be in **Possible-SP**(Ord,X) if and only if $V \neq X$ and there is an undirected path U in π between V and X such that every vertex on U except for X is a predecessor of X in Ord, and no vertex on U except for the endpoints is a definite-noncollider on U. For a partially oriented inducing path graph π over $\mathbf{O}$ and ordering Ord acceptable for π, V is in **Definite-SP**(Ord,X) if and only if $V \neq X$ and there is an undirected path U in π between V and X such that every vertex on U except for X is a predecessor of X in Ord, and every vertex on U except for the endpoints is a collider on U.

THEOREM 7.2: If $P(\mathbf{O})$ is the marginal of a distribution faithful to G over $\mathbf{V}$, π is a partially oriented inducing path graph of G over $\mathbf{O}$, and Ord is an ordering of variables in $\mathbf{O}$ acceptable for some inducing path graph over $\mathbf{O}$ with partially oriented inducing path graph π, then there is a minimal I-map G_{Min} of $P(\mathbf{O})$ in which **Definite-SP**(Ord,X) in π is included in **Parents**(G_{Min},X) which is included in **Possible-SP**(Ord,X) in π.

We can use theorem 7.2 as a heuristic for searching for a minimal I-map of $P(\mathbf{O})$. The procedure is only a heuristic for the following reason. While from π we can identify orderings that are not acceptable for any inducing path graph over $\mathbf{O}$ with partially oriented inducing path graph π, we cannot always definitely tell that some ordering acceptable for π is acceptable for some inducing path graph over $\mathbf{O}$ with partially oriented inducing path graph π. For orderings not acceptable for any such inducing path graph over $\mathbf{O}$, it is possible that making $\mathbf{SP}(Ord,X)$ the parents of X in G_{Min} does not make G_{Min} a minimal I-map, in which case it may be that no set $\mathbf{M}$ including **Definite-SP**(Ord,X) and included in **Possible-SP**(Ord,X) makes **Predecessors**$(Ord,V)\backslash\mathbf{M}$ independent of X given $\mathbf{M}$. If that is the case, we must conduct a wider search.

Step C2: If P satisfies the Markov condition for directed acyclic graph G, the following lemma shows how to determine an expression E for $P(\mathbf{Y}|\mathbf{Z})$. (For a related result see Geiger, Verma, and Pearl 1990)

LEMMA 3.3.5: If P satisfies the Markov condition for directed acyclic graph G over $\mathbf{V}$, then

$$P(\mathbf{Y}|\mathbf{Z}) = \frac{\sum_{\overrightarrow{\mathbf{IV}(\mathbf{Y},\mathbf{Z})}} \prod_{W \in \mathbf{IV}(\mathbf{Y},\mathbf{Z}) \cup \mathbf{IP}(\mathbf{Y},\mathbf{Z}) \cup \mathbf{Y}} P(W|\mathbf{Parents}(W))}{\sum_{\overrightarrow{\mathbf{IV}(\mathbf{Y},\mathbf{Z}) \cup \mathbf{Y}}} \prod_{W \in \mathbf{IV}(\mathbf{Y},\mathbf{Z}) \cup \mathbf{IP}(\mathbf{Y},\mathbf{Z}) \cup \mathbf{Y}} P(W|\mathbf{Parents}(W))}$$

for all values of $\mathbf{V}$ for which the conditional distributions in the factorization are defined, and for which $P(\mathbf{z}) \neq 0$.

Step C3: We use theorems 7.3 and 7.4 below to determine from π whether a given conditional distribution is invariant under a direct manipulation of X in G_{Unman}. If π is a partially oriented inducing path graph over $\mathbf{O}$, then a vertex B on an undirected path U in a partially oriented inducing path graph π over $\mathbf{O}$ is a **definite noncollider** on U if and only if B is an endpoint of U or there are edges A *-* $\underline{B}$ *-* C, A *-* $B \rightarrow C$, or $A \leftarrow B$ *-* * C on U. If $A \neq B$, and A and B are not in $\mathbf{Z}$, then an undirected path U between A and B in a partially oriented inducing path graph π over $\mathbf{O}$ is a **possibly d-connecting** path between A and B given $\mathbf{Z}$ if and only if every collider on U is the source of a semidirected path to a member of $\mathbf{Z}$, and every definite noncollider is not in $\mathbf{Z}$. If $\mathbf{Y}$ and $\mathbf{Z}$ are disjoint, then X is in **Possibly-IP**$(\mathbf{Y},\mathbf{Z})$ if and only if X is in $\mathbf{Z}$, and there is a possibly d-connecting path between X and some Y in $\mathbf{Y}$ given $\mathbf{Z}\backslash\{X\}$ that is not out of X. If $\mathbf{Y}$ and $\mathbf{Z}$ are disjoint, X is in **Possibly-IV**$(\mathbf{Y},\mathbf{Z})$ if and only if X is not in $\mathbf{Z}$, there is a possibly d-connecting path between X and some Y in $\mathbf{Y}$ given $\mathbf{Z}$, and there is a semidirected path from X to a member of $\mathbf{Y} \cup \mathbf{Z}$. Note that theorems 7.3 and 7.4 also entail that if there is a directed acyclic graph G for which an ordering of variables is acceptable that makes steps C1 and C2 successful, then so does the minimal I-map for which that ordering is acceptable.

THEOREM 7.3: If G is a directed acyclic graph over $\mathbf{V} \cup \mathbf{W}$, $\mathbf{W}$ is exogenous with respect to $\mathbf{V}$ in G, $\mathbf{O}$ is included in $\mathbf{V}$, G_{Unman} is the subgraph of G over $\mathbf{V}$, π is the FCI partially oriented inducing path graph over $\mathbf{O}$ of G_{Unman}, $\mathbf{Y}$ and $\mathbf{Z}$ are included in $\mathbf{O}$, $\mathbf{X}$ is included in $\mathbf{Z}$, $\mathbf{Y}$ and $\mathbf{Z}$ are disjoint, and no X in $\mathbf{X}$ is in **Possibly-IP(Y,Z)** in π, then $P(\mathbf{Y}|\mathbf{Z})$ is invariant under direct manipulation of $\mathbf{X}$ in G by changing the value of $\mathbf{W}$ from $\mathbf{w_1}$ to $\mathbf{w_2}$.

THEOREM 7.4: If G is a directed acyclic graph over $\mathbf{V} \cup \mathbf{W}$, $\mathbf{W}$ is exogenous with respect to $\mathbf{V}$ in G, $\mathbf{O}$ is included in $\mathbf{V}$, G_{Unman} is the subgraph of G over $\mathbf{V}$, π is the FCI partially oriented inducing path graph over $\mathbf{O}$ of G_{Unman}, $\mathbf{X}$, $\mathbf{Y}$ and $\mathbf{Z}$ are included in $\mathbf{O}$, $\mathbf{X}$, $\mathbf{Y}$ and $\mathbf{Z}$ are pairwise disjoint, and no X in $\mathbf{X}$ is in **Possibly-IV(Y,Z)** in π, then $P(\mathbf{Y}|\mathbf{Z})$ is invariant under direct manipulation of $\mathbf{X}$ in G by changing the value of $\mathbf{W}$ from $\mathbf{w_1}$ to $\mathbf{w_2}$.

The Prediction Algorithm is based upon the construction of a partially oriented inducing path graph from $P_{Unman(\mathbf{W})}(\mathbf{V})$. Consider the model in figure 7.14, where the relationships among X, Z, and T are linear in graph G_1, and W is a policy variable.

Although the distribution over X, Z, and T is not faithful to G_1 when $W = w_1$ if $a = -bc$, the distribution over X and Z is faithful to G_1'. In effect, although the distribution over X and Z when $W = w1$ is faithful to a directed acyclic graph, it is not faithful to the graph of the causal process that generated the distribution. Graph G_2 depicts the model when X is directly manipulated by changing the value of W from w_1 to w_2; this makes the coefficient of T in the equation for X equal to 0, and imposes some new distribution upon X. The manipulated distribution over X and Z does not satisfy the Markov condition for G_1'; rather it satisfies the Markov condition for graph G_2', which contains an edge between X and Z that G_1' does not contain. If we were to construct a partially oriented inducing path graph from the unmanipulated distribution over X and Z it would contain no edges, and make the prediction that the distribution of Z would be the same in the manipulated and unmanipulated distributions, we would be wrong. Hence the Prediction Algorithm is only guaranteed to be correct when the unmanipulated distribution is faithful to the unmanipulated graph (which includes the $X \rightarrow Z$ edge because the combined graph contains the $X \rightarrow Z$ edge.)

Graph G_1 (with $W = w_1$): edges $X \xrightarrow{a} Z$, $T \xrightarrow{b} X$, $T \xrightarrow{c} Z$

Graph G_1': X and Z disconnected

$$a = -bc$$

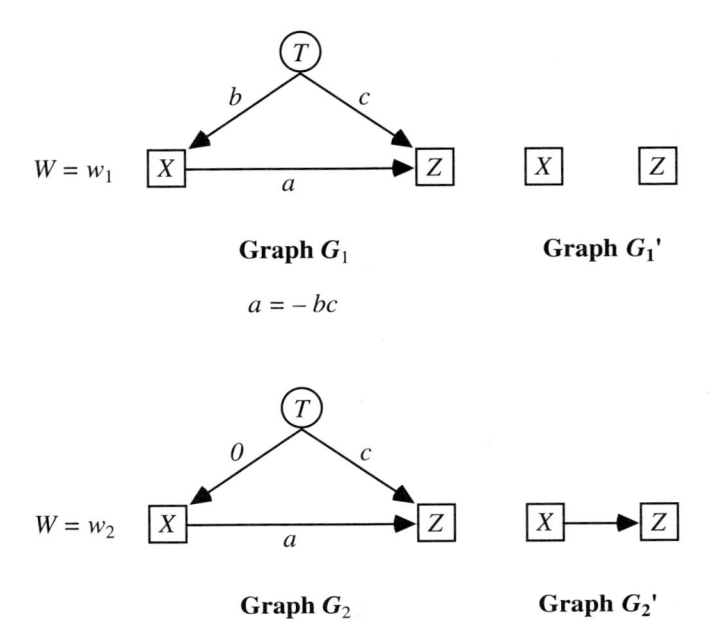

Graph G_2 (with $W = w_2$): edges $X \xrightarrow{a} Z$, $T \xrightarrow{0} X$, $T \xrightarrow{c} Z$

Graph G_2': $X \rightarrow Z$

Figure 7.14

This assumption is not as restrictive as it might first appear. Suppose that we perform an experiment of the effects of *Smoking* upon *Cancer*. We decide to assign each subject a number of cigarettes smoked per day in the following way. For each subject in the experiment, we roll a die: if the die comes up 1, they are assigned to smoke no cigarettes, if the die comes up 2, they are assigned to smoke 10 cigarettes per day, etc. Let **W** = {*Experiment*} and **V** = {*Die, Smoking, Drinking, Cancer*}. Figure 7.15 shows the causal graph for the combined population of experimental and non-experimental subjects, and G_{Unman}. The policy variable is *Experiment*: it has the same value (0) for everyone in the non-experimental population, and the same value (1) for everyone in the experimental population. *Die* is not a policy variable because it takes on different values for members of the experimental population.

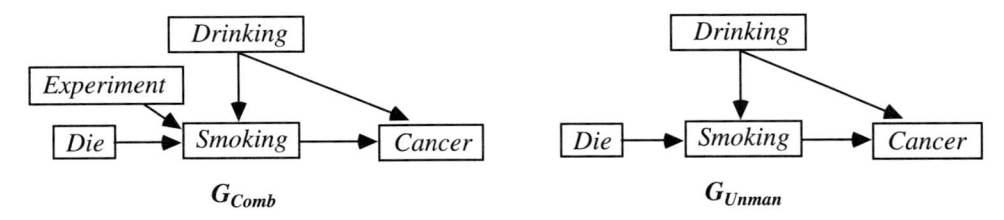

G_{Comb}: *Experiment* → *Die* → *Smoking* → *Cancer*; *Drinking* → *Smoking*; *Drinking* → *Cancer*

G_{Unman}: *Die* → *Smoking* → *Cancer*; *Drinking* → *Smoking*; *Drinking* → *Cancer*

Figure 7.15

Prediction 181

In this case, the assumption that $P_{Unman}(V)$ is faithful to G_{Unman} is clearly false because the outcome of the roll of a die and the number of cigarettes smoked by a subject are independent in the non-experimental population, but there is an edge between them in G_{Unman}. Suppose, however that we consider the subset of variables $V' = \{Smoking, Drinking, Cancer\}$. The causal graphs that result from marginalizing over V' are shown in figure 7.16. In this case, $P_{Unman}(V')$ is faithful to G_{Unman}. Since variables that are causes of *Smoking* in the manipulated population but not in the unmanipulated population complicate the analysis, we will in general simply not consider them. There is no problem in leaving them out of the causal graphs, as long as relative to the set of measured variables they are direct causes only of the manipulated variable. This guarantees that the set of variables that remain after they are removed is causally sufficient.

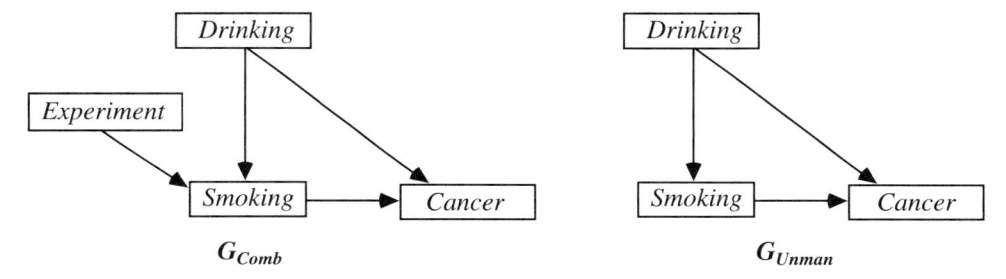

Figure 7.16

THEOREM 7.5: If G is a directed acyclic graph over $V \cup W$, W is exogenous with respect to V in G, G_{Unman} is the subgraph of G over V, $P_{Unman(W)}(V) = P(V|W = w_1)$ is faithful to G_{Unman}, and changing the value of W from w_1 to w_2 is a direct manipulation of X in G, then the Prediction Algorithm is correct.

The Prediction Algorithm is not complete; it may say that $P_{Man}(Y|Z)$ is unknown when it is calculable in principle.

7.6 Examples

First we consider our hypothetical example from the previous chapter, with the directed acyclic graph depicted in figure 7.17, and the partially oriented inducing path graph π over $\mathbf{O} = \{Income, Parents' smoking habits, Smoking, Cilia damage, Heart disease, Lung capacity, Measured breathing dysfunction\}$ depicted in figure 7.18. We assume that P_{Unman} is faithful to G_{Unman}, and that in the manipulated graph that *Income* and *Parents' smoking habits* are not parents of *Smoking*. We will use the Prediction Algorithm to draw our conclusions.

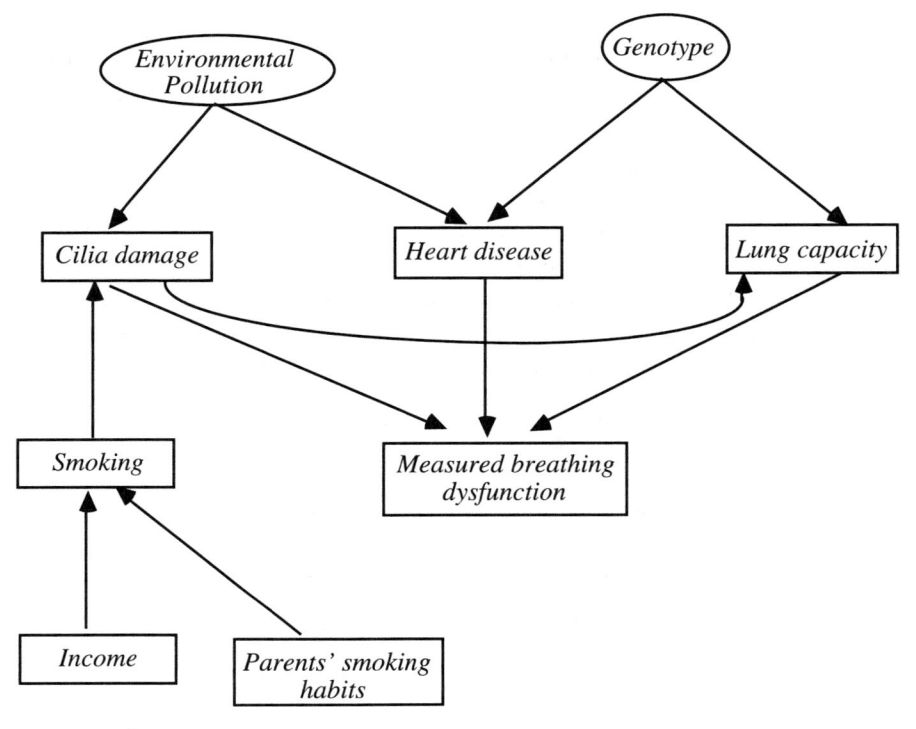

Figure 7.17

We will show in some detail the process of determining that the entire joint distribution of {*Income*, *Parents' smoking habits*, *Heart disease*, *Lung capacity* and *Measured breathing dysfunction*} is predictable given a direct manipulation of *Smoking*. Let us abbreviate the names of the variables in the following way:

Income	*I*
Parents' Smoking Habits	*PSH*
Smoking	*S*
Cilia damage	*C*
Heart disease	*H*
Measured breathing dysfunction	*M*
Lung capacity	*L*

Prediction 183

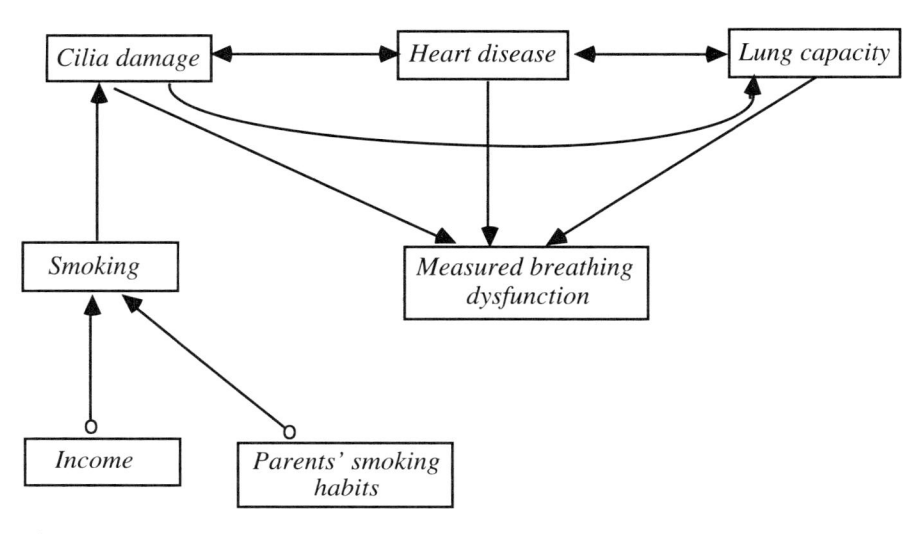

Figure 7.18

We begin by choosing an ordering for the variables. There are two constraints we impose upon the orderings. First, the only variables that precede S are those variables that are in **Definite-Nondescendant**(S), and second, the ordering is acceptable for the partially oriented inducing path graph. That means that I, PSH, and H precede S. Second, in order to be acceptable for the partially oriented inducing path graph, S, C, L, and M have to occur in that order. We arbitrarily choose one ordering Ord compatible with these constraints: I, PSH, H, S, C, L M. (Note that the ordering among the variables that are predecessors of the directly manipulated variable never matters because each term containing only variables that are predecessors of the directly manipulated variable is always invariant.)

We generate a directed graph for which $P_{Unman}(I,PSH,S,C,H,M,LC)$ satisfies the Minimality and Markov conditions. In this case we can determine that any ordering acceptable for the partially oriented inducing path graph in figure 7.18 is also an ordering acceptable for the inducing path graph. Hence, we can apply theorem 7.2. The resulting factorization is $P_{Unman}(I) \times P_{Unman}(PSH) \times P_{Unman}(H) \times P_{Unman}(S|I,PSH) \times P_{Unman}(C|S,H) \times P_{Unman}(L|C,H,S) \times P_{Unman}(M|C,H,L)$.

We now determine which terms in the factorized distribution are needed in order to predict the conditional distribution under consideration. Because we are predicting the entire joint distribution, it is trivial that we need every term in the factorized distribution.

Finally, we use the partially oriented inducing path graph to test whether each of the terms except $P_{Unman}(S|I,PSH)$ in the factorized distribution is invariant under direct manipulation of S in G_{Unman}. $P_{Unman}(I)$, $P_{Unman}(PSH)$, and $P_{Unman}(H)$ are invariant by theorem 7.4 because there are no semidirected paths from S to I, H, or PSH. $P_{Unman}(C|S,H)$ is invariant by theorem 7.3 because every path possibly d-connecting path

S to C given H is out of S. $P_{Unman}(L|C,S,H)$, is invariant by theorem 7.3 because every path possibly d-connecting path between S and L given C and H is out of S. Finally $P_{Unman}(M | C,H,L)$ is invariant by theorem 7.4 because there is no possible d-connecting path between S and M given C, H, and L.

Hence, $P_{Man}(I,PSH,H,S,C,L,M) = P_{Unman}(I) \times P_{Unman}(PSH) \times P_{Unman}(H) \times P_{Man}(S) \times P_{Unman}(C | S,H) \times P_{Unman}(L|C,H,S) \times P_{Unman}(M | C,H,L)$.

In this case, the search was simple because for the given ordering of variables, every term in the expression for $P_{Unman}(I,PSH,H,S,C,L,M)$ except for $P_{Man}(S)$ is invariant under direct manipulation of *Smoking* in G_{Unman}. If the expression had failed this test we would have repeated the process by generating different orderings of variables, until we had found a factorized expression of $P(I,PSH,H,S,C,L,M)$ in which each term except $P_{Man}(S)$ was invariant or we ran out of orderings.

For the next example, consider three alternative models of the relationship between *Smoking* and *Lung cancer* depicted in figure 7.19. In G_1, *Smoking* causes *Lung cancer*, and there is a common cause of *Smoking* and *Lung cancer*; in G_2, *Smoking* does not cause *Lung cancer*, but there is a common cause of *Lung cancer* and *Smoking*; and in G_3, *Smoking* causes *Lung cancer*, but there is no common cause of *Smoking* and *Lung cancer*.

The maximally informative partially oriented inducing path graph of G_1, G_2, and G_3 over **O** = {*Smoking*, *Lung cancer*} is shown in figure 7.20.

From this partially oriented inducing path graph it is impossible to determine whether *Smoking* causes *Lung cancer* (as in G_3) or *Smoking* does not cause *Lung cancer* but there is a common cause of *Smoking* and *Lung cancer* (as in G_2), or *Smoking* causes *Lung cancer* and there is also a common cause (as in G_1). In addition, we cannot predict the distribution of *Lung cancer* when *Smoking* is directly manipulated. If we try the ordering of variables <*Smoking,Lung cancer*> then in order to apply the Prediction Algorithm, we need to show that $P(Lung cancer|Smoking)$ is invariant under direct manipulation of *Smoking* in G_{Unman}. But we cannot use theorem 7.3 to show that $P(Lung cancer|Smoking)$ is invariant because the *Smoking* o-o *Lung cancer* edge guarantees that there is a possibly d-connecting path between *Smoking* and *Lung cancer* given the empty set that is not out of *Smoking*. This is a quite general feature of the method; it cannot be used to predict a conditional distribution of Y whenever there is an edge between the variable X being directly manipulated and Y that has a "o" at the X end. Of course, this feature does not of itself show that $P(Lung cancer)$ is not predictable by some other method (although in this example it clearly is not.)

Suppose, however, that **O** = {*Smoking*, *Lung cancer*, *Income*}. If the true graph is G_2, the partially oriented inducing path graph is shown in figure 7.21.

Prediction

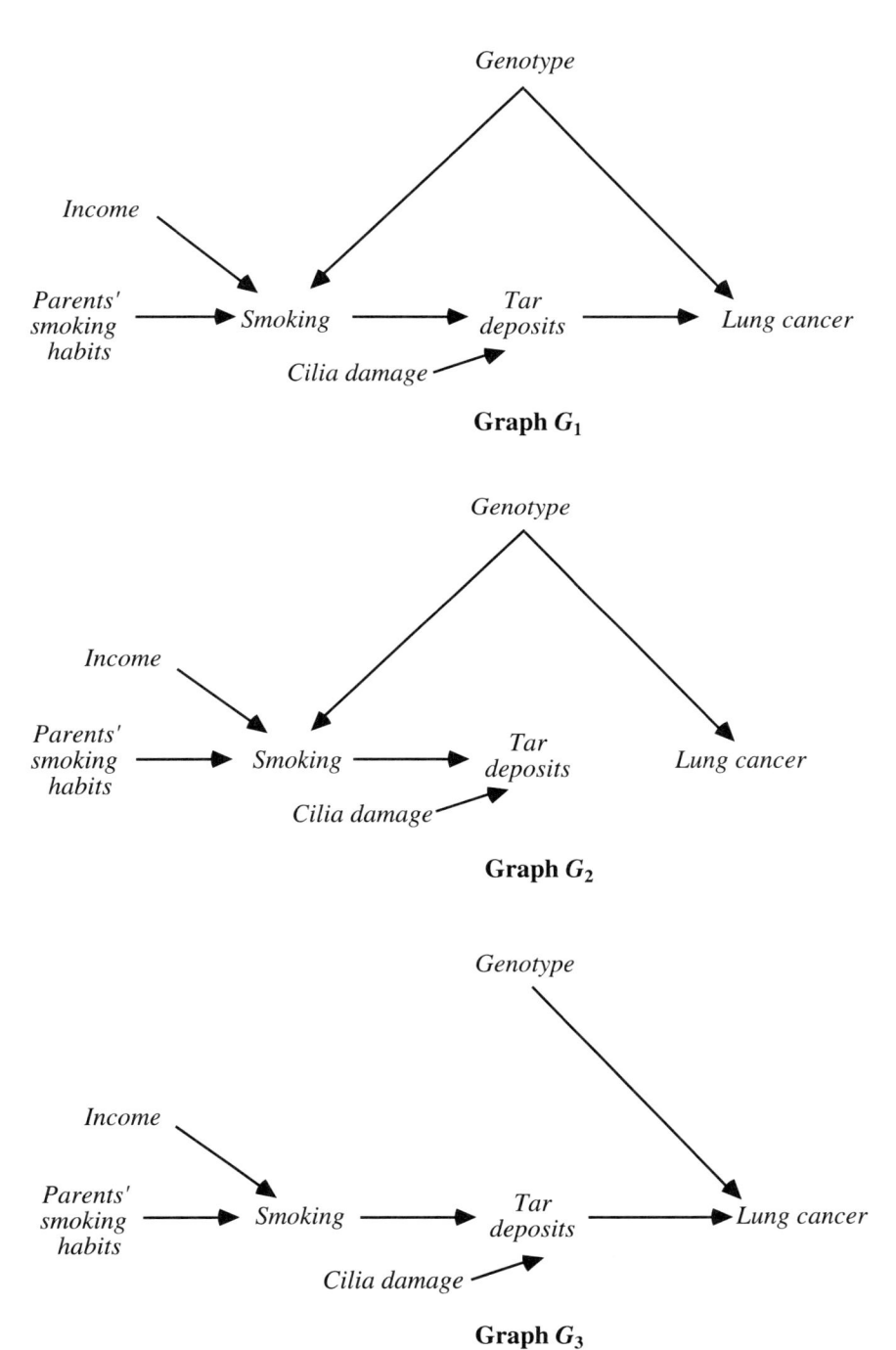

Graph G_1

Graph G_2

Graph G_3

Figure 1.19

[Smoking]o————o[Lung cancer]

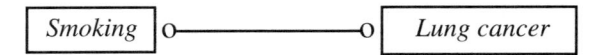

Figure 7.20

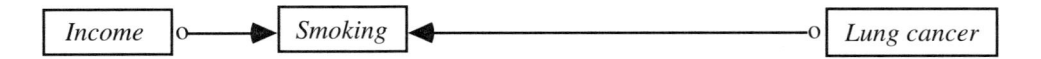

[Income]o——▶[Smoking]◀————o[Lung cancer]

Figure 7.21

By the results of the previous chapter, we can conclude that *Smoking* does not cause *Lung cancer*, because there is no semidirected path from *Smoking* to *Lung cancer*. In this case *P(Lung cancer)* is invariant under direct manipulation of *Smoking* in G_{Unman}, so $P_{Man}(Lung\ cancer)$ is predictable.

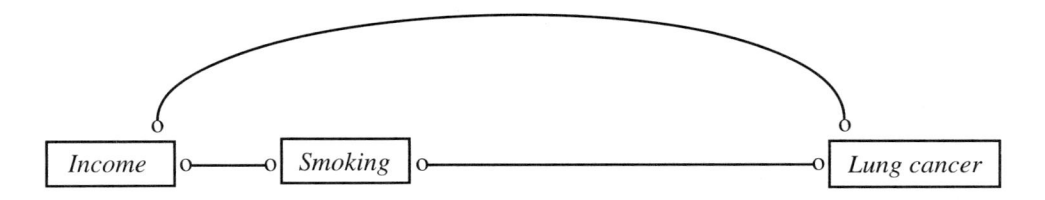

[Income]o——o[Smoking]o————o[Lung cancer]

Partially Oriented Inducing Path Graph of G_1 over **O** = {Lung Cancer, Smoking, Income}

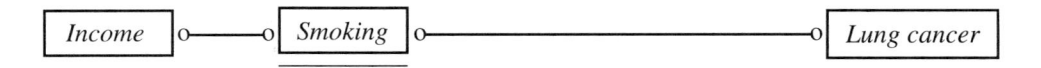

[Income]o——o[Smoking]o————o[Lung cancer]

Partially Oriented Inducing Path Graph of G_3 over **O** = {Lung Cancer, Smoking, Income}

Figure 7.22

The partially oriented inducing path graphs for G_1 and G_3 over **O** = {*Lung cancer, Smoking, Income*} (shown in figure 7.22) do not contain enough information in order to determine whether *Smoking* causes *Lung cancer*. Because in each case there is a *Smoking* o-o *Lung cancer* edge it follows that we cannot use the Prediction Algorithm to predict $P_{Man}(Lung\ cancer)$.

If the true graph is G_3 it is possible to determine that *Smoking* causes *Lung cancer* by also measuring two causes of *Smoking* that are not directly connected in the partially oriented inducing path graph, as in figure 7.23. Because there is a directed path from

Smoking to *Lung cancer* in the partially oriented inducing path graph, by the results of the preceding chapter there is a directed path from *Smoking* to *Lung cancer* in the causal graph of the process that generated the data, and *Smoking* causes *Lung cancer*. The output of the Prediction Algorithm is:

$$P_{Man}(Lung\ Cancer) = \sum_{Smoking} \overrightarrow{P_{Man}(Smoking)} P_{Unman}(Lung\ Cancer|Smoking)$$

Note that it is not necessary that *Parents' Smoking Habits* and *Income* be uncorrelated, or direct parents of *Smoking*. The *Smoking* to *Lung cancer* edge is oriented by any pair of variables that have edges that collide at a third variable V, that are not adjacent in the partially oriented inducing path graph, and such that there is a directed path U from V to *Smoking* and for every subpath $<X,Y,Z>$ of U, X, Y, and Z do not form a triangle.

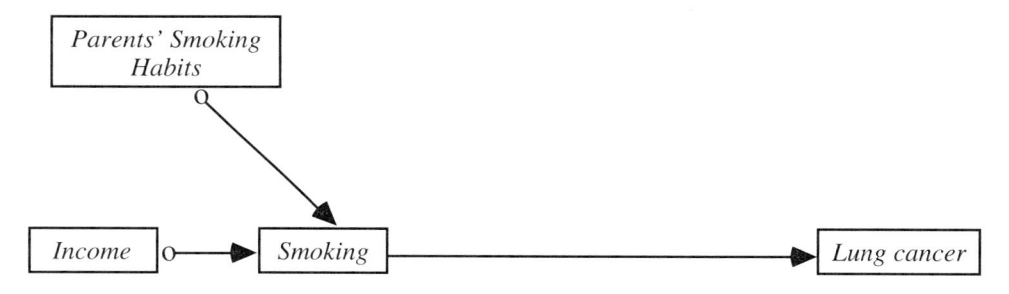

Figure 7.23

Unfortunately, it is more difficult to determine whether *Smoking* is a cause of *Lung cancer* if G_1 is the true causal graph. If **O** = {*Smoking, Lung cancer, Income, Parents' Smoking Habits*} and G_1 is the true causal graph, without further background knowledge we cannot determine whether *Smoking* causes *Lung cancer*. Figure 7.24 shows that in the partially oriented inducing path graph the *Smoking* to *Lung cancer* edge is in triangles with *Income* and *Parents' smoking habits* and hence is oriented with an 'o' at each end. It follows from the existence of the *Smoking* o-o *Lung cancer* edge that we cannot use the Prediction Algorithm to predict $P(Lung\ cancer)$ when *Smoking* is directly manipulated.

It is plausible that *Income* does not cause *Lung cancer* directly. If we know from background knowledge that if there is a causal connection between *Income* and *Lung cancer* it contains a causal path from *Smoking* to *Lung cancer*, then we can conclude from the partially oriented inducing path graph that *Smoking* does cause *Lung cancer*.

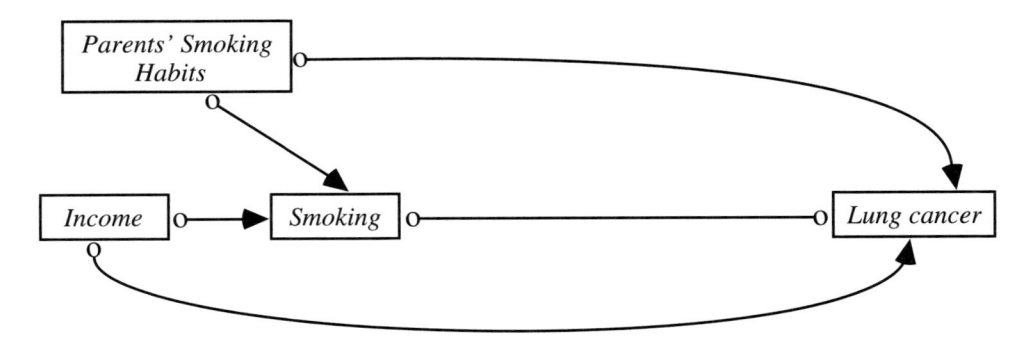

Figure 7.24

Alternatively, if G_1 is the correct model, we might try to determine that *Smoking* is a cause of cancer by measuring a variable such as *Tar deposits*, that is causally between *Smoking* and *Lung cancer*. While there is still an induced edge between *Income* and *Lung Cancer* in the partially oriented inducing path graph, *Income*, *Smoking*, and *Tar deposits* are not in a triangle, and the edge from *Smoking* to *Tar deposits* can be oriented. Unfortunately, as figure 7.25 illustrates, this now leaves one end of the edge between *Tar deposits* and *Lung cancer* oriented with a "o" at one end, so the partially oriented inducing path graph still does not entail that *Smoking* causes *Lung cancer*. And because there is a *Smoking* o-o *Lung cancer* edge, P_{Man}(*Lung cancer*) is not predictable using the Prediction Algorithm.

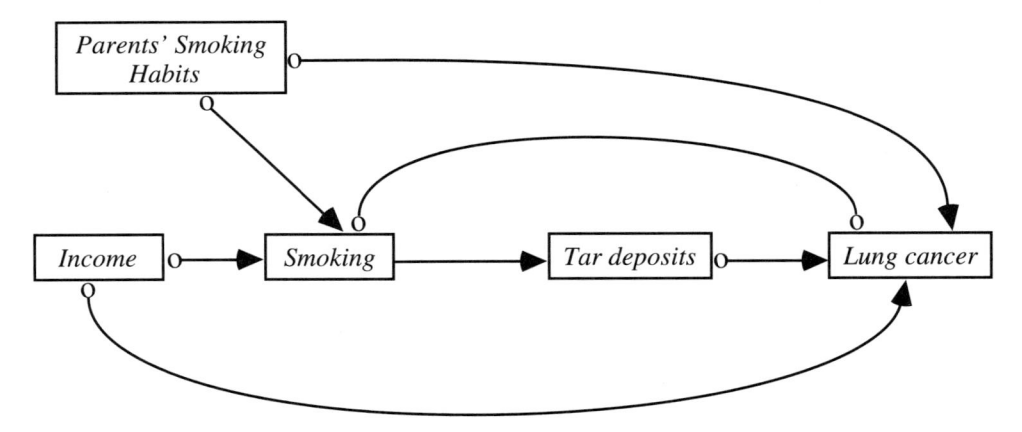

Figure 7.25

However, if G_1 is the correct model, and we measure a variable between *Smoking* and *Lung cancer*, such as *Tar deposits*, and another cause of *Tar deposits*, such as *Cilia damage*, we can determine that *Smoking* causes *Lung cancer*. (See figure 7.26.) However,

Prediction 189

we cannot predict $P_{man}(Lung\ cancer)$ using the Prediction Algorithm because of the *Smoking* o→ *Lung cancer* edge.

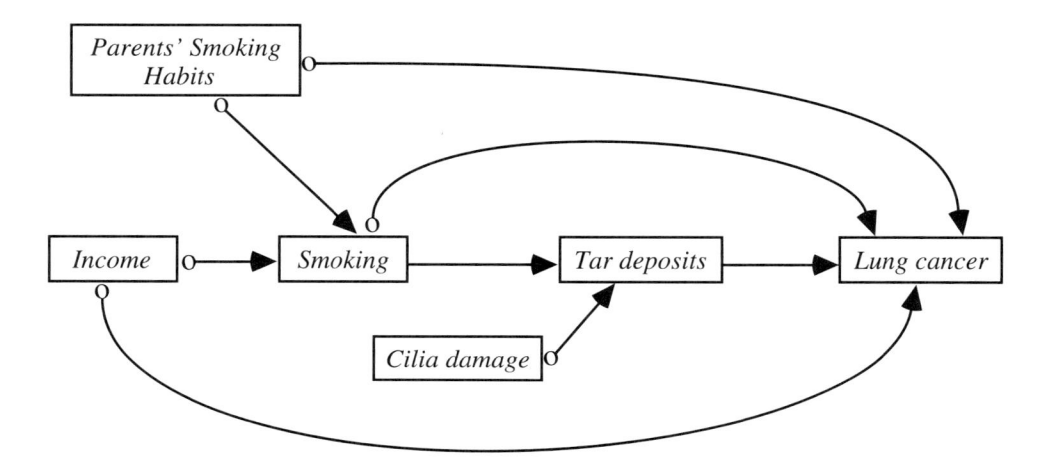

Figure 7.26

We can also determine that *Smoking* is a cause of *Lung cancer* by breaking the *Income-Smoking-Lung cancer* triangle by measuring all of the common causes of *Smoking* and *Lung cancer* (in this case, *Genotype*). By measuring all of the common causes of *Smoking* and *Lung cancer*, the edge between *Income* and *Lung cancer* is removed from the partially oriented inducing path graph. This breaks triangles involving *Income*, *Smoking*, and *Lung cancer*, so that the *Smoking* to *Lung cancer* edge can be oriented by the edge between *Income* and *Smoking*, as in figure 7.27. In addition, $P_{Man}(Lung\ cancer)$ is predictable. The output of the Prediction Algorithm is:

$$P_{Man}(Lung\ Cancer) =$$

$$\sum_{Smoking, Genotype} \vec{P}_{Man}(Smoking) P_{Unman}(Genotype) P_{Unman}(Lung\ Cancer|Smoking, Genotype)$$

Of course, measuring *all* of the common causes of *Smoking* and *Lung cancer* may be difficult both because of the number of such common causes, and because of measurement difficulties (as in the case of *Genotype*). So long as even one common cause remains unmeasured, the inducing path graph has an *Income - Smoking - Lung cancer* triangle, and the edge between *Smoking* and *Lung cancer* cannot be oriented.

Although we cannot determine from the partially oriented inducing path graph in figure 7.27 whether *Genotype* is a common cause of *Smoking* and *Lung cancer*, we can determine that there is *some* common cause of *Smoking* and *Lung cancer*.

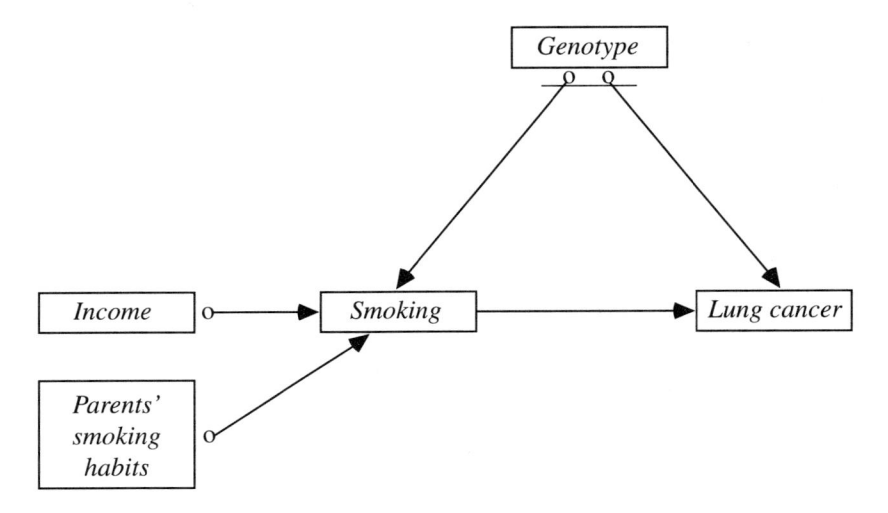

Figure 7.27

7.7 Conclusion

The results developed here show that there exist possible cases in which predictions of the effects of manipulations can be obtained from observations of unmanipulated systems, and predictions of experimental outcomes can be made from uncontrolled observations. Some examples from real data analysis problems will be considered in the next chapter. We do not know whether our sufficient conditions for prediction are close to maximally informative, and a good deal of theoretical work remains to be done on the question.

7.8 Background Notes

Anticipations of the theory developed in this chapter can be found in Strotz and Wold 1960, in Robins 1986, and in the tradition of work inaugurated by Rubin. The special case of the Manipulation Theorem that applies when an intervention makes a single directly manipulated variable X independent of its parents was independently conjectured by Fienberg in a seminar in 1991. Subsequently, Pearl (1995) has given rules for calculating predictions from interventions. The rules, which follow from theorem 7.1, are discussed in chapter 12.

8 Regression, Causation, and Prediction

Regression is a special case, not a special subject. The problems of causal inference in regression studies are instances of the problems we have considered in the previous chapters, and the solutions are to be found there as well. What is singular about regression is only that a technique so ill suited to causal inference should have found such wide employment to that purpose.

8.1 When Regression Fails to Measure Influence

Regression models are commonly used to estimate the "influence" that regressors **X** have on an outcome variable, Y.[1] If the relations among the variables are linear then for each X_i the expected change in Y that would be produced by a unit change in X_i if all other **X** variables are forced to be constant can be represented by a parameter, say α_i. It is obvious and widely noted (see, e.g., Fox 1984) that the regression estimate of α_i will be incorrect if X_i and Y have one or more unmeasured common causes, or in more conventional statistical terminology, the estimate will be biased and inconsistent if the error variable for Y is correlated with X_i. To avoid such errors, it is often recommended (Pratt and Schlaifer 1988) that investigators enlarge the set of potential regressors and determine if the regression coefficients for the original regressors remain stable, in the hope that confounding common causes, if any, will thereby be measured and revealed. Regression estimates are known often to be unstable when the number of regressors is enlarged, because, for example, additional regressors may be common causes of previous regressors and the outcome variable (Mosteller and Tukey 1977). The stability of a regression coefficient for X when other regressors are added is taken to be evidence that X and the outcome variable have no common cause.

It does not seem to be recognized, however, that when regressors are statistically dependent, the existence of an unmeasured common cause of regressor X_i and outcome variable Y may bias estimates of the influence of *other* regressors, X_k; variables having no influence on Y whatsoever, nor even a common cause with Y, may thereby be given significant regression coefficients. The error may be quite large. The strategy of regressing on a larger set of variables and checking stability may compound rather than remedy this problem. A similar difficulty may arise if one of the measured candidate regressors is an *effect*, rather than a cause, of Y, a circumstance that we think may sometimes occur in uncontrolled studies.

To illustrate the problem, consider the linear structures in figure 8.1, where for concreteness we specify that exogenous and error variables are all uncorrelated and jointly normally distributed, the error variables have zero means, and linear coefficients are not zero. Only the X variables and Y are assumed to be measured. Each set of linear equations is accompanied by a directed graph illustrating the assumed causal and functional dependencies among the non-error variables. In large samples, for data from each of these structures linear multiple regression will give all variables in the set $\{X_1, X_2, X_3, X_5\}$ nonzero regression coefficients, even though X_2 has no direct influence on Y in any of these structures, and X_3 has no influence direct or indirect on Y in structures (i), and (ii), and the effect of X_3 in structures (iii) and (iv) is confounded by an unmeasured common cause. The regression estimates of the influences of X_2 and X_3 will in all four cases be incorrect. If a specification search for regressors had selected X_1 alone or X_1 and X_5 in (i) or (ii), or X_5 alone in (i), (ii), (iii), or (iv), a regression on these variables would give consistent, unbiased estimates of their influence on Y. But the textbook procedures in commercial statistical packages will in all of these cases fail to identify $\{X_1\}$ or $\{X_5\}$ or $\{X_1, X_5\}$ as the appropriate subset of regressors.

It is easy to produce examples of the difficulty by simulation. Using structure (i), twenty sets of values for the linear coefficients were generated, half positive and half negative, each with absolute value greater than .5. For each system of coefficient values a random sample of 5,000 units was generated by substituting those values for the coefficients of structure (i) and using uncorrelated standard normal distributions for the exogenous variables. Each sample was given to MINITAB, and in all cases MINITAB found that $\{X_1, X_2, X_3, X_5\}$ is the set of regressors with significant regression coefficients. The STEPWISE procedure in MINITAB selected the same set in approximately half the cases and in the others added X_4 to boot; selection by the lowest value of Mallow's C_p or adjusted R^2 gave results similar to the STEPWISE procedure.

The difficulty can be remedied if one measures all common causes of the outcome variable and the candidate regressors, but unfortunately nothing in regression methods informs one as to when that condition has been reached. And the addition of extra candidate regressors may create the problem rather than remedy it; in structures (i) and (ii), if X_3 were not measured the regression estimate of X_2 would be consistent and unbiased.

The problem we have illustrated is quite general; it will lead to error in the estimate of the influence of any regressor X_k that directly causes or has a common direct unmeasured common cause with any regressor X_i such that X_i and Y have an unmeasured common cause (or X_i is an effect of Y). Depending on the true structure and coefficient values the error may be quite large. It is easy to construct cases in which a variable with no influence on the outcome variable has a standardized regression coefficient larger than any other single regressor. Completely parallel problems arise for categorical data. Recall theorem 3.4:

Regression, Causation, and Prediction

$Y = a_1 X_1 + a_2 X_5 + \varepsilon_Y$

$X_1 = a_3 X_2 + a_4 X_4 + \varepsilon_1$

$X_3 = a_5 X_2 + a_6 Y + \varepsilon_3$

$Y = a_1 X_1 + a_2 X_5 + a_3 T + \varepsilon_Y$

$X_1 = a_4 X_2 + a_5 X_4 + \varepsilon_1$

$X_3 = a_6 X_2 + a_7 T + \varepsilon_3$

(i) **(ii)**

$Y = a_1 X_1 + a_2 X_5 + a_3 T_2 + a_4 X_3 + \varepsilon_Y$

$X_1 = a_5 X_2 + a_6 X_4 + \varepsilon_1$

$X_2 = a_7 T_1 + \varepsilon_2$

$X_3 = a_8 T_1 + a_9 T_2 + \varepsilon_3$

$Y = a_1 X_1 + a_2 X_5 + a_3 T_2 + a_4 X_3 + \varepsilon_Y$

$X_1 = a_5 X_2 + a_6 X_4 + \varepsilon_1$

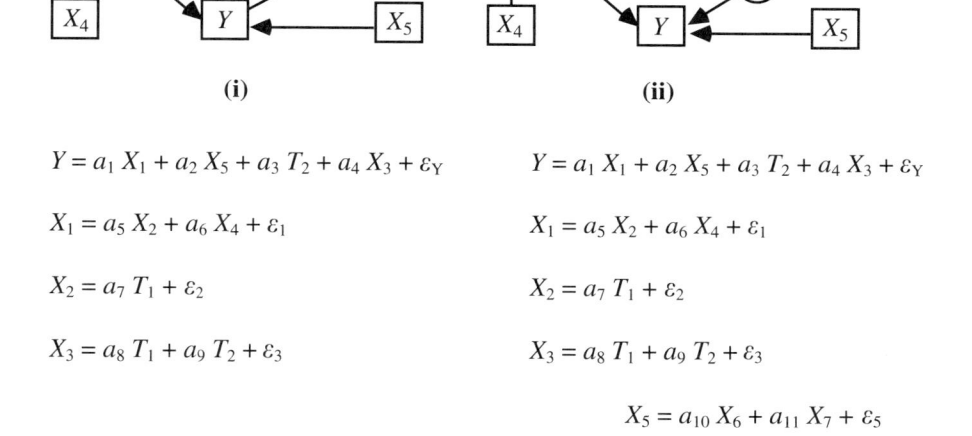

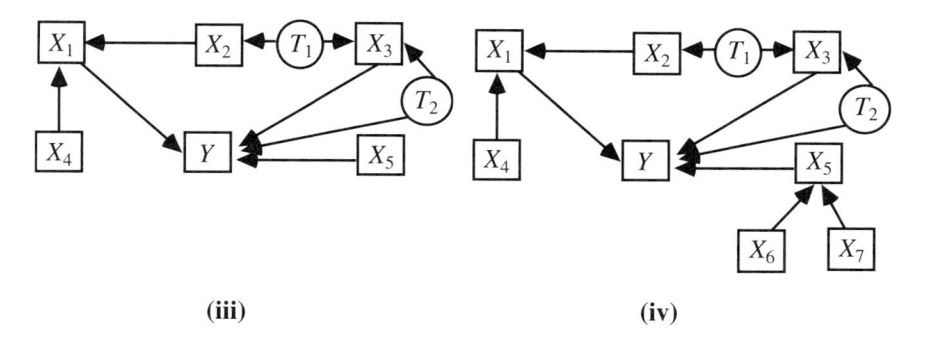

$X_2 = a_7 T_1 + \varepsilon_2$

$X_3 = a_8 T_1 + a_9 T_2 + \varepsilon_3$

$X_5 = a_{10} X_6 + a_{11} X_7 + \varepsilon_5$

(iii) **(iv)**

Figure 8.1

THEOREM 3.4: If P is faithful to some graph, then P is faithful to G if and only if (i) for all vertices, X, Y of G, X and Y are adjacent if and only if X and Y are dependent conditional on every set of vertices of G that does not include X or Y; and (ii) for all vertices X, Y, Z such that X is adjacent to Y and Y is adjacent to Z and X and Z are not adjacent, $X \rightarrow Y \leftarrow Z$ is a subgraph of G if and only if X, Z are dependent conditional on every set containing Y but not X or Z.

Consideration of the first part of this theorem explains why in structure (i) in figure 8.1 regression procedures incorrectly select X_2 as a variable directly influencing Y: The structure and distribution satisfy the Markov and Faithfulness conditions, but linear regression takes a variable X_i to influence Y provided the partial correlation of X_i and Y controlling for *all* of the other **X** variables does not vanish. Part (i) of theorem 3.4 shows that the regression criterion is insufficient. It follows immediately from theorem 3.4 that, assuming the Markov and Faithfulness Conditions, regression of Y on a set **X** of variables will only yield an unbiased or consistent estimate of the influences of the **X** variables provided in the true structure no **X** variable is the effect of Y or has a common unmeasured cause with Y.

Since typical empirical data sets to which multiple regression methods are applied have some correlated regressors, and in uncontrolled studies it is rare to know that unmeasured common causes are not acting on both the outcome variable and the regressors, the problem is endemic. One of the most common uses of statistical methods thus appears to be little more than elaborate guessing.

8.2 A Solution and Its Application
Assuming the right variables have been measured, there is a straightforward solution to these problems: apply the PC, FCI, or other reliable algorithm, and appropriate theorems from the preceding chapters, to determine which X variables influence the outcome Y, which do not, and for which the question cannot be answered from the measurements; then estimate the dependencies by whatever methods seem appropriate and apply the results of the previous chapter to obtain predictions of the effect of manipulating the X variables. No extra theory is required. We will give a number of illustrations, both empirical and simulated.

We begin by noting that for the twenty samples from structure (i), in every case our implementation of the PC algorithm—which of course assumes there are no latent variables—selects $\{X_1, X_5\}$ as the variables that directly influence Y. Our implementation of the FCI algorithm, which makes no such assumption, in every case says that X_1 directly influences Y, that X_5 may, and that the other variables do not.

In each of the other three structures in figure 8.1 with sufficiently large samples multiple regression methods will make errors, always including X_2 and X_3 among the "significant" or "best" or "important" variables. In contrast the FCI algorithm together with theorems 6.5 through 6.8 give the following results:

Regression, Causation, and Prediction

Structure	Direct Influence	No Direct Influence	Undetermined
(ii)	X_1	X_2, X_3, X_4	X_5
(iii)	X_1	X_4	X_2, X_3, X_5
(iv)	X_1, X_5	X_4, X_6, X_7	X_2, X_3

In all of these cases the FCI procedure either determines definitely that X_2 and X_3 have no direct influence on Y, or determines that it cannot be decided whether they have any unconfounded direct influence.

8.2.1 Components of the Armed Forces Qualification Test

The *AFQT* is a test battery used by the United States armed forces. It has a number of component tests, including those listed below:

Arithmetical Reasoning (*AR*)
Numerical Operations (*NO*)
Word Knowledge (*WK*)

In addition a number of other tests, including those listed below, are not part of the *AFQT* but are correlated with it and with its components:

Mathematical Knowledge (*MK*)
Electronics Information (*EI*)
General Science (*GS*)
Mechanical Comprehension (*MC*)

Given scores for these 8 measures on 6224 armed forces personnel, a linear multiple regression of AFQT on the other seven variables gives significant regression coefficients to all except EI and MC and thus fails to distinguish the tests that are in fact linear components of *AFQT*. The covariance matrix is shown below.

$n = 6224$

AFQT	NO	WK	AR	MK	EI	MC	GS
253.9850							
29.6490	51.7649						
60.3604	6.2931	41.967					
57.6566	14.5143	16.0226	40.9329				
29.3763	18.2701	13.2055	20.6052	40.7386			
36.2318	2.10733	22.6958	16.3664	12.1773	63.1039		
35.8244	4.45539	17.4155	20.3952	16.459	35.1981	62.9647	
38.2510	5.61516	27.1492	14.7402	14.8442	29.9095	26.6842	48.9300

Given the prior information that *AFQT* is not a cause of any of the other variables, the PC algorithm in TETRAD II correctly picks out {*AR*, *NO*, *WK*} as the only variables adjacent to *AFQT*, and hence the only variables that can be components of *AFQT*. (Spirtes, Glymour, Scheines, and Sorensen 1990).[2]

8.2.2 The Causes of Spartina Biomass

A recent textbook on regression (Rawlings 1988) skillfully illustrates regression principles and techniques for a biological study in which it is reasonable to think there is a causal process at work relating the variables. The question at issue is plainly causal: among a set of 14 variables, which have the most influence on an outcome variable, the weight of Spartina grass? Since the example is the principal application given for an entire textbook on regression, the reader who reaches the 13th chapter may be surprised to find that the methods yield almost no useful information about that question.

According to Rawlings, Linthurst (1979) obtained five samples of Spartina grass and soil from each of nine sites on the Cape Fear Estuary of North Carolina. Besides the mass of Spartina (*BIO*), fourteen variables were measured for each sample:

Free Sulfide (H_2S)
Salinity (*SAL*)
Redox potentials at pH 7 (EH_7)
Soil pH in water (*PH*)
Buffer acidity at pH 6.6 (*BUF*)
Phosphorus concentration (*P*)
Potassium concentration (*K*)
Calcium concentration (*CA*)
Magnesium concentration (*MG*)
Sodium concentration (*NA*)
Manganese concentration (*MN*)
Zinc concentration (*ZN*)
Copper concentration (*CU*)
Ammonium concentration (NH_4)

The correlation matrix is as follows[3]:

```
     BIO  H2S  SAL  EH7  PH   BUF  P    K    CA   MG   NA   MN   ZN   CU   NH4
     1.0
     .33  1.0
    -.10  .10  1.0
     .05  .40  .31  1.0
     .77  .27 -.05  .09  1.0
    -.73 -.37 -.01 -.15 -.95  1.0
    -.35 -.12 -.19 -.31 -.40  .38  1.0
    -.20  .07 -.02  .42  .02 -.07 -.23  1.0
     .64  .09  .09 -.04  .88 -.79 -.31 -.26  1.0
    -.38 -.11 -.01  .30 -.18  .13 -.06  .86 -.42  1.0
    -.27  .00  .16  .34 -.04 -.06 -.16  .79 -.25  .90  1.0
    -.35  .14 -.25 -.11 -.48  .42  .50 -.35 -.31 -.22 -.31 1.00
    -.62 -.27 -.42 -.23 -.72  .71  .56  .07 -.70  .35  .12  .60  1.0
     .09  .01 -.27  .09  .18 -.14 -.05  .69 -.11  .71  .56 -.23  .21  1.0
    -.63 -.43 -.16 -.24 -.75  .85  .49 -.12 -.58  .11 -.11  .53  .72  .01  1.0
```

The aim of the data analysis was to determine for a later experimental study which of these variables most influenced the biomass of Spartina in the wild. Greenhouse experiments would then try to estimate causal dependencies out of the wild. In the best case one might hope that the statistical analyses of the observational study would correctly select variables that influence the growth of Spartina in the greenhouse. In the worst case, one supposes, the observational study would find the wrong causal structure, or would find variables that influence growth in the wild (e.g., by inhibiting or promoting growth of a competing species) but have no influence in the greenhouse.

Using the SAS statistical package, Rawlings analyzed the variable set with a multiple regression and then with two stepwise regression procedures. A search through all possible subsets of regressors was not carried out, presumably because the candidate set of regressors is too large. The results were as follows:

(i) a multiple regression of *BIO* on all other variables gives only *K* and *CU* significant regression coefficients;
(ii) two stepwise regression procedures[4] both yield a model with *PH, MG, CA,* and *CU* as the only regressors, and multiple regression on these variables alone gives them all significant coefficients;
(iii) simple regressions one variable at a time give significant coefficients to *PH, BUF, CA, ZN,* and *NH4*.

What is one to think? Rawling's reports that "None of the results was satisfying to the biologist; the inconsistencies of the results were confusing and variables expected to be biologically important were not showing significant effects" (p. 361). This analysis is supplemented by a ridge regression, which increases the stability of the estimates of coefficients, but the results for the point at issue—identifying the important variables—are much the same as with least squares. Rawlings also provides a principal components factor analysis and various geometrical plots of the components. These calculations provide no information about which of the measured variables influence Spartina growth.

Noting that *PH*, for example, is highly correlated with *BUF*, and using *BUF* instead of *PH* along with *MG*, *CA* and *CU* would also result in significant coefficients, Rawlings effectively gives up on this use of the procedures his book is about:

> Ordinary least squares regression tends either to indicate that none of the variables in a correlated complex is important when all variables are in the model, or to arbitrarily choose one of the variables to represent the complex when an automated variable selection technique is used. A truly important variable may appear unimportant because its contribution is being usurped by variables with which it is correlated. Conversely, unimportant variables may appear important because of their associations with the real causal factors. It is particularly dangerous in the presence of collinearity to use the regression results to impart a "relative importance," whether in a causal sense or not, to the independent variables. (p. 362)

Rawling's conclusion is correct about multiple regression and about conventional methods for choosing regressors, but it is not true of more reliable inference procedures. If we apply the PC algorithm to the Linthurst data then there is one robust conclusion: the only variable that may *directly* influence biomass in this population[5] is *PH*; *PH* is distinguished from all other variables by the fact that the correlation of every other variable (except *MG*) with *BIO* vanishes or vanishes when *PH* is controlled for.[6] The relation is not symmetric; the correlation of *PH* and *BIO*, for example, does not vanish when *BUF* is controlled. The algorithm finds *PH* to be the only variable adjacent to *BIO* no matter whether we use a significance level of .05 to test for vanishing partial correlations, or a level of 0.1, or a level of 0.2. In all of these cases, the PC algorithm or the FCI algorithm yield the result that *PH* and only *PH* can be directly connected with *BIO*. If the system is linear normal and the Causal Markov Condition obtains, then in this population any influence of the other regressors on *BIO* would be blocked if *PH* were held constant. Of course, over a larger range of values of the variables there is little reason to think that *BIO* depends linearly on the regressors, or that factors that have no influence in producing variation within this sample would continue to have no influence. Nor can the analysis determine whether the relationship between *PH* and *BIO* is confounded by one or more unmeasured common causes, but the principles of the theory in this case suggest otherwise. If *PH* and *BIO* have a common unmeasured cause *T*, say, and any other variable, *Z*, among the 13 others either causes *PH* or has a common unmeasured cause with *PH*, then *Z* and *BIO* should be correlated conditional on *PH*, which appears not to be the case.

The program and theory lead us to expect that if *PH* is forced to have values like those in the sample—which are almost all either below *PH* 5 or above *PH* 7—then manipulations of other variables within the ranges evidenced in the sample will have no effect on the growth of Spartina. The inference is a little risky, since growing plants in a greenhouse under controlled conditions may not be a direct manipulation of the variables relevant to growth in the wild. If for example, in the wild variations in *PH* affect Spartina growth chiefly through their influence on the growth of competing species not present in the greenhouse, a greenhouse experiment will not be a direct manipulation of *PH* for the system.

The fourth chapter of Linthurst's thesis partly confirms the PC algorithm's analysis. In the experiment Linthurst describes, samples of Spartina were collected from a salt marsh creekbank (presumably at a different site than those used in the observational study). Using a 3 x 4 x 2 (*PH* x *SAL* x *AERATION*) randomized complete block design with four blocks, after transplantation to a greenhouse the plants were given a common nutrient solution with varying values *PH* and *SAL* and *AERATION*. The *AERATION* variable turned out not to matter in this experiment. Acidity values were *PH* 4, 6 and 8. *SAL* for the nutrient solutions was adjusted to 15, 25, 35 and 45 %.

Linthurst found that growth varied with *SAL* at *PH* 6 but not at the other *PH* values, 4 and 8, while growth varied with *PH* at all values of *SAL* (p. 104). Each variable was correlated with plant mineral levels. Linthurst considered a variety of mechanisms by which extreme *PH* values might control plant growth:

> At pH 4 and 8, salinity had little effect on the performance of the species. The pH appeared to be more dominant in determining the growth response. However, there appears to be no evidence for any causal effects of high or low tissue concentrations on plant performance unless the effects of pH and salinity are also accounted for. (p. 108)
>
> The overall effect of pH at the two extremes is suggestive of damage to the root directly, thereby modifying its membrane permeability and subsequently its capacity for selective uptake. (p. 109)

A comparison of the observational and experimental data suggests that the PC Algorithm result was essentially correct and can be extrapolated through the variation in the populations sampled in the two procedures, but cannot be extrapolated through *PH* values that approach neutrality. The result of the PC search was that in the non-experimental sample, observed variations in aerial biomass were perhaps caused by variations in *PH*, but were not caused by variations in other variables. In the observational data Rawlings reports (p. 358) almost all *SAL* measurements are around 30—the extremes are 24 and 38. Compared to the experimental study rather restricted variation was observed in the wild sample. The observed values of *PH* in the wild, however, are clustered at the two extremes; only four observations are within half a *PH* unit of 6, and no observations at all occurred at *PH* values between 5.6 and 7.1. For the observed values of *PH* and *SAL*, the experimental results appear to be in very good agreement with our results from the observational study: small variations in *SAL* have no effect on Spartina growth if the *PH* value is extreme. The is further discussion of this case in chapter 12, section 5.10.

8.2.3 The Effects of Foreign Investment on Political Repression

Timberlake and Williams (1984) used regression to claim foreign investment in third-world countries promotes dictatorship. They measured political exclusion (*PO*) (i.e., dictatorship), foreign investment penetration in 1973 (*FI*), energy development in 1975 (*EN*), and civil liberties (*CV*). Civil liberties was measured on an ordered scale from 1 to 7, with lower values indicating *greater* civil liberties. Their correlations for 72 "noncore" countries are:

PO	FI	EN	CV
1.0			
-.175	1.0		
-.480	.330	1.0	
.868	-.391	-.430	1.0

Their inference is unwarranted. Their model and the model obtained from the SGS algorithm using a .12 significance level to test for vanishing partial correlations) are shown in figure 8.2.[7]

Regression Model **SGS Algorithm Model**

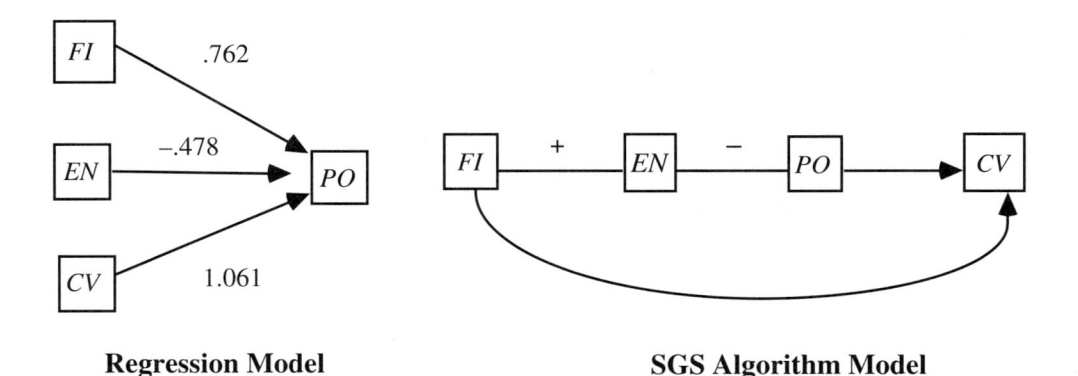

Figure 8.2

The SGS Algorithm will not orient the *FI-EN* and *EN-PO*, edges, or determine whether they are due to at least one unmeasured common cause. Maximum likelihood

Regression, Causation, and Prediction 201

estimates of any of the SGS Algorithm models require that the influence of *FI* on *PO* (if any) be negative, and the models easily pass a likelihood ratio test with the EQS program. If one of the SGS Algorithm models is correct, Timberlake and William's regression model appears to be a case in which an effect of the outcome variable is taken as a regressor, as in structure (i) of figure 8.1.

This analysis of the data assumes their are no unmeasured common causes. If we run the correlations through the FCI algorithm using the same significance level, we obtain the partially oriented inducing path graph shown in figure 8.3.

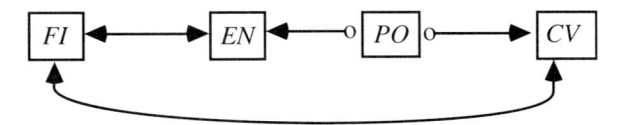

Figure 8.3

The graph together with the required signs of the dependencies, says that foreign investment and energy consumption have a common cause, as do foreign investment and civil liberties, that energy development has no influence on political exclusion, but political exclusion may have a negative effect on energy development, and that foreign investment has no influence, direct or indirect, on political exclusion.

8.2.4 More Simulation Studies

In the following simulation study we used data generated from the graph of figure 8.4, which illustrates some of the confusions that seem to be present in the regression produced by Timberlake and Williams.

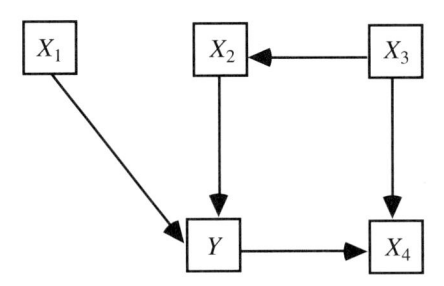

Figure 8.4

For both the linear and the discrete cases with binary variables, one hundred trials were run at each of sample sizes 2,000 and 10,000 using the SGS algorithm. A similar set was run using the PC algorithm for linear and ternary variables. (Each of these algorithms assumes causal sufficiency.) Results were scored separately for errors concerning the existence and the directions of edges, and for correct choice of regressors. Let us call the pattern of the graph in figure 8.4 the true pattern. Recall that an edge existence error of commission (Co) occurs when any pair of variables are adjacent in the output pattern but not in the true pattern. An edge direction error of commission occurs when in an edge occurring in both the true pattern and the output pattern there is an arrowhead in the output pattern but not the true pattern. Errors of omission (Om) are defined analogously in each case. The results are tabulated as the average over the trial distributions of the ratio of the number of actual errors to the number of possible errors of each kind. The proportion of trials in which both (Both) actual causes of Y were correctly identified (with no incorrect causes), and in which one (One) but not both causes of Y were correctly identified (again with no incorrect causes) were recorded for each sample size:

Variable Type	#trials	n	%Edge Existence		%Edge Direction		%Both Correct	%One Correct
			Co	Om	Co	Om		
SGS								
Linear	100	2000	1.4	3.6	3.0	5.4	85.7	3.6
Linear	100	10000	1.6	1.0	2.7	2.2	90.0	7.0
Binary	100	2000	0.6	16.6	29.5	21.8	38.0	34.0
Binary	100	10000	1.2	7.4	30.0	9.1	60.0	25.0
PC								
Linear	100	2000	6.0	2.0	1.0	6.2	80.0	15.0
Linear	100	10000	0.0	1.0	2.5	2.9	95.0	0.0
Ternary	100	2000	3.0	1.0	29.1	8.3	65.0	35.0
Ternary	100	10000	3.0	2.0	10.8	1.2	85.0	15.0

The differences in the results with the SGS and PC algorithms for discrete data are due to the choice of binary variables in the former case and ternary variables in the latter case. The tests for statistical independence with discrete variables appear to have more power when variables can have more than two values.

For purposes of prediction and policy, the numbers in the last two columns suggest that the procedure quite reliably finds real causes of the outcome variable when the statistical assumptions of the simulations are met, the sample is large and a causal structure like that in figure 8.4 obtains. Regression will in these cases find that all of the regressors influence the outcome variable.

8.3 Error Probabilities for Specification Searches

We have shown that various algorithms for specifying causal structure from the data are correct if the requisite statistical decisions are correctly made, but we have given no results about the probability of various sorts of errors in small and medium size samples. The Neyman-Pearson account of testing has made popular two measures of error: the probability of rejecting the null hypothesis when it is true (type I), and the probability of not rejecting the null hypothesis when an alternative is true (type II). Correspondingly, when a search procedure yields a model M from a sample, we can ask for the probability that, were the model M true, the procedure would not find it on samples of that size, and given an alternative M', we can ask for the probability that were M' true the search procedure would find M on samples of that size. We shall also refer to the error probabilities for the outcomes of search procedures as probabilities of type I and type II errors respectively. Especially in small samples, the significance levels and powers of the tests used in deciding conditional independence may not be reliable indicators of the probabilities of corresponding errors in the search procedure.

Error probabilities for search procedures are nearly impossible to obtain analytically, and we have recommended that Monte Carlo methods be used instead. (A discussion of assumptions under which confidence intervals for search do or do not exist is given in chapter 12.) When a procedure yields M from a sample of size n, estimate M and use the estimated model to generate a number of samples of size n, run the search procedure on each and count the frequency with which something other than M is found. For plausible or interesting alternative models M', estimate M', use the estimated model to generate a number of samples of size n, run the search procedure on each and count the frequency with which M is found. We will illustrate the determination of error probabilities for specification searches with a case in which probability of type II error is quite high.

Weisberg (1985) illustrates a procedure for detecting outliers with an experimental study for which regression produces anomalous results:

> An experiment was conducted to investigate the amount of a particular drug present in the liver of a rat. Nineteen rats were randomly selected, weighed, placed under light ether anesthesia and given an oral dose of the drug. Because it was felt that large livers would absorb more of a given dose than smaller liver, the actual dose an animal received was approximately determined as 40 mg of the drug per kilogram of body weight.
>
> The experimental hypothesis was that, for the method of determining the dose, there is no relationship between the percentage of the dose in the liver (Y) and the body weight (X_1), liver weight (X_2), and relative dose (X_3). (pp. 121–124)

Regressing Y on (X_1, X_2, X_3) gives a result not in agreement with the hypothesis; the coefficients of Y on body weight (X_1) and dose (X_3) are both significant, even though one is determined by the other. We find the following regression values for Weisberg's data (standard errors are parenthesized and below the coefficients, and t statistics are shown just below the standard errors):

$$Y = -3.902*X_1 + .197*X_2 + 3.995*X_3 + \varepsilon$$
    (1.345)   (.218)   (1.336)
    −2.901   .903   2.989

A multiple regression not including X_2 also yields significant regressors for X_1 and X_3 at the .05 level. Yet, Weisberg observes that no individual regression of Y on any one of the X variables is significant at that level. The results of the several statistical decisions are therefore inconsistent; we have, for example, that $\rho_{X_1 Y} = 0$, $\rho_{X_2 Y} = 0$, $\rho_{X_3 Y} = 0$ but $\rho_{X_1 Y . X_3} \neq 0$. One might take any of several views about such inconsistencies. One is that it is largely an artifact of the particular significance level used. If the .01 level were used to reject hypotheses of vanishing correlations and partial correlations, the correlations with Y would vanish and so would the partial correlations controlling for one other variable. But the partial correlation of X_1 with Y controlling for both of the other regressors could not be rejected, and an inconsistency would remain. Another view is that inconsistencies in the outcomes of sequential statistical decisions are to be expected, especially in small samples, and where possible, inferences should be based on the statistical decisions that are most reliable. In the case at hand the power of any of the statistical tests is low because of the sample size, but the lower the order of the partial correlation the greater the power. The rule of thumb is that to control for an extra variable is to throw away a data point. Thus in this case the PC algorithm never considers the partial correlations and concludes solely from the vanishing correlations that none of the X variables cause the Y variable. Weisberg instead recommends excluding one of the 19 data points, after which a multiple regression using the remaining data gives no (.05) significant regression coefficients.

From the experimental setup we can assume that body weight and liver weight are causally prior to dose, which itself is prior to the outcome, that is, the amount of the drug found in the rat's liver. Applying the PC algorithm to the original data set with this background knowledge, and using the .05 significance level in the program's tests, we get the pattern in figure 8.5.

The PC algorithm gives the supposed correct result in this case because no correlation of an X variable with Y is significant, and that is all the program needs to decide absence of influence. The regression of Y against each individual X variable alone is an essentially equivalent test. To estimate the type I error of the PC search, we obtained a maximum likelihood estimate (assuming normal distributions) of the model shown in figure 8.5 and used it to generate 100 simulated data sets each of size 19. The PC algorithm was then applied to each data set. In four of the 100 samples the procedure erroneously introduced an edge between Y and one or another of the X variables.

Regression, Causation, and Prediction

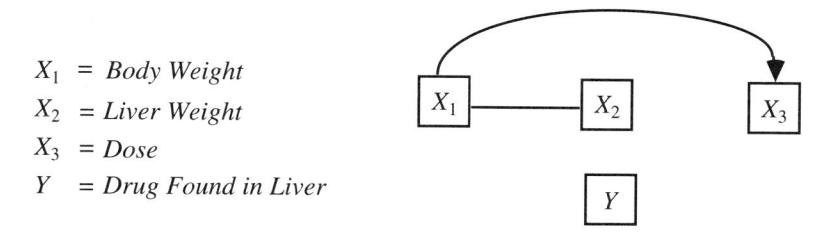

X_1 = Body Weight
X_2 = Liver Weight
X_3 = Dose
Y = Drug Found in Liver

Model Output by the PC Algorithm

Figure 8.5

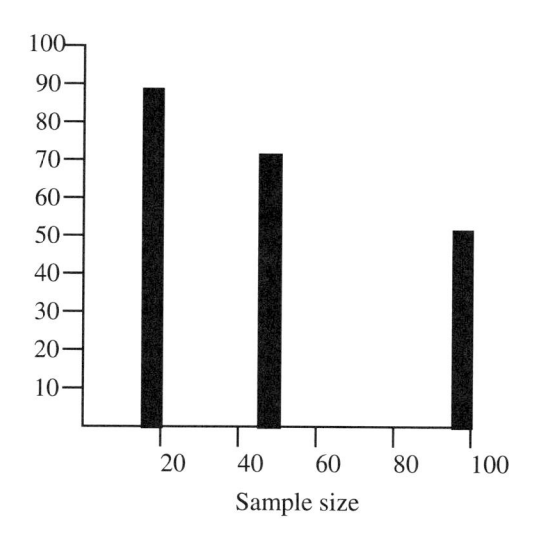

Figure 8.6

To investigate the power of the procedure against alternatives, we consider three models in which Y is connected to at least one X variable. The first is simply the regression model with correlations among the regressors estimated from the sample. With a correlation of about .99 between X_1 and X_3, the regression model with correlated errors is nearly unfaithful, and we should expect the search to be liable not to find the structure. We generated 100 samples at each of the sizes 19, 50, 100, and 1000. We then ran the PC algorithm on each sample, counting the output as a type 2 error if it included no edge between Y and some X variable. Figure 8.6 gives the results for the first three sample sizes.

In 100 trials at sample size 1,000 PC never makes a type 2 error against this alternative.

The second alternative is an elaboration of the original PC output. We add an edge from body weight to the outcome, giving the graph in figure 8.7.

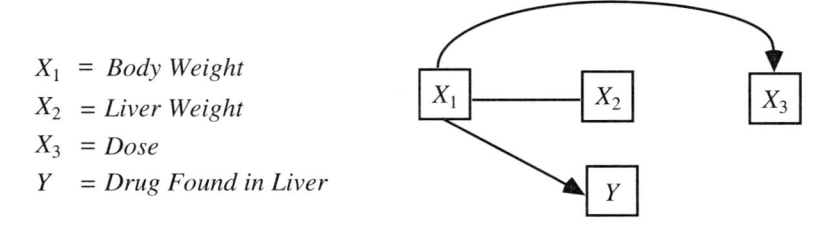

X_1 = Body Weight
X_2 = Liver Weight
X_3 = Dose
Y = Drug Found in Liver

Figure 8.7

We estimated this model with the EQS program, which found a value of .228 for the linear coefficient associated with the $X_1 \rightarrow Y$ edge, and then used the estimated model to again generate 100 samples at each of the four sample sizes. The results for the first three sample sizes are shown in figure 8.8.

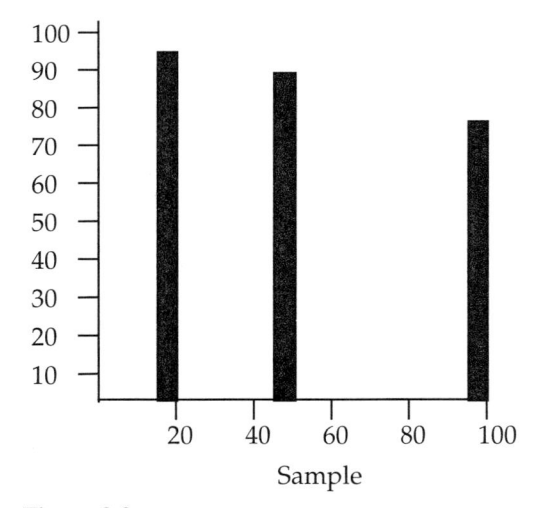

Figure 8.8

Even at sample size 1,000 the search makes an error of type 2 against this alternative in 55% of the cases. "Small" influences of body weight on Y cannot be detected. We would expect the same to be true of dose.

In the third case we increased the linear coefficient connecting X_1 and Y in the model in figure 8.7 to 1.0.

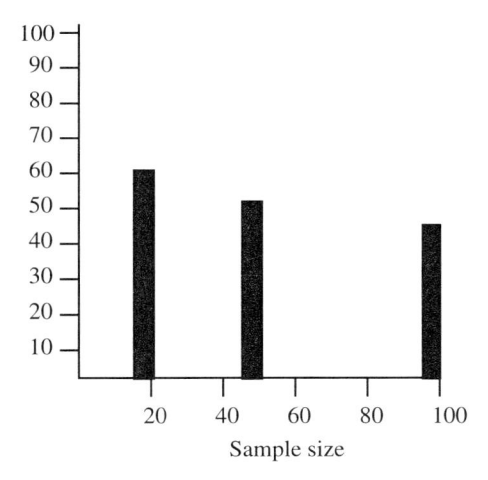

[bar chart: y-axis 0–100, x-axis Sample size 20, 40, 60, 80, 100; bars approximately 60 at 20, 52 at 40, 45 at 80]

Figure 8.9

At sample size 1,000 the search makes an error against this alternative in 2% of the cases.

For this problem at small sample sizes the search has little power against some alternatives, and little power even at large sample sizes against alternatives that may not be implausible.

8.4 Conclusion

In the absence of very strong prior causal knowledge, multiple regression should not be used to select the variables that influence an outcome or criterion variable in data from uncontrolled studies. So far as we can tell, the popular automatic regression search procedures should not be used at all in contexts where causal inferences are at stake. Such contexts require improved versions of algorithms like those described here to select those variables whose influence on an outcome can be reliably estimated by regression. In applications, the power of the specification searches against reasonable alternative explanations of the data is easy to determine by simulation and ought to be investigated.

It should be noted that the present state of the algorithm is scarcely the last word on selecting direct causes. There are cases in which a partially oriented inducing path graph of a directed acyclic graph G over O contains a directed edge from X to Y even though X is not a *direct* cause of Y relative to O (although of course there is a directed path from X to Y in G). However, theorem 6.8 states a sufficient condition for a directed edge in a partially oriented inducing path graph to entail that X is a direct cause of Y. In some cases tests based on constraints such as Verma and Pearl's, noted in section 6.9, would help with the problem, but they have not been developed or implemented.

9 The Design of Empirical Studies

Simple extensions of the results of the preceding chapters are relevant to the design of empirical studies. In this chapter we consider only a few fundamental issues. They include a comparison of the powers of observational and experimental designs, some implications for sampling and variable selection, and some considerations regarding ethical experimental design. We conclude with a reconsideration from the present perspective of the famous dispute over the causal conclusions that could legitimately be drawn from epidemiological studies of smoking and health.

9.1 Observational or Experimental Study?

There are any number of practical issues about both experimental and non-experimental studies that will not concern us here. Questions of the practical difficulty of obtaining an adequate random sample aside, when can alternative possible causal structures be distinguished without experiment and when only by experiment?

Suppose that one is interested in whether a treatment T causes an outcome O. According to Fisher (1959) one important advantage of a randomized experiment is that it eliminates from consideration several alternatives to the causal hypothesis to be tested. If the value of T is assigned randomly, then the hypothesis that O causes T or that there is an unmeasured common cause of O and T can be eliminated. Fisher argues that the elimination of this alternative hypothesis greatly simplifies causal inference; the question of whether T causes O is reduced to the question of whether T is statistically dependent on O. (This assumes, of course, instances of the Markov and Faithfulness Conditions.)

Critics of randomized experiments, for example, Howson and Urbach (1989), have correctly questioned whether randomization in all cases does eliminate this alternative hypothesis. The treatments given to people are typically very complex and change the values of many random variables. For example, suppose one is interested in the question of whether inhaling tobacco smoke from cigarettes causes lung cancer. Imagine a randomized experiment in which one group of people is randomly assigned to a control group (not allowed to smoke) and another group is randomly assigned to a treatment group (forced to smoke 20 cigarettes a day). Further imagine that the experimenter does not know that an unrecorded feature of the cigarettes, such as the presence of a chemical in some of the paper wrappings of the cigarettes, is the actual cause of lung cancer, and inhaling tobacco smoke does not cause lung cancer. In that case lung cancer and inhaling tobacco smoke from cigarettes are statistically dependent even though inhaling tobacco smoke from cigarettes does not cause lung cancer. They are dependent because assignment to the treatment group is a common cause of inhaling tobacco smoke from cigarettes and of lung cancer.

Fisher (1951, p. 20) suggests that "the random choice of the objects to be treated in different ways would be a complete guarantee of the validity of the test of significance, if these treatments were the last in time of the stages in the physical history of the objects which might affect their experimental reaction." But this does not explain how an experimenter who does not even suspect that cigarette paper might be treated with some cancer causing chemical could know that he had not eliminated all common causes of lung cancer and inhaling tobacco smoke from cigarettes, even though he had randomized assignment to the treatment group. This is an important and difficult question about randomization, made more difficult by the fact that randomization often produces deterministic relationships between such variables as drug dosage and treatment group, producing violations of the Faithfulness Condition.

In this section we will put aside this question, and simply assume that an experimenter has some method that correctly eliminates the possibility that O causes T or that there are common causes of O and T. In general, causal inferences from experiments are based on the principles described in chapters 6 and 7. The theory applies uniformly to inferences from experimental and from non-experimental data. Inferences to causal structure are often more informative when experimental data is available, not because causation is somehow logically tied to experimental manipulations, but because the experimental setup provides relevant background causal knowledge that is not available about non-experimental data. (See Pearl and Verma 1991 for a similar point.)

There are, of course, besides the argument that randomization eliminates some alternative causal hypotheses from consideration, a variety of other arguments that have been given for randomization. It has been argued that it reduces observer bias; that it warrants the experimenter assigning a probability distribution to features of the outcomes conditional on the null (causal) hypothesis being true, thereby allowing him to perform a statistical test and calculate the probability of type I error; that for discrete random variables it can increase the power of a test by simulating continuity; and that by bypassing 'nuisance factors' it provides a basis for precise confidence levels. We will not address these arguments for randomization here; for a discussion of these arguments from a Bayesian perspective see, for example, Kadane and Seidenfeld 1990.

Consider three alternative causal structures, and let us suppose for the moment that they exhaust the possibilities and are mutually exclusive: (i) A causes C, (ii) some third variable B causes both A and C, or (iii) C causes A. If by experimental manipulation we can produce a known distribution on A not caused by B or C, and if we can produce a known distribution on C not caused by A or B, we can distinguish these causal structures. In the experiment, all of the edges into A in the causal graph of the non-experimental population are broken, and replaced by an edge from U to A; furthermore there is no non-empty undirected path between U and any other variable in the graph that does not contain the edge from U to A. Any procedure in which A is caused only by a variable U with these properties we will call a **controlled experiment**. In a controlled experiment we know three useful facts about U: U causes A, there is no common cause of U and C,

The Design of Empirical Studies 211

and if U causes C it does so by a mechanism that is blocked if A is held constant (i.e., in the causal graph if there is a directed path from U to C it contains A.). As we noted in chapter 7, U is not a policy variable and is not included in the combined, manipulated or unmanipulated causal graphs.

The controlled experimental setups for the three alternative causal structures are shown in figure 9.1, where an A-experiment represents a manipulation of A breaking the edges into A, and a C-experiment represents a manipulation of C breaking edges into C. If we do an A-experiment and find partially oriented inducing path graph (ia*) over $\{A,C\}$ then we know that A causes C because we know that we have broken all edges into A. (We do not include U (or V) in the partially oriented inducing path graphs in figure 9.1 because including them does not strengthen the conclusions that can be drawn in this case, but does complicate the analysis because of the possible deterministic relationships between U and A.) Similarly, if we perform a C-experiment and find partially oriented inducing path graph (iiic*) then we know that C causes A. If we perform an A-experiment and get (iia*) and a C-experiment and get (iic*) then we know that there is a latent common cause of A and C (assuming that A and C are dependent in the non-experimental population.)

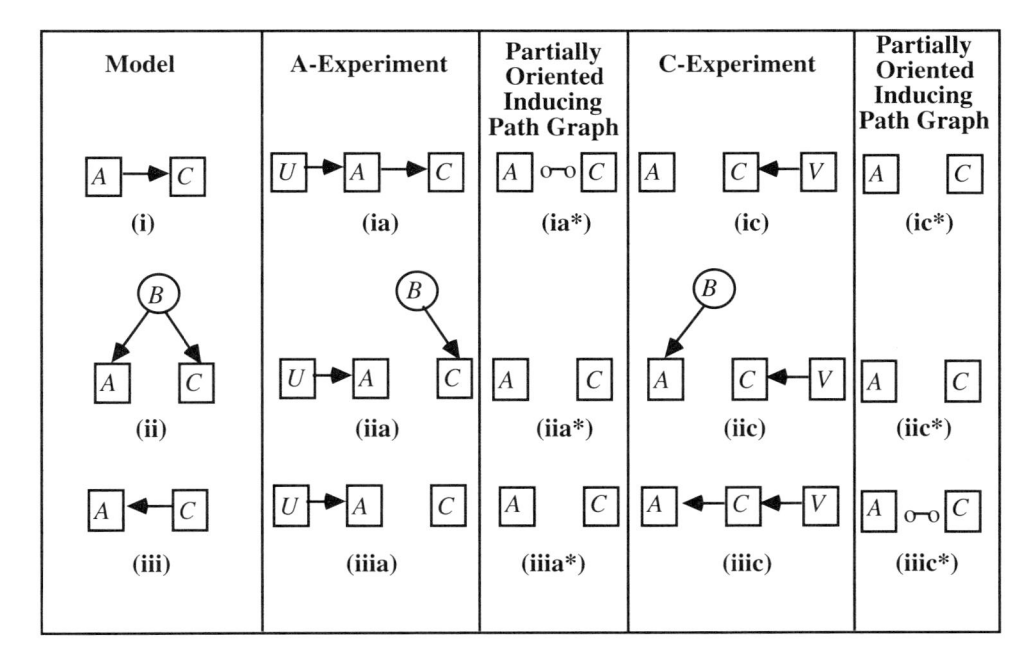

Figure 9.1

Now suppose that in the non-experimental population there are variables U and V *known* to bear the same relations to A and C respectively as in the experimental setup. (We assume in the non-experimental population that A is not a deterministic function of U, and C is not a deterministic function of V.) That is, U causes A, there is no common cause of U and C, and if there is any directed path from U to C it contains A; also, V causes C, there is no common cause of V and A, and if there is any directed path from V to A it contains C. Can we still distinguish (i), (ii), and (iii) from each other without an experiment? The answer is yes. In figure 9.2, (io*), (iio*), and (iiio*) are the partially oriented inducing path graphs corresponding to (io), (iio), and (iiio) respectively. Suppose the FCI algorithm constructs (io*). If it is known that U causes A, then from the fact that the edge between U and A and the edge between A and C do not collide, we can conclude that the edge between A and C is oriented as $A \rightarrow C$ in the inducing path graph. It follows that A causes C. Similarly, if the FCI algorithm constructs (iiio*) ideally we can conclude that C causes A. The partially oriented inducing path graph in (iio*) indicates by theorem 6.9 that there is a latent common cause of A and C, and by theorem 6.6 that A does not cause C, and C does not cause A.

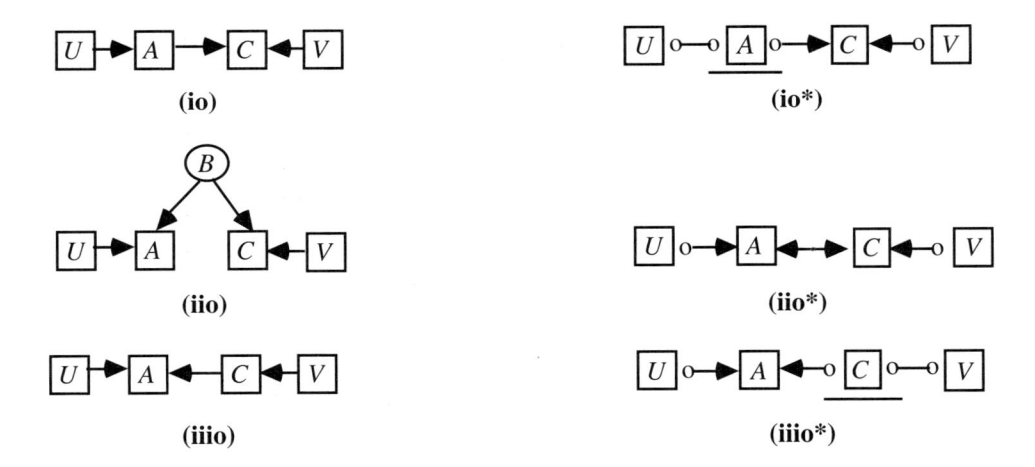

Figure 9.2

The Design of Empirical Studies

Note that if we had measured variables such as *W*, *U*, *V*, and *X* in figure 9.3 then the corresponding partially oriented inducing path graphs would enable us to distinguish (i), (ii), and (iii) without experimentation and without the use of any prior knowledge about the causal relations among the variables.

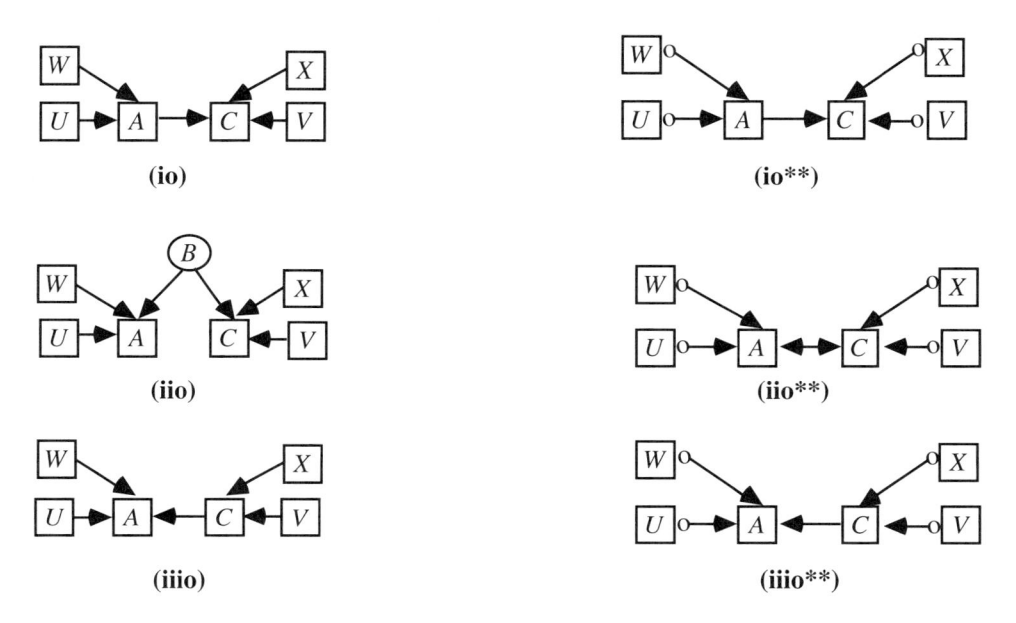

Figure 9.3

214   Chapter 9

Consider now the more complex cases in which the possibilities are (i) *A* causes *C and* there is a latent common cause *B* of *A* and *C*, (ii) there is a latent common cause *B* of *A* and *C*, and (iii) *C* causes *A and* there is a latent common cause *B* of *A* and *C*. Each of the structures (i), (ii), and (iii) can be distinguished from the others by experimental manipulations in which for a sample of systems we break the edges into *A* and impose a distribution on *A* and for another sample we break the edges into *C* and impose a distribution on *C*. The corresponding graphs are presented in figure 9.4, and the analysis of the experiment is essentially the same as in the previous case.

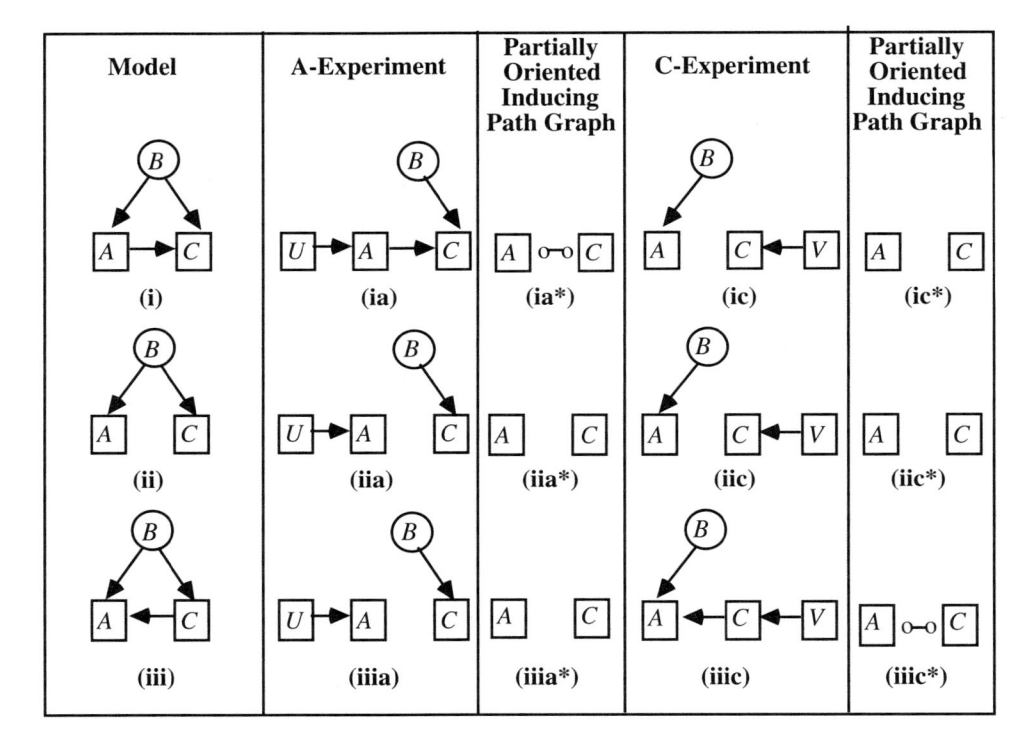

Figure 9.4

The analysis of the corresponding non-experimental case is more complicated. Assume that there is a variable *U* and it is known that *U* causes *A*, there is no common cause of *U* and *A*, and if there is any directed path from *U* to *C* it contains *A*, and that there is a variable *V* and it is known that *V* causes *C*, there is no common cause of *V* and *A*, and if there is any directed path from *V* to *A* it contains *C*. The directed acyclic graphs and their corresponding partially oriented inducing path graphs are shown in figure 9.5. Now suppose that the directed acyclic graphs are true of an observed non-experimental population. Can we still distinguish (i), (ii), and (iii) from each other?

Once again the answer is yes. For example, suppose that an application of the FCI algorithm produces (io*). The existence of the $U \circ\!\!\to C$ edge entails that either there is a common cause of U and C or a directed path from U to C. By assumption, there is no common cause of U and C, so there is a directed path from U to C. Also by assumption, all directed paths from U to C contain A, so there is a directed path from A to C. Given that there is an edge between U and C in the partially oriented inducing path graph, and the same background knowledge, it also follows that there is a latent common cause of A and C. (The proof is somewhat complex and we have placed it in an Appendix to this chapter.) Similarly, if we obtain (iiio*) then we know that C causes A and there is a latent common cause of A and C. If we obtain (iio*) then we know that A and C have a latent common cause but that A does not cause C and C does not cause A. It is also possible to distinguish (i), (ii), and (iii) from each other without any prior knowledge of particular causal relations, but it requires a more complex pattern of measured variables, as shown in figure 9.6. If we obtain (io**) then we know without using any such prior knowledge about the causal relationships between the variables that A causes C and that there is a latent common cause of A and C, and similarly for (iio**) and (iiio**).

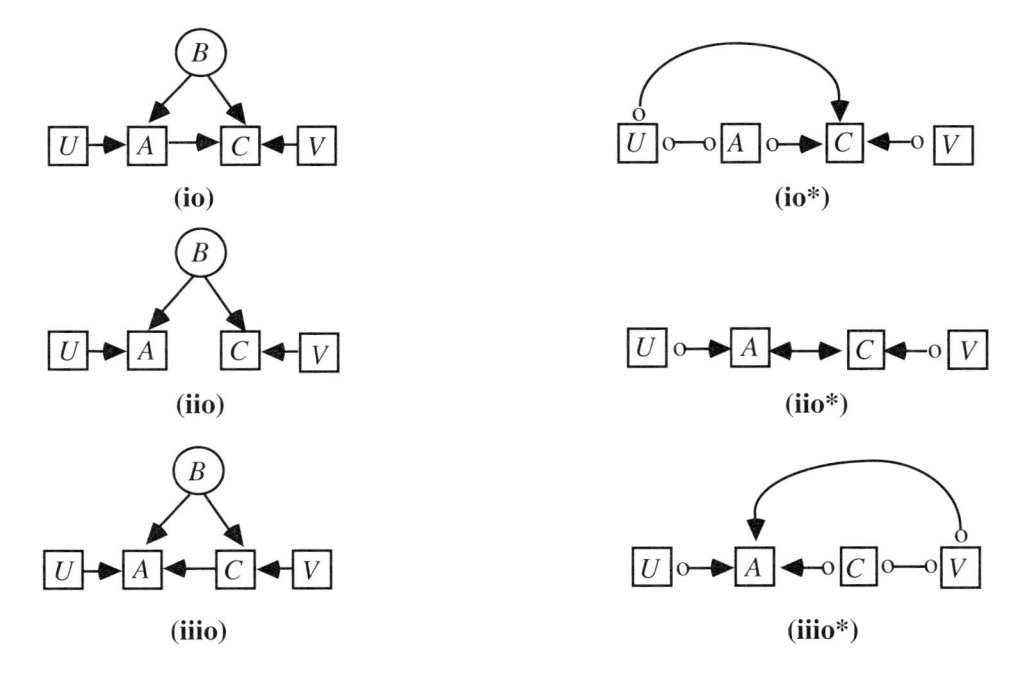

Figure 9.5

There is an important advantage to experimentation over passive observation in one of these cases. By performing an experiment we can make a quantitative prediction about the consequences of manipulating A in (i), (ii), and (iii). But if (i) is the correct causal model, we cannot use the Prediction Algorithm to make a quantitative prediction of the effects of manipulating A. (In the linear case, a prediction could be made because U serves as an "instrumental variable.")

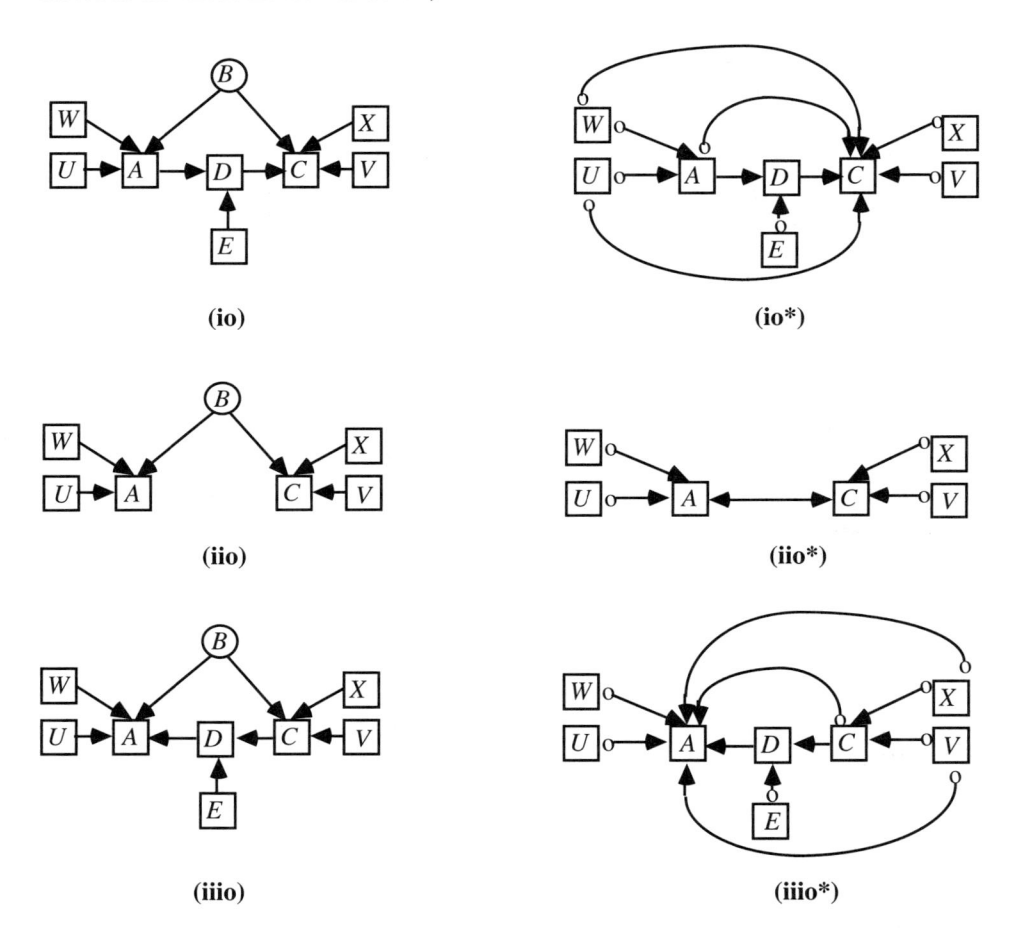

Figure 9.6

The Design of Empirical Studies 217

Suppose finally that we want to know whether there are two causal pathways that lead from *A* to *C*. More specifically, suppose we want to distinguish which of (i), (ii), and (iii) in figure 9.7 obtains, remembering again that *B* is unmeasured.

The question is fairly close to Blyth's version of Simpson's paradox. By experimental manipulation that breaks the edges directing into *A* and imposes a distribution on *A*, we can distinguish structures (i) and (iii) from structure (ii) but not from one another. Note that in figure 9.7 the partially oriented inducing path graph (ia*) is identical to (iiia*) and (ic*) is identical to (iiic*).

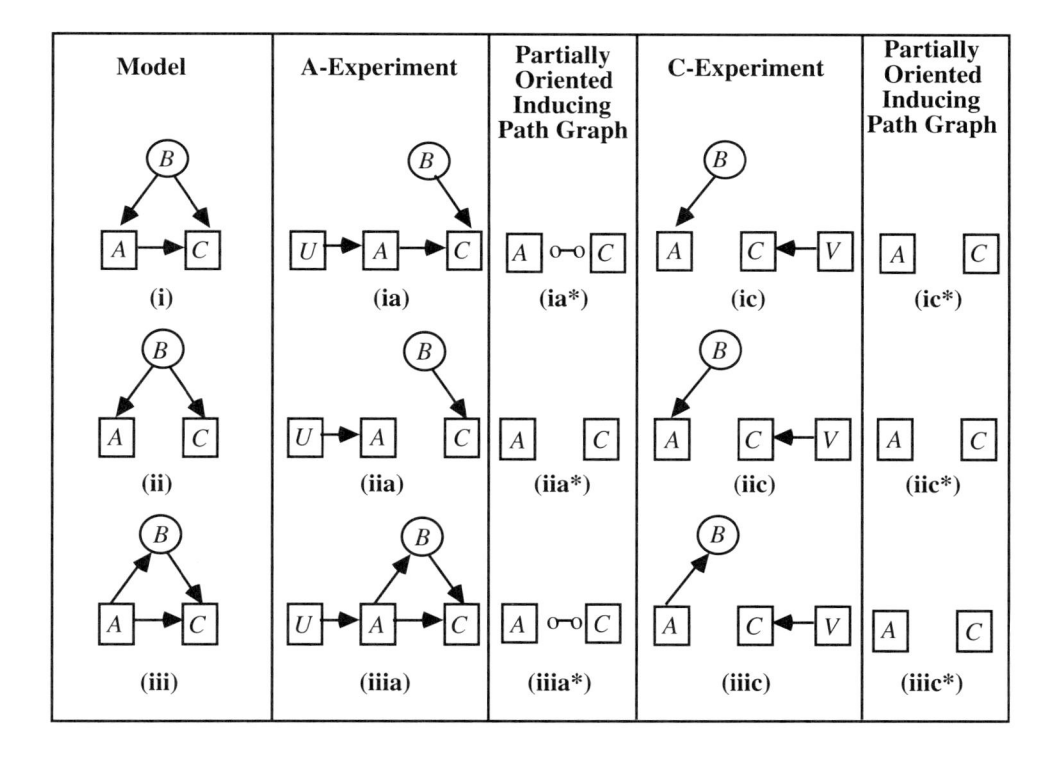

Figure 9.7

Assume once again that in a non-experimental population it is known that U causes A, there is no common cause of U and C, and if there is any path from U to C it contains A, and V causes C, there is no common cause of V and A, and if there is any path from V to A it contains C. The directed acyclic graphs and their corresponding partially oriented inducing path graphs are shown in figure 9.8.

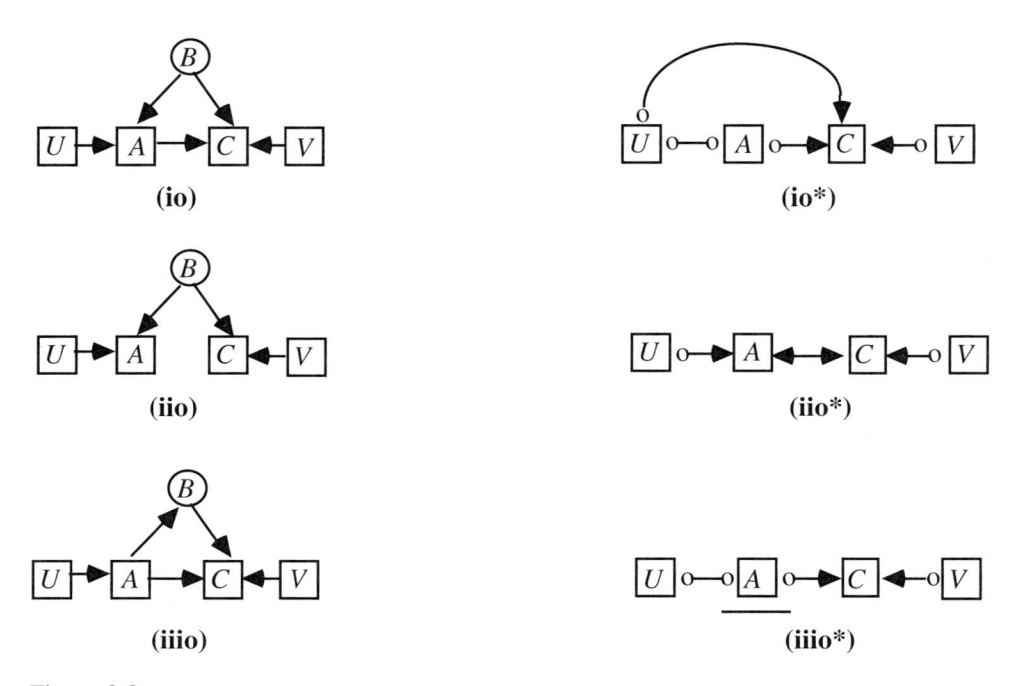

Figure 9.8

Unlike the controlled experimental case, where (i) and (iii) cannot be distinguished, in the non-experimental case they *can* be distinguished. Suppose we obtain (iiio*). We know from the background knowledge that U causes A, and from (iiio*) that the edge between U and A does not collide with the edge between A and C. Hence in the corresponding inducing path graph there is an edge from A to C and in the corresponding directed acyclic graph there is a path from A to C. (Of course we cannot tell how many paths from A to C there are; (iiio*) is compatible with a graph like (iiio) but in which the <A,B,C> path does not exist.) We also know that there is no latent common cause of A and C because (iiio*) together with our background knowledge entails that there is no path in the inducing path graph between A and C that is into A. Suppose on the other hand that we obtain (io*). Recall that the background knowledge together with the partially oriented inducing path graph entail that A is a cause of C and that there is a latent common cause of A and C. (We have placed the proof in an appendix to this chapter.)

Once again if more variables are measured, it is also possible to distinguish these three cases without any background knowledge about the causal relationships among the variables, as shown in figure 9.9.

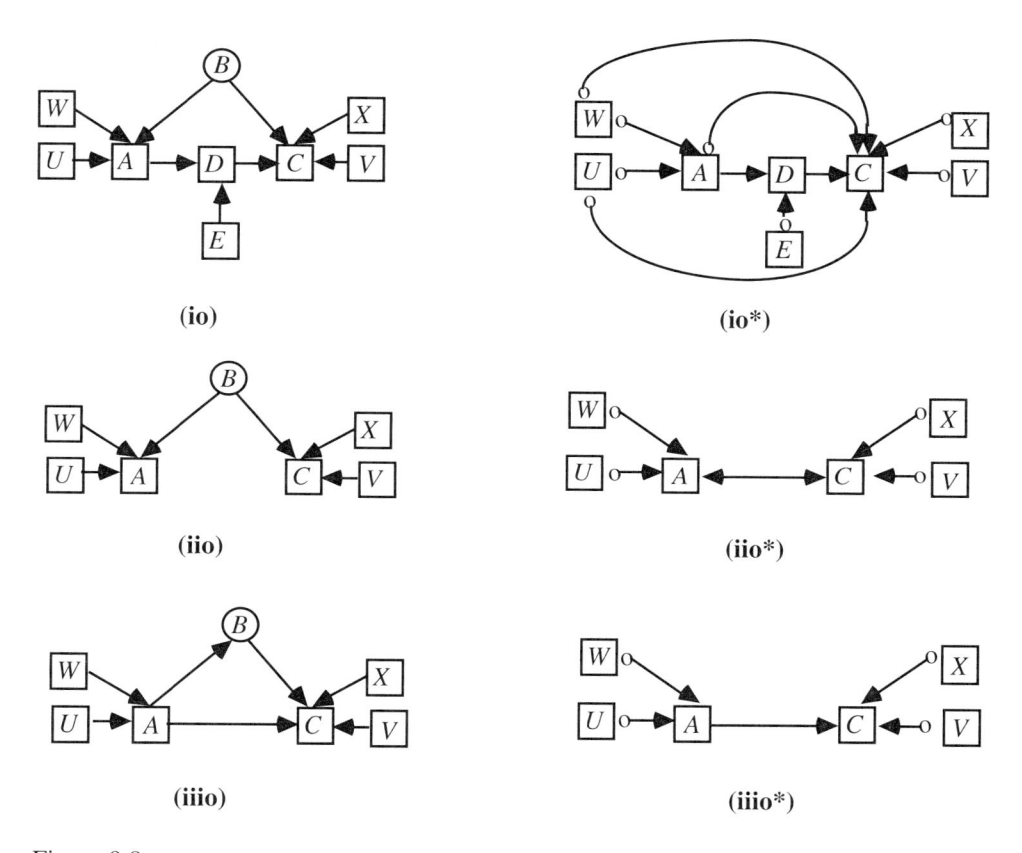

(io)

(io*)

(iio)

(iio*)

(iiio)

(iiio*)

Figure 9.8

Thus all three structures can be distinguished without experimental manipulation or prior knowledge.

It may seem extraordinary to claim that structure (i) in figure 9.7 cannot be distinguished from structure (iii) by a controlled experiment, but can be distinguished without experimental control if the structure is appropriately embedded in a larger structure whose variables are measured. It runs against common sense to claim that when A causes C, a controlled experiment cannot distinguish A and C also having an unmeasured common cause from A also having a second mechanism through which it effects B, but that observation without experiment sometimes can distinguish these situations. But controlled experimental manipulation that forces a distribution on A

breaks the dependency (in the experimental sample) of *A* on *B* in structure (i), and thus information that is essential to distinguish the two structures is lost.

While a controlled experiment alone cannot distinguish (i) from (iii) in figure 9.7 the combination of a simple observational study and controlled experimentation can distinguish (i) from (iii). We can determine from an *A*-experiment that there is a path from *A* to *C*, and hence no path from *C* to *A*. We know if $P(C|A)$ is not invariant under manipulation of *A* then there is a trek between *C* and *A* that is into *A*. Hence if $P(C|A)$ is different in the non-experimental population and the *A*-experimental population we can conclude that there is a common cause of *A* and *C*. If $P(C|A)$ is invariant under manipulation of *A* then we know that either there is no common cause of *A* and *C* or the particular parameter values of the model "coincidentally" produce the invariance. By combining information from an observational study and an experimental study it is sometimes possible to infer causal relations that cannot be inferred from either alone. This is often done in an informal way. For example, suppose that in both an *A*-experiment and a *C*-experiment *A* and *C* are independent. This indicates that there is no directed path from *A* to *C* or *C* to *A*. But it does not distinguish between the case where there is no common cause of *A* and *C* (i.e., there is no trek at all between *A* and *C*) and the case where there is a common cause of *A* and *C*. Of course in practice these two models are distinguished by determining whether *A* and *C* are independent in the non-experimental population; assuming faithfulness, there is a trek between *A* and *C* if and only if *A* is not independent of *C*.

In view of these facts the advantages of experimental procedures in identifying (as distinct from measuring) causal relations need to be recast. There are, of course, well known practical difficulties in obtaining adequate non-experimental random samples without missing values but we are interested in issues of principle. One disadvantage of non-experimental studies is that in order to make the distinctions in structure just illustrated either one has to know something in advance about some of the causal relations of some of the measured variables, or else one must be lucky in actually measuring variables that stand in the right causal relations. The chief advantage of experimentation is that we sometimes know how to *create* the appropriate causal relations. A further advantage to experimental studies is in identifying causal structures in mixed samples. In the experimental population the causal relation between a manipulating variable and a manipulated variable is known to be common to every system so treated. Mixing different causal structures acts like the introduction of a latent variable, which makes inferences about other causal relations from a partially oriented inducing path graph more difficult. Similar conclusions apply to cases in which experimental and statistical controls are combined.

In the "controlled" experiments we have discussed thus far, we have assumed that the experimental manipulation breaks all of the edges into *A* in the causal graph of the non-experimental population, and that the variable *U* used to manipulate the value of *A* has no common cause with *C*. However, it is possible to do informative experiments that satisfy

The Design of Empirical Studies

neither of these assumptions. Suppose, for example, the causal graph of figure 9.10 describes a non-experimental population.

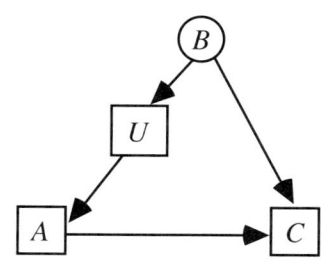

Figure 9.10

Suppose that in an experiment in which we manipulate *A*, we force a distribution upon *P(A|U)*. In this case the causal graph of the experimental population is the same as the causal graph of the non-experimental population, although of course the parametrization of the graph is different in the two populations. This kind of experiment does not break the edges into *A*. More generally, we assume that there is a set of variables **U** used to influence the value of *A* such that any direct cause *V* of *A* that is not in **U** is connected to some outcome variable *C* only by undirected paths that contain *A* as a definite noncollider. (This may occur for example, if **U** is a proper subset of the variables used to fix the value of *A*, and the other variables used to fix the value of *A* are directly connected *only* to *A*.) These are just the conditions that we need in order to guarantee the invariance of the distribution of a variable *C* given **U** and *A*, and hence allows the use of the Prediction Algorithm. (A more extensive discussion of this kind of experiment is given in section 9.4.) With an experiment of this kind, it is possible to distinguish model (i) from model (iii) in figure 9.7. Of course, with the same background knowledge assumptions it is also possible to distinguish (i) from (iii) in a non-experimental study in which the distribution of *P(A|U)* is not changed. Indeed with this kind of experiment, the only difference between the analysis of the experimental population and a non-experimental population lies in the background knowledge employed.

9.2 Selecting Variables

The selection of variables is the part of inference that at present depends almost entirely on human judgment. We have seen that poor variable selection will usually not of itself lead to incorrect causal conclusions, but can very well result in a loss of information. Discretizing continuous variables and using continuous approximations for discrete variables both risk altering the results of statistical decisions about conditional independence.

One fundamental new consideration in the selection of variables is that in the absence of prior knowledge of the causal structure, empirical studies that aim to measure the influence, if any, of *A* on *C* or of *C* on *A*, should try to measure at least two variables correlated with *A* that are thought to influence *C*, if at all, only through *A*, and likewise for *C*. As the previous section illustrates, variables with these properties are especially informative about whether *A* causes *C*, or *C* causes *A*, or neither causes the other but there is a common cause of *A* and *C*.

The strategy of measuring every variable that might be a common cause of *A* and *B* and conditioning on all such variables is hazardous. If one of the additional variables is in fact an effect of *B*, or shares a common cause with *A* and a common cause with *B*, conditioning on that variable will produce a spurious dependency between *A* and *B*. That is not to say that extra variables should not be measured if it is thought that they may be common causes; but if they are measured, they should be analyzed by the methods of chapters 5 and 6 rather than by multiple regression.

Finally, if methods like those described in chapters 5, 6, and 7 are to be employed, we offer the obvious but unusual suggestion that variables be selected for which good conditional independence tests are available.

9.3 Sampling

We can view many sampling designs as procedures that specify a property *S*, which may have two values or several, and from subpopulations with particular *S* values draw a sample in which the distribution of values of the ith unit drawn is distributed independently of and identically to the distribution of all other sample places from that subpopulation. In the simplest case *S* can be viewed as a binary variable with the value 1 indicating that a unit has the sample property, which of course does not mean that the unit occurs in any particular actual sample. We distinguish the sample property *S* from any treatments that might be applied to members of the sample. In sampling according to a property *S* we obtain information directly not about the general population but about the segments of the population that have various values of *S*. Our general questions therefore concern when conditioning on any value of *S* in the population leaves unaltered the conditional probabilities or conditional independence relations for variables in the causal graph *G* describing the causal structure of each unit in the population. That is, suppose there is a population in which the causal structure of all units is described by a directed graph *G*, and let the values of the variables be distributed as *P*, where *P* is faithful to *G*. What are the causal and statistical constraints a sampling property *S* must satisfy in order that a sub-population consisting of all units with a given value of *S* will accurately reflect the conditional independence relations in *P*—and thus the causal structure *G*—and under what conditions will the conditional probabilities for such sub-populations be as in *P*? The answers to these questions bear on a number of familiar questions about sampling, including the appropriateness of retrospective versus prospective sampling and of random sampling as against other sampling arrangements. We will not consider questions about the sampling distributions obtained by imposing various constraints on the distribution of

The Design of Empirical Studies 223

values of *S* in a sample. Our discussion assumes that *S* (which may be identical to one of the variables in *G*) is not determined by any subset of the other variables in *G*.

We assume in our discussion that *S* is defined in such a way that if the sampling procedure necessarily excludes any part of the population from occurring in a sample, then the excluded units have the same *S* value. For example, if a sample is to be drawn from the sub-population of people over six feet tall, then we will assume that $S = 0$ corresponds to people six feet tall or under and $S = 1$ corresponds to people over 6 feet tall.

The causal graph *G* relating the variables of interest can be expanded to a graph *G(S)* that includes *S* and whatever causal relations *S* and the other variables realize. We assume a distribution *P(S)* faithful to *G(S)* whose marginal distribution summing over *S* values will of course be *P*. We suppose that the sampling distribution is determined by the conditional distribution $P(\ |S)$. Our questions are then, more precisely, when this conditional distribution has the same conditional probabilities and conditional independence relations as *P*. We require, moreover, that the answer be given in terms of the properties of the graph *G(S)*. The following theorem is obvious and will not be proved.

THEOREM 9.1: If *P(S)* is faithful to *G(S)*, and **X** and **Y** are sets of variables in *G(S)* not containing S, then $P(\mathbf{Y}|\mathbf{X}) = P(\mathbf{Y}|\mathbf{X},S)$ if and only if **X** d-separates **Y** and *S* in *G(S)*.

Our sampling property should not be the direct or indirect cause or effect of **Y** save through a mechanism blocked by **X**, and **X** should not be the effect, direct or indirect of both **Y** and the sampling property. (The second clause in effect guarantees that Simpson's paradox is avoided in a faithful distribution). The theorem is essentially the observation that $P(\mathbf{Y}|\mathbf{X} \cup \mathbf{Z}) = P(\mathbf{Y}|\mathbf{X} \cup \mathbf{Z} \cup \{S\})$ if and only if in *P* **Y** and *S* are independent conditional on $\mathbf{X} \cup \mathbf{Z}$. It entails, for example, that if we wish to estimate the conditional probability of **Y** on **X** from a sample of units with an *S* property (say, $S = 1$), we should try to ensure that there is

(i) no direct edge between any *Y* in **Y** and *S*,
(ii) no trek between any *Y* in **Y** and *S* that does not contain some *X* in **X**, and
(iii) no pair of directed paths from any *Y* in **Y** to an *X* in **X** and from *S* to *X*.

Figure 9.11 illustrates some of the ways estimation from the sampling property can bias estimates of the conditional probability of **Y** given **X** and **Z**.

Cases (i) and (iii) are typical of retrospective designs. In case (ii) the sampling property biases estimates of *P(Y|X,Z)* because *Y* and the sample property *S* are dependent conditional on $\{X,Z\}$. Theorem 9.1 amounts to a (very) partial justification of the notion that: "prospective" sampling is more reliable than "retrospective" sampling, if by the former is meant a procedure that selects by a property that causes or is caused by *Y*, the

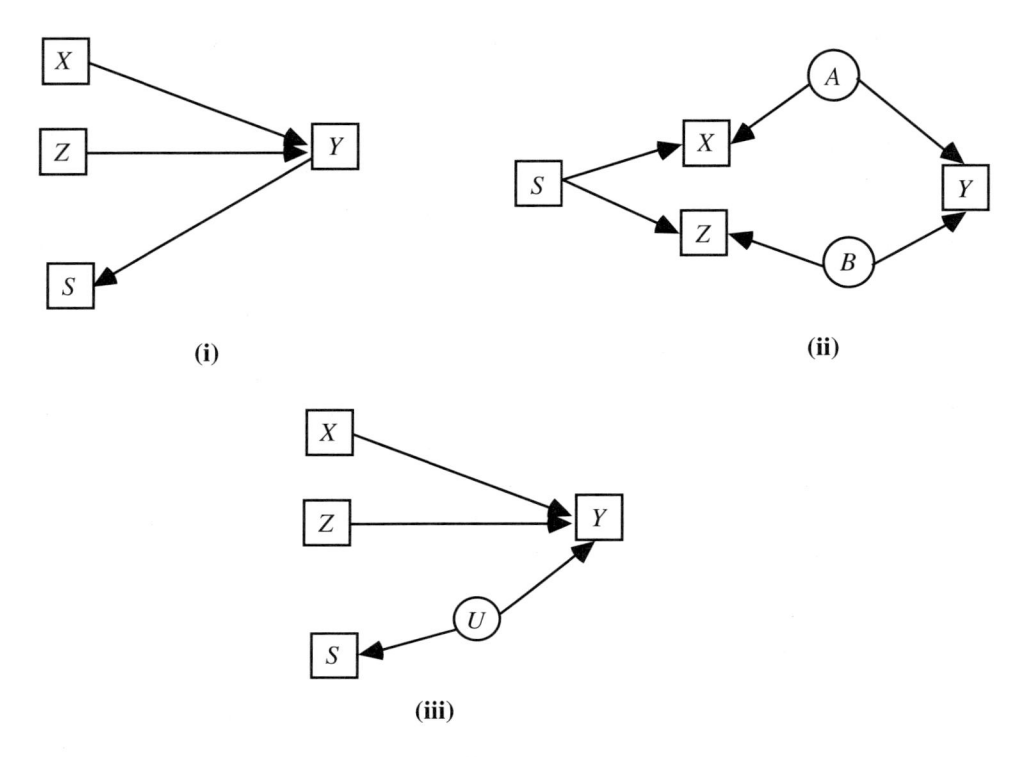

Figure 9.11

effect, if at all only through X, the cause, and by the latter is meant a procedure that selects by a property that causes or is caused by X only through Y. In a prospective sampling design in which X is the only direct cause of S, and S does not cause any variable, the estimate of $P(Y|X,Z)$ is not biased. But case (ii) shows that under some conditions prospective samples can bias estimates as well.

Similar conclusions should be drawn about random sampling. Suppose as before that the goal is to estimate the conditional probability $P(Y|X)$ in distribution P. In drawing a random sample of units from P we attempt to sample according to a property S that is entirely disconnected with the variables of interest in the system. If we succeed in doing that then we ensure that S has no causal connections that can bias the estimate of the conditional probability. Of course even a random sample may fail if the very property of being selected for a study (a property, it should be noted, different from having some particular value of S) affects the outcome, which is part of the reason for blinding treatments. Further, *any* property that has the same causal disconnection will do as a basis for sampling; there is nothing special in this respect about randomization, except that a random S is believed to be causally disconnected from other variables.

When the aim is only to determine the causal structure, and not to estimate the distribution P or the conditional probabilities in P, the asymmetry between prospective and retrospective sampling vanishes.

In model (iii) of figure 9.11, which is an example of retrospective design, for any three disjoint sets of variables **A**, **B**, and **C** not containing S, **A** is d-separated from **B** given **C** if and only if **A** is d-separated from **B** given $\mathbf{C} \cup S$. So these cases in which conditional probability in P cannot be determined from S samples are nonetheless cases in which conditional independence in P, and hence causal structure, can be determined from S samples.

Theorem 9.2 states conditions under which the set of conditional independence relations true in the population as a whole is different from the set of conditional independence relations true in a subpopulation with a constant value of S. In theorem 9.2 let **Z** be any set of variables in G not including X and Y.

THEOREM 9.2: For a joint distribution P, faithful to graph G, exactly one of $<Y \perp\!\!\!\perp X | \mathbf{Z};$ $Y \perp\!\!\!\perp X | \mathbf{Z} \cup \{S\}>$ is true in P if and only if the corresponding member and only that member of $<\mathbf{Z}$ d-separates $X, Y; \mathbf{Z} \cup \{S\}$ d-separates $X, Y>$ is true in G.

Although theorem 9.2 is no more than a restatement of theorem 3.3, its consequences are rather intricate. Suppose that **X** and **Y** are independent conditional on **Z** in distribution P. When will sample property S make it appear that **X** and **Y** are instead dependent conditional on **Z**? The answer is exactly when **X**, **Y** are dependent conditional on $\mathbf{Z} \cup S$ in $P(S)$. This circumstance—conditional independence in P and conditional dependence in $P(S)$—can occur for faithful distributions when and only when there exists an undirected path U from X in **X** to Y in **Y** such that

i. no noncollider on U is in $\mathbf{Z} \cup \{S\}$;
ii. every collider on U has a descendant in $\mathbf{Z} \cup \{S\}$;
iii. some collider on U does not have a descendant in **Z**.

The converse error involves conditional dependence in P and conditional independence in $P(S)$. That can happen in a faithful distribution when and only when there exists an undirected path U from X to Y such that
i. every collider on U has a descendant in **Z**;
ii. no noncollider in U is in **Z**;
and S is a noncollider on every such path. Again, asymptotically both of these errors can be avoided by sampling randomly, or by any property S that is unconnected with the variables of interest.

In experimental designs the aim is sometimes to sample from an ambient population, apply a spectrum of treatments to the sampled units, and then infer from the outcome the effect a policy of treatment would have if applied to the general population. In the next section we consider some relations between experimental design, policy prediction, and causal reasoning.

9.4 Ethical Issues in Experimental Design

Clinical trials of alternative therapies have at least two ethical problems. (1) In the course of the trials (or even beforehand) suspicion may grow to near certainty that some treatments are better than others; is it ethical to assign people to treatments, or to continue treatments, found to be less efficacious? (2) In clinical trials, whether randomized or other, patients are generally *assigned* to treatment categories; if the patients were not part of an experimental design, presumably they would be free to choose their treatment (free, that is, if they could pay for it, or persuade their insurer to); is it ethical to ask or induce patients to forego choosing? Suppose the answer one gives to each of these questions is negative. Are there experimental designs for clinical trials that avoid or mitigate the ethical problems but still permit reasonable predictions of the effects of treatment throughout the population from which the experimental subjects are obtained?

Kadane and Sedransk (1980) describes a design (jointly proposed by Kadane, Sedransk, and Seidenfeld) to meet the first problem. Their design has been used in trials of drugs for post-operative heart patients and in other applications. The inferences Kadane and Seidenfeld (1990) make in explaining the design are in accord with the Markov Condition, and indeed follow from it, and the case nicely illustrates the role of causal reasoning in experimental design. Furthermore, combining the Markov and Faithfulness Conditions, and using the Manipulation Theorem and fheorem 7.1 leads to two novel conclusions:

1. The efficacy of treatments in an experiment can be reliably assessed in an experimental procedure that takes patient preference into account in allocating treatment, but except in special cases the knowledge so acquired could not be used to predict the effects of a general policy of treatment.

2. Perhaps of more practical interest, given comparatively weak causal knowledge on the part of the experts, another design in which treatment allocation depends on patient preference can be used to determine whether or not patient self-assignment in the experiment will confound prediction of the outcome of a treatment policy in the general population. When all influences are linear, the effects of treatment policy can be predicted even if confounding occurs.

9.4.1 The Kadane/Sedransk/Seidenfeld Design

In the Kadane/Sedransk/Seidenfeld experimental design (described in Kadane and Seidenfeld 1990), for each member of a panel of experts, degrees of belief are elicited about the outcome O of each treatment T conditional on each profile of values of $X_1...X_n$.

The elicited judgments are used to specify some prior distribution over parameters in a model of the treatment process. For each experimental subject, the panel of experts receives information on the variables $X_1, ..., X_n$. Nothing else about the patient is known to the experts. Based on the values of $X_1, ..., X_n$ each expert i recommends a preferred treatment $p_i(X)$ to the patient, and the patient is assigned to treatment by some rule $T = h(X, p_1,...,p_k)$ that is a function of the X values and the experts' treatment preferences $(p_1,...p_k)$ for patients described by X, and perhaps some random factor. The rule guarantees that no patient is given a treatment unless at least one expert recommends it for patients with that profile. The model determines the likelihood, for each vector of parameter values, of outcomes conditional on X and T values. As data are collected on patients, the prior distribution over the parameters is updated by conditioning. If the evidence reaches a stage at which all the experts agree that some treatment T for patients with profile X is not the best treatment for the patient, then treatment T is suspended for such patients. As evidence accrues, the experts' degrees of belief about the parameter values of the likelihood model should converge.

Let X_j be a vector of observed characteristics of the jth patient, "including those that are used as a basis for deciding what treatment each patient is to receive, and possibly other characteristics as well." (We do not place X_j in boldface in order to match Kadane and Seidenfeld's notation.) Let T_j be the treatment assigned to patient j. Let O_j be the outcome for patient j. Let $P_j = (O_j, T_j, X_j, O_{j-1}, T_{j-1}, X_{j-1}, ..., X_1)$ be the past evidence up to and including what is known about patient j. Let θ be a vector of the parameters of interest, those that determine the probabilities of outcomes O_j for a patient j given characteristics X_j and treatment T_j. For example, the degrees of belief of an expert might be represented by a mixture of linear models parametrized by exogenous variances, means and linear coefficients. A unique value of these parameters then "determines" the probability of an outcome given X values. For reasons that will become clear, it is essential to the definition of θ that alternative values for the parameter not give alternative specifications of the distribution of X variables.

The expression $f_\theta(P_J)$ represents the expert's conditional degree of belief, given θ, that the total evidence is P_J. Kadane and Seidenfeld add that "It is part of the definition of θ as the parameter that

$$f_\theta(O_j|T_j, X_j, P_{j-1}) = f_\theta(O_j|T_j, X_j) \quad (1 \leq j \leq J) \qquad (1)$$

What this means is that θ contains all the information contained in P_{j-1} that might be useful for predicting O_j from T_j and X_j." The factorization of degree of belief,

$$f_\theta(P_J) = \underbrace{\left[\prod_{j=1}^{J} f_\theta(O_j|T_j, X_j, P_{j-1})\right]}_{1} \underbrace{\left[\prod_{j=1}^{J} f_\theta(T_j|X_j, P_{j-1})\right]}_{2} \underbrace{\left[\prod_{j=1}^{J} f_\theta(X_j|P_{j-1})\right]}_{3}$$

follows by the definition of conditional probability. The terms are marked 1, 2, and 3. Kadane and Seidenfeld claim that term 3 does not depend on θ if one believes that the features, treatments and outcomes for earlier subjects in the experimental trial have no influence on "the kinds of people" who subsequently become subjects. (Recall that parameters relevant to the distribution of $X_1...X_n$ are not included in θ.). Kadane and Seidenfeld also say that term 2 does not depend on θ because there is a fixed rule for treatment assignment as a function of X values and the history of the experimental outcomes.

It follows from (1) that

$$\prod_{j=1}^{J} f_\theta(O_j|T_j, X_j, P_{j-1}) = \prod_{j=1}^{J} f_\theta(O_j|T_j, X_j)$$

Kadane and Seidenfeld say the proportionality given by this term:

$$f_\theta(P_J) \propto \prod_{j=1}^{J} f_\theta(O_j|T_j, X_j)$$

"is the form that we use to evaluate the results of a clinical trial of the kind considered here." That is for each value θ_i of θ, multiplying $f_{\theta_i}(P_J)$ by the prior density of θ_i gives a quantity proportional to the posterior density of θ_i. The ratios of the posterior densities of two values of θ_i can therefore be found.

Now for a new case, each value of θ determines a probability of treatment outcome given an X profile and a treatment T, and so the posterior distribution of θ yields, for any one expert, degrees of belief in the outcomes of various treatment regimes to various classes of patients. Although Kadane and Seidenfeld say nothing explicit about predicting the effects of policies of treatment, these degrees of belief may be transformed into expected values if outcome is somehow quantified. In any case, an expert who began believing that a rule of treatment given by $T = k(X)$ would most often result in a successful outcome, may come instead to predict that a different rule of treatment, say $T = g(X)$ will more often be successful.

Why can't the experiment let the subjects simply choose their own treatments, and seek any advice that they want? Kadane and Seidenfeld give two reasons. One is that if patients were to determine their own treatment, the argument that term 2 in the

The Design of Empirical Studies 229

factorization does not depend on θ would no longer hold. The other is that "It would now be necessary to explain statistically the behavior of patients in choosing their treatments, and there might well be contamination between these choices and the effect of the treatment itself." We will consider the force of these considerations in the next subsection.

9.4.2 Causal Reasoning in the Experimental Design

What is it that the experts believe that warrants this analysis of the experiment, or the derivation of any predictions? The expert surely entertains the possibility that some unknown common causes U may influence both the X features of a patient and the outcome of the patient's treatment. And yet the analysis assumes that in the expert's degrees of belief, treatment is independent of any such U conditional on X values. That is implicit in the claim that term 2 in the factorization is independent of θ. *Why should treatment T and unknown causes U be conditionally independent given X?* The reason, clearly, is that in the experiment the only factors that influence the treatment a patient receives are the X values for that patient and P_{j-1}; any such U, should it exist, has no influence on T except through X. A causal fact is the basis for independence of probabilities.

Aspects of the expert's understanding of the experimental set-up are pictured in figure 9.12 (where we have made X_j a single variable.)

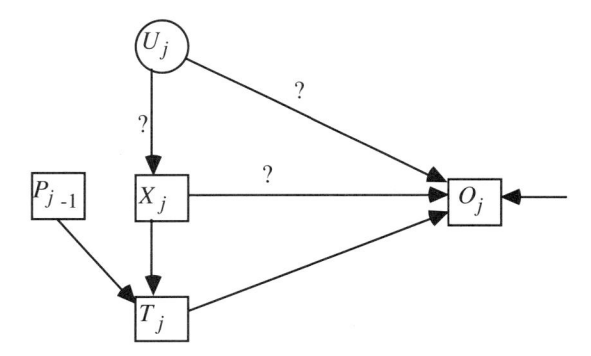

Figure 9.12

The expert may not be at all sure that the edges with "?" correspond to real influences, but she is sure there is *no* influence of the kind in figure 9.13 in boldface.

The experimental design, which makes treatment assignment a function of the X variables and P_{j-1} only, is contrived to exclude such influences. The expert's thought seems to be that if U influences T only through X, then U and T are independent conditional on X. That thought is an instance of the Markov Condition. The probabilities in the Markov Condition can be understood either objectively or subjectively. But in the

Kadane/Sedransk/Seidenfeld design the probability in the Markov Condition cannot be the expert's unconditional degrees of belief, because those probabilities are mixtures over θ of distributions conditional on θ, and mixtures of distributions satisfying the Markov Condition do not always satisfy the Markov Condition. We will assume the distributions conditional on θ do so.

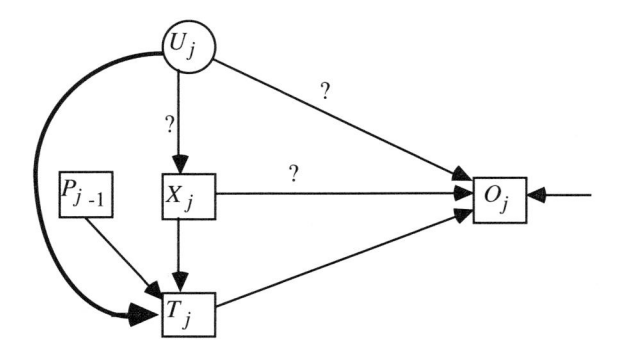

Figure 9.13

Consider another feature of the idealized expert belief. An idealized expert in Kadane and Seidenfeld's experiment changes his probability distribution for a parameter whose values specify a model of the experimental process. At the end of the experiment the expert has a view not only about the outcome to be expected for a new patient with profile X if that patient were assigned treatment according to the rule $T = h(X, P_{j-1})$ used in the experiment, but also about the outcome to be expected for a new patient with profile X if that patient were assigned treatment according to the rule $T = g(X)$ that the expert now, in light of the evidence, prefers. In principle, the expert's probabilities for outcomes if the new patient were treated by the experimental rule $T = h(X, P_{j-1})$ is easy to compute because that probability is determinate for each value of θ, and we know the expert's posterior distribution of θ. *But what determines the expert's probability for outcomes if the patient with profile X is now treated according the preferred rule, $T = g(X)$?* Why doesn't *changing* the rule change the dependence of O on X and T? The sensible answer, implicit in Kadane and Seidenfeld's analysis, is that the outcome for any patient depends on the X profile of the patient and the treatment given to the patient, but not on the "rule" by which treatments are assigned. Changing the assignment rule changes the probability of treatment T given profile X, but has no effect on other relevant conditional probabilities; the probability O given T and X is unaltered. We can derive this more formally in the following way.

If for a fixed value of θ the distribution $f_\theta(O_j, T_j, X_j, P_{j-1})$ satisfies the Markov condition for graphs of the type in figure 9.12, then theorem 7.1 entails that $f_\theta(O_j | T_j, X_j)$ is invariant under a manipulation of T_j. According to theorem 7.1, in a distribution that satisfies the Markov condition for graphs of the type in figure 9.12, f_θ ($O_j|T_j,X_j$) is invariant under a manipulation of T_j if there is no path that d-connects O_j and X_j given T_j that is into T_j. Every undirected path between T_j and O_j that contains some X_j variable satisfies this condition because some member of X_j is a noncollider on such a path. There are no undirected paths between T_j and O_j that contain P_{j-1}. Hence f_θ ($O_j|T_j,X_j$) is invariant under manipulation of T_j.

But does the Markov condition reasonably apply in an experiment designed according to the Kadane/Sedransk/Seidenfeld specifications in which f_θ (O_j,T_j,X_j,P_{j-1}) does not represent frequencies but an expert's opinions? We are concerned with the circumstance in which the experiment is concluded, and the expert's degrees of belief, we will suppose, have converged so far as they will. The expert is uncertain as to whether there are common causes of X and outcome, or how many there are, but all of the causal structures she entertains are like figure 9.12 and none are like figure 9.13. Conditional on θ and any particular causal hypothesis we suppose the Markov condition is satisfied, but the expert's actual degrees of belief are some mixture over different causal structures. Should f_θ ($O_j|T_j,X_j$) be invariant under manipulation of T_j in the opinion of the expert when her distribution for a given value of θ is a mixture of several different causal hypotheses? The answer is yes, as the following argument shows.

Let us call the experimental (unmanipulated in the sense of chapter 7) population *Exp*, and the hypothetical population subjected to some policy based on the results of the experiment *Pol*. Let $f_{\theta,Exp}(O_j,T_j,X_j,P_{j-1})$ represent the expert's degrees of belief about O_j, T_j, X_j, and P_{j-1} conditional on θ in the experimental population, and $f_{\theta,Pol}(O_j,T_j,X_j,P_{j-1})$ represent the expert's degrees of belief about O_j, T_j, X_j, and P_{j-1} conditional on θ in the hypothetical population subjected to some policy. Let *CS* be a random variable that denotes a causal structure. We have already noted that it follows from theorem 7.1 that

$$f_{\theta,Exp}(O_j|T_j,X_j,CS) = f_{\theta,Pol}(O_j|T_j,X_j,CS)$$

Because θ determines the density of O_j conditional on T_j and X_j,

$$f_{\theta,Exp}(O_j|T_j,X_j,CS) = f_{\theta,Exp}(O_j|T_j,X_j)$$
$$f_{\theta,Pol}(O_j|T_j,X_j,CS) = f_{\theta,Pol}(O_j|T_j,X_j)$$

Hence,

$$f_{\theta,Exp}(O_j|T_j,X_j) = f_{\theta,Pol}(O_j|T_j,X_j)$$

Consider next the question of "bias" raised by Kadane and Seidenfeld. The very notion requires us to consider not just degrees of belief but also some facts and some potential facts. We suppose there is really a correct (or nearly correct) value for the parameters in the likelihood model, and the true values describe features of the process that go on in the experiment. We suppose the expert converges to the truth, so that her posterior distribution is concentrated around the true values. What the public that pays for these experiments cares about is whether the expert's views about the best treatment are correct: *Would a policy that puts in place the expert's preferred rule of treatment, say T = g(X), result in better outcomes than alternative policies under consideration?* One way to look at that question is to ask if the expert's expected values for outcome conditional on X profile and treatment roughly equal the population mean for outcome under these conditions. If degrees of belief accord with population distributions, that is just to ask when the frequency of O conditional on T and X that would result if every relevant person in the population were treated on the basis of the experimental assignment rule $T = h(X, P_{j-1})$ would be roughly the same as the frequency of O conditional on T and X for the general population if a revised rule $T = g(X)$ for assigning treatments were used. In other words: *Will the frequency of O conditional on T and X be invariant under a direct manipulation of T?*

As we have just seen for this case, the Markov Condition and theorem 7.1 *entail* that for the graph in figure 9.12, and all others like it (those in which every trek whose source is a common cause of O_j and T_j contains an X_j variable, O_j does not cause T_j, and every common cause of an X_j variable and T_j is an X_j variable) the probability of O_j on T_j and X_j is invariant under a direct manipulation of T_j. No other assumptions are required. The example is a very special case of a general sufficient condition for the invariance of conditional probabilities under a direct manipulation.

Consider next why the experimental design forbids that treatment assignment depend directly on "unrecorded" features of the patients, such as Y. Suppose such assignments were allowed; Kadane and Seidenfeld say the outcome might be "contaminated" which we understand to mean that some unrecorded causes of the patient preference, and hence of T_j, might also be causes of O_j. So that we have the causal picture in figure 9.14.

The question marks indicate that, we (or the expert), are uncertain about whether the corresponding causal influences exist. Suppose the directed edges from U_j to Y and from U_j to O_j exist. Then there is an undirected path between O_j and T_j that contains Y. In this case the Markov condition entails that the probability of O_j conditional on T_j and the X_j variables is *not* invariant under a direct manipulation of T_j except for "coincidental" parameter values. So if unrecorded Y values were allowed to influence assignments then for all we or the experts know, the expert's prediction of the effects of his proposed rule $T = g(X)$ would be wrong.

The Design of Empirical Studies 233

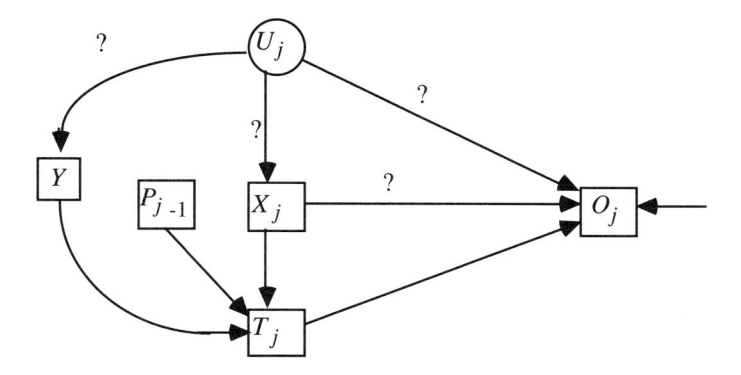

Figure 9.14

Let us now return to the question of why patient preference cannot be used to influence treatment. The reason why patient preferences cannot be used to determine treatment assignment in the experiment is not only because there may be, for all one knows, a causal interaction between patient preference and treatment outcome. It is true that if it were known that no such confounding occurs, then patient preference could be used in treatment assignment, but why cannot such assignment be used even if there is confounding? In order to make treatment assignments depend on patient preference (and presumably also on other features, such as the X_j variables) the patients' preferences must be ascertained. If the preferences are known, why not conditionalize outcome on *Preference*, *T* and the X_j variables, just as we have conditionalized outcome on *T* and the X_j variables? The probability of O_j conditional on *Preference*, the X_j variables, and T_j has no formally different role than the probability of O_j conditional on the X_j variables and T_j. If in figure 9.14 we make $Y = Preference$, then according to the Markov condition and theorem 7.1, the probability of O_j conditional on T_j, X_j and *Preference is* invariant under a manipulation of T_j. Of course some precautions would have to be taken in the course of an experiment that allows patient preference to influence treatment assignment. There is not much point to allowing a patient to choose his or her treatment unless the choice is informed. In the experimental setting, it would be necessary to standardize the information and advice that each patient received.

Could this design actually be used to predict the effects of a policy of treatment? At the end of the study the expert has a density function of outcomes given *T*, *X*, and *Preference*; subsequent patients' preferences for treatment would have to be recorded (but not necessarily used in determining treatment). If the announced results of the study alter *Preference*, the probability of outcome conditional on *Preference*, *X* and *T* depends on whether influences represented by the edges adjacent to *U* in figure 9.12 exist. But if patients are informed of the experimental results we must certainly expect in many cases that their preferences will be changed, that is, announcing the result of the experiment is a direct manipulation of *Preference*. Reliable predictions could only be made if the experimental outcome were kept secret!

The reason that patient preferences cannot be used for assigning treatment is therefore, not *just* because their preferences might have complicated interactions with the outcome of the treatment—so might the *X* variables. No analysis that does not also consider how a policy changes variables that were relevant to outcome in an experimental study can give a complete account of when predictions can be relied upon. As Kadane has pointed out,[1] it is more likely that the causes of *Preference* in the experimental population are different from the causes of *Preference* in the non-experimental population than it is that the causes of *X* in the experimental population are different from the causes of *X* in the non-experimental population. In the case we are considering, announcements of experimental results (or of recommendations) about policies that use patient preferences for assigning treatment can generally be expected to directly change those very preferences—whereas policies that use the *X* variables for assigning treatment do not generally change the values of the *X* variables for people in the population. (Of course, there may be instances where the results of a study that does not base treatment on patient preference also directly manipulates the distribution of the *X* variables, in which case the prediction of outcome conditional on treatment and the *X* variables would also be unreliable. Suppose, fancifully, in experimental trials that assign treatment as a function of cholesterol levels, it were found that a certain drug is very effective against cancer for subjects with low cholesterol.)

The Kadane/Sedransk/Seidenfeld design thus reveals an ethical conundrum that conventional methodological prejudices against nonrandomized trials has hidden. There is an obligation to find the most effective and cost effective treatments, and an obligation to take into account in treatment the preferences of people who participate as subjects in clinical trials. Both can be satisfied. But there is also an obligation fully to inform patients about the relevant scientific results that bear on decisions about their treatment. This obligation is incompatible with the others.

9.4.3 Toward Ethical Trials

Finally, we can use the causal analysis to obtain some more optimistic results about patient selection of treatment in experimental trials. Suppose in an experiment treatment assignment is a function $T = h(X_j, Preference, P_{j-1})$, and every undirected path between *Preference* and O_j contains some member of X_j as a noncollider. (If this is the case then we will say that *Preference* is not confounded with *O*.) Then it can be shown strictly as a consequence of the Markov and Faithfulness Conditions that the probability of O_j conditional on T_j and X_j is invariant under a direct manipulation of T_j; we may or may not conditionalize on *Preference*, or take *Preference* into account in the treatment rule used in policy, and in that case whether or not the announced experimental results changes the distribution of preferences is irrelevant to the accuracy of predictions. Now in some cases it may very well be that patient preference is not confounded with treatment outcome. If investigators could in fact discover that *Preference* is not confounded with O, then they could let such preferences be a factor both in treatment assignments in the experimental protocol and in the recommended policy. Kadane and Seidenfeld say that such dependencies, if they exist, are undetectable. If the experts are completely ignorant about what factors do not influence the outcome of treatment, Kadane and Seidenfeld are right; but if the experts know something, anything, that varies with patients and that has no effect on outcome except through treatment and has no common cause with outcome, we disagree. The something could be the phase of the moon on the patient's last birthday, the angular separation of the sun and Jupiter on the celestial sphere on the day of the patient's mother's day of birth, or simply the output of some randomizing device assigned to each patient. How is that?

In any distribution faithful to the graph of figure 9.15, E and C are dependent conditional on B. The relation is necessary in linear models, without assuming faithfulness.

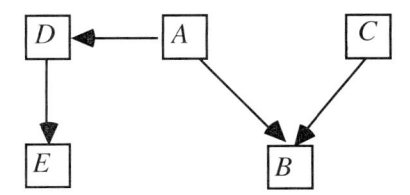

Figure 9.15

Now, returning to our problem, let Z be any feature whatsoever that varies from patient to patient, and that the experts agree in regarding as independent of patients' preferences for treatments and as affecting outcome only through treatment. Adopt a rule in the experiment that makes treatment a function of *Preference*, the patient's X profile, P_{j-1}, and the patient's Z value. Then the expert view of the causal process in the experiment looks something like figure 9.16.

If O_j and Z_j are independent conditional on T_j and X_j then there is (assuming faithfulness) no path d-connecting $Preference_j$ and O_j given X_j that is into $Preference_j$. A confounding relation between $Preference_j$ and O_j, if it exists, can be discovered from the experimental data. (Similarly, if T_j and O_j are dependent given X_j because the experimental population consists of a mixture of causal structures, then O_j and Z_j are dependent conditional on T_j and X_j unless some particular set of parameter values produces a "coincidental" independence.) Indeed, on the rather brave assumption that all dependencies are linear, Z_j is an instrumental variable (Bowden and Turkington 1984) and the linear coefficient representing the influence of T conditional on O and X can be calculated from the correlations and partial correlations.

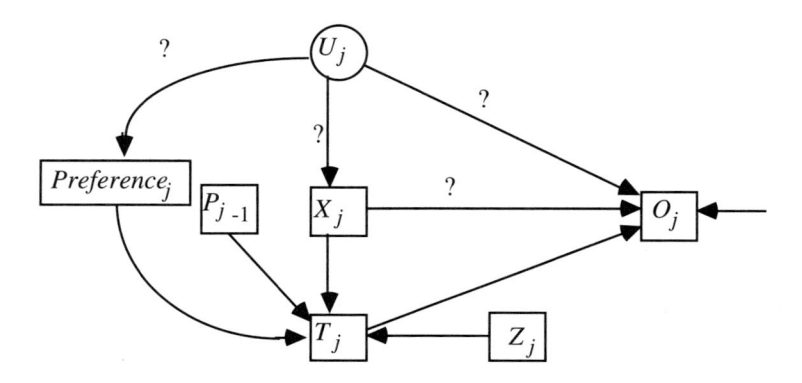

Figure 9.16

This suggests that it is possible to do a pilot study to determine whether *Preference* is confounded with *O* in the experimental population. In the pilot study, *Preference* can be a factor influencing, but not completely determining, *T*. If the results of the pilot study indicate that *Preference* is not confounded with *O* in the experimental population, a larger study in which *Preference* completely determines *T* can be done; otherwise, the Kadane/Sedransk/Seidenfeld design can be employed.

The goal of a medical experiment might be to predict outcomes in a population where a policy of assigning treatments without consulting patient preference is adopted. For example, the question might be "What would the death rate be if only halothane were used as a general anesthetic?" In this case, patient preference has little or nothing to do with the assigned treatment. If patient preference is not used to assign treatment in the policy population there is no reason to think that predictions of $P(O|X,T)$ in the policy population will be inaccurate when based upon experiments in which patients choose (or at least influence) their treatment, and *Preference* and *O* are not confounded.

It might be, however, that the goal of the experiment is to predict $P(O|X,T)$ in the policy population, and to let the patients choose or at least influence the choice of the treatment they receive. For example, in choosing between lumpectomy and mastectomy, patient preference may be the deciding factor. In this case there are a number of reasons to question the accuracy of a prediction of $P(O|X,T)$ in the policy population based upon the design we propose. But in this case every design meets the same difficulties, whether or not patients have assigned themselves in experimental treatments. These are equally good reasons for questioning the accuracy of a prediction of $P(O|X,T)$ (interpreted as frequencies or propensities and not as degrees of belief) in the policy population based upon the Kadane/Sedransk/Seidenfeld or a classical randomized design. The fundamental problem is that there are any number of plausible ways in which the causal relationships among preference and other variables in the experimental population may be different from the causal relationships in the policy population.[2] For example, in the experimental population the assignment of treatment will not depend on the patient's income.

However, in the actual population, the choice of treatment may very well depend upon income. There could easily be a common causal pathway connecting income and outcome that does not contain any variable in the patient's X profile. Again, in the experimental population the information and advice patients receive can be standardized. We can also try and ensure that the advice given is a function only of the patient's X profile. In the policy population, however, the advice and information that patients receive cannot be controlled in this way. If this is the case, the determination of preference may be a mixture of different causal structures in the policy population. Finally, the determination of patient preference in the policy population could easily be unstable. There are fads and fashions among patients, and also fads and fashions among doctors. New information could be released, or an intensive advertising campaign introduced. Any of these might create a trek between *Preference* and O, and hence between T and O, that does not contain any member of the X profile as a noncollider.

So even if in the experimental population *Preference* and O are not confounded, they very well might be in the policy population. If they are confounded in the policy population then $P(O|X,T)$ will not be the same in the experimental and policy populations (unless the parameters of the different causal structures coincidentally have values that make them equal.) Note that the same is true of predictions of $P(O|X,T)$ based on the Kadane/Sedransk/Seidenfeld design or of a prediction of $P(O|T)$ based upon a randomized experiment. This does not mean that no useful predictions can be made in situations where patient preference will be used to influence treatment in the policy population. It is still possible to inform the patient what $P(O|X,T)$ would be if a particular treatment were given *without* patient choice The patient can use this information to help make an informed decision. And with the design we have proposed, this (counterfactual) prediction is accurate as long as *Preference* and O are not confounded in the *experimental* population, regardless of how *Preference* is causally connected to O in the policy population.

Suppose then that we are merely trying to predict $P(O|X,T)$ in a population in which everyone is *assigned* a treatment. Is the design we have suggested practical? One potential problem is the obligation to give patients who are experimental subjects advice and information about their treatments; if this were not done the experiment would be unethical. If patients have access to advice from physicians, the advice is likely to be based upon their X profile. Even if the only information subjects receive is that all of the experts agree that they should not choose treatment T_1, their X profile is a cause of their preference. Will giving this advice and information make it unlikely that $P(O|X,T)$ is invariant under manipulation of T? It is true that the variables in the X profiles were chosen to be variables thought to be causes of O or to have common causes with O. Hence advice of this kind is very likely to *create* a common cause of *Preference* and O in the experimental population. Hence, it is likely that in the experimental population there will be a trek between T and O that is into T and contains *Preference*. However, such a trek would *not* d-connect T and O given X because it would also contain some member of

X as a noncollider. (See figure 9.17.) Hence such a trek does not invalidate the invariance of $P(O|X,T)$ under manipulation of T. Moreover, there is no problem with changing the advice as the experiment progresses, under the assumption that P_{j-1} is causally connected to O_j only through $Preference_j$.

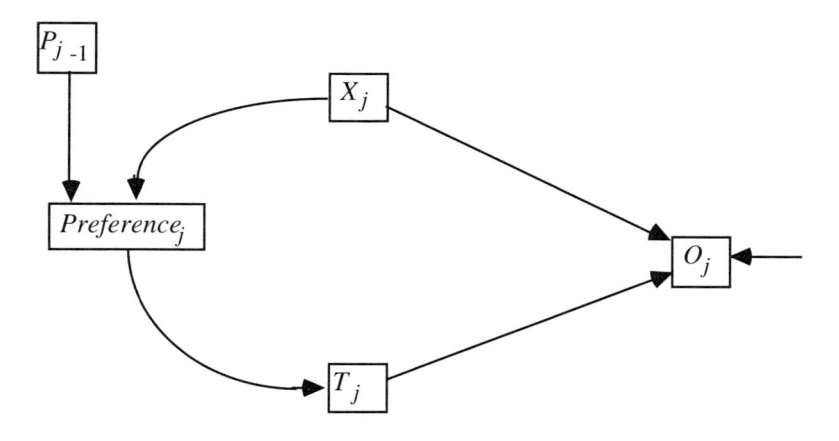

Figure 9.17

Can we let patients choose their own treatment, or merely influence the choice of treatment? As long as we are merely trying to predict $P(O|X,T)$ in a population in which everyone is *assigned* a treatment we can let patients choose their own treatment, as long as this doesn't result in all patients with a particular X profile always failing to choose some treatment T_1. In that case, $P(O|X,T = T_1)$ is undefined in the experimental population and cannot be used to predict $P(O|X,T = T_1)$ in a population where that quantity is defined.

In summary, so long as the goal is to predict $P(O|X,T)$ in a population where everyone is assigned a treatment, and there is enough variation of choice of treatments among the patients, and the experimental population is not a mixture of causal structures, and *Preference* and O are not confounded in the experimental population (an issue that must be decided empirically) it is possible to make accurate predictions from an experimental population in which informed patients choose their own treatment. If it is really important to let patient preferences influence their treatment in experiments, then it is worth risking some cost to realize that condition if it is possible to do so consistent with reliable prediction. How much it is worth, either in money or in degradation in confidence about the reliability of predictions, is not for us to say. But a simple modification of the Kadane/Sedransk/Seidenfeld design which has initial trials base treatment assignment on X_j, P_{j-1}, Z_j, and $Preference_j$, and then allows patient self-assignment if $Preference_j$ and O_j are discovered to be unconfounded, would in some cases permit investigators to conduct clinical trials that conform to ethical requirements of autonomy and informed consent.

9.5 An Example: Smoking and Lung Cancer

The fascinating history of the debates over smoking and lung cancer illustrates the difficulties of causal inference and prediction from policy studies, and also illustrates some common mistakes. Perhaps no other hypothetical cause and effect relationship has been so thoroughly studied by non-experimental methods or has so neatly divided the professions of medicine and statistics into opposing camps. The theoretical results of this and the preceding chapters provide some insight into the logic and fallacies of the dispute.

The thumbnail sketch is as follows: In the 1950s a retrospective study by Doll and Hill (1952) found a strong correlation between cigarette smoking and lung cancer. That initial research prompted a number of other studies, both retrospective and prospective, in the United States, the United Kingdom, and soon after in other nations, all of which found strong correlations between cigarette smoking and lung cancer, and more generally between cigarette smoking and cancer and between cigarette smoking and mortality. The correlations prompted health activists and some of the medical press to conclude that cigarette smoking causes death, cancer, and most particularly, lung cancer. Sir Ronald Fisher took very strong exception to the inference, preferring a theory in which smoking behavior and lung cancer are causally connected only through genetics. Fisher wrote letters, essays, and eventually a book against the inference from the statistical dependencies to the causal conclusion. Neyman ventured a criticism of the evidence from retrospective studies. The heavyweights of the statistical profession were thus allied against the methods of the medical community. A review of the evidence containing a response to Fisher and Neyman was published in 1959 by Cornfield, Haenszel, Hammond, Lilienfeld, Shimkin, and Wynder. The Cornfield paper became part of the blueprint for the Report of the Surgeon General on Smoking and Health in 1964, which effectively established that as a political fact smoking would be treated as an unconfounded cause of lung cancer, and set in motion a public health campaign that is with us still. Brownlee (1965) reviewed the 1964 report in the *Journal of the American Statistical Association* and rejected its arguments as statistically unsound for many of the reasons one can imagine Fisher would have given. In 1979, the Surgeon General published a second report on smoking and health, repeating the arguments of the first report but with more extensive data, but offering no serious response to Brownlee's criticisms. The report made strong claims from the evidence, in particular that cigarette smoking was the largest preventable cause of death in the United States. The foreword to the report, by Joseph Califano, was downright vicious, and claimed that any criticism of the conclusions of the report was an attack on science itself. That did not stop P. Burch (1983), a physicist turned theoretical biologist turned statistician, from publishing a lengthy criticism of the second report, again on grounds that were detailed extensions of Fisher's criticisms, but buttressed as well by the first reports of randomized clinical trials of the effects of smoking intervention, all of which were either null or actually suggested that intervention programs increased mortality. Burch's remarks brought a reply by A. Lilienfeld (1983), which began and ended with an *ad hominem* attack on Burch.

Fisher's criticisms were directed against the claim that uncontrolled observations of a correlation between smoking and cancer, no matter whether retrospective or prospective, provided evidence that smoking causes lung cancer, as against the alternative hypothesis that there are one or more common causes of smoking and lung cancer. His strong views can be understood in the light of features of his career. Fisher had been largely responsible for the introduction of randomized experimental designs, one of the very points of which was to obtain statistical dependencies between a hypothetical cause and effect that could not be explained by the action of unmeasured common causes. Another point of randomization was to insure a well-defined distribution for tests of hypotheses, something Fisher may have doubted was available in observational studies. Throughout his adult life Fisher's research interests had been in heredity, and he had been a strong advocate of the eugenics movement. He was therefore disposed to believe in genetic causes of very detailed features of human behavior and disease. Fisher thought a likely explanation of the correlation of lung cancer and smoking was that a substantial fraction of the population had a genetic predisposition both to smoke and to get lung cancer.

One of Fisher's (1959, p. 8) fundamental criticisms of these epidemiological arguments was that correlation underdetermines causation: besides smoking causing cancer, wrote Fisher "there are two classes of alternative theories which any statistical association, observed without the precautions of a definite experiment, always allows— namely, (1) that the supposed effect is really the cause, or in this case that incipient cancer, or a precancerous condition with chronic inflammation, is a factor in inducing the smoking of cigarettes, or (2) that cigarette smoking and lung cancer, though not mutually causative, are both influenced by a common cause, in this case the individual genotype." Not even Fisher took (1) seriously. To these must be added others Fisher did not mention, for example that smoking and lung cancer have several distinct unmeasured common causes, or that while smoking causes cancer, something unmeasured also causes both smoking and cancer.

If we interpret "statistical association" as statistical dependence, Fisher is correct that given observation only of a statistical dependence between smoking and lung cancer in an uncontrolled study, the possibility that smoking does not cause lung cancer cannot be ruled out. However, he does not mention the possibility that this hypothesis, if true, could have been established without experimentation by finding a factor associated with smoking but independent, or conditionally independent (on variables other than smoking) of cancer. By the 1960s a number of personal and social factors associated with smoking had been identified, and several causes of lung cancer (principally associated with occupational hazards and radiation) potentially independent of smoking had been identified, but their potential bearing on questions of common causes of smoking and lung cancer seems to have gone unnoticed. The more difficult cases to distinguish are the hypotheses that smoking is an unconfounded cause of lung cancer versus the joint hypotheses that smoking causes cancer and that there is also an unmeasured common cause—or causes—of smoking and cancer.

Fisher's hypothesis that genotype causes both smoking behavior and cancer was speculative, but it wasn't a will-o-the-wisp. Fisher obtained evidence that the smoking behavior of monozygotic twins was more alike than the smoking behavior of dizygotic twins. As his critics pointed out, the fact could be explained on the supposition that monozygotic twins are more encouraged by everyone about them to do things alike than are dizygotic twins, but Fisher was surely correct that it could also be explained by a genetic disposition to smoke. On the other side, Fisher could refer to evidence that some forms of cancer have genetic causes.

The paper by Cornfield et al. (including Lilienfeld) argued that while lung cancer may well have other causes besides, cigarette smoking causes lung cancer. This view had already been announced by official study groups in the United States and Great Britain. Cornfield's paper is of more scientific interest than the Surgeon General's report five years later, in part because the former is not primarily a political document. Cornfield et al. claimed the existing data showed several things:

1. Carcinomas of the lung found at autopsy had systematically increased since 1900, although different studies gave different rates of increase. Lung cancers are found to increase monotonically with the amount of cigarette smoking and to be higher in current than in former cigarette smokers. In large prospective studies diagnoses of lung cancer may have an unknown error rate, but the total death rate also increases monotonically with cigarette smoking.

2. Lung cancer mortality rates are higher in urban than in rural populations, and rural people smoke less than city people, but in both populations smokers have higher death rates from lung cancer than do nonsmokers.

3. Men have much higher death rates from lung cancer than women, especially among persons over 55, but women smoked much less and as a class had taken up the habit much later than men.

4. There are a host of causes of lung cancer, including a variety of industrial pollutants and unknown circumstances associated with socioeconomic class, with the poorer and less well off more likely than the better off to contract the disease, but no more likely to smoke. Cornfield et al. emphasize that "The population exposed to established industrial carcinogens is small, and these agents cannot account for the increasing lung-cancer risk in the remainder of the population. Also, the effects associated with socioeconomic class and related characteristics are smaller than those noted for smoking history, and the smoking class differences cannot be accounted for in terms of these other effects" (p. 179). This passage states that the difference in cancer rates for smokers and nonsmokers

could not be explained by socioeconomic differences. While this claim was very likely true, no analysis was given in support of it, and the central question of whether smoking and lung cancer were independent or nearly independent conditional on all subsets of the known risk factors *that are not effects* of smoking and cancer—area of residence, exposure to known carcinogens, socioeconomic class, and so on, was not considered. Instead, Cornfield et al. note that different studies measured different variables and "The important fact is that in all studies when other variables are held constant, cigarette smoking retains its high association with lung cancer."

5. Cigarette smoking is not associated with increased cancer of the upper respiratory tract, the mouth tissues or the fingers. Carcinoma of the trachea, for example, is a rarity. But, Cornfield et al. point out, "There is no a priori reason why a carcinogen that produces bronchogenic cancer in man should also produce neoplastic changes in the anspharynx or in other sites" (p. 186).

6. Experimental evidence shows that cigarette smoke inhibits the action of the cilia in cows, rats and rabbits. Inhibition of the cilia interferes with the removal of foreign material from the surface of the bronchia. Damage to ciliated cells is more frequent in smokers than in nonsmokers.

7. Application of cigarette tar directly to the bronchia of dogs produced changes in the cells, and in some but not all other experiments applications of tobacco tar to the skin of mice produced cancers. Exposure of mice to cigarette smoke for up to 200 days produced cell changes but no cancers.

8. A number of aromatic polycyclic compounds have been isolated in tobacco smoke, and one of them, the α form of benzopyrene, was known to be a carcinogen.

Perhaps the most original technical part of the argument was a kind of sensitivity analysis of the hypothesis that smoking causes lung cancer. Cornfield et al. considered a single hypothetical binary latent variable causing lung cancer and statistically dependent on smoking behavior. They argued such a latent cause would have to be almost perfectly associated with lung cancer and strongly associated with smoking to account for the observed association. The argument neglected, however, the reasonable possibility of multiple common causes of smoking and lung cancer, and had no clear bearing on the hypothesis that the observed association of smoking and lung cancer is due both to a direct influence and to common causes.

In sum, Cornfield et al. thought they could show a mechanism for smoking to cause cancer, and claimed evidence from animal studies, although their position in that regard tended to trip over itself (compare items 5 and 7). They didn't put the statistical case entirely clearly, but their position seems to have been that lung cancer is also caused by a number of measurable factors that are not plausibly regarded as effects of smoking but which may cause smoking, and that smoking and cancer remain statistically dependent conditional on these factors. Against Fisher they argued as follows:

> The difficulties with the constitutional hypothesis include the following considerations: (a) changes in lung-cancer mortality over the last half century; (b) the carcinogenicity of tobacco tars for experimental animals; (c) the existence of a large effect from pipe and cigar tobacco on cancer of the buccal cavity and larynx but not on cancer of the lung; (d) the reduced lung-cancer mortality among discontinued cigarette smokers. No one of these considerations is perhaps sufficient by itself to counter the constitutional hypothesis, ad hoc modification of which can accommodate each additional piece of evidence. A point is reached, however, when a continuously modified hypothesis becomes difficult to entertain seriously. (p. 191)

Logically, Cornfield et al. visited every part of the map. The evidence was supposed to be inconsistent with a common cause of smoking and lung cancer, but also consistent with it. Objections that a study involved self-selection—as Fisher and company would object to (d)—was counted as an "ad hoc modification" of the common cause hypothesis. The same response was in effect given to the unstated but genuine objections that the time series argument ignored the combined effects of dramatic improvements in diagnosis of lung cancer, a tendency of physicians to bias diagnoses of lung cancer for heavy smokers and to overlook such a diagnosis for light smokers, and the systematic increase in the same period of other factors implicated in lung cancer, such as urbanization. The rhetoric of Cornfield et al. converted reasonable demands for sound study designs into *ad hoc* hypotheses. In fact none of the evidence adduced was inconsistent with the "constitutional hypothesis."

A reading of the Cornfield paper suggests that their real objection to a genetic explanation was that it would require a very close correlation between genotypic differences and differences in smoking behavior and liability to various forms of cancer. Pipe and cigar smokers would have to differ genotypically from cigarette smokers; light cigarette smokers would have to differ genotypically from heavy cigarette smokers; those who quit cigarette smoking would have to differ genotypically from those who did not. Later the Surgeon General would add that Mormons would have to differ genotypically from non-Mormons and Seventh Day Adventists from nonseventh Day Adventists. The physicians simply didn't believe it. Their skepticism was in keeping with the spirit of a time in which genetic explanations of behavioral differences were increasingly regarded as politically and morally incorrect, and the moribund eugenics movement was coming to be viewed in retrospect as an embarrassing bit of racism.

In 1964 the Surgeon General's report reviewed many of the same studies and arguments as had Cornfield, but it added a set of "Epidemiological Criteria for Causality," said to be sufficient for establishing a causal connection and claimed that smoking and cancer met the criteria. The criteria were indefensible, and they did not promote any good scientific assessment of the case. The criteria were the "consistency" of the association, the "strength" of the association, the "specificity" of the association, the temporal relationship of the association and the "coherence" of the association.

All of these criteria were left quite vague, but no way of making them precise would suffice for reliably discriminating causal from common causal structures. Consistency meant that separate studies should give the "same" results, but in what respects results should be the same was not specified. Different studies of the relative risk of cigarette smoking gave very different multipliers depending on the gender, age and nationality of the subjects. The results of most studies were the same in that they were all positive; they were plainly not nearly the same in the seriousness of the risk. Why stronger associations should be more likely to indicate causes than weaker associations was not made clear by the report. Specificity meant the putative cause, smoking, should be associated almost uniquely with the putative effect, lung cancer. Cornfield et al. had rejected this requirement on causes for good reason, and it was palpably violated in the smoking data presented by the Surgeon General's report. "Coherence" in the jargon of the report meant that no other explanation of the data was possible, a criterion the observational data did not meet in this case. The temporal issue concerned the correlation between increase in cigarette smoking and increase in lung cancer, with a lag of many years. Critics pointed out that the time series were confounded with urbanization, diagnostic changes and other factors, and that the very criterion Cornfield et al. had used to avoid the issue of the unreliability of diagnoses, namely total mortality, was, when age-adjusted, uncorrelated with cigarette consumption over the century.

Brownlee (1965) made many of these points in his review of the report in the *Journal of the American Statistical Association*. His contempt for the level of argument in the report was plain, and his conclusion was that Fisher's alternative hypothesis had not been eliminated or even very seriously addressed. In Brownlee's view, the Surgeon General's report had only two arguments against a genetic common cause: (a) the genetic hypothesis would allegedly have to be very complicated to explain the dose/response data, and (b) the rapid historical rise in lung cancer following by about 20 years a rapid historical rise in cigarette smoking. Brownlee did not address (a), but he argued strongly that (b) is poor evidence because of changes in diagnostics, changes in other factors of known and unknown relevance, and because of changes in the survival rate of weak neonates whom, as adults, might be more prone to lung cancer.

One of the more interesting aspects of the review was Brownlee's "very simplified" proposal for a statistical analysis of "E_2 causes E_1" which was that E_1 and E_2 be dependent conditional on every possible vector of values for all other variables of the system. Brownlee realized, of course, that his condition did not separate "E_2 causes E_1" from E_1 causes E_2," but that was not a problem with smoking and cancer. But even ignoring the direction of causation, Brownlee's condition—perhaps suggested to him by the fact that the same principle is used (erroneously) in regression—is quite wrong. It

The Design of Empirical Studies 245

would be satisfied, for example, if, E_1 and E_2 had no causal connection whatsoever provided some measured variable E_j were a direct effect of both E_1 and E_2.

Brownlee thought his way of considering the matter was important for prediction and intervention:

> If the inequality holds only for, say, one particular subset $E_j,..., E_k$, and for all other subsets equality holds, and if the subset $E_j,...,E_k$ occurs in the population with low probability, then $\Pr\{E_1|E_2\}$, while not strictly equal to $\Pr\{E_1|E_2^c\}$, will be numerically close to it, and then E_2 as a cause of E_1 may be of small practical importance. These considerations are related to the Committee's responsibility for assessment of the *magnitude* of the health hazard (page 8). Further complexities arise when we distinguish between cases in which one of the required secondary conditions $E_j,...,E_k$ is, on the one hand, presumably controllable by the individual, e.g., the eating of parsnips, or uncontrollable, e.g., the presence of some genetic property. In the latter case, it further makes a difference whether the genetic property is identifiable or non-identifiable: for example it could be brown eyes which is the significant subsidiary condition E_j, and we could tell everybody with not-brown eyes it was safe for *them* to smoke. (p. 725)

No one seems to have given any better thought than this to the question of how to predict the effects of public policy intervention against smoking. Brownlee regretted that the Surgeon General's report made no explicit attempt to estimate the expected increase in life expectancy from not smoking or from quitting after various histories.

Fifteen years later, in 1979, the second Surgeon General's Report on Smoking and Health was able to report studies that showed a monotonic increase in mortality rates with virtually every feature of smoking practice that increased smoke in the lungs: number of cigarettes smoked per day, number of years of smoking, inhaling versus not inhaling, low tar and nicotine versus high tar and nicotine, length of cigarette habitually left unsmoked. The monotonic increase in mortality rates with cigarette smoking had been shown in England, the continental United States, Hawaii, Japan, Scandinavia and elsewhere, for whites and blacks, for men and women. The report dismissed Fisher's hypothesis in a single paragraph by citing a Scandinavian study (Cederlof, Friberg, and Lundman 1977) that included monozygotic and dizygotic twins:

> When smokers and nonsmokers among the dizygotic pairs were compared, a mortality ratio of 1.45 for males and 1.21 for females was observed. Corresponding mortality ratios for the monozygotic pairs were 1.5 for males and 1.222 for females. Commenting on the constitutional hypothesis and lung cancer, the authors observed that "the constitutional hypothesis as advanced by Fisher and still supported by a few, has here been tested in twin studies. The results from the Swedish monozygotic twin series speak strongly against the constitutional hypothesis."

The second Surgeon General's report claimed that tobacco smoking is responsible for 30% of all cancer deaths; cigarette smoking is responsible for 85% of all lung cancer deaths.

A year before the report appeared, in a paper for the British Statistical Association P. Burch (1978) had used the example of smoking and lung cancer to illustrate the problems of distinguishing causes from common causes without experiment. In 1982 he published a full fledged assault on the second Surgeon General's report. The criticisms of the argument of the report were similar to Brownlee's criticisms of the 1964 report, but Burch was less restrained and his objections more pointed. His first criticism was that while all of the studies showed a increase in risk of mortality with cigarette smoking, the degree of increase varied widely from study to study. In some studies the age adjusted multiple regression of mortality on cigarettes, beer, wine and liquor consumption gave a smaller partial correlation with cigarettes than with beer drinking. Burch gave no explanation of why the regression model should be an even approximately correct account of the causal relations. Burch thought the fact that the apparent dose/response curve for various culturally, geographically, and ethnically distinct groups were very different indicated that the effect of cigarettes was significantly confounded with environmental or genetic causes. He wanted the Surgeon General to produce a unified theory of the causes of lung cancer, with confidence intervals for any relevant parameter estimates: Where, he asked, did the 85% figure come from?

Burch pointed out, correctly, that the cohort of 1487 dizygotic and 572 monozygotic twins in the Scandinavian study born between 1901 and 1925 gave no support at all to the claim that the constitutional explanation of the connection between smoking and lung cancer had been refuted, despite the announcements of the authors of that study. The study showed that of the dizygotes exactly 2 nonsmokers or infrequent smokers had died of lung cancer and 10 heavy smokers had died of lung cancer; of the monozygotes, 2 low non smokers and 2 heavy smokers had died of the disease. The numbers were useless, but if they suggested anything, it was that if genetic variation was controlled there is no difference in lung cancer rates between smokers and nonsmokers. The Surgeon General's report of the conclusion of the Scandinavian study was accurate, but not the less misleading for that.

Burch also gave a novel discussion of the time series data, arguing that it virtually refuted the causal hypothesis. The Surgeon General and others had used the time series in a direct way. In the U.K. for example, male cigarette consumption per capita had increased roughly a hundredfold between 1890 and 1960, with a slight decrease thereafter. The age-standardized male death rate from lung cancer began to increase steeply about 1920, suggesting a thirty-year lag, consistent with the fact that people often begin smoking in their twenties and typically present lung cancer in their fifties. According to Burch's data, the onset of cigarette smoking for women lagged behind males by some years, and did not begin until the 1920s. The Surgeon General's report noted that the death rate from lung cancer for women had also increased dramatically between 1920 and 1980. Burch pointed out that the autocorrelations for the male series and female series didn't mesh: there was no lag in death rates for the women. Using U.K. data, Burch plotted the *percentage change* in the age-standardized death rate from lung cancer for both men and women from 1900 to 1980. The curves matched perfectly until 1960. Burch's conclusion is that whatever caused the increase in death rates from lung cancer affected both men and women at the same time, from the beginning of the century on, although whatever it is had a smaller absolute effect on women than on men. But then the whatever-it-was could not have been cigarette smoking, since increases in women's consumption of cigarettes lagged twenty to thirty years behind male increases.

Burch was relentless. The Surgeon General's report had cited the low occurrence of lung cancer among Mormons. Burch pointed out that Mormon's in Utah not only have lower age-adjusted incidences of cancer than the general population, but also have higher incidences than non-Mormon nonsmokers in Utah. Evidently their lower lung cancer rates could not be simply attributed to their smoking habits.

Abraham Lilienfeld, who only shortly before had written a textbook on epidemiology and who had been involved with the smoking and cancer issue for more than twenty years, published a reply to Burch that is of some interest. Lilienfeld gives the impression of being at once defensive and disdainful. His defense of the Surgeon General's report began with an *ad hominem* attack, suggesting that Burch was so out of fashion as to be a crank, and ended with another *ad hominem*, demanding that if Burch wanted to criticize others' inferences from their data he go get his own. The most substantive reply Lilienfeld offered is that the detailed correlation of lung cancer with smoking habits in one subpopulation after another makes it seem very implausible that the association is due to a common cause. Lilienfeld said, citing himself, that the conclusion that 85% of lung cancer deaths are due to cigarettes is based on the relative risk for cigarette smokers and the frequency of cigarette smoking in the population, predicting, in effect, that if cigarette smoking ceased the death rate from lung cancer would decline by that percentage. (The prediction would only be correct, Burch pointed out in response, provided cigarette smoking is a completely unconfounded cause of lung cancer.) Lilienfeld challenged the source of Burch's data on female cigarette consumption early in the century, which Burch subsequently admitted were estimates.

Both Burch and Lilienfeld discussed a then recent report by Rose et al. (1982) on a ten-year randomized smoking intervention study. The Rose study, and another that appeared at nearly the same time with virtually the same results, illustrates the hazards of prediction. Middle-aged male smokers were assigned randomly to a treatment or nontreatment group. The treatment group was encouraged to quit smoking and given counseling and support to that end. By self-report, a large proportion of the treatment group either quit or reduced cigarette smoking. The difference in self-reported smoking levels between the treatment and nontreatment groups was thus considerable, although the difference declined toward the end of the ten-year study. To most everyone's dismay, Rose found that there was no statistically significant difference in lung cancer between the two groups after ten years (or after five), but there was a difference in overall mortality—the group that had been encouraged to quit smoking, and had in part done so, suffered higher mortality.

Fully ignoring their own evidence, the authors of the Rose study concluded nonetheless that smokers should be encouraged to give up smoking, which makes one wonder why they bothered with a randomized trial. Burch found the Rose report unsurprising; Lilienfeld claimed the numbers of lung cancer deaths in the sample are too small to be reliable, although he did not fault the Surgeon General's report for using the Scandinavian data, where the numbers are even smaller, and he simply quoted the conclusion of the report, which seems almost disingenuous. To Burch's evident delight, as Lilienfeld's defense of the Surgeon General appeared so did yet further experimental evidence that intervening in smoker's behavior has no benign effect on lung cancer rates. The Multiple Risk Factor Intervention Trial Research Group (1982) reported the results after six years of a much larger randomized experimental intervention study producing roughly three times the number of lung cancer deaths as in the Rose study. But the intervention group showed more lung cancer deaths than the usual care group! The absolute numbers were small in both studies but there could be no doubt that nothing like the results expected by the epidemiological community had materialized.

The results of the controlled intervention trials illustrate how naive it is to think that experimentation always produces unambiguous results, or frees one from requirements of prior knowledge. One possible explanation for the null effects of intervention on lung cancer, for example, is that the reduced smoking produced by intervention was concentrated among those whose lungs were already in poor health and who were most likely to get lung cancer in any case. (Rose et al. gave insufficient information for an analysis of the correlation of smoking behavior and lung cancer within the intervention group.) This possibility could have been tested by experiments using blocks more finely selected by health of the subjects.

In retrospect the general lines of the dispute were fairly simple. The statistical community focused on the want of a good scientific argument against a hypothesis given prestige by one of their own; the medical community acted like Bayesians who gave the "constitutional" hypothesis necessary to account for the dose/response data so low a prior that it did not merit serious consideration. Neither side understood what uncontrolled studies could and could not determine about causal relations and the effects of interventions. The statisticians pretended to an understanding of causality and correlation they did not have; the epidemiologists resorted to informal and often irrelevant criteria, appeals to plausibility, and in the worst case to *ad hominem*.

Fisher's prestige as well as his arguments set the line for statisticians, and the line was that uncontrolled observations cannot distinguish among three cases: smoking causes cancer, something causes smoking and cancer, or something causes smoking and cancer and smoking causes cancer. The most likely candidate for the "something" was genotype. Fisher was wrong about the logic of the matter, but the issue never was satisfactorily clarified, even though some statisticians, notably Brownlee and Burch, tried unsuccessfully to characterize more precisely the connection between probability and causality. While the statisticians didn't get the connection between causality and probability right, the Surgeon General's "epidemiological criteria for causality" were inadequate and arguments in defense of the conclusions of the Surgeon General's Report were flawed. The real view of the medical community seems to have been that it was just too implausible to suppose that genotype strongly influenced how much one smoked, whether one smoked at all, whether one smoked cigarettes as against a cigar or pipe, whether one was a Mormon or a Seventh day Adventist, and whether one quit smoking or not. After Cornfield's survey the medical and public health communities gave the common cause hypothesis more invective than serious consideration. And, finally, in contrast to Burch, who was an outsider and maverick, leading epidemiologists, such as Lilienfeld, seem simply not to have understood that if the relation between smoking and cancer is confounded by one or more common causes, the effects of abolishing smoking cannot be predicted from the "risk ratios," that is, from sample conditional probabilities. The subsequent controlled smoking intervention studies gave evidence of how very bad were the expectations based on uncontrolled observation of the relative risks of lung cancer in those who quit smoking compared to those who did not.

9.6 Appendix

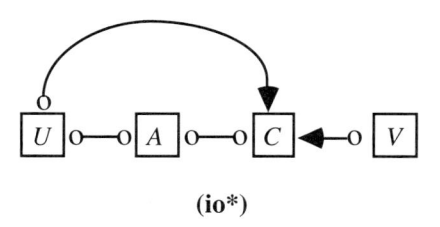

Figure 9.18

We will prove that the partially oriented inducing path graph (io*) in figure 9.18, together with the assumptions that U causes A, that there is no common cause of U and C, and that every directed path from U to C contains A, entail that A causes C and that there is a latent common cause of A and C. We assume that A is not a deterministic function of U.

Let $\mathbf{O} = \{A,C,U,V\}$, and G be the directed acyclic graph that generated (io*). The U o$\rightarrow$ C edge in (io*) entails that in the inducing path graph of G either $U \rightarrow C$ or $U \leftrightarrow C$. If there is a $U \leftrightarrow C$ edge, then there is a latent common cause of U and C, contrary to our assumption. Hence the inducing path graph contains a $U \rightarrow C$ edge. It follows that in G there is a directed path from U to C. Because every directed path from U to C contains A, there is a directed path from A to C in G. Hence A causes C.

The $U \to C$ edge in the inducing path graph of G entails that there is an inducing path Z relative to **O** that is out of U and into C in G. If Z does not contain a collider then Z is a directed path from U to C and hence it contains A. But then A is a noncollider on Z, and Z is not an inducing path relative to **O** (because Z contains a member of **O**, namely A, that is a noncollider on Z) contrary to our assumption. Hence Z contains a collider.

We will now show that no collider on Z is an ancestor of U. Suppose, on the contrary that there is a collider on Z that is an ancestor of U; let M be the closest such collider on Z to C. No directed path from M to U contains C, because there is a directed path from U to C, and hence no directed path from C to U. There are two cases.

Suppose first that there is no collider between M and C. Then there is a variable Q on Z, such that $Z(Q,C)$ is a directed path from Q to C and $Z(Q,M)$ is a directed path from Q to M. (As in the proofs in chapter 13, we adopt the convention that on an acyclic path Z containing Q and C, $Z(Q,C)$ represents the subpath of Z between Q and C.) $U \neq M$ because M is a collider on Z and U is not. U does not lie on $Z(Q,C)$ or $Z(Q,M)$ because Z is acyclic. The concatenation of $Z(Q,M)$ and a directed path from M to U contains a directed path from Q to U that does not contain C. $Z(Q,C)$ is a directed path from Q to C that does not contain U. Q is a noncollider on Z, and because Z is an inducing path relative to **O**, Q is not in **O**. Hence Q is a latent common cause of U and C, contrary to our assumption.

Suppose next that there is a collider between M and C, and N is the collider on Z closest to M and between M and C. Then there is a variable Q on Z, such that $Z(Q,N)$ is a directed path from Q to N and $Z(Q,M)$ is a directed path from Q to M. $U \neq M$ because M is a collider on Z and U is not. U does not lie on $Z(Q,N)$ or $Z(Q,M)$ because Z is acyclic. The concatenation of $Z(Q,M)$ and a directed path from M to U contains a directed path from Q to U that does not contain C. There is a directed path from N to C, and by hypothesis no such directed path contains U. The concatenation of $Z(Q,N)$ and a directed path from N to C contains a directed path from Q to C that does not contain U. Q is a noncollider on Z, and because Z is an inducing path relative to **O**, Q is not in **O**. Hence Q is a latent common cause of U and C, contrary to our assumption.

It follows that no collider on Z is an ancestor of U.

Let X be the collider on Z closest to U. There is a directed path from X to C. $Z(U,X)$ is a directed path from U to X. The concatenation of $Z(U,X)$ and a directed path from X to C contains a directed path from U to C. By assumption, such a path contains A. A does not lie between U and X on Z, because every vertex between U and X on Z is a noncollider, and if A occurs on Z it is a collider on Z. Hence A lies on every directed path from X to C. Hence there exists a collider on Z that is the source of a directed path to C that contains A. Let R be the collider on Z closest to C such that there is a directed path D from R to C that contains A. There are again two cases.

If there is no collider between R and C on Z, then there is a vertex Q on Z such that $Z(Q,C)$ is a directed path from Q to C and $Z(Q,R)$ is a directed path from Q to R. A does not lie on $Z(Q,C)$ because no vertex on $Z(Q,C)$ is a collider on Z. C does not lie on

$D(R,A)$ because the directed graph is acyclic. $C \neq Q$ because Z has an edge into C but not Q. $C \neq R$ because R is a collider on Z and C is not. Hence, C does not lie on $Z(Q,R)$ because Z is acyclic. The concatenation of $Z(Q,R)$ and $D(R,A)$ contains a directed path from Q to A that does not contain C. Q is not a collider on Z, so it not in **O**. Hence Q is a latent common cause of A and C.

Suppose next that there is a collider between R and C on Z, and N is the closest such collider to R on Z. Then there is a vertex Q on Z such that $Z(Q,N)$ is a directed path from Q to N and $Z(Q,R)$ is a directed path from Q to R. $Q \neq N$ because by hypothesis there is a path from N to C that does not contain A. A does not lie on $Z(Q,N)$ because no vertex on $Z(Q,N)$ except N is a collider on Z. There is a directed path from N to C, but it does not contain A by hypothesis. Hence the concatenation of $Z(Q,N)$ and a directed path from N to C contains a directed path that does not contain A. C does not lie on $D(R,A)$ because the directed graph is acyclic. $C \neq Q$ because Z has an edge into C but not Q. $C \neq R$ because R is a collider on Z and C is not. Hence, C does not lie on $Z(Q,R)$ because Z is acyclic. The concatenation of $Z(Q,R)$ and $D(R,A)$ contains a directed path from Q to A that does not contain C. Q is not a collider on Z, so it not in **O**. Hence Q is a latent common cause of A and C.

Hence in either case, A and C have a latent common cause in G.

10 The Structure of the Unobserved

10.1 Introduction

Many theories suppose there are variables that have not been measured but that influence measured variables. In studies in econometrics, psychometrics, sociology and elsewhere the principal aim may be to uncover the causal relations among such "latent" variables. In such cases it is usually assumed that one knows that the measured variables (e.g., responses to questionnaire items) are not themselves causes of unmeasured variables of interest (e.g., attitude), and the measuring instruments are often designed with fairly definite ideas as to which measured items are caused by which unmeasured variables. Survey questionnaires may involve hundreds of items, and the very number of variables is ordinarily an impediment to drawing useful conclusions about structure. Although there are a number of procedures commonly used for such problems, their reliability is doubtful. A common practice, for example, is to form aggregated scales by averaging measures of variables that are held to be proxies for the same unmeasured variable, and then to study the correlations of the scales. The correlations thus obtained have no simple systematic connection with causal relations among the unmeasured variables.

What can a mixture of substantive knowledge about the measured indicators and statistical observations of those indicators reveal about the causal structure of the unobserved variables? And under what assumptions about distributions, linearity, etc.? This chapter begins to address these questions. The procedures for forming scales, or "pure measurement models," that we will describe in this chapter have found empirical application in the study of large psychometric data sets (Callahan and Sorensen 1992).

10.2 An Outline of the Algorithm

Consider the problem of determining the causal structure among a set of unmeasured variables of interest in linear pseudoindeterministic models, commonly called "structural equation models with latent variables." Assume the distributions are linearly faithful. Structural equation models with latent variables are sometimes presented in two parts: the "measurement model," and the "structural model" (see figure 10.1). The structural model involves only the causal connections among the latent variables; the remainder is the measurement model. From a mathematical point of view, the distinction marks only a difference in the investigator's interests and access and not any distinction in formal properties. The same principles connecting graphs, probabilities and causes apply to the measurement model as to the structural model. In figure 10.1 we give an example of a latent variable model in which the measured variables (Q_1-Q_{12}) might be answers to survey questions.

Let **T** be a set of latent variables and **V** a set of measured variables. We will assume that **T** is causally sufficient, although that is clearly not the general case. We let **C** denote the set of "nuisance" latent common causes, that is, unobserved common causes, not in **T**, of two or more variables in **T** ∪ **V**. Call a subgraph of G that contains all of the edges in G except for edges between members of **T** a **measurement model** of G.

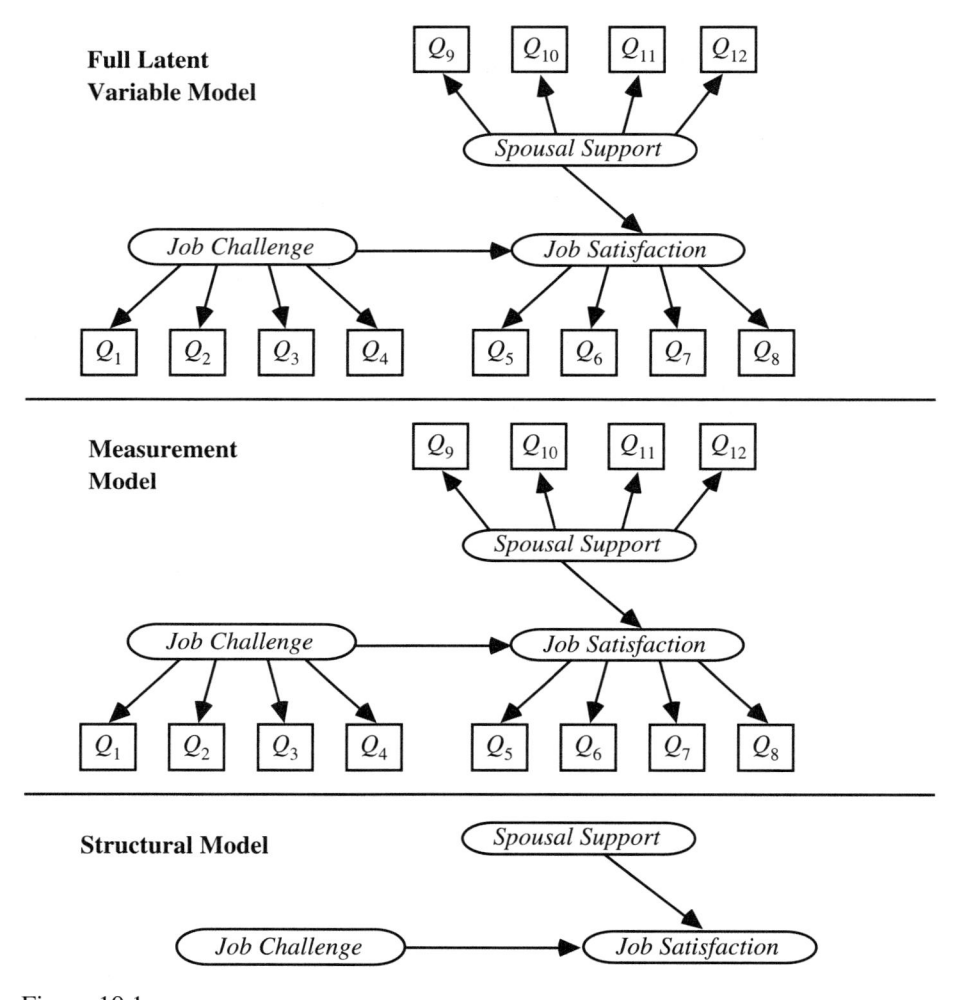

Figure 10.1

In actual research the set **V** is often chosen so that for each T_i in **T**, a subset of **V** is intended to measure T_i. In Kohn's (1969) study of class and attitude in America, for example, various questionnaire items where chosen with the intent of measuring the same attitude; factor analysis of the data largely agreed with the clustering one might expect on intuitive grounds. Accordingly, we suppose the investigator can partition **V** into **V**(T_i), such that for each i the variables in **V**(T_i) are direct effects of T_i. We then seek to eliminate those members of **V**(T_i) that are impure measures of T_i, either because they are

The Structure of the Unobserved

also the effects of some other unmeasured variable in **T**, because they are also causes or effects of some other measured variable, or because they share an unmeasured common cause in **C** with another measured variable.

In the class of models we are considering, a measured variable can be an impure measure for four reasons, which are exhaustive:

(i) If there is a directed edge from some T_i in **T** to some V in $\mathbf{V}(T_i)$ but also a trek between V and T_j that does not contain T_i or any member of **V** except V then V is **latent-measured impure**.
(ii) If there is a trek between a pair of measured variables V_1, V_2 from the same cluster $\mathbf{V}(T_i)$ that does not contain any member of **T** then V_1 and V_2 are **intra-construct impure**.
(iii) If there is a trek between a pair of measured variables V_1, V_2 from distinct clusters $\mathbf{V}(T_i)$ and $\mathbf{V}(T_j)$ that does not contain any member of **T** then we say V_1 and V_2 are **cross-construct impure**. (iv) If there is a variable in **C** that is the source of a trek between T_i and some member V of $\mathbf{V}(T_i)$ we say V is **common cause impure**.

In figure 10.2, for example, if $\mathbf{V}(T_1) = \{X_1, X_2, X_3\}$ and $\mathbf{V}(T_2) = \{X_4, X_5, X_6\}$ then X_4 is latent-measured impure, X_1 and X_2 are intra-construct impure, X_2 and X_5 are cross-construct impure, and X_6 is common cause impure. Only X_3 is a pure measure of T_1.

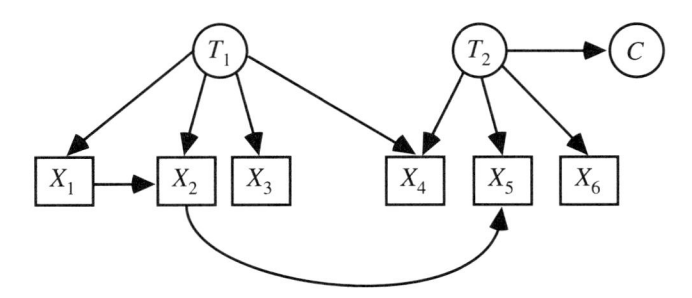

Figure 10.2

We say that a measurement model is **almost pure** if the only kind of impurities among the measured variables are common cause impurities. An **almost pure latent variable graph** is a directed acyclic graph with an almost pure measurement model. In an almost pure latent variable graph we continue to refer throughout the rest of this chapter to the set of measured variables as **V**, a subset of the latent variables as **T**, and the "nuisance" latent variables that are common causes of members of **T** and **V** as **C**.

The strategy that we employ has three steps:

(i) Eliminate measured variables until the variables that remain form the largest almost pure measurement model with at least two indicators for each latent variable.

(ii) Use vanishing tetrad differences among variables in the measurement model from (i) to determine the zero and first order independence relations among the variables in **T**.
(iii) Use the PC algorithm to construct a pattern from the zero and first order independence relations among the variables in **T**.

The next section describes a procedure for identifying the appropriate measured variables. The details are rather intricate; the reader should bear in mind that the procedures have all been automated, that they work very well in simulation tests, and they all derive from fundamental structural principles. Given the population correlations the inference techniques would be reliable (in large samples) for any conditions under which the Tetrad Representation Theorem holds. The statistical decisions involve a substantial number of joint tests, and no doubt could be improved. We occasionally resort to heuristics for cases in which each latent variable has a large number of measured indicators.

10.3 Finding Almost Pure Measurement Models

If G is the true model over $\mathbf{V} \cup \mathbf{T} \cup \mathbf{C}$ with measurement model G_M, then our task in this section is to find a subset **P** of **V** (the larger the better) such that the sub-model of G_M on vertex set $\mathbf{P} \cup \mathbf{T} \cup \mathbf{C}$ is an almost pure measurement model, if one exists with at least two indicators per latent variable. Our strategy is to use different types of foursomes of variables to sequentially eliminate impure measures.

As in figure 10.3, we call four measured variables an **intra-construct foursome** if all four are in $\mathbf{V}(T_i)$ for some T_i in **T**; otherwise call it a **cross-construct foursome.**

10.3.1 Intra-Construct Foursomes

In this section we discuss what can be learned about the measurement model for T_i from $\mathbf{V}(T_i)$ alone. We take advantage of the following principle, which is a consequence of the Tetrad Representation Theorem.

(P–1) If a directed acyclic graph linearly implies all tetrad differences among the variables in $\mathbf{V}(T_i)$ vanish, then no pair of variables in $\mathbf{V}(T_i)$ is intra-construct impure.

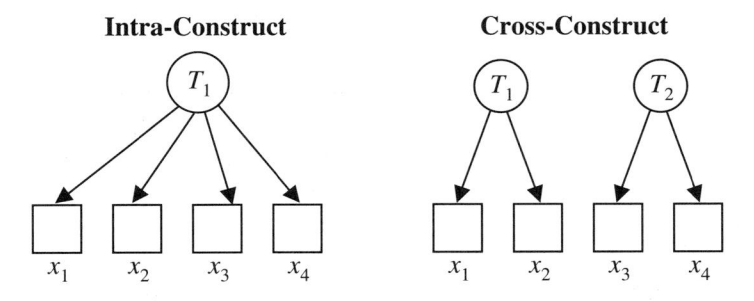

Figure 10.3

So given a set, $\mathbf{V}(T_i)$, of variables that measure T_i, we seek the largest subset, $\mathbf{P}(T_i)$, of $\mathbf{V}(T_i)$ such that all tetrad differences are judged to vanish among $\mathbf{P}(T_i)$. The number of subsets of $\mathbf{V}(T_i)$ is $2^{|\mathbf{V}(T_i)|}$, so it is not generally feasible to examine each of them. Further, in realistic samples we won't find a sizable subset in which all tetrad differences are judged to vanish. A more feasible strategy is to prune the set iteratively, removing at each stage the variable that improves the performance of the remaining set $\mathbf{P}(T_i)$ on easily computable heuristic criteria derived from principle P–1. In practice, if the set $\mathbf{V}(T_i)$ is large, some small subset of $\mathbf{V}(T_i)$ may by chance do well on these two criteria. For example, if $\mathbf{V}(T_i)$ has 12 variables, then there are 495 subsets of size 4, each of which has only 3 possible vanishing tetrad differences. There are 792 subsets of size 5, but there are 15 possible tetrad differences that must all be judged to vanish among each set instead of 3. Because the larger the size of $\mathbf{P}(T_i)$ the more unlikely it is that all tetrad differences among $\mathbf{P}(T_i)$ will be judged to vanish by chance, and because we might eliminate variables from $\mathbf{P}(T_i)$ later in the process, we want $\mathbf{P}(T_i)$ to be as large as possible. On the other hand, no matter how well a set $\mathbf{P}(T_i)$ does on the first criterion above, some subset of it will do at least as well or better. Thus, in order to avoid always choosing the smallest possible subsets we have to penalize smaller sets.

We use the following simple algorithm. We initialize $\mathbf{P}(T_i)$ to $\mathbf{V}(T_i)$. If the *set* of tetrad differences among variables in $\mathbf{P}(T_i)$ passes a statistical test, we exit. (We count a set of n tetrad differences as passing a statistical test at a given significance level *Sig* if each individual tetrad difference passes a statistical test at significance level *Sig/n*. The details of the statistical tests that we employ on individual tetrad differences are described in chapter 11.) If the set does not pass a statistical test, we look for a variable to eliminate from $\mathbf{P}(T_i)$. We score each measured variable X in the following way. For each tetrad difference t among variables in $\mathbf{P}(T_i)$ in which X appears we give X credit if t passes a statistical test, and discredit if t fails a statistical test. We then eliminate the variable with the lowest score from $\mathbf{P}(T_i)$. We repeat this process until we arrive at a set $\mathbf{P}(T_i)$ that passes the statistical test, or we run out of variables.

10.3.2 Cross-Construct Foursomes
Having found, for each latent variable T_i, a subset $\mathbf{P}(T_i)$ of $\mathbf{V}(T_i)$ in which no variables are intra-construct impure, we form a subset $\mathbf{P}$ of $\mathbf{V}$ such that

$$\mathbf{P} = \bigcup_{T_i \in \mathbf{T}} \mathbf{P}(T_i).$$

We next eliminate members of $\mathbf{P}$ that are cross-construct impure.

2x2 foursomes involve two measured variables from $\mathbf{P}(T_i)$ and two from $\mathbf{P}(T_j)$, where i and j are distinct. A 2x2 foursome in a pure latent variable model linearly implies exactly one tetrad equation, *regardless of the nature of the causal connection between T_i and T_j in the structural model*. For example, the graph in figure 10.4 linearly implies the vanishing tetrad difference $\rho_{XY}\rho_{WZ} - \rho_{XW}\rho_{YZ} = 0$. Graphs in which T_j causes T_i and graphs in which T_i and T_j are not causally connected (i.e., there is no trek between them) also linearly imply $\rho_{XY}\rho_{WZ} - \rho_{XW}\rho_{YZ} = 0$.

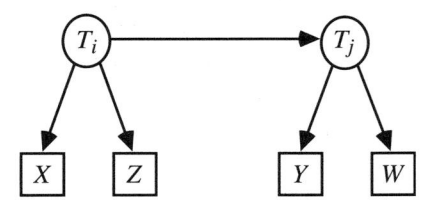

Figure 10.4

If one variable in $\mathbf{V}(T_i)$ is latent-measured impure because of a trek containing T_j, and one variable in $\mathbf{V}(T_j)$ is latent-measured impure because of a trek containing T_i, then the tetrad differences among the foursome are not linearly implied to vanish by the graph. If T_i and T_j are connected by some trek and a pair of variables in $\mathbf{V}(T_i)$ and $\mathbf{V}(T_j)$ respectively are cross-construct impure then again the tetrad difference is not linearly implied to vanish by the graph. (The case where T_i and T_j are not connected by some trek is considered below.) In figure 10.5, for example, model (i) implies the tetrad equation $\rho_{XY}\rho_{WZ} = \rho_{XW}\rho_{YZ}$ but models (ii) and (iii) do not.

So if we test a 2x2 foursome F_1 and the hypothesis that the appropriate tetrad difference vanishes can be rejected, then we know that in at least one of the four pairs in which there is a measured variable from each construct, both members of the pair are impure. We don't yet know which pair. We can find out by testing other 2x2 foursomes that share variables with F_1. Suppose the largest subgraph of the true model containing $\mathbf{P}(T_1)$ and $\mathbf{P}(T_2)$ is the graph in figure 10.6.

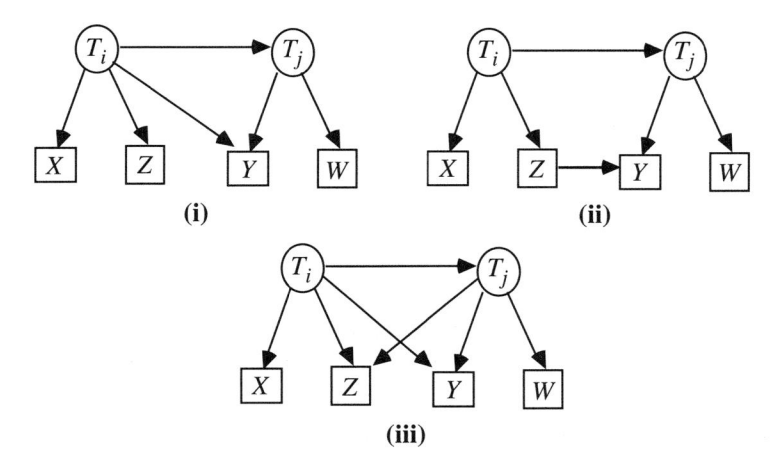

Figure 10.5

The Structure of the Unobserved

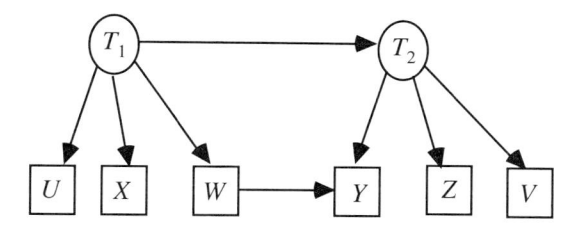

Figure 10.6

Only 2x2 foursomes involving the pair <W,Y> will be recognizably impure. When we test vanishing tetrad differences in the foursome F_1 = <X,W,Y,Z>, we won't know which of the pairs <W,Z>, <X,Y>, <X,Z>, <W,Y> is impure. When we test the foursome F_2 = <X,W,Z,V>, however, we find that no pair among <X,Z>, <X,V>, <W,Z>, or <W,V> is impure. We know therefore that the pairs <X,Z> and <W,Z> are not impure in F_1. By testing the foursome F_3 = <U,X,Y,Z> we find that <X,Y> is not impure, entailing <W,Y> is impure in F_1. If there are at least two pure indicators within each construct, then we can detect exactly which of the other indicators are impure in this way.

By testing all the 2x2 foursomes in **P**, we can in principle eliminate all variables that are cross-construct impure. We cannot yet eliminate all the variables that are latent-measured impure, because if there is only one such variable it is undetectable from 2x2 foursomes.

Foursomes that involve three measured variables from $\mathbf{P}(T_i)$ and one from $\mathbf{P}(T_j)$, where *i* and *j* are distinct, are called **3x1 foursomes**. All 3x1 foursomes in a pure measurement model linearly imply all three possible vanishing tetrad differences (see model (i) in figure 10.7 for example), *no matter what the causal connection between T_i and T_j*. If the variable from $\mathbf{P}(T_j)$ in a 3x1 foursome is impure because it measures both latents (model (ii) in figure 10.7), then T_i is still a choke point and all three equations are linearly implied. If a variable Z from $\mathbf{P}(T_i)$ is impure because it measures both latents (model (iii) in figure 10.7), however, then the latent variable model does not linearly imply that the tetrad differences containing the pair <Z,W> vanish. This entails that a nonvanishing tetrad differences among the variables in a 3x1 foursome can identify a unique measured variable as latent-measured impure.

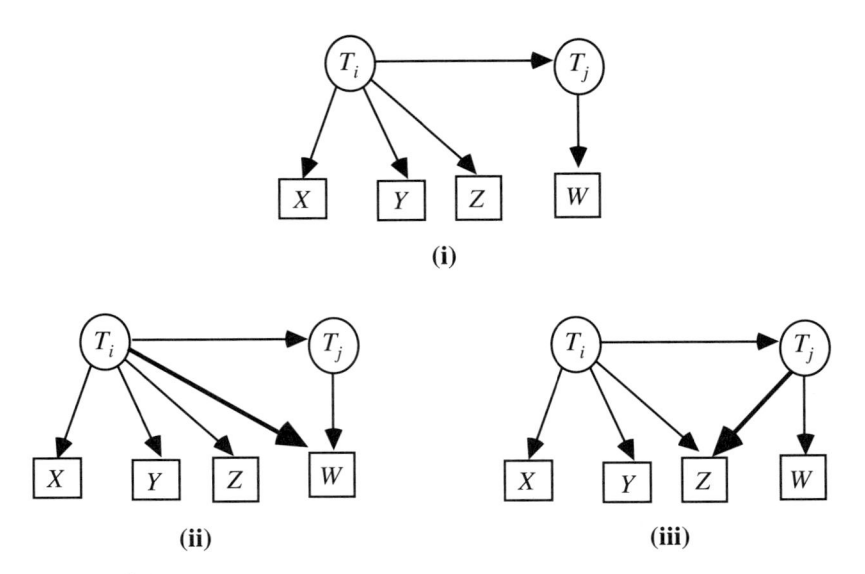

Figure 10.7

Also if T_i and T_j are not trek-connected and a pair of variables V_1 and V_2 in $\mathbf{P}(T_i)$ and $\mathbf{P}(T_j)$ respectively are cross-construct impure, then the correlation between V_1 and V_2 does not vanish, and a tetrad difference among a 3x1 foursome that contains V_1 and V_2 is not linearly implied to vanish; hence the impure member of $\mathbf{P}(T_j)$ will be recognized.

If there are least three variables in $\mathbf{P}(T_i)$ for each i, then when we finish examining all 3x1 foursomes we will have a subset $\mathbf{P}$ of $\mathbf{V}$ such that the sub-model of the true measurement model over $\mathbf{P}$ (which we call G_P) is an almost pure measurement model.

10.4 Facts about the Unobserved Determined by the Observed

In an almost pure latent variable model constraints on the correlation matrix among the *measured* variables determine
(i) for each pair A, B, of latent variables, whether A, B are uncorrelated,
(ii) for each triple A, B, C of latent variables, whether A and B are d-separated given $\{C\}$.

Part (i) is obvious: two measured variables are uncorrelated in an almost pure latent variable model if and only if they are effects of distinct unmeasured variables that are not trek connected (i.e., there is no trek between them) and hence are d-separated given the empty set of variables. Part (ii) is less obvious, but in fact certain d-separation facts are determined by vanishing tetrad differences among the measured variables.

Theorem 10.1 is a consequence of the Tetrad Representation Theorem:

THEOREM 10.1: If G is an almost pure latent variable graph over $\mathbf{V} \cup \mathbf{T} \cup \mathbf{C}$, $\mathbf{T}$ is causally sufficient, and each latent variable in $\mathbf{T}$ has at least two measured indicators,

The Structure of the Unobserved   261

then latent variables T_1 and T_3, whose measured indicators include J and L respectively, are d-separated given latent variable T_2, whose measured indicators include I and K, if and only if G linearly implies $\rho_{JI}\rho_{LK} = \rho_{JL}\rho_{KI} = \rho_{JK}\rho_{IL}$.

For example, in the model in figure 10.8, the fact that T_1 and T_3 are d-separated given T_2 is entailed by the fact that for all m, n, o, and p between 1 and 3, where o and p are distinct:

$$\rho_{A_m D_n}\rho_{B_o B_p} = \rho_{A_m B_o}\rho_{D_n B_p} = \rho_{A_m B_p}\rho_{D_n B_o}$$

By testing for such vanishing tetrad differences we can test for first order d-separability relations among the unmeasured variables in an almost pure latent variable model. (If **A** and **B** are d-separated given **D**, we call the number of variables in **D** the **order** of the d-separability relation.) These zero and first order d-separation relations can then be used as input to the PC algorithm, or to some other procedure, to obtain information about the causal structure among the latent variables. In the ideal case, the pattern among the latents that is output will always contain the pattern that would result from applying the PC algorithm directly to d-separation facts among the latents, but it may contain extra edges and fewer orientations.

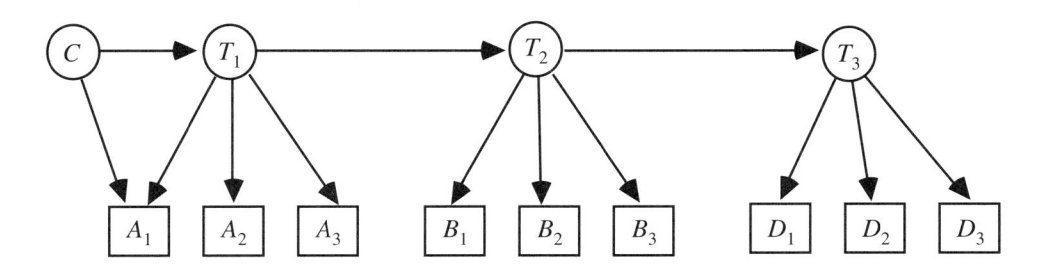

Figure 10.8

10.5 Unifying the Pieces
Suppose the true but unknown graph is shown in figure 10.9.

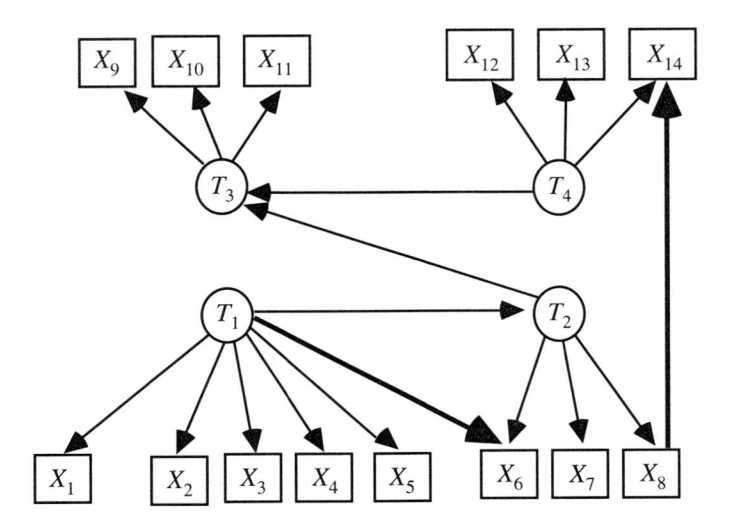

Figure 10.9. True causal structure

We assume that a researcher can accurately cluster the variables in the specified measurement model, for example, figure 10.10.

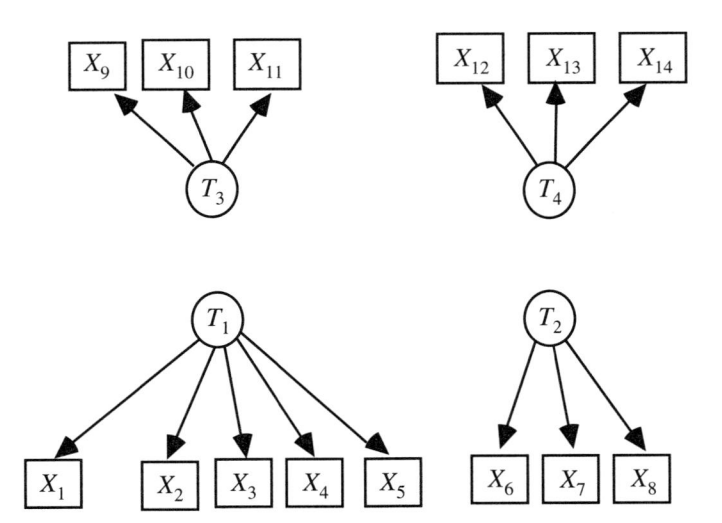

Figure 10.10. Specified measurement model

The actual measurement model is then the graph in figure 10.11.

The Structure of the Unobserved 263

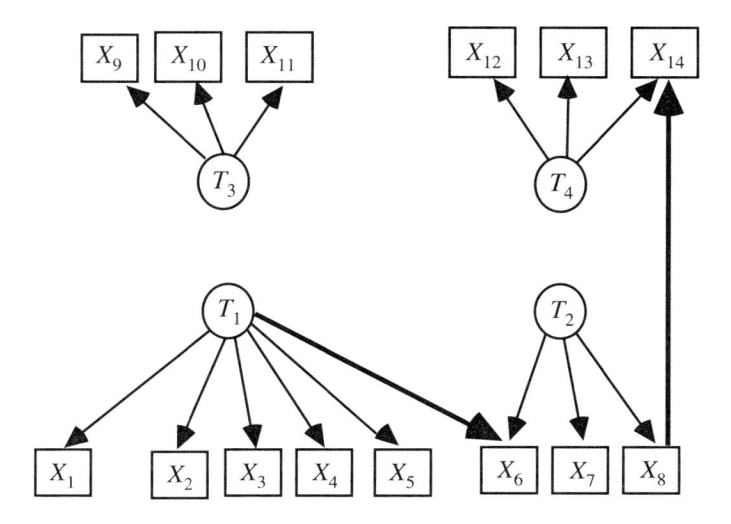

Figure 10.11. Actual measurement model

Figure 10.12 shows a subset of the variables in G (one that leaves out X_1, X_6, and X_{14}) that do form an almost pure measurement model.

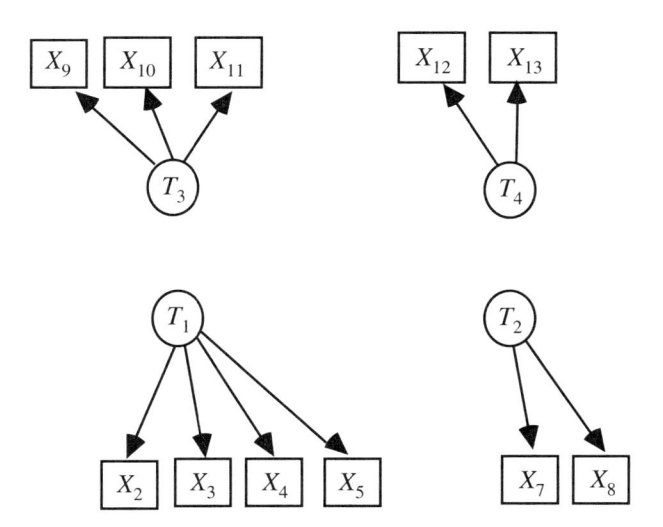

Figure 10.12. Almost pure measurement model

Assuming the sequence of vanishing tetrad difference tests finds such an almost pure measurement model, a sequence of tests of 1x2x1 vanishing tetrad difference tests then decides some d-separability facts for the PC or other algorithm through theorem 10.1. Since in figure 10.12 there are many 1x2x1 tetrad tests with measured variables drawn respectively from the clusters for T_1, T_2 and T_3, the results of the tests must somehow be aggregated. For each 1x2x1 tetrad difference among variables in $\mathbf{V}(T_1)$, $\mathbf{V}(T_2)$, and $\mathbf{V}(T_3)$ we give credit if the tetrad difference passes a significance test and discredit if it fails a significance test; if the final score is greater than 0, we judge that T_1 and T_3 are d-separated by T_2.

With two slight modifications, the PC algorithm can be applied to the zero and first order d-separation relations determined by the vanishing tetrad differences. The first modification is of course that the algorithm never tries to test any d-separation relation of order greater than 1 (i.e., in the loop in step B) of the PC Algorithm the maximum value of n is 1.) The second is that in step D) of the PC algorithm we do not orient edges to avoid cycles.

Without all of the d-separability facts available, the PC algorithm may not find the correct pattern of the graph. It may include extra edges and fail to orient some edges. However, it is possible to recognize from the pattern that some edges are definitely in the graph that generated the pattern, while others may or may not be. We add the following step to the PC algorithm to label with a "?" edges that may or may not be in the graph. Y is a **definite noncollider** on an undirected path U in pattern Π if and only if either X *-* $Y \rightarrow Z$, or $X \leftarrow Y$ *-* Z are subpaths of U, or X and Z are not adjacent and not $X \rightarrow Y \leftarrow Z$ on U.

E.) Let $\mathbf{P}$ be the set of all undirected paths in Π between X and Y of length ≥ 2. If X and Y are adjacent in Π, then mark the edge between X and Y with a "?" unless either

(i) no paths are in $\mathbf{P}$, or
(ii) every path in $\mathbf{P}$ contains a collider, or
(iii) there exists a vertex Z such that Z is a definite noncollider on every path in $\mathbf{P}$, or
(iv) every path in $\mathbf{P}$ contains the same subpath $<A,B,C>$.

We refer to the combined procedure as the Multiple Indicator Model Building (MIMBuild) Algorithm.

THEOREM 10.2: If G is an almost pure latent variable graph over $\mathbf{V} \cup \mathbf{T} \cup \mathbf{C}$, $\mathbf{T}$ is causally sufficient, each variable in $\mathbf{T}$ has at least two measured indicators, the input to MIMBuild is a list of all vanishing zero and first order correlations among the latent variables linearly implied by G, and Π is the output of the MIMBuild Algorithm then:

A–1) If X and Y are not adjacent in Π, then they are not adjacent in G.
A–2) If X and Y are adjacent in Π and the edge is not labeled with a "?," then X and Y are adjacent in G.

O–1) If $X \to Y$ is in Π, then every trek in G between X and Y is into Y.
O–2) If $X \to Y$ is in Π and the edge between X and Y is not labeled with a "?," then $X \to Y$ is in G.

The algorithm's complexity is bounded by the number of tetrad differences it must test, which in turn is bounded by the number of foursomes of measured variables. If there are n measured variables the number of foursomes is $O(n^4)$. We do not test each possible foursome, however, and the actual complexity depends on the number of latent variables and how many variables measure each latent. If there are m latent variables and s measured variables for each, then the number of foursomes is $O(m^3 \times s^4)$. Since $m \times s = n$, this is $O(n^3 \times s)$.

10.6 Simulation Tests

The procedure we have sketched has been fully automated in the TETRAD II program, with sensible but rather arbitrary weighting principles where required. To test the behavior of the procedure we generated data from the causal graph in figure 10.13, which has 11 impure indicators.

The distribution for the exogenous variables is standard normal. For each sample, the linear coefficients were chosen randomly between .5 and 1.5.

We conducted 20 trials each at sample sizes of 100, 500, and 2000. We counted errors of commission and errors of omission for detecting uncorrelated latents (0-order d-separation) and for detecting 1st-order d-separation. In each case we counted how many errors the procedure could have made and how many it actually made. We also give the number of samples in which the algorithm identified the d-separations perfectly. The results are shown in table 10.1, where the proportions in each case indicate the number of errors of a given kind over all samples divided by the number of possible errors of that kind over all samples.

Extensive simulation tests with a variety of latent topologies for as many as six latent variables, and 60 normally distributed measured variables, show that for a given sample size the reliability of the procedure is determined by the number of indicators of each latent and the proportion of indicators that are confounded. Increased numbers of almost pure indicators make decisions about d-separability more reliable, but increased proportions of confounded variables makes identifying the almost pure indicators more difficult. For large samples with ten indicators per latent the procedure gives good results until more than half of the indicators are confounded.

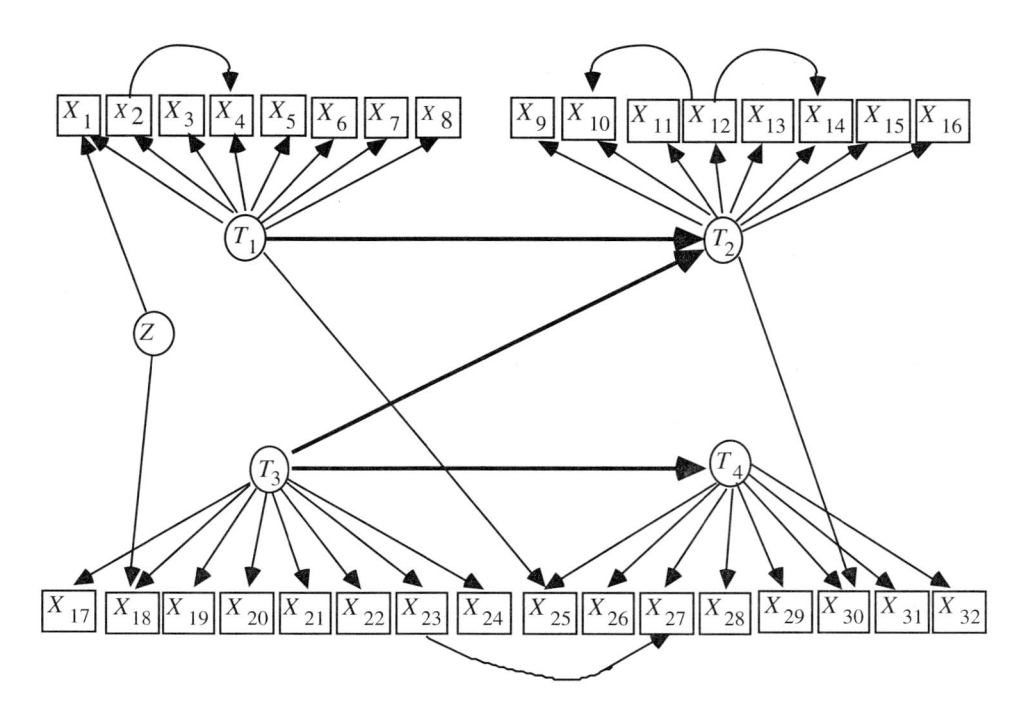

Figure 10.13. Impure Indicators =
$\{X_1, X_2, X_4, X_{10}, X_{12}, X_{14}, X_{18}, X_{23}, X_{25}, X_{27}, X_{30}\}$

Sample Size	0-order Commission	0-order Omission	1st-Order Commission	1st-Order Omission	Perfect
100	2.50%	0.00%	3.20%	5.00%	65.00%
500	1.25%	0.00%	0.90%	0.00%	95.00%
2000	0.00%	0.00%	0.00%	0.00%	100.00%

Table 10.1

10.7 Conclusion

Alternative strategies are available. One could, for example, purify the measurement sets, and specify a "theoretical model" in which each pair of latent variables is directly correlated. A maximum likelihood estimate of this structure will then give an estimate of the correlation matrix for the latents. The correlation matrix could then be used as input to the PC or FCI algorithms. The strategy has two apparent disadvantages. One is that these estimates typically depend on an assumption of normality. The other is that in preliminary simulation studies with normal variates and using LISREL to estimate the latent correlations, we have found the strategy less reliable than the procedure described in this chapter. Decisions about d-separation facts among latent variables seem to be more reliable if they are founded on a weighted average of a number of decisions about vanishing tetrad differences based on measured correlations than if they are founded on decisions about vanishing partial correlations based on estimated correlations.

The MIMBuild algorithm assumes **T** is causally sufficient; an interesting open question is whether there are reliable algorithms that do not make this assumption. In addition, although the algorithm is correct, it is incomplete in a number of distinct ways. There is further orientation information linearly implied by the zero and first order vanishing partial correlations. Further, we do not know whether there is further information about which edges definitely exist (i.e., should not be marked with a "?") that is linearly implied by the vanishing zero and first order partial correlations. Moreover, it is sometimes the case that for each edge labeled with a "?" in the MIMBuild output there exists a pattern compatible with the vanishing zero and first order partial correlations that does not contain that edge, but no pattern compatible with the vanishing zero and first order partial correlations that does not contain two or more of the edges so labeled.

Finally, and most importantly, the strategy we have described is not very informative about latent structures that have multiple causal pathways among variables. An extension of the strategy might be more informative and merits investigation. In addition to tetrad differences, one could test for higher-order constraints on measured correlations (e.g., algebraic combinations of five or more correlations) and use the resulting decisions to determine higher-order d-separation relations among the latent variables. The necessary theory has not been developed.

11 Elaborating Linear Theories with Unmeasured Variables[1]

11.1 Introduction

In many cases investigators have a causal theory in which they place some confidence, but they are unsure whether the model contains all important causal connections, or they believe it to be incomplete but don't know which dependencies are missing. How can further unknown causal connections be discovered? The same sort of question arises for the output of the PC or FCI algorithms when, for example, two correlated variables are disconnected in the pattern; in that case we may think that some mechanism not represented in the pattern accounts for the dependency, and the pattern needs to be elaborated. In this chapter we consider a special case of the "elaboration problem," confined to linear theories with unmeasured common causes each having one or more measured indicators. The general strategy we develop for addressing the elaboration problem can be adapted to models without latent variables, and also to models for discrete variables. Other strategies than those we consider here are also promising; the Bayesian methods of Cooper and Herskovits, in particular, could be adapted to the elaboration problem.

The problem of elaborating incomplete "structural equation models" has been addressed in at least two commercial computer packages, the LISREL program (Joreskog and Sorbom 1984) and the EQS program (Bentler 1985). We will describe detailed tests of the reliabilities of the automated search procedures in these packages. Generally speaking, we find them to be very unreliable, but not quite useless, and the analysis of why they fail when other methods succeed suggests an important general lesson about computerized search in statistics: in specification search computation matters, and in large search spaces it matters far more than does using tests that would, were computation free, be optimal.

We will compare the EQS and LISREL searches with a search procedure based on tests of vanishing tetrad differences. In principle, the collection of tetrad tests is less informative than maximum likelihood tests of an entire model used by the LISREL and EQS searches. In practice, this disadvantage is overwhelmed by the computational advantages of the tetrad procedure. Under some general assumptions, the procedure we describe gives correct (but not necessarily fully informative) answers if correct decisions are made about vanishing tetrad differences in the population. We demonstrate that for many problems the procedure obtains very reliable conclusions from samples of realistic sizes.

11.2. The Procedure

The procedure we will describe is implemented in the TETRAD II program. It takes as input:

(i) a sample size,
(ii) a correlation or covariance matrix, and
(iii) the directed acyclic graph of an initial linear structural equation model.

A number of specifications of internal parameters can also be input. The graph is given to the program simply by specifying a list of paired causes and effects. The algorithm can be divided into two parts, a scoring procedure and a search procedure.

11.2.1 Scoring

The procedure uses the following methodological principles.

Falsification Principle: Other things being equal, prefer models that do not linearly imply constraints that are judged not to hold in the population.

Explanatory Principle: Other things being equal, prefer models that linearly imply constraints that are judged to hold in the population.

Simplicity Principle: Other things being equal, prefer simpler models.

The intuition behind the Explanatory Principle is that an explanation of a constraint based on the causal structure of a model is superior to an explanation that depends upon special values of the free parameters of a model. This intuition has been widely shared in the natural sciences; it was used to argue for the Copernican theory of the solar system, the General Theory of Relativity, and the atomic hypothesis. A more complete discussion of the Explanatory Principle can be found in Glymour et al. 1987, Scheines 1988, and Glymour 1983. As with vanishing partial correlations, the set of values of linear coefficients associated with the edges of a graph that generate a vanishing tetrad difference not linearly implied by the graph has Lebesgue measure zero.

Unfortunately, the principles can conflict. Suppose, for example, that model M' is a modification of model M, formed by adding an extra edge to M. Suppose further that M' linearly implies fewer constraints that are judged to hold in the population, but also linearly implies fewer constraints that are judged not to hold in the population. Then M' is superior to M with respect to the Falsification Principle, but inferior to M with respect to the Simplicity and Explanatory Principles. The procedure we use introduces a heuristic scoring function that balances these dimensions.[2]

In order to calculate the *Tetrad-score* we first calculate the **associated probability** $P(t)$ of a vanishing tetrad difference, which is the probability of obtaining a tetrad difference as large or larger than the one actually observed in the sample, under the assumption that the tetrad difference vanishes in the population. Assuming normal variates, Wishart (1928) showed that the variance of the sampling distribution of the vanishing tetrad difference $\rho_{IJ}\rho_{KL} - \rho_{IL}\rho_{JK}$ is equal to

$$\frac{D_{12}D_{34}(N+1)}{(N-1)(N-2)} - D$$

where D is the determinant of the population correlation matrix of the four variables I, J, K, and L, D_{12} is the determinant of the two-dimensional upper left-corner submatrix, D_{34} is the determinant of the lower right-corner submatrix and I, J, K, and L, have a joint normal distribution. In calculating $P(t)$ we substitute the sample covariances for the corresponding population covariances in the formula. $P(t)$ is determined by lookup in a chart for the standard normal distribution. An asymptotically distribution free test has been described by Bollen (1990).

Among any four distinct measured variables I, J, K, and L we compute three tetrad differences:

$t_1 = \rho_{IJ}\rho_{KL} - \rho_{IL}\rho_{JK}$
$t_2 = \rho_{IL}\rho_{JK} - \rho_{IK}\rho_{JL}$
$t_3 = \rho_{IK}\rho_{JL} - \rho_{IJ}\rho_{KL}$

and their associated probabilities $P(t_i)$ on the hypothesis that the tetrad difference vanishes. If $P(t_i)$ is larger than the given significance level, the procedure takes the tetrad difference to vanish in the population. If $P(t_i)$ is smaller than the significance level, but the other two tetrad differences have associated probabilities higher than the significance level, then t_i is ignored. Otherwise, if $P(t_i)$ is smaller than the significance level, the tetrad difference is judged not to vanish in the population.

Let **Implied$_H$** be the set of vanishing tetrads linearly implied by a model M that are judged to hold in the population and **Implied$_{\sim H}$** be the set of vanishing tetrads linearly implied by M that are judged not to hold in the population. Let **Tetrad-score** be the score of model M for a given significance level assigned by the algorithm, and let **weight** be a parameter (whose significance is explained below). Then we define

$$T = \sum_{t \in \textbf{Implied}_H} P(t) - \sum_{t \in \textbf{Implied}_{\sim H}} weight * (1 - P(t))$$

The first term implements the explanatory principle while the second term implements the falsification principle. The simplicity principle is implemented by preferring, among models with identical *Tetrad-scores*, those with fewer free parameters—which amounts to preferring graphs with fewer edges. The weight determines how conflicts between the explanatory and falsification principles are resolved by determining the relative importance of explanation relative to residual reduction.

The scoring function is controlled by two parameters. The *significance level* is used to judge when a given tetrad difference is zero in the population. The *weight* is used to determine the relative importance of the Explanatory and Falsification Principles. The scoring function has several desirable asymptotic properties, but we do not know whether the particular value for *weight* we use is optimal.

11.2.2 Search

The TETRAD II procedure searches a tree of elaborations of an initial model. The search is comparatively fast because there is an easy algorithm for determining the vanishing tetrad differences linearly implied by a graph (using the Tetrad Representation Theorem), because most of the computational work required to evaluate a model can be stored and reused to evaluate elaborations, and because the scoring function is such that if a model can be conclusively eliminated from consideration because of a poor score, so can any elaboration of it.

The search generates each possible one-edge elaboration of the initial model, orders them by the tetrad score, and eliminates any that score poorly. It then repeats this process recursively on each model generated, until no improvements can be made to a model.

The search is guided by a quantity called **T-maxscore**, which for a given model M represents the maximum *Tetrad-score* that could possibly be obtained by any elaboration of M. *T-maxscore* is equal to:

$$T\text{-}maxscore = \sum_{t \in \mathbf{Implied}_H} P(t)$$

The use we make of this quantity is justified by the following theorem.

THEOREM 11.1. If G is a subgraph of directed acyclic graph G', than the set of tetrad equations among variables of G that are linearly implied by G' is a subset of those linearly implied by G.

In order to keep the following example small, suppose that there are just 4 edges, e_1, e_2, e_3, or e_4 which could be added to the initial model. The example illustrates the search procedure in a case where each possible elaboration of the initial model is considered. Node 1 in figure 11.1 represents the initial model. Each node in the graph represents the model generated by adding the edge next to the node to its parent. For example, node 2 represents the initial model + e_1; node 7 represents node 2 + e_4, which is the initial model + e_1 + e_4. We will say that a program **visits** a node when it creates the model M corresponding to the node and then determines whether any elaboration of M has a higher *Tetrad-score* than M. (Note that the algorithm can generate a model M without visiting M if it generates M but does not determine whether any elaboration of M has a higher

Tetrad-score than *M*.) The numbers inside each node indicate the order in which the models are visited. Thus for example, when the algorithm visits node 2, it first generates all possible one edge additions of the initial model + e_1, and orders them according to their *T-maxscore*. It then first visits the one with the highest *T-maxscore* (in this case, node 3 that represents the initial model + e_1 + e_2). Note that the program does not visit the initial model + e_2 (node 10) until after it has visited all elaborations of the initial model + e_1.

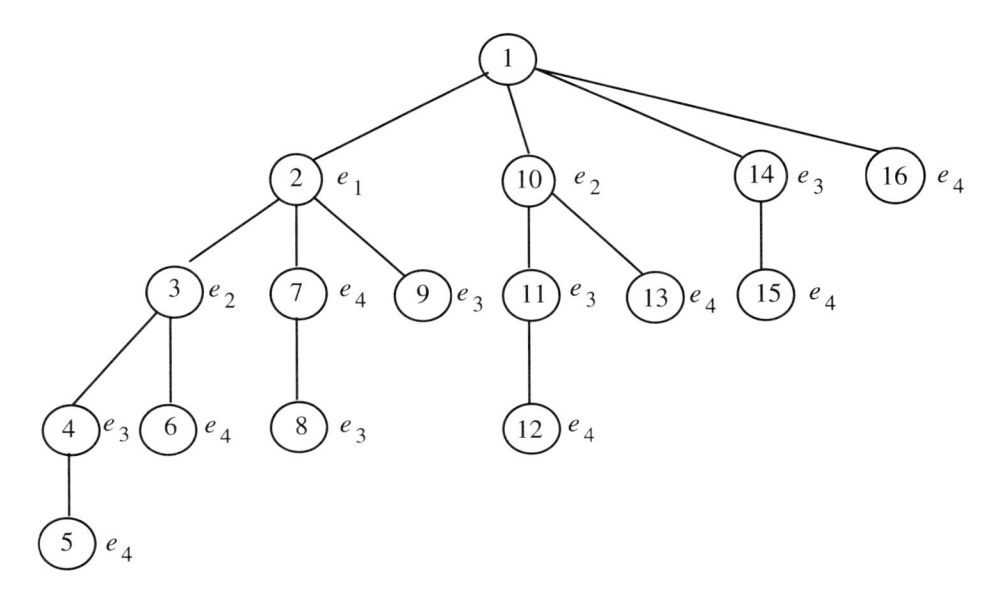

Figure 11.1

In practice, this kind of complete search could not possibly be carried out in a reasonable amount of time. Fortunately we are able to eliminate many models from consideration without actually visiting them. Addition of edges to a graph may defeat the linear implication of tetrad equations, but in view of theorem 11.1 will never cause more tetrad equations to be linearly implied by the resulting graph. If the *T-maxscore* of a model *M* is less than the *Tetrad-score* of some model *M'* already visited, then we know that neither *M* nor any elaboration of *M* has a *Tetrad-score* as high as that of *M'*. Hence we need never visit *M* or any of its elaborations. This is illustrated in figure 11.2. If we find that *T-maxscore* of the initial model + e_4 is lower than the *Tetrad-score* of the initial model + e_1, we can eliminate from the search all models that contain the edge e_4.

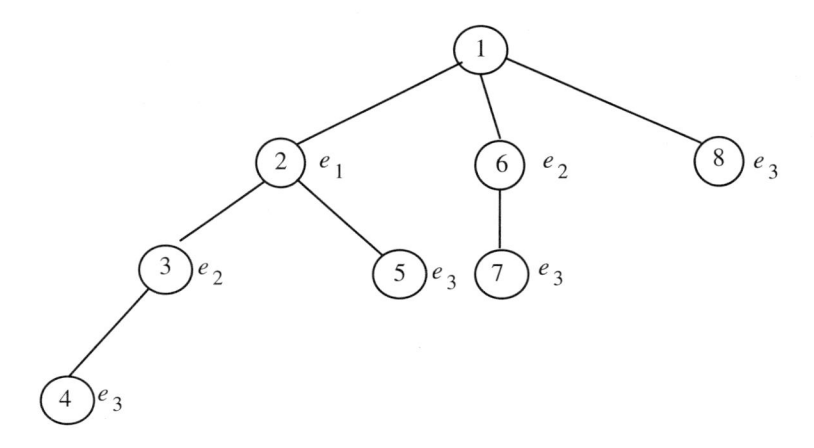

Figure 11.2

In some cases in the simulation study described later, the procedure described here is too slow to be practical. In those cases the time spent on a search is limited by restricting the depth of search. (We made sure that in every case the depth restriction was large enough that the program had a chance to err by overfitting.) The program adjusts the search to a depth that can be searched in a reasonable amount of time; in many of the Monte Carlo simulation cases no restriction on depth was necessary.[3]

11.3. The LISREL and EQS Procedures

LISREL VI and EQS are computer programs that perform a variety of functions, such as providing maximum likelihood estimates of the free parameters in a structural equation model. The feature we will consider automatically suggests modifications to underspecified models.

11.3.1 Input and Output

Both programs take as input:

(i) a sample size,
(ii) a sample covariance matrix,
(iii) initial estimates of the variances of independent variables,
(iv) initial estimates of the linear coefficients,
(v) an initial causal model (specified by fixing at zero the linear coefficient of A in the equation for B if and only if A is not a direct cause of B), in the form of equations (EQS) or a system of matrices (LISREL VI)
(vi) a list of parameters not to be freed during the course of the search,
(vii) a significance level, and
(viii) a bound on the number of iterations in the estimation of parameters.

The output of both programs includes a *single* estimated model that is an elaboration of the initial causal model, various diagnostic information as well as a χ^2 value for the suggested revision, and the associated probability of the χ^2 measure.

11.3.2 Scoring

LISREL VI and EQS provide maximum likelihood estimates of the free parameters in a structural equation model. More precisely, the estimates are chosen to minimize the fitting function

$$F = \log|\Sigma| + tr(S\Sigma^{-1}) - \log|S| - t$$

where S is the sample covariance matrix, Σ is the predicted covariance matrix, t is the total number of indicators, and if A is a square matrix then $|A|$ is the determinant of A and $tr(A)$ is the trace of A. In the limit, the parameters that minimize the fitting function F also maximize the likelihood of the covariance matrix for the given causal structure.

After estimating the parameters in a given model, LISREL VI and EQS test the null hypothesis that Σ is of the form implied by the model against the hypothesis that Σ is unconstrained. If the associated probability is greater than the chosen significance level, the null hypothesis is accepted, and the discrepancy is attributed to sample error; if the probability is less than the significance level, the null hypothesis is rejected, and the discrepancy is attributed to the falsity of M. For a "nested" series of models $M_1,...,M_k$ in which for all models M_i in the sequence the free parameters of M_i are a subset of the free parameters of M_{i+1}, asymptotically, the *difference* between the χ^2 values of two nested models also has a χ^2 distribution, with degrees of freedom equal to the difference between the degrees of freedom of the two nested models.

11.3.3 The LISREL VI Search[4]

The LISREL VI search is guided by the "modification indices" of the fixed parameters. Each modification index is a function of the derivatives of the fitting function with respect to a given fixed parameter. More precisely, the modification index of a given fixed parameter is defined to be $N/2$ times the ratio between the squared first-order derivative and the second-order derivative (where N is the sample size). Each modification index provides a lower bound on the decrease in the χ^2 obtained if that parameter is freed and all previously estimated parameters are kept at their previously estimated values.[5] (Note that if the coefficient for variable A in the linear equation for B is fixed at zero, then freeing that coefficient amounts to adding an edge from A to B to the graph of the model.) LISREL VI first makes the starting model the current best model in its search. It then calculates the modification indices for all of the fixed parameters[6] in the starting model. If LISREL VI estimates that the difference between the χ^2 statistics of M, the current best model, and M', the model obtained from M by freeing the parameter with the largest modification index, is not significant, then the search ends, and LISREL VI suggests model M. Otherwise, it makes M' the current best model and repeats the process.

11.3.4 The EQS Search
EQS computes a Lagrange Multiplier statistic, which is asymptotically distributed as χ^2.[7] EQS performs univariate Lagrange Multiplier tests to determine the approximate separate effects on the χ^2 statistic of freeing each fixed parameter in a set specified by the user. It frees the parameter that it estimates will result in the largest decrease in the χ^2 value. The program repeats this procedure until it estimates that there are no parameters that will significantly decrease the χ^2. Unlike LISREL VI, when EQS frees a parameter it does not reestimate the model.[8]

It should be noted that both LISREL VI and EQS are by now quite complicated programs. An understanding of their flexibility and limitations can only be obtained through experimentation with the programs.

11.4. The Primary Study
Eighty data sets, forty with a sample size of 200 and forty with a sample size of 2,000, were generated by Monte Carlo methods from *each* of nine different structural equation models with latent variables. The models were chosen because they involve the kinds of causal structures that are often thought to arise in social and psychological scientific work. In each case part of the model used to generate the data was omitted and the remainder, together in turn with each of the data sets for that model, was given to the LISREL VI, EQS, and TETRAD II programs. A variety of specification errors are represented in the nine cases. Linear coefficient values used in the true models were generated at random to avoid biasing the tests in favor of one or another of the procedures. In addition, a number of ancillary studies were suggested by the primary studies and bear on the reliability of the three programs.

11.4.1 The Design of Comparative Simulation Studies
To study the reliability of automatic respecification procedure under conditions in which the general structural equation modeling assumptions are met, the following factors should be varied independently:
(i) the causal structure of the true model;
(ii) the magnitudes and signs of the parameters of the true model;
(iii) how the starting model is misspecified;
(iv) the sample size.

In addition, an ideal study should:

(i) Compare fully algorithmic procedures, rather than procedures that require judgment on the part of the user. Procedures that require judgment can only adequately be tested by carefully blinding the user to the true model; further, results obtained by one user may not transfer to other users. With fully algorithmic procedures, neither of these problems arises.

(ii) Examine causal structures that are of a kind postulated in empirical research, or that there are substantive reasons to think occur in real domains.

(iii) Generate coefficients in the models randomly. Costner and Herting showed that the size of the parameters affects LISREL's performance. Further, the reliability of TETRAD II depends on whether vanishing tetrad differences hold in a sample because of the particular numerical values of the coefficients rather than because of the causal structure, and it is important not to bias the study either for or against this possibility.

(iv) Ensure insofar as possible that all programs compared must search the same space of alternative models.

11.4.2 Study Design
11.4.2.1 Selection of Causal Structures

The nine causal structures studied are illustrated in figures 3, 4, and 5. For simplicity of depiction we have omitted uncorrelated error terms in the figures, but such terms were included in the linear models. The heavier directed or undirected lines in each figure represent relationships that were included in the model used to generate simulated data, but were omitted from the models given to the three programs; that is, they represent the dependencies that the programs were to attempt to recover. The starting models are shown in figure 11.6. The models studied include a one factor model with five measured variables, seven multiple indicator models each with eight measured variables and two latent variables, and one multiple indicator model with three latent variables and eight measured variables.

One factor models commonly arise in psychometric and personality studies (see Kohn 1969); two latent factor models are common in longitudinal studies in which the same measures are taken at different times (see McPherson et al. 1977), and also arise in psychometric studies; the triangular arrangement of latent variables is a typical geometry (see Wheaton et al. 1977).

The set of alternative structures determines the search space. Each program was forced to search the same space of alternative elaborations of the initial model, and the set of alternatives was chosen to be as large as possible consistent with that requirement.

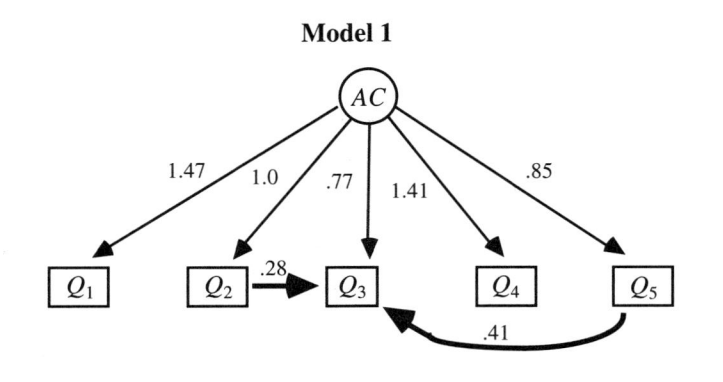

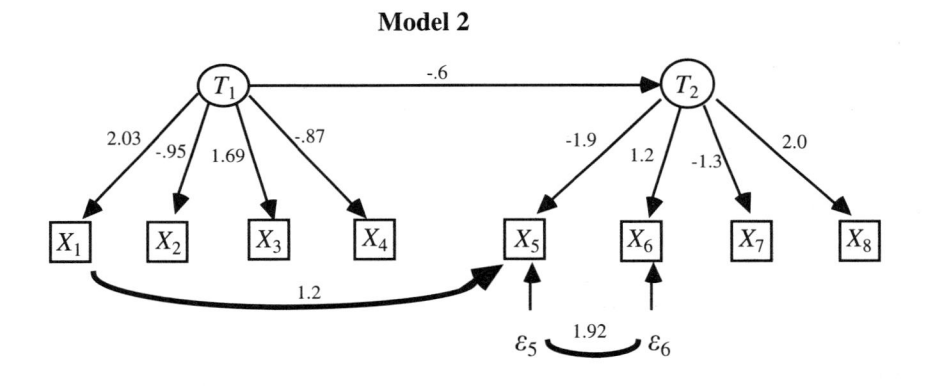

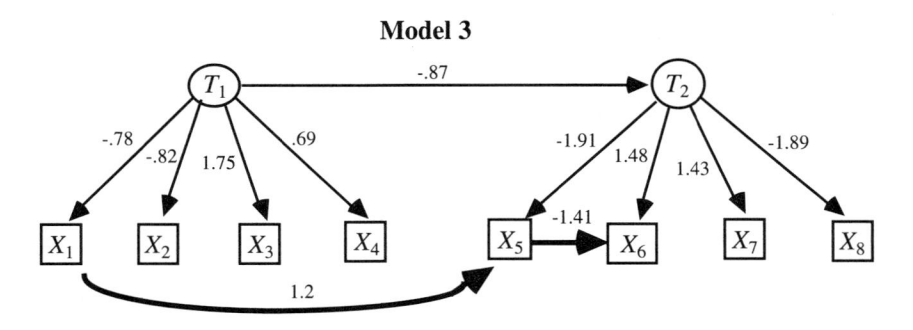

Figure 11.3

Model 4

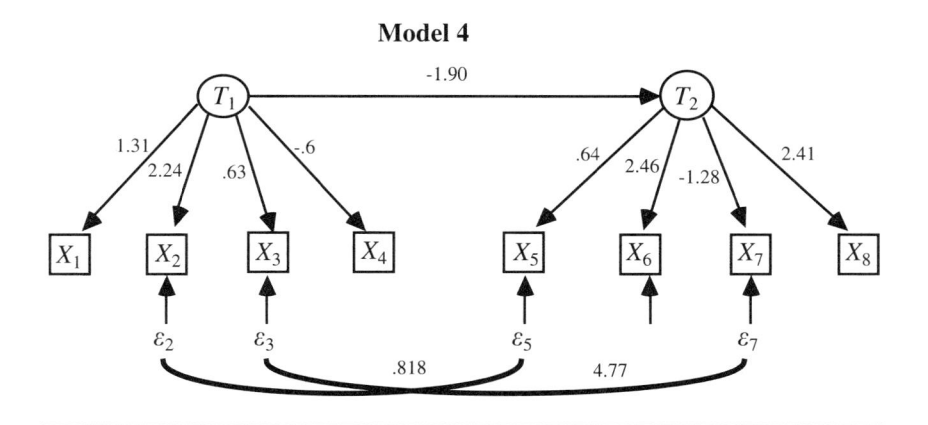

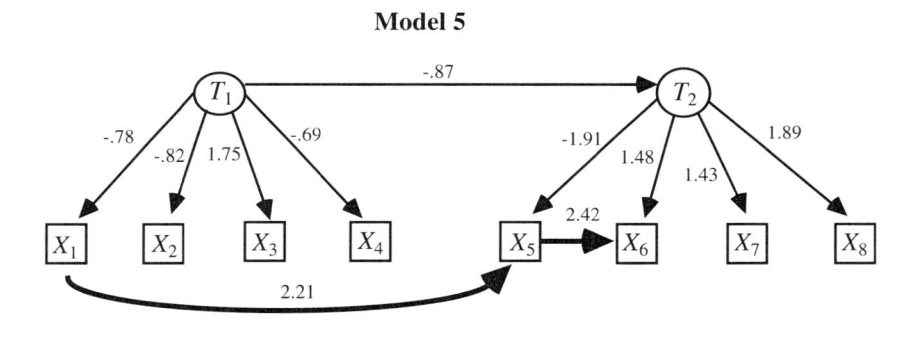

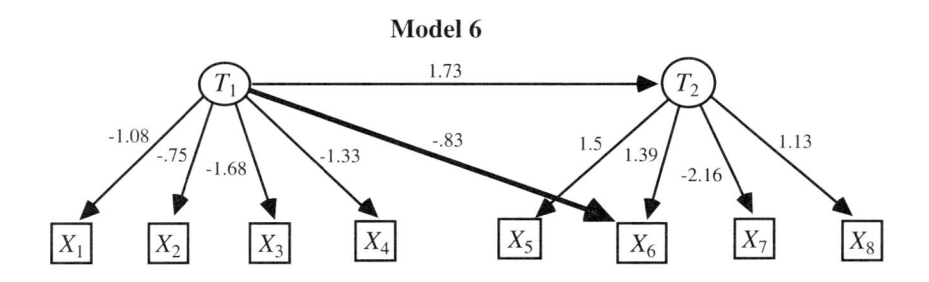

Model 5

Model 6

Figure 11.4

Model 7

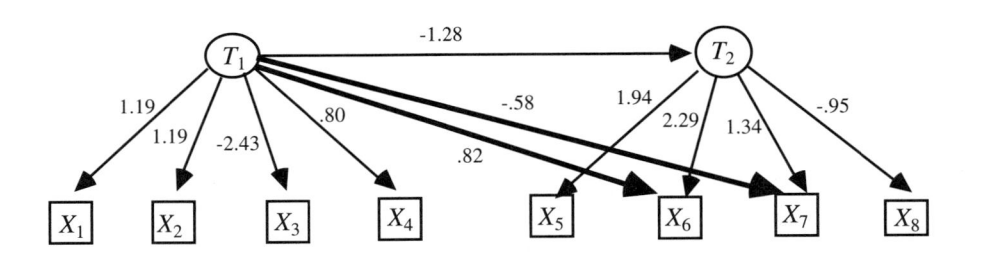

Model 8

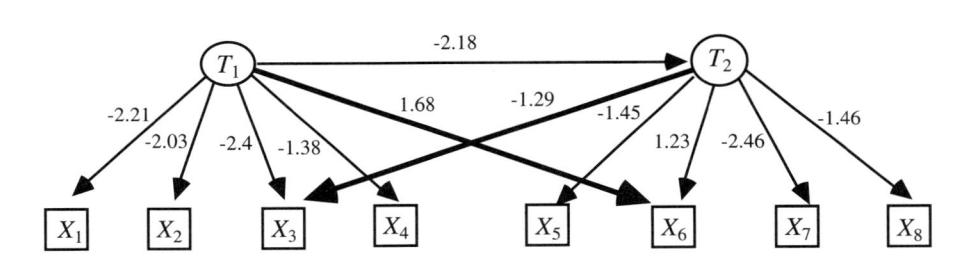

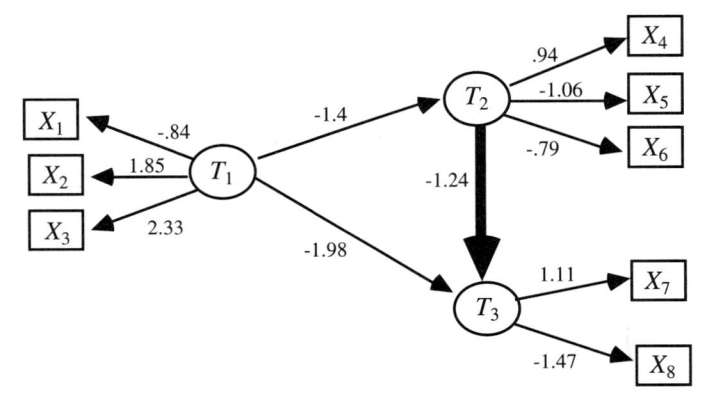

Model 9

Figure 11.5

Elaborating Linear Theories with Unmeasured Variables 281

Start for Model 1

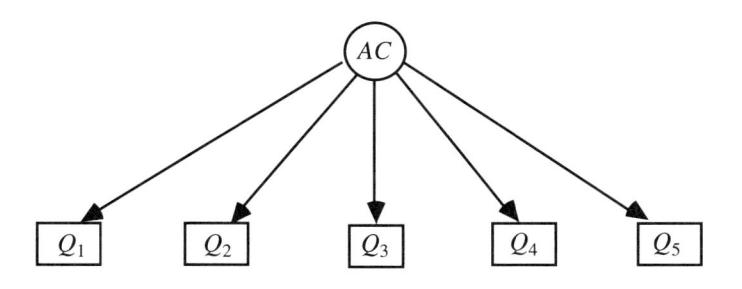

Start for Models 2 - 8

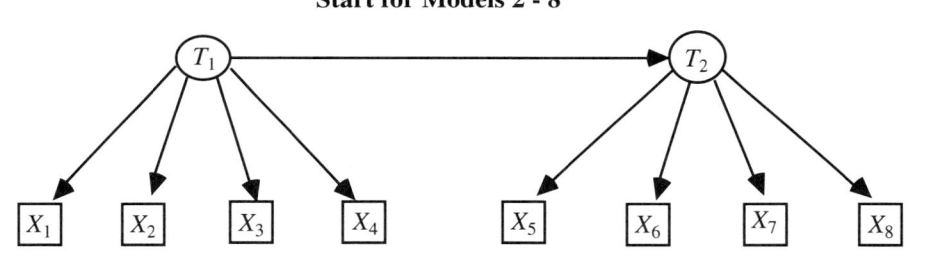

Start for Model 9

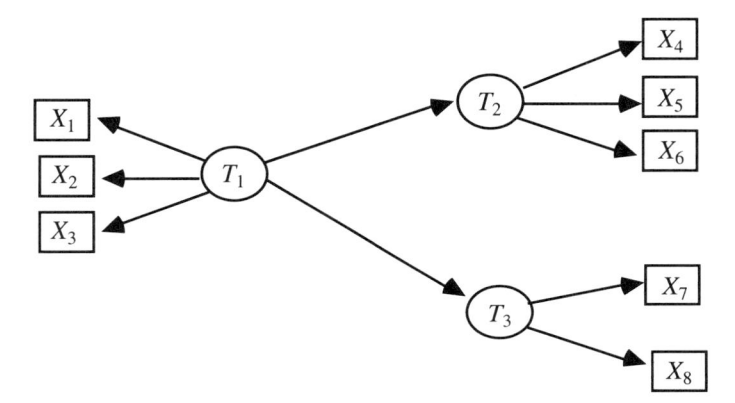

Figure 11.6

1.4.2.2 Selection of Connections to Be Recovered
The connections to be recovered include:

(i) Directed edges from latent variables to latent variables; relations of this kind are often the principal point of empirical research. See Maruyama and McGarvey (1980) for an example.
(ii) Edges from latent variables to measured variables; connections of this kind may arise when measures are impure, and in other contexts. See Costner and Schoenberg (1973) for an example.
(iii) Correlated errors between measured variables; relationships of this kind are perhaps the most frequent form of respecification.
(iv) Directed edges from measured variables to measured variables. Such relations cannot obtain, for example, between social indices, but they may very well obtain between responses to survey or psychometric instruments (see Campbell et al. 1966), and of course between measured variables such as interest rates and housing sales.

We have not included cases that we know beforehand cannot be recovered by one or another of the programs. Details are given in a later section.

11.4.2.3 Selection of Starting Models
Only three starting models were used in the nine cases. The starting models are, in causal modeling terms, pure factor models or pure multiple indicator models. In graph theoretic terms they are trees.

11.4.2.4 Selection of Parameters
In the figures showing the true models the numbers next to directed edges represent the values given to the associated linear coefficients. The numbers next to undirected lines represent the values of specified covariances. In all cases, save for models 1 and 5, the coefficients were chosen by random selection from a uniform distribution between .5 and 2.5. The value obtained was then randomly given a sign, positive or negative.

In model 1, all linear coefficients were made positive. The values of the causal connections between indicators were specified nonrandomly. The case was constructed to simulate a psychometric or other study in which the loadings on the latent factor are known to be positive, and in which the direct interactions between measured variables are comparatively small.

Model 5 was chosen to provide a comparison with model 3 in which the coefficients of the measured-measured edges were deliberately chosen to be large relative to those in model 3.

11.4.2.5 Generation of Data

For each of the nine cases, twenty data sets with sample size 200 and twenty data sets with sample 2,000 were generated by Monte Carlo simulation methods.

Pseudorandom samples were generated by the method described in chapter 5. In order to optimize the performance of each of the programs, we assumed that all of the exogenous variables had a standard normal distribution. This assumption made it possible to fix a value for each exogenous variable for each unit in the population by pseudo random sampling from a standard normal distribution. Correlated errors were obtained in the simulation by introducing a *new* exogenous common cause of the variables associated with the error terms.

11.4.2.6 Data Conditioning

The entire study we discuss here was performed twice. In the original study, we gave LISREL VI and EQS positive starting values for all parameters. If either program had difficulty estimating the starting model, we reran the case with the initial values set to the correct sign.

LISREL and EQS employ iterative procedures to estimate the free parameters of a model. These procedures are sensitive to "poorly conditioned" variables and will not perform optimally unless the data are transformed. For example, it is a rule of thumb with these procedures that no two variances should vary by more than an order of magnitude in the measured variables. After generating data in the way we describe above, a small but significant percentage of our covariance matrices were ill conditioned in this way.

To check the possibility that the low reliability we obtained in the first study for the LISREL VI and EQS procedures was due to "ill-conditioned" data, the entire study was repeated. Sample covariances were transformed into sample correlations by dividing each cell $[I,J]$ in the covariance matrix by $s_I s_J$, where s_I is the *sample* standard deviation of I. To avoid sample variances of widely varying magnitudes, we transformed each cell $[I,J]$ in the sample covariance matrix by dividing it by $\sigma_I \sigma_J$ where σ_I is the *population* standard deviation of I[9]. We call the result of this transformation the *pseudocorrelation* matrix. The transformation makes all of the variances of the measured variables almost equal, without using a data-dependent transformation. Of course in empirical research, this transformation could not be performed, since the population parameters would not be known.

In practice, we found that conditioning the data and giving the population parameters as starting values did little to change the performance of LISREL VI or EQS. The performance of the TETRAD II procedure was essentially the same in both cases. Conditioning the data improved LISREL VI's reliability very slightly for small samples, and degraded it slightly for large samples.

11.4.2.7 Starting Values for the Parameters

We selected the linear coefficients for our models randomly, allowing some to be negative and some to be positive. Models with negative parameters actually represent a harder case for the TETRAD procedures. If a model implies a vanishing tetrad difference then the signs of its parameters make no difference. If a model does not imply that a tetrad difference vanishes, however, but instead implies that the tetrad difference is equal to the sum of two or more terms, then it is possible, if not all of the model's parameters are positive, that these terms sum to zero. Thus, in data generated by a model with negative parameters, we are more likely to observe vanishing tetrad differences that are *not* linearly implied by the model.

The iterative estimation procedures for LISREL and EQS begin with a vector of parameters θ. They update this vector until the likelihood function converges to a local maximum. Inevitably, the iterative procedures are sensitive to starting values. Given the same model and data, but two different starting vectors θ^i and θ^j, the procedures might converge for one but not for the other. This is especially true when the parameters are of mixed signs. To give LISREL and EQS the best chance possible in the second study, we set the starting values of each parameter to its actual value whenever possible. For the linear coefficients that correspond to edges in the generating model left out of the starting model, we assigned a starting value of 0. For all other parameters, however, we started LISREL and EQS with the *exact value in the population.*[10]

11.4.2.8 Significance Levels

EQS and LISREL VI continue to free parameters as long as the associated probability of the decrease in χ^2 exceeds the user-specified significance level. For both LISREL and EQS, we set the significance level to .01. (This is the default value for LISREL; the default value for EQS is .05.) The lower the significance level, the fewer the parameters that each program tends to set free. Since both LISREL and EQS both tend to overfit even at .01, we did not attempt to set the significance level any higher. (It may appear in our results that LISREL VI and EQS both underfit more than they overfit, but almost all of the "underfitting" was due to aborted searches that did not employ the normal stopping criterion.)

11.4.2.9 Number of Iterations

The default number of maximum iterations for estimating parameters for LISREL VI on a personal computer is 250. We set the number of maximum iterations to 250 for both our LISREL VI and EQS tests.

11.4.2.10 Specifying Starting Models in LISREL VI

LISREL VI, like previous editions of the program, requires the user to put variables into distinct matrices according to whether they are exogenous, endogenous but unmeasured, measured but dependent on exogenous latent, measured but dependent on endogenous latent, and so forth. Variables in certain of these categories cannot have effects on variables in other categories. When formulated as recommended in the LISREL manual, LISREL VI would be in principle unable to detect many of the effects considered in this study. However, these restrictions can in most cases be overcome by a system of substitutions of phantom variables in which measured variables are actually represented as endogenous latent variables.[11] In the current study, we were not able to get LISREL VI to accept changing ζ variables, which are exogenous and latent, to η variables, which are endogenous and latent. This had the unfortunate effect that LISREL would not consider adding any edges into T_1 (represented by the ζ variable). To ensure a comparable search problem, we restricted TETRAD II and EQS in the same way.

11.4.2.11 Implementation
The LISREL VI runs were performed with the personal computer version of the program, run on a Compaq 386 computer with a math coprocessor. EQS runs were performed on an IBM XT clone with a math coprocessor. All TETRAD II runs were performed on Sun 3/50 workstations. For TETRAD II, which also runs on IBM clones, the processing time for the Compaq 386 and the Sun 3/50 are roughly the same.

11.4.2.12 Specification of TETRAD II Parameters
TETRAD II requires that the user set a value of the weight parameter, a value for the significance level used in the test for vanishing tetrad differences, and a value for a percentage parameter that bounds the search. In all cases we set the significance level at 0.05. At sample size 2000, we set the weight to .1 and the percentage to 0.95.

At smaller sample size the estimates of the population covariances are less reliable, and more tetrad differences are incorrectly judged to vanish in the population. This makes judgments about the Explanatory Principle less reliable. For this reason, at sample size 200, we set the weight to 1, in order to place greater importance upon the Falsification Principle. Less reliable judgments about the Explanatory Principle also make lowering the percentage for small sample sizes helpful. At sample size 200, we set the percentage to 0.90. We do not know if these parameter settings are optimal.

11.5 Results
For each data set and initial model, TETRAD II produces a set of best alternative elaborations. In some cases that set consists of a single model; typically it consists of two or three alternatives. EQS and LISREL VI, when run in their automatic search mode, produce as output a single model elaborating the initial model. The information provided by each program is scored "correct" when the output contains the true model. But it is important to see how the various programs err when their output is not correct, and we have provided a more detailed classification of various kinds of error. We have classified the output of TETRAD II as follows (where a model is in TETRAD's top group if and

only if it is tied for the highest *Tetrad score*, and no model with the same *Tetrad-score* has fewer edges):

Correct—the true model is in TETRAD's top group.
Width—the average number of alternatives in TETRAD's top group.

Errors:
Overfit—TETRAD's top group does not contain the true model but contains a model that is an elaboration of the true model.
Underfit—TETRAD's top group does not contain the true model but does contain a model of which the true model is an elaboration.
Other—none of the previous categories apply to the output.

We have scored the output of the LISREL VI and EQS programs as follows:

Correct—the true model is recommended by the program.

Errors:
In TETRAD's Top Group—the recommended model is not correct, but is among the best alternatives suggested by the TETRAD II program for the same data.
Overfit—the recommended model is an elaboration of the true model.
Underfit—the true model is an elaboration of the recommended model.
Right Variable Pairs—the recommended model is not in any of the previous categories, but it does connect the same pairs of variables as were connected in the omitted parts of the true model.
Other—none of the previous categories apply to the output.

In most cases no estimation problems occurred for either LISREL VI or EQS. In a number of data sets for cases 3 and 5, LISREL VI and EQS either issued warnings about estimation problems or aborted the search due to computational problems. Since our input files were built to minimize convergence problems, we ignored such warnings in our tabulation of the results. If either program recommended freeing a parameter, we counted that parameter as freed regardless of what warnings or estimation problems occurred before or after freeing it. If either program failed to recommend freeing any parameters because of estimation problems in the starting model, we counted it as an underfit. The results are shown in the next table and figure.

For a sample size of 2000, TETRAD II's set included the correct respecification in 95% of the cases. LISREL VI found the right model 18.8% of the time and EQS 13.3%. For a sample size of 200, TETRAD II's set included the correct respecification 52.2% of the time, while LISREL VI corrected the misspecification 15.0% of the time, and EQS corrected the misspecification 10.0 % of the time. A more detailed characterization of the errors is given in figure 11.8.

Width, n=2000									
Case	1	2	3	4	5	6	7	8	9
LISREL VI	1	1	1	1	1	1	1	1	1
EQS	1	1	1	1	1	1	1	1	1
TETRAD	4	2.1	2	1	1.1	3	7.1	11.3	2.9
Width, n=200									
Case	1	2	3	4	5	6	7	8	9
LISREL VI	1	1	1	1	1	1	1	1	1
EQS	1	1	1	1	1	1	1	1	1
TETRAD	1.9	3.5	1.5	1	1	3.2	5.9	8.4	3

Table 11.1

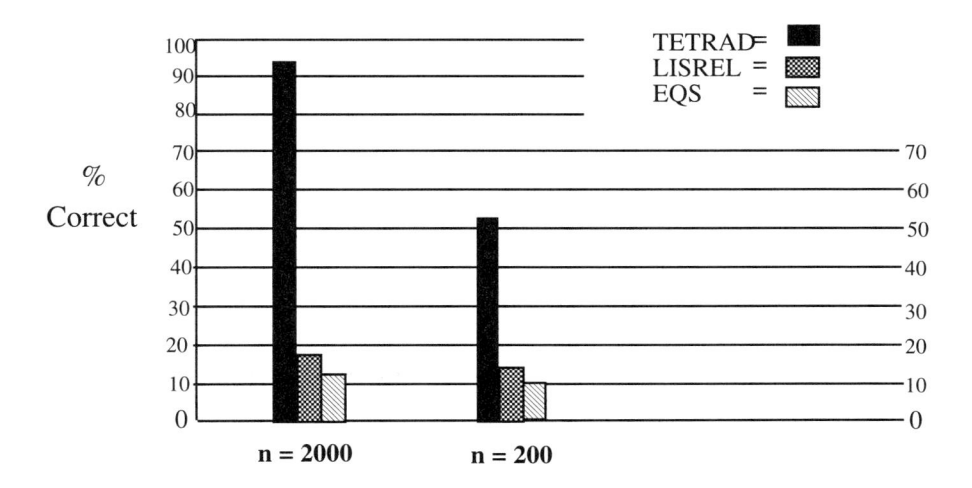

Figure 11.7

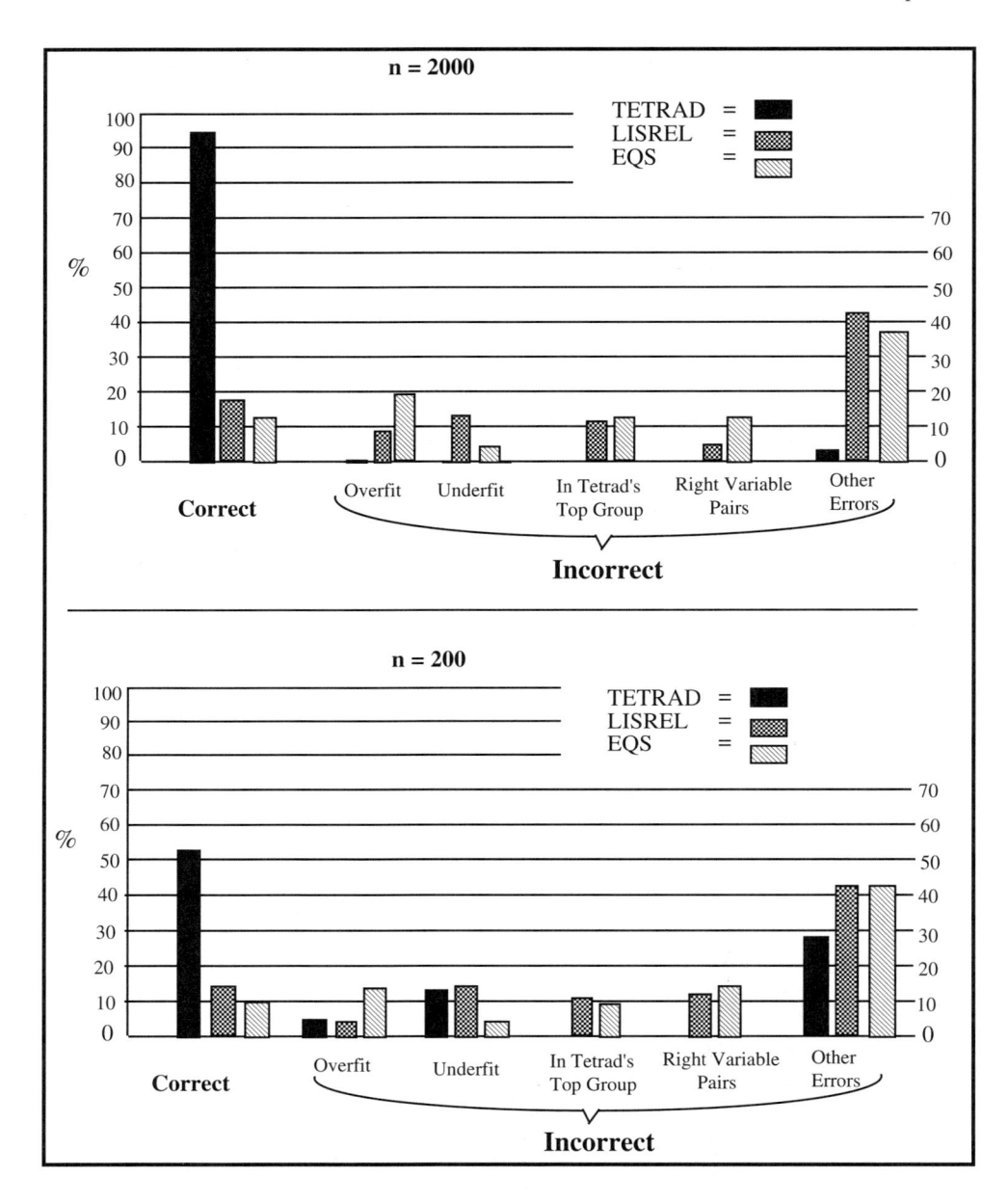

Figure 11.8

11.6 Reliability and Informativeness

There are two criteria by which the suggestions of each of these programs can be judged. The first is reliability. Let the **reliability** of a program be defined as the probability that its set of suggested models includes the correct one. In these cases, the TETRAD search procedures are clearly more reliable than either LISREL VI and EQS. One can achieve higher reliability simply by increasing the number of guesses. A program that outputs the top million models might be quite reliable, but its suggestions would be uninformative. Thus we call the second criterion boldness. Let the **boldness** of a program's suggestions be the reciprocal of the number of models suggested. On this measure, our procedure does worse than LISREL VI or EQS in seven of the nine cases considered.

Since neither our procedure nor the modification index procedures dominate on both of these criteria, it is natural to ask whether the greater reliability of the former is due simply to reduced boldness. This question can be interpreted in at least two ways:

(i) If TETRAD II were to increase its boldness to match LISREL VI and EQS, that is., if it were to output a single model, would it be more or less reliable than LISREL VI or EQS?

(ii) If LISREL VI or EQS were to decrease their boldness to match TETRAD II, that is, were they to output a set of models as large as does TETRAD II, would they be more or less reliable than TETRAD II?

If we have no reason to believe that any one model in the TETRAD II output is more likely than any other to be correct, we could simply choose a model at random. We can calculate the **expected single model reliability** of our procedure in the following way. We assume that when TETRAD II outputs a list of n models for a given covariance matrix, the probability of selecting any particular one of the models as the best guess is $1/n$. So instead of counting a list of length n that contains the correct model as a single correct answer, we would count it as $1/n$ correct answers.[12] Then simply divide the expected number of correct answers by the number of trial runs.

Were TETRAD II to be as bold as LISREL VI or EQS, its single model reliability at sample size 2000 would drop from 95% to about 42.3%. On our data, LISREL VI has a reliability of 18.8% and EQS has a reliability of 13.3%. At sample size 200 the TETRAD II single model reliability is 30.2% LISREL has a reliability of 15.0% for sample size 200 and EQS 10.0%. In a more realistic setting one might have substantive reasons to prefer one model over another. If substantive knowledge is worth anything, and we use it to select a single model M, then M is more likely to be true than a model selected at random from TETRAD II's set of suggested models. Thus, in a sense the numbers given in the paragraph above are worst case.

An alternative strategy is to cut down the size of the set before one picks a model. We can often eliminate some of the TETRAD II suggestions by running them through EQS or LISREL VI and discarding those that were not tied for the highest associated probability. There is little effect. We raise the (worst case) single model reliability of TETRAD II at sample size 2000 from 42.3% to about 46%, and at sample size 200 from 30.2 to approximately 32%.

There are a number of good reasons to want a list of equally good suggestions rather than a single guess. All have to do with the reliability and informativeness of the output.

First, it is important for the user of a program to have a good idea of how reliable the output of a program is. At sample size 2000, in the range of cases that we considered, the reliability of the TETRAD II output was very stable, ranging from a low of 90% to a high of 100%. For reasons explained below, the single model output by LISREL VI and EQS is at best in effect a random selection from a list of models that contains all of the models whose associated probabilities are equal to that of the true model (and possibly others of lower associated probabilities as well). Unfortunately, the size of the list from which the suggested model is randomly selected varies a great deal depending on the structure of the model, and is not known to the user. Thus, even ignoring the cases where LISREL VI had substantial computational difficulties, the reliability of LISREL VI's output at sample size 2000 ranged from 0 out of 20 to 11 out of 20. So it is rather difficult for a user of LISREL VI or EQS to know how much confidence to have in the suggested models.

Second, more than one model in a suggested set might lead to the same conclusion. For example, many of the models suggested by TETRAD II might overlap, that is, they might agree on a substantial number of the causal connections. If one's research concerns are located within those parts of the models that agree, then choosing a single model is not necessary. In this case one need not sacrifice reliability by increasing boldness, because all competitors agree.

Finally, having a well-defined list of plausible alternatives is more useful than a single less reliable suggestion for guiding further research. In designing experiments and in gathering more data it is useful to know exactly what competing models have to be eliminated in order to establish a conclusive result. For example, consider case 3. The correct model contains edges from X_1 to X_5 and X_5 to X_6. TETRAD II suggests the correct model, as well as a model containing edges from X_1 to X_5 and X_1 to X_6. An experiment which varied X_1, and examined the effect on X_6 would not distinguish between these two alternatives (since both predict that varying X_1 would cause X_6 to change), but an experiment which varied X_5 and examined the effect on X_6 would distinguish between these alternatives. Only by knowing the plausible alternatives can we decide which of these experiments is more useful.

If LISREL VI or EQS were to output a set of models as large as does our procedure, would they be as reliable? The answer depends upon how the rest of the models in the set were chosen. In many cases LISREL VI and EQS find several parameters tied, or almost tied, for the highest modification index. Currently both programs select one, and only one, of these parameters to free, on the basis of an arbitrary ordering of parameters. For example, if after evaluating the initial model it found that $X_3 \rightarrow X_5$ and $X_3 \, C \, X_5$[13] were tied for the highest modification indices, LISREL VI or EQS would choose one of them (say $X_3 \to X_5$) and continue until the search found no more parameters to free. Then they would suggest the single model that had the highest associated probability. If LISREL VI or EQS searched all branches corresponding to tied modification indices, instead of arbitrarily choosing one, their reliability would undoubtedly increase substantially. For example, after freeing $X_3 \to X_5$ and then freeing parameters until no more should be freed, LISREL VI or EQS could return to the initial model, free X_3 C X_5, and again continue freeing parameters until no more should be freed. They could then suggest all of the models tied for the highest associated probability. This is essentially the search strategy followed by the TETRAD II program.

If the LISREL VI search were expanded in this way on case 1 at sample size 2000, it would increase the number of correct outputs from 3 to 16 out of 20. In other cases, this strategy would not improve the performance of LISREL VI or EQS much at all. For example, in case 5 at sample size 2000, LISREL VI was incorrect on every sample in part because of a variety of convergence and computational problems, while TETRAD II was correct in every case. In case 4 at sample size 2000, LISREL VI missed the correct answer on nine samples (while TETRAD II missed the correct answer on only two samples) for reasons having nothing to do with the method of breaking ties.

LISREL VI and EQS would pay a substantial price for expanding their searches; their processing time would increase dramatically. A branching procedure that retained three alternatives at each stage and which stopped on all branches after freeing two parameters in the initial model, would increase the time required by about a factor of 7. In general, the time required for a branching search increases exponentially as the number of alternatives considered at each stage. Could such a search be run in a reasonable amount of time? Without a math coprocessor, a typical LISREL VI run on a Compaq 386 took roughly 20 minutes; with a math coprocessor it took about 4 minutes. EQS runs were done on a LEADING EDGE (an IBM XT clone that is considerably slower than the COMPAQ 386) with a math coprocessor and the average EQS run was about 5 minutes. This suggests that a branching strategy is possible for LISREL VI even for medium-sized models only on relatively fast machines; a branching search is practical on slower machines for the faster, but less reliable EQS search.

11.7 Using LISREL and EQS as Adjuncts to Search

There are two ways in which the sort of search TETRAD II illustrates can profitably be used in conjunction with LISREL VI or EQS. A procedure such as ours can be used to generate a list of alternative revisions of an initial model, which can then be estimated by LISREL or EQS, discarding those alternatives that have very low, or comparatively low associated probabilities.[14] We found that in only three cases could the associated probabilities distinguish among models suggested by TETRAD II. In case 6, one of the three models suggested by TETRAD II had a lower associated probability that the other two. In case 7, one of the six models suggested by TETRAD II had a lower associated probability that the other five. The largest reduction in TETRAD II's suggestions came in case 8, where 8 of the 12 models suggested by TETRAD II had associated probabilities lower than the top four. These results were obtained when LISREL VI was given the correct starting values for all of the edges in the true model, and a starting value of zero for edges not in the true model; in previous tests when LISREL VI was not given the true parameters as initial values, it often suffered convergence problems.

It is also instructive to run the both the automatic searches of TETRAD II and LISREL VI or EQS together. When LISREL VI and TETRAD II agree (that is when the model suggested by LISREL VI is in TETRAD II's top group) both programs are correct a higher percentage of times than their respective averages; conversely when they disagree, both programs are wrong a higher percentage of times than their average. The same holds true of EQS when used in conjunction with TETRAD II. Indeed, at sample size 2000, neither EQS nor LISREL VI was *ever* correct when it disagreed with TETRAD II. In contrast, at sample size 2000 LISREL VI was correct 61.8% of the time when it agreed with TETRAD II, and EQS was correct 53.3% of the time when it agreed with TETRAD II. Again, at sample size 2000, TETRAD II was *always* correct when it agreed with either LISREL VI or EQS. At sample size 200, while TETRAD II was correct on average 52.2% of the time, when it agreed with LISREL VI it was correct 75.7% of the time, and when it agreed with EQS it was correct 75.0% of the time. These results are summarized below:

Sample size 2000:

P(TETRAD correct)	95.0
P(LISREL VI correct)	18.8
P(EQS correct)	13.3
P(TETRAD correct | LISREL VI agree)	100.0
P(TETRAD correct | LISREL VI disagree)	92.1
P(TETRAD correct | EQS agree)	100 0
P(TETRAD correct | EQS disagree)	92.6
P(LISREL VI correct | TETRAD II agree)	61.8
P(LISREL VI correct | TETRAD II disagree)	0.0
P(EQS correct | TETRAD II agree)	53.3
P(EQS correct | TETRAD II disagree)	0.0

Sample size 200:

P(TETRAD correct)	52.2
P(LISREL VI correct)	15.0
P(EQS correct)	10.0
P(TETRAD correct | LISREL VI agree)	75.7
P(TETRAD correct | LISREL VI disagree)	46.9

P(TETRAD correct | EQS agree)	75.0
P(TETRAD correct | EQS disagree)	47.2
P(LISREL VI correct | TETRAD II agree)	39.4
P(LISREL VI correct | TETRAD II disagree)	9.5
P(EQS correct | TETRAD II agree)	43.7
P(EQS correct | TETRAD II disagree)	2.7

11.8 Limitations of the TETRAD II Elaboration Search

The TETRAD II procedure cannot find the correct model if there are a large number of vanishing TETRAD differences that are not linearly implied by the true model, but hold because of coincidental values of the free parameters. Our study indicates that this occurrence is unusual, at least given the uniform distribution that we placed on the linear coefficients in the models that generated our data, but it certainly does occur. The same results can be expected for any other "natural" distribution on the parameters. Further, the search does not guarantee that it will find all of the models that have the highest *Tetrad-score*. But in many cases, depending upon the size of the model, the amount of background knowledge, the structure of the model, and the sample size, the search space is so large that a search that *guarantees* finding the models with the highest *Tetrad-score* is not practical. One way the procedure limits search is through the application of the simplicity principle. This is a substantive assumption that may be false. The simplicity assumption is not needed for some small models, but in many problems with more variables there may be a large number of models that have maximal scores but contain many redundant edges that do not contribute to the score. Without the use of the simplicity principle, it is often difficult to search this space of models and if it is searched, there may be so many models tied for the highest score that the output is uninformative. If a model with "redundant" edges is correct, then our procedure will not find it. Typically these structures are underidentified, and so they could not be found by either LISREL VI or EQS.

The search procedure we have described here is practical for no more than several dozen variables. However, for larger numbers of variables, the MIMBuild algorithm described in chapter 10 may be applicable.

Finally, there exist many latent variable models that cannot be distinguished by the vanishing tetrad differences they imply, but are nonetheless in principle statistically distinguishable. More reliable versions of the LISREL or EQS procedures might succeed in discovering such structures when the TETRAD procedures fail.

11.9 Some Morals for Statistical Search

There were three reasons why the TETRAD II procedure proved more reliable over the problems considered here than either of the other search procedures.

(i) TETRAD II, unlike LISREL VI or EQS, does not need to estimate any parameters in order to conduct its search. Because the parameter estimation must be performed on an initial model that is wrong, LISREL VI and EQS often failed to converge, or calculated highly inaccurate parameter estimates. This in turn, led to problems in their respective searches.

(ii) In the TETRAD II search, when the scores of several different models are tied, the program considers elaborations of each model. In contrast, LISREL VI and EQS arbitrarily chose a single model to elaborate.

(iii) Both LISREL VI and EQS are less reliable than TETRAD II in deciding when to stop adding edges.

The morals for statistical search are evident: avoid iterative numerical procedures wherever possible; structure search so that it is feasible to branch when alternative steps seem equally good; find structural properties that permit reliable pruning of the search tree; for computational efficiency use local properties whenever possible; don't rely on statistical tests as stopping criteria without good evidence that they are reliable in that role.

Statistical searches cannot be adequately evaluated without clarity about the goals of search. We think in the social, medical and psychological uses of statistics the goals are often to find and estimate causal influence. The final moral for search is simple: once the goals are clearly and candidly given, if theoretical justifications of reliability are unavailable for the short run or even the long run, the computer offers the opportunity to subject the procedures to experimental tests of reliability under controlled conditions.

12 Prequels and Sequels

12.1 Graphical Representations, Independence, and Data Generating Processes

A variety of graphical objects have been introduced to represent both constraints on probability distributions and aspects of data generating processes. Each family of objects is accompanied by one or more principles relating graphical structure to conditional independence properties, just as undirected graphs are paired with separability and directed acyclic graphs are paired with the Markov Condition or with d-separation. Lauritzen et al. (1990) describe various Markov properties for different kinds of graphical models, and the relationships between the Markov properties. In their terminology, the Markov Condition of chapter 2 is a "local" Markov property, while d-separation is a "global" Markov property. Graphical objects consist of vertices, edges, and marks on edges or edge pairs (chapter 2), and families of such objects may restrict the possibilities in various ways. For example, undirected graphs (chapter 2, section 4) contain only undirected edges, and the natural global undirected Markov property for such objects specifies that if disjoint sets $\mathbf{X}$, $\mathbf{Y}$, $\mathbf{Z}$ are such that $\mathbf{Y}$ separates $\mathbf{X}$ and $\mathbf{Z}$, in the sense that every path connecting a member of $\mathbf{X}$ with a member of $\mathbf{Z}$ contains a member of $\mathbf{Y}$, then $\mathbf{X} \perp\!\!\!\perp \mathbf{Z} \mid \mathbf{Y}$.

In some cases—directed cyclic graphs for example—the representations have been in use for many years, without any general articulation either of the principle that relates graphical structure to independence properties, or of the data generating processes such structures describe. In this section we will consider directed acyclic graphs (DAGs), directed cyclic graphs (DCGs), partial ancestral graphs (PAGs), mixed ancestral graphs (MAGs), and chain graphs (CGs). The set of directed graphs (DGs) is the union of DAGs and DCGs. These representations are studied not merely because they represent one or another family of conditional independence relations, but because they describe the relations between causal hypotheses and conditional independence relations in a variety of models commonly used in applied statistics. For a discussion of these and other structures, as well as other distributional families representable by graphs, see Lauritzen 1996, Shafer 1996, and Edwards 1995. For a discussion of causal inference from graphical models see Lauritzen 2000.

12.1.1 Markov Conditions

As presented in chapter 3, the Causal Markov Condition gives a causal interpretation to a formal condition usually known as the local Markov property. The Causal Markov Condition is necessarily true of any system representable by a DAG in which the exogenous variables—those represented by vertices of zero indegree—are independently distributed, and each variable is any (measurable, deterministic) function of its parents (direct causes) and unique, jointly independent noises or "errors." It is also necessarily true of the subgraph and marginal probability distribution obtained by eliminating any subset of vertices with zero indegree and unit outdegree, and marginalizing accordingly.

It is a matter of some debate whether it applies to quantum mechanical systems (chapter 3 and Maudlin 1994). The Causal Markov Condition does not apply to systems of variables in which some variables are defined in terms of other variables, nor to systems with interunit causation (e.g., epidemics, where the units are people), although if the units are redefined so there is no interunit causation, it will apply (e.g., in an epidemic among a group of people, the group can be taken as a single unit). As we emphasized in chapter 6, even when it is true of the population described by some data-generating process, it may not characterize the conditional independence relations found for measured variables in a sample due to:

1. sampling error;
2. causal relations between the sampling mechanism and the observed variables (chapter 9, section 12.1.3);
3. lack of causal sufficiency among the measured variables (chapter 6);
4. aggregation of variable values (chapter 3, for example, representing blood pressure by "low," "medium," or "high," instead of two real numbers);
5. when one variable is a function of another variable by definition (e.g., X and X^2);
6. samples in which for some units A causes B and for other units B causes A;
7. reversible systems.

Sober (1987) criticized a consequence of the Causal Markov Condition on the grounds that two time series, such as the price of bread in England and the sea level in Venice may both be rising, and hence correlated, even though there is no causal connection between them. However, in this example it is not clear what the units are, and what the variables are. If the variables are bread price and sea level, then the units are years, and there is interunit causation (since the sea level at one year affects the sea level at another year). If one removes the interunit causation by taking differences of bread prices and differences of sea levels, there is no reason to believe the differences are correlated. On the other hand, if sea levels in different years are distinct variables, and bread prices in different years are distinct variables, then there is only one unit, and hence no correlation.

The Causal Markov Condition may not apply to samples from feedback systems (section 12.1.2) which are generated by time series, because depending upon what the process is and what the units are taken to be, there is interunit causation, or mixing units for which A causes B and units for which B causes A, or aggregation of variable values (e.g., by time averaging.)

The constituent implications of causal claims have been carefully analyzed by Hausman (1998) and the Causal Markov Condition has been defended at length in an interesting essay by Hausman and Woodward (in press) which emphasizes the close connections between the condition and the relations between interventions and mechanisms. The condition or its consequences have also been criticized (even for systems not in the list of exceptions noted above) by several writers (Lemmer 1996, Cartwright 1993, Artzenius

1992 Humphreys and Freedman 1996, which also criticizes some of the models in Chapter 5); replies to these criticisms are in Hausman and Woodward (in press), Spirtes et al. 1997, and Korb and Wallace 1997. A qualitative version of the Causal Markov Condition (not using probabilities) has been proposed in Goldszmidt and Pearl 1992.

12.1.2 Directed Cyclic Graphs

The models that we called pseudoindeterministic causal structures in chapters 2 and 3 are special cases of what are generally called structural equation models (SEMs). The variables in a SEM can be divided into two sets, the "error variables" or "error terms," and the substantive variables. Corresponding to each substantive variable X_i is an equation with X_i on the left hand side of the equation, and the direct substantive causes of X_i plus the error term ε_i on the right hand side of the equation, where ε_i represents the combined effect of all of causes other than the substantive ones. (We write the equation "$X_i = \varepsilon_i$" for an exogenous substantive variable X_i; this is nonstandard but serves to give the error terms a unified and special status as providing all the exogenous source of stochastic variation for the system.) Associated with each SEM is a graph ("path diagram" in the SEM literature.) There is a directed edge from X_i to X_j in the associated path diagram if and only if X_j is a function of X_i in the corresponding structural equation. Directed cycles are allowed in path diagrams. A distribution is associated with a SEM by assigning a probability distribution to the exogenous variables (which in turn determines the joint distribution over all of the variables.) An error term is generally not included in the path diagram of a SEM unless the error term is dependent upon some other error term. If two error terms are dependent, then they are included in the path diagram, and they are connected by a double-headed edge ("↔"). In other words, all error terms are assumed independent unless they are explicitly connected by double-headed edges in the path diagram. A SEM in which each vertex is a linear function of its associated error term and its parents in the associated path diagram is a linear SEM. (See figure 12.1 for an example of a linear SEM and its associated path diagram.) A good introduction to SEMs is Bollen 1989.

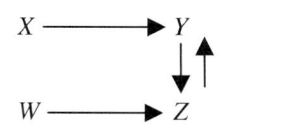

$X \longrightarrow Y$
$W \longrightarrow Z$
(with vertical double arrows between Y and Z)

$X = \varepsilon_X$ $Y = a \times X + b \times Z + \varepsilon_Y$
$W = \varepsilon_W$ $Z = c \times W + d \times Y + \varepsilon_Z$

$\varepsilon_X, \varepsilon_Y, \varepsilon_Z, \varepsilon_W$ are jointly independent standard Gaussians

Figure 12.1

The distribution associated with the DCG in figure 12.1 does not in general satisfy the natural extension of the local Markov property to DCGs. It follows from the linear equations associated with this DCG, and from the assumed joint independence of the exogenous variables, that $X \perp\!\!\!\perp W$ and $X \perp\!\!\!\perp W \mid \{Y, Z\}$, but, contrary to the natural extension of the local Markov property to DCGs, X is not independent of Z conditional on $\{Y, W\}$, the set of parents of Z. There is however a straightforward extension of the d-separation relation to cyclic directed graphs; the definition for d-separation in DAGs can be carried over unchanged. Spirtes (1994, 1995), and separately, Koster (1995, 1996), show that if **X** and **Z** are d-separated given **Y** in the DCG corresponding to a linear SEM, then the linear SEM entails that $\mathbf{X} \perp\!\!\!\perp \mathbf{Z} \,|\, \mathbf{Y}$. Spirtes (1995) shows that if a linear SEM (without dependent errors) entails that $\mathbf{X} \perp\!\!\!\perp \mathbf{Z} \,|\, \mathbf{Y}$ for all values of the free parameters then **X** and **Z** are d-separated given **Y** in the corresponding DCG. Spirtes (1994) also provides a sufficient condition for entailed conditional independence in nonlinear SEMs. For linear SEMs with dependent errors, Spirtes et al. (1998) proved that if each double-headed arrow between dependent errors is replaced with an independent latent common cause of the errors, the conditional independence relations among the substantive variables are still characterized by d-separation. Thus d-separation characterizes the independence relations entailed by path diagrams associated with linear SEMs generally (which is also shown in Koster forthcoming). Koster (1996) also generalizes chain graphs to include cycles.

Naive attempts to extend factorization conditions for DAGs (where the joint distribution is equal to the product over the vertices of the distribution of each vertex conditional on its parents) to DCGs can lead to absurdities. For example, with binary variables one might try to represent distributions for the graph in figure 12.2 by the factorization $P(Y,Z) = P(Y|Z)P(Z|Y)$. However, the factorization implies that Y and Z are independent.

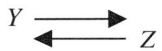

$Y \rightleftarrows Z$

Figure 12.2

Pearl and Dechter (1996) have shown that in structual equation models of discrete variables in which (i) the exogenous variables (including the error terms) are jointly independent, and (ii) the values of the exogenous variables uniquely determine the values of the endogenous variables, if **X** and **Z** are d-separated given **Y** then $\mathbf{X} \perp\!\!\!\perp \mathbf{Z} \,|\, \mathbf{Y}$ even if the associated path diagram is cyclic. It is not always the case, however, that if a graph is cyclic, and each vertex is a function of its parents in the graph and its associated error term, that the non-error term variables are functions of the error terms alone. Neal (2000) shows that in order to derive their result, Pearl and Dechter actually need the stronger assumption that each variable is a function of its ancestral error terms.

The data generating processes that are appropriately described by DCGs are still not well understood. Consider a population composed of two subpopulations, one with the causal DAG (i) of figure 12.3, and the other with the causal DAG (ii).

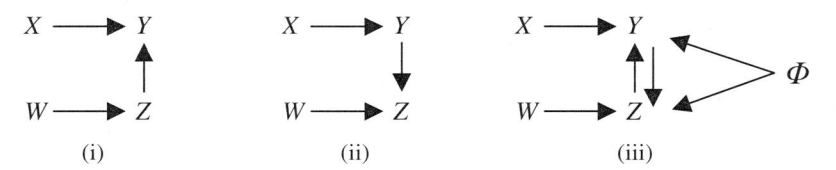

Figure 12.3

Assume the joint distribution of X, W is the same in both sub-populations. Then the independencies and the causal structure in the combined population can be represented by the DG in figure 12.3 (iii). For each unit in the sample, the value of Φ codes which pathway, $Y \rightarrow Z$, or $Z \leftarrow Y$ obtains.

Certain DCGs can describe aspects of the causal structure and conditional independencies in corresponding feedback systems represented by time series, but there does not exist a recipe for writing an interesting time series for an arbitrary DCG, or vice-versa. Particular cases are known (Fisher 1970, Richardson 1996a, Wermuth et al. 1999).

A theory of intervention for linear simultaneous equation models was given by Strotz and Wold (1960), and consists in simply replacing a manipulated variable in an equation by the value given it by an intervention. This account fits nicely with Fisher's time series model. There is not a developed theory of intervention for DCGs whose variables take a finite set of values. The importance of such a theory depends largely on whether there is an interesting class of data generating processes described by such DCGs.

12.1.3 Partial Ancestral Graphs

Any graphical model inevitably leaves out interesting aspects of the causal system it tries to describe. A DAG, for example, may specify $X \rightarrow Y$, but the mechanism referred to by $X \rightarrow Y$ is unspecified; it might, for example, contain a feedback loop in unrecorded variables, or it might not. Nor do DAGs or DCGs say anything about the time required for a variation in a cause to result in a variation in an effect, a feature that is often important in understanding dynamical systems.

Similarly, patterns can be viewed as descriptions of a class of causal processes described by various DAGs, or as an incomplete description of the process represented by some specific DAG. Again, partially ordered inducing path graphs, described in chapter 6, represent both a (generally infinite) class of DAGs or, alternatively, incompletely describe a particular DAG.

Search is often based on data from a marginal distribution that omits causally relevant variables. Variables that are not observed for any unit in a sample we will call **latent** or **hidden** variables; otherwise they are **observed**. Observational data is often obtained by conditioning on some variable (e.g., we do observations on *hospitalized* pneumonia patients). We associate with each measured variable X in a DAG a **selection variable** S_X that has the value 1 for each unit in the sample for which the value of X has been measured, and 0 otherwise. We do not place restrictions on how the selection variables are causally related to each other or to the other variables. **Selection bias** occurs when a selection variable is causally related to the observed (nonselection) variables. For a given DG G and a partition of the variable set **V** of G into observed (**O**), selection (**S**), and latent (**L**) variables, we will write $G(\mathbf{O},\mathbf{S},\mathbf{L})$. When every selection variables equals 1 ($\mathbf{S = 1}$) for a given unit there is no missing data for the measured variables for that unit. If **X**, **Y**, and **Z** are included in **O**, and $\mathbf{X} \perp\!\!\!\perp \mathbf{Z}|(\mathbf{Y} \cup (\mathbf{S = 1}))$, then we say it is an **observed** conditional independence relation.

Recall that in chapter 4 we said that two DAGs G_1 and G_2 are faithfully indistinguishable when the set of distributions that satisfied the Markov and Faithfulness conditions for G_1 was the same set of distributions that satisfied the Markov and Faithfulness conditions for G_2. This is equivalent to saying that G_1 and G_2 have the same set of d-separation relations. Faithful indistinguishability is more commonly called **Markov equivalence** now, so henceforth we will adopt that terminology. Markov equivalence extends straightforwardly to DGs as well as DAGs. We now extend the concept of Markov equivalence to DGs that may have latent variables or selection bias. Say that two graphs $G_1(\mathbf{O,S,L})$ and $G_2(\mathbf{O,S,L'})$ are **O-Markov equivalent** if and only if for $\mathbf{X}$, $\mathbf{Y}$, and $\mathbf{Z} \subseteq \mathbf{O}$, $G_1(\mathbf{O,S,L})$ entails that $\mathbf{X} \perp\!\!\!\perp \mathbf{Z}|(\mathbf{Y} \cup (\mathbf{S = 1}))$ if and only if $G_2(\mathbf{O,S,L'})$ entails that $\mathbf{X} \perp\!\!\!\perp \mathbf{Z}|(\mathbf{Y} \cup (\mathbf{S = 1}))$.

Richardson (1996a, 1996b) introduces a class of objects, Partial Ancestral Graphs (PAGs), which represents features common to Markov equivalence classes of DGs (that is DGs without selection bias or latent variables.) Spirtes et al. (1996, 1998, 1999) and Scheines et al. (1998) extend the structure to represent **O**-Markov equivalence classes of DAGs with latent variables and selection bias. One important feature of PAGs is that they give an uniform representation to the Markov equivalence classes of DGs and the **O**-Markov equivalence class of DAGs.

PAGs may contain directed edges (→), double-headed edges (↔), semidirected edges with an "o" symbol at the tail (o→), or undirected edges with "o" symbols at both ends (o—o). The symbol "*" does not occur in PAGs, but we use it as a meta-symbol to stand for any kind of endpoint (i.e., "o," "<," or "—.") For example, "*→" stands for "o→," or "↔," or "→." Let Δ be a subset of an **O**-Markov equivalence class.

DEFINITION 12.1.1: Γ is a **partial ancestral graph** (**PAG**) that represents class Δ if and only if

(1) Every vertex in Γ is in **O**.
(2) If A and B are in **O**, there is an edge between A and B in Γ if and only if for every $\mathbf{W} \subseteq \mathbf{O}\setminus\{A,B\}$, A and B are d-connected given $\mathbf{W} \cup \mathbf{S}$ in every graph in Δ.
(3) If there is an edge in Γ, $A -\!\!* B$, out of A (not necessarily into B), then in every graph in Δ, A is an ancestor of B or **S**.
(4) If there is an edge in Γ, $A *\!\!\to B$, into B, then in every graph in Δ, B is not an ancestor of A or of **S**.
(5) If there is an underlining $A *\!\!-\!\!\underline{*B}*\!\!-\!\!*C$ in Γ then B is an ancestor of (at least one of) A or C or **S** in every graph in Δ.
(6) If there are edges in Γ, from A to B and from C to B, ($A \to B \leftarrow C$), then the arrow heads at B are joined by dotted underlining ($A \to \underset{...}{B} \leftarrow C$), only if in every graph in Δ, B is not a descendant of a common child of A and C.
(7) Any edge endpoint not marked in one of the above ways is left with a small circle thus o–*

Prequels and Sequels

If a DG $G(\mathbf{O},\mathbf{S},\mathbf{L})$ is in the class Δ represented by a PAG Γ, we also say that the PAG represents $G(\mathbf{O},\mathbf{S},\mathbf{L})$. When the output of the FCI algorithm is interpreted as a PAG that represents an **O**-Markov equivalence class of DAGs, assuming a zero probability for unfaithful distributions, and an extension of the Causal Markov Condition to cases where there may be selection bias, in the large sample limit the algorithm is correct with probability 1 even when there are latent variables and selection bias (Spirtes et al. 1995, 1999). A PAG output by the FCI algorithm has enough orientations to represent a unique **O**-Markov equivalence class of DAGs with latent variables and selection bias. Similarly, the output of the cyclic discovery algorithm described in Richardson (1996a, 1996b) is a PAG with respect to the Markov equivalence class of DGs (without latent variables or selection bias), and represents a unique Markov equivalence class.

For example, there is a Markov equivalence class of DGs that contains only G_1 and G_2 of figure 12.4, and which is represented by the PAG in figure 12.4. The undirected edge between X and Y in the PAG indicates that X is an ancestor of Y, and Y is an ancestor of X in every member of the Markov equivalence class represented by the PAG, and hence no DAG has the same set of d-separation relations as G_1 and G_2.

As with POIPGs, a graph may have several PAGs, all sharing the same adjacencies but some with more orientation information than others. Not every graphical object written with the marks and underlinings of PAGs is a PAG that represents an **O**-Markov equivalence class of DAGs. While there are consistency tests, there is no available direct algorithm for determining whether, for an arbitrary PAG-like structure, there exists an **O**-Markov equivalence class of DAGs represented by the PAG-like structure. Applications of PAGs are given in section 12.5.7.

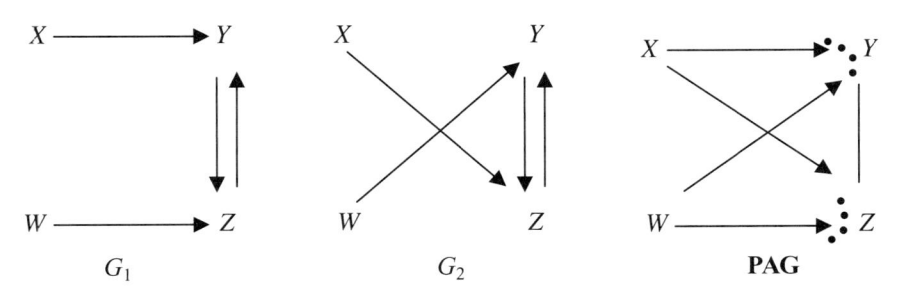

Figure 12.4

12.1.4 Mixed Ancestral Graphs

Mixed ancestral graphs were introduced in Spirtes and Richardson 1996 and investigated in Spirtes, Richardson, and Meek 1996 for two technical reasons connected with search. First, mixed ancestral graphs provide a direct means to decide whether any two DAGs imply (by the local Markov property) the same conditional independencies in any distribution obtained by marginalizing latent variables and conditioning on selection variables.

Second, DAGs with latent variables imply nonindependence constraints, illustrated by Verma's example in chapter 6. Other constraints of this sort have been investigated in Desjardins 1999, Settimi and Smith 1999, and Geiger et al. 1996. Nonindependence constraints make it difficult to determine the dimensionality of the marginal distribution over the observed variables of a latent variable model. Indeed, the marginal of a latent variable model often has no well defined dimension (Geiger et al. 1999). The dimension is a parameter that is used in many methods (BIC, AIC, MDL) of assigning data based scores to models. (For a description of the BIC and MDL scores, see section 12.5.5.2.) Since scores for models are desirable for several reasons (section 12.5), it is important to find an appropriate representation for scoring models with latent variables from data with selection bias. MAGs describe aspects of the causal relations of such structures, but they imply *only* independence and conditional independence constraints on the observed variables, and have a well defined dimension that in the Gaussian case can be easily calculated.

MAGs may contain directed edges (→), double-headed edges (↔), semidirected edges with an "o" symbol at the tail (o→), or undirected edges with "o" symbols at both ends (o—o). The symbol "*" does not occur in MAGs, but we use it as a meta-symbol to indicate any kind of endpoint (i.e., "o," "<," or "–.") For example "*→" stands for "o→," or "↔," or "→").

DEFINITION 12.1.2: MAG Γ represents DAG $G(\mathbf{O},\mathbf{S},\mathbf{L})$ if and only if:

1. If A and B are in $\mathbf{O}$, there is an edge between A and B in Γ if and only if for any subset $\mathbf{W} \subseteq \mathbf{O}\backslash\{A,B\}$, A and B are d-connected given $\mathbf{W} \cup \mathbf{S}$ in $G(\mathbf{O},\mathbf{S},\mathbf{L})$.
2. There is an edge $A \to B$ in Γ if and only if A is an ancestor of B but not of $\mathbf{S}$ in $G(\mathbf{O},\mathbf{S},\mathbf{L})$.
3. There is an edge $A \leftarrow\!* B$ in Γ if and only if A is not an ancestor of B or $\mathbf{S}$ in $G(\mathbf{O},\mathbf{S},\mathbf{L})$.
4. There is an edge A o—$*$ B in Γ if and only if A is an ancestor of $\mathbf{S}$ in $G(\mathbf{O},\mathbf{S},\mathbf{L})$.(Note that "o" has a different meaning in PAGS.)

There is a natural extension of d-separation to MAGs which is called **m-separation**. The definition requires extending the notions of collider and directed path to graphs with selection bias and latent variables. A **path** from X_1 to X_n in MAG M is a sequence of distinct vertices $<X_1,\ldots X_n>$ such that for every $i < n$, there is an edge (of any kind) between X_i and X_{i+1} in M. A **directed path** from X_1 to X_n in MAG M is a sequence of distinct vertices $<X_1,\ldots X_n>$ such that for every $i < n$, there is a directed edge from X_i to X_{i+1} in M. A vertex V is an **ancestor** of X_i if and only if $V=X_i$, or there is a directed path from V to X_i. X_i is a **collider** on a path U in M, if there are edges $X_{i-1}*\!\to X_i \leftarrow\!* X_{i+1}$ on U. For disjoint sets of vertices $\mathbf{X}$, $\mathbf{Y}$, and $\mathbf{Z}$ in MAG M, $\mathbf{X}$ is **m-connected** to $\mathbf{Y}$ if there is a path U between some $X \in \mathbf{X}$ and some $Y \in \mathbf{Y}$ such that every collider on U is an ancestor of a member of $\mathbf{Z}$, and no noncollider on U is in $\mathbf{Z}$; otherwise $\mathbf{X}$ is **m-separated** from $\mathbf{Y}$ given $\mathbf{Z}$. This entails that m-separation (m-connection) when applied to a DAG is identical to d-separation (d-connection). Applications of MAGs are given in section 12.5.7.

The problem of representing DAGs that may have latent variables and selection bias in a graphical structure that contains only observed variables was posed by Wermuth et al. (1994, 1998). The representation they propose is called a summary graph. Several differences between MAGs and summary graphs are (1) in MAGs, but not in summary graphs, a missing edge entails a conditional independence relation; (2) in summary graphs, but not in MAGs, there can be multiple edges between a pair of observed variables; (3) a Gaussian MAG is always identifiable, but a Gaussian summary graph is not always identifiable (4) a MAG entails only conditional independence constraints, while a summary graph may entail nonconditional independence constraints (which means a summary graph may contain more information about the DAGs it represents than a MAG does.) For further details see Cox and Wermuth et al. 1994.

12.1.5 Chain Graphs

Chain graphs are a much-studied (see Cox and Wermuth 1996, Lauritzen 1996) class of graphical objects introduced to represent situations in which there are "symmetric associations" between variables. Chain graphs may contain both directed and undirected edges, but may not contain partially directed cycles, that is, they do not contain a sequence of n distinct edges with endpoints X_1, X_{n+1}, such that $X_1 = X_{n+1}$ and for all i, $1 \leq i < n+1$, either $X_i - X_{i+1}$ or $X_i \to X_{i+1}$, and for some j, $1 \leq j < n+1$, $X_j \to X_{j+1}$.

Two different Markov Conditions have been proposed for chain graphs, one by Lauritzen, Wermuth and Frydenberg (Lauritzen and Wermuth 1989; Frydenberg 1990), and another by Andersson, Madigan, and Perlman (1996). The conditions are not equivalent to one another, although for undirected graphs both reduce to separation and for DAGs both reduce to d-separation. The respective Markov properties determine whether a conditional independence relation $\mathbf{X} \perp\!\!\!\perp \mathbf{Z}|\mathbf{Y}$ is entailed by a chain graph in a two step process. First, they associate a chain graph with an undirected graph. Second, $\mathbf{X} \perp\!\!\!\perp \mathbf{Z}|\mathbf{Y}$ is entailed by the chain graph if $\mathbf{X}$ is separated from $\mathbf{Z}$ by $\mathbf{Y}$ in the associated undirected graph. But the undirected graphs constructed by the two methods differ in their separation properties. The following summary is based on Richardson 1998.

A vertex V in a chain graph is **anterior** to a set $\mathbf{W}$ of vertices if there is a path P from V to some W in $\mathbf{W}$ and for all directed edges $X \to Y$ on P, Y is between X and W. **Ant(W)** is the set of vertices anterior to $\mathbf{W}$. For chain graph CG, with vertex set $\mathbf{V}$ and $\mathbf{W} \subseteq \mathbf{V}$, the induced subgraph $CG(\mathbf{W})$ is obtained by removing all vertices in $\mathbf{V}\backslash\mathbf{W}$ and all edges with an endpoint in $\mathbf{V}\backslash\mathbf{W}$. A **complex** is an induced subgraph of the form: $X \to V_1 - \ldots - V_n \leftarrow Y$, $n \geq 1$. **Moral(CG)** is the undirected graph obtained by connecting X, Y with an undirected edge if they are the endpoints of a complex, and then replacing each directed edge with an undirected edge. The Lauritzen-Wermuth-Frydenberg (LWF) global Markov Property says that CG entails $\mathbf{X} \perp\!\!\!\perp \mathbf{Z}|\mathbf{Y}$ if $\mathbf{X}$ is separated from $\mathbf{Z}$ by $\mathbf{Y}$ in the undirected graph $Moral(CG(\mathbf{Ant}(\mathbf{Z} \cup \mathbf{Y} \cup \mathbf{X}))$.

The Andersson-Madigan-Perlman chain graph global Markov property is defined as follows. In a chain graph CG, vertices V and W are **connected** if there is a path between

V and W containing only undirected edges. **Con(W)** = $\{V \mid V$ is connected to some $W \in$ **W**$\}$. **Ext**$(CG,\mathbf{W})$ contains the vertex set **Con(W)**, and all the directed edges in $CG(\mathbf{W})$, and all undirected edges in $CG(\mathbf{Con(W)})$. V is an **ancestor** of **W** if there is a path from V to $W \in \mathbf{W}$ such that all edges on the path are directed ($X \to Y$) and are such that Y is between X and W on the path. **Anc(W)** = $\{V \mid V$ is an ancestor of some $W \in \mathbf{W}\}$. A triple of vertices $<X, Y, Z>$ is a **triplex** if $CG(\{X,Y,Z\})$ is either $X \to Y \longleftarrow Z$, $X \to Y \leftarrow Z$, or $X \longleftarrow Y \leftarrow Z$. A triplex is **augmented** by adding the $X \longleftarrow Z$ edge. Four vertices $<X, A, B, Y>$ form a **bi-flag** if the edges $X \to A$, $Y \to B$, and $A \longleftarrow B$ occur in the induced subgraph over $\{X, A, B, Y\}$. A bi-flag is **augmented** by adding an $X \longleftarrow Y$ edge. **Aug**(CG) is the undirected graph that is formed by augmenting all triplexes and bi-flags in CG, and replacing all of the directed edges with undirected edges. Let Aug$[CG; \mathbf{X}, \mathbf{Y}, \mathbf{Z}]$ = Aug(**Ext**$(CG,\mathbf{Anc}$ (**X** $\cup$ **Y** $\cup$ **Z**))). The Andersson-Madigan-Perlman (AMP) global Markov property is that CG entails that $\mathbf{X} \perp\!\!\!\perp \mathbf{Z}|\mathbf{Y}$ if **X** is separated from **Z** by **Y** in the undirected graph Aug$[CG; \mathbf{X},\mathbf{Y},\mathbf{Z}]$.

An interesting discussion has developed about what data generating processes are explained by the extra structure allowed by chain graph Markov properties. For example, two simple chain graphs among four variables are shown in figure 12.5.

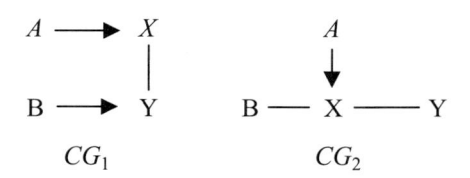

Figure 12.5

Richardson (1998) shows that the local Lauritzen-Wermuth-Frydenberg Markov Property applied to CG_1 implies a different set of independence and conditional independence relations than do any of the known ways of producing symmetrical associations by causal processes (marginalizing out a latent common cause, conditioning on a common effect, feedback) representable by DGs, and similarly for CG_2 and the AMP Markov property. (The set of conditional independencies entailed by the LWF intepretation of CG_1 is $\{A \perp\!\!\!\perp B, A \perp\!\!\!\perp Y|\{B,X\}, B \perp\!\!\!\perp X|\{A,Y\}\}$; the set of conditional independencies entailed by the AMP interpretation of CG_2 is $\{A \perp\!\!\!\perp B, A \perp\!\!\!\perp B|\{Y\}, A \perp\!\!\!\perp Y, A \perp\!\!\!\perp Y|\{B\}, B \perp\!\!\!\perp Y|\{A,X\}\}$.)

In as yet unpublished work, Lauritzen has proposed that chain graph models (with the LWF global Markov property) such as CG_1 give the independencies and conditional independencies in the limiting distribution of certain dynamical systems. The procedure is as follows: Specify $P(A)$, $P(B)$, $P(X|A,Y)$ and $P(Y|X,B)$. For each unit in a population, at $t = 0$ draw a value A_0 of A from $P(A)$ and a value B_0 of B from $P(B)$. Pick an arbitrary starting value for Y, say Y_0. Now draw X_1 from $P(X|Y_0, A_0)$ and draw Y_1 from $P(Y|X_1, B_0)$. Repeat many times, drawing X_{i+1} from $P(X|Y_i, A_0)$ and Y_{i+1} from $P(Y|X_{i+1}, B_0)$. After sufficiently long, (A_0, B_0, X_n, Y_n) is, with some further restrictions, a sample from a distribution that satisfies the LWF global Markov property for CG_1 above. The further restrictions are required because X and Y are treated asymmetrically, which implies that some restriction on the transition probabilities is required to generate a distribution that satisfies the LWF global Markov property.[1]

Cox and Wermuth (1999), and Wermuth, Cox, Richardson, and Glonek (1999) also consider what data-generating processes might lead to distributions represented by chain graphs.

12.2 Equivalence

Equivalence of models is always with respect to some selected set of variables, representing either a set **O** of observed variables, or a set **S** of selection variables, or both, and features of distributions obtained by conditioning on the selection variables and marginalizing out the variables that are unobserved. The distributional features in question may be the independence and conditional independence relations in the conditional marginal distributions, or other constraints such as vanishing tetrad differences, or, most generally, the entire conditional marginal distributions. Say that $P(\mathbf{O}|\mathbf{S}=1)$ is an **observed distribution that satisfies the Markov condition** for $G(\mathbf{O},\mathbf{S},\mathbf{L})$ if it is formed by conditionalization and marginalization from a distribution $P(\mathbf{O},\mathbf{S},\mathbf{L})$ that satisfies the Markov condition for $G(\mathbf{O},\mathbf{S},\mathbf{L})$. Two DAGs G_1 and G_2 with vertex set **V** are **distribution equivalent** if and only if $P(\mathbf{V})$ satisfies the local Markov property for G_1 if and only if $P(\mathbf{V})$ satisfies the local Markov property for G_2. Two DAGs $G_1(\mathbf{O},\mathbf{S},\mathbf{L})$ and $G_2(\mathbf{O},\mathbf{S},\mathbf{L}')$ are **O-distribution equivalent** when an observed distribution $P(\mathbf{O}|\mathbf{S}=1)$ satisfies the local Markov property for $G_1(\mathbf{O},\mathbf{S},\mathbf{L})$ if and only if $P(\mathbf{O}|\mathbf{S}=1)$ satisfies the local Markov property for $G_2(\mathbf{O},\mathbf{S},\mathbf{L}')$. Distribution equivalence and **O**-distribution equivalence can be defined similarly for restricted families of distributions (e.g., Gaussian, or multinomial.) Similar notions apply to DGs and to chain graphs.

If the family of distributions represented by the DAG is multivariate Gaussian, multinomial, or unrestricted, two DAGs without latent variables or selection bias are **O**-distribution equivalent if and only if they are **O**-Markov equivalent That relation does not, however, obtain in general if the DAGs contain latent variables or if there is sample selection bias.

The equivalence relation with respect to a data feature essentially characterizes a limit of resolution for search procedures that exclusively use estimates of that feature from the data. For example, **O**-Markov equivalence characterizes the limits of algorithms such as FCI that depend on conditional independence relations. From a Bayesian perspective, equivalence results are of less theoretical interest, since even asymptotically, **O**-Markov equivalent models need not have the same posterior probabilities. However, for searches such as those discussed in section 12.5, Bayesian search procedures that attempt to distinguish between latent variable models that are **O**-Markov equivalent face some difficult theoretical and computational problems.

Using the following result, Spirtes and Verma (1992) showed there is a more or less feasible (depending on the structure of the graphs) decision procedure for the equivalence of two DAGS which may contain unobserved (latent) variables, but no selection bias. Say that FCI uses a DAG G with vertices **V** as an **O-oracle** when it only tests d-separation relations among variables in $\mathbf{O} \subseteq \mathbf{V}$, and uses the d-separation relations in G among the variables in **O** to decide questions of d-separation.

THEOREM 12.2.1: (Spirtes and Verma): Two DAGs G, H, entail the same independence and conditional independence relations among variables in a common subset **O** of variables in G and H if and only if the output of the FCI algorithm using G as an **O**-oracle is equal to the output of the FCI algorithm using H as an **O**-oracle.

Using MAGs, Spirtes and Richardson (1996) provides a polynomial time decision procedure for **O**-Markov equivalence of models with latent variables and selection variables. Richardson (1996c) shows there is a polynomial time algorithm ($O(n^5)$), where n is the number of vertices) for deciding the Markov equivalence of DGs (without selection bias or latent variables.)

Geiger and Meek (1999) have obtained theoretically fascinating, but as yet impractical, results about distributional equivalence and other "structural" features of graphical models, such as the identification problem—the problem of deciding whether a model parameter can be uniquely estimated from the marginal probability distribution over observed variables. Their results show a remarkable sequence of connections between mathematical logic, probability theory, and methodology.

Tarski axiomatized ordinary real algebra—the theory of real closed fields, RCF—and proved that the theory is complete, hence decidable, and admits elimination of quantifiers. That is, for every formula F in the language of RCF there is a formula H without quantifiers such that RCF $\models$ F $\Leftrightarrow$ H. One can use the theory to test distributional equivalence of two linear, Gaussian structural equation models M and N as follows. The variance/covariance matrix of the observed variables in Model M can be written as polynomial functions of the model parameters, which are real variables. Model M asserts that there exist values of the parameters such that each covariance of the observed variables equals the specified function of the parameters. That claim is a sentence S_M of a simple extension of RCF, for which Tarski's theorem holds. Hence there is a sentence Q_M *without quantification over the model parameters and without names of values of the model parameters* such that RCF $\models Q_M \Leftrightarrow S_M$. For model N with the same observables there is likewise a sentence S_N asserting the existence of values of parameters in N such that covariances of observed variables are specified functions of the parameters, and likewise an equivalent sentence Q_N without quantifiers. Models M and N are therefore distribution equivalent if and only if RCF $\models Q_M \Leftrightarrow Q_N$.

Since RCF is decidable, there is an algorithm to decide distribution equivalence in linear Gaussian structural equation models—no matter whether acyclic ("recursive") or cyclic ("nonrecursive"), and with or without latent variables. Identification problems are solvable by a similar strategy, since the identifiability of a parameter corresponds to an RCF formula saying that if two values of a parameter result in equal values of the polynomial functions for the covariances, then the two values are identical. Quantifier elimination then results in a sentence, using only the vocabulary for observable correlations, which is a theorem of RCF if and only if the parameter is identifiable.

The same argument works for any family of distributions whose marginal distribution over observables can be described by a finite set of polynomial functions of real valued model parameters. Graphical models with categorical variables can therefore be treated in the same way, since the marginal distributions over measured variables are sums of products of conditional probabilities, and the latter are real valued variables with a restricted range.

But the solution is not yet practical. Tarski's decision procedure is hyper-exponential. Although faster algorithms have since been found, they are still hyper-exponential, and Geiger and Meek are able to work out an example for only three variables. Even for these faster algorithms, a problem with six variables is hopeless. Since, however, decisions about equivalence, identifiability, and bounding of parameter values require deciding only formulas of special logical forms, there is still hope that more efficient algorithms may be possible for these special cases.

12.3 Prediction and Manipulation
12.3.1 Causation and Subjunctives

The Rubin approach to making predictions about manipulations of causal models (discussed in chapter 7) introduces subjunctive variables, such as $Y_{X=0}$, to represent the value that Y would have if X were manipulated to have the value 0. Rubin's approach also uses judgments about the conditional independence of subjunctive variables from occurrent variables. Two problems arise with this approach; interpreting what it means to have a joint distribution over subjunctive and occurrent variables, and whether people can make judgments about the independence of subjunctive and occurrent variables (especially in light of the fact that that without using graphical methods people are poor at making judgments about conditional independence even for occurrent variables alone.)

In contrast, in chapter 7, we used DAGs to make predictions from causal models. Instead of introducing a new subjunctive variable to represent the value that Y would have if X were manipulated to 0, we added a policy variable and an edge from the policy variable to X, and took the value that Y would have if X were *manipulated* to 0 to be the value of Y *conditional* on the policy variable equaling 1 (i.e., the manipulation had occurred). Two advantages of the DAG approach are that it does not require a joint probability distribution over subjunctive and occurrent random variables (since we always condition on a value of the policy variable in all of our calculations), and it uses causal DAGs to calculate conditional independence relations. This approach led to theorem 7.1 (equivalent to what Pearl later called the "Calculus of Interventions" in Pearl 1995) which gives sufficient graphical conditions for conditional probabilities to remain invariant under a manipulation. It is not known whether the conditions of theorem 7.1 are also necessary. This theorem has interesting applications discussed in section 12.3.2.

While the DAG supplemented with policy variables does not require joint distributions of subjunctive and occurent variables, it also does not *allow* representation of a joint distribution over subjunctive and occurent variables, or a joint distribution over subjunctive variables corresponding to different manipulations, which in some cases is desirable. For example, suppose a patient who is not on drug treatment presents with high blood pressure. The physician believes the causal relations are those shown in figure 12.6.

Suppose that *Drug therapy* = 1 represents being given drug therapy, and *Arterial disease* = 1 represents the occurrence of arterial disease. Consider the probability that a patient would have *Arterial disease* if *Drug therapy* were to be manipulated to be present (a subjunctive variable) conditional on the patient's actual *Blood pressure*. Here *Blood pressure* is actually measured (without intervention on the causal process), but *Drug therapy* and *Arterial disease* are subjunctive variables in this case, that is they are features that are actual only if the intervention subsequently occurs. This is in general not equal to the probability a patient would have *Arterial disease* if *Drug therapy* were manipulated to 1 (a subjunctive variable) conditional on the *Blood pressure* a patient would have if *Drug therapy* were manipulated to 1 (another subjunctive variable). In the language of chapter 7 the latter probability is $P_{Man(Drug)}$ (*Arterial disease*|*Blood pressure*), and can be calculated from theorem 7.1. There is no way in the language of chapter 7 to express the former probability and it cannot be directly calculated by an application of Theorem 7.1. In this section we consider how Balke and Pearl (1994), Pearl (1999), and Galles and Pearl (1998a) use structural equation semantics to clarify the meaning of a joint distribution over subjunctive and occurent variables, and use DAGs to calculate the required conditional independence relations between subjunctive and occurent variables.

Arterial disease   *Drug therapy*

Blood pressure

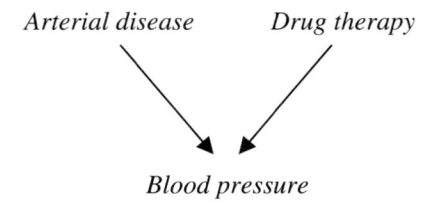

Figure 12.6

For the sake of illustration, suppose that the statistical model associated with the DAG in figure 12.6 is a linear structural equation model. Suppose the structural equation for *Blood pressure* is of the following form:

Blood pressure = a × *Drug therapy* + b × *Arterial disease* + 100 + ε_{bp}

Assume that *Arterial disease* is binary (1 representing having the disease) and *Drug therapy* is binary (1 representing being given the drug), that the probabilities of *Arterial disease* and *Drug therapy* are given, ε_{bp} follows a standard Gaussian distribution, and *Arterial disease*, *Drug therapy*, and ε_{bp} are mutually independent.

We need a notation to express the probability *Arterial disease* would have if *Drug therapy* were to be manipulated to the value 1, conditional on the actual value of *Blood pressure*. In order to do that we (following Rubin as in chapter 7, Balke and Pearl 1994, and Pearl 1999) split *Drug therapy*, and all its descendants into two variables, one variable representing the value that would occur if *Drug therapy* were manipulated to the value 1, and the other variable representing the unmanipulated value of *Drug therapy*. In this example there is *Drug therapy*$_{Man(Drug)}$ and *Drug therapy*$_{Unman}$, and *Blood pressure*$_{Man(Drug)}$ and *Blood pressure*$_{Unman}$. Note that because *Arterial disease* and ε_{bp} are not descendants of *Drug therapy* and hence (by the Manipulation Theorem) are unaffected by the manipulation, the manipulated and unmanipulated values of *Arterial disease* and ε_{bp} have the same distribution, so we do not need to split these variables. Using the structural equation model, we can write:

Blood pressure$_{Man(Drug)}$ = $a \times$ *Drug therapy*$_{Man(Drug)}$ + $b \times$ *Arterial disease* + 100 + ε_{bp}
Blood pressure$_{Unman}$ = $a \times$ *Drug therapy*$_{Unman}$ + $b \times$ *Arterial disease* + 100 + ε_{bp}

If the assumption is made that the manipulated value of *Drug therapy* does not depend on the unmanipulated value of *Drug therapy*, then by the Causal Markov Condition *Drug therapy*$_{Unman}$ and *Drug therapy*$_{Man(Drug)}$ are independent of each other. The joint distribution over the subjunctive and occurent variables then follows from this assumption, the joint distribution over the exogenous occurrent variables, and the structural equations. (Balke and Pearl [1994], and Pearl [1999] use a DAG with latent variables rather than double-headed arrows. Madigan [1999] also considers graphical representations of subjunctive variables.)

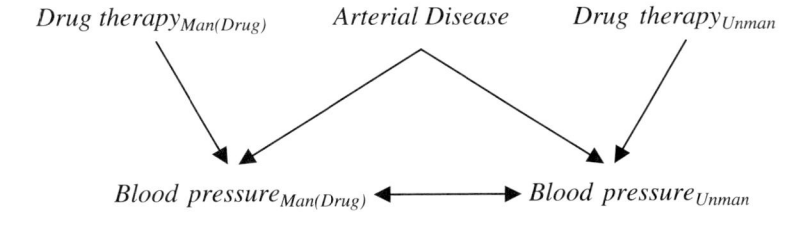

Figure 12.7

When we make the modifications to the causal DAG in figure 12.6, the result is the MAG in figure 12.7. There is a correlated error between *Blood pressure*$_{Man(Drug)}$ and *Blood pressure*$_{Unman}$ because ε_{bp} is a cause of both, as seen in their respective equations. *Blood* *pressure*$_{Unman}$ is not caused by *Drug therapy*$_{Man(Drug)}$ according to its structural equation.

Now m-separation applied to the causal MAG of figure 12.7 shows that *P(Arterial disease|Blood pressure*$_{Unman}$, *Drug therapy*$_{Man(Drug)}$) = *P(Arterial disease|Blood pressure*$_{Unman}$), that is the drug has no effect among a group of people with a given actual *Blood pressure*.

There are equality constraints among the parameters in the causal MAG (e.g., the distribution of *Blood pressure*$_{Man(Drug)}$ conditional on its parents *Drug therapy*$_{Man(Drug)}$ and *Arterial disease* equals the distribution of *Blood pressure*$_{Unman}$ conditional on its parents *Drug therapy*$_{Unman}$ and *Arterial disease*). Hence there may be an equality between a conditional probability among the unmanipulated variables (e.g., *P(Blood pressure*$_{Man(Drug)}$| *Arterial disease,Drug therapy*$_{Man(Drug)}$) and the corresponding conditional probability among the manipulated variables (e.g., *P(Blood pressure*$_{Unman}$|*Arterial disease,Drug therapy*$_{Unman}$), an equality that is not entailed by the m-separation relations in the causal MAG, but is entailed by d-separation in the causal DAG representation of chapter 7 using policy variables. So there are advantages to using the causal DAG representation of chapter 7 as long as the quantities of interest are not mixtures of subjunctive and occurrent variables.

Graphs similar in structure to figure 12.7 can also be used to represent *Drug therapy* at different times, instead of unmanipulated and manipulated *Drug therapy*, where the variables are indexed by time, instead of whether or not *Drug therapy* has been manipulated. Theorem 7.1 can be applied directly to such temporal graphs. See Boyen et al. (1999) for one representation of dynamic systems.

It is also possible to use (a minor modification of) the Balke-Pearl graphical representation (using MAGs instead of DAGs) to calculate some conditional probabilities of subjunctive and occurent variables in the following special case that is of particular interest for reasons described below. A special case of a manipulation arises when the manipulated variable is set to a constant value. Interpreting the causal DAG as a structural equation model gives an especially clear interpretation to subjunctive variables in this case. (This view also seems implicit in the use of subjunctives by some analyses based on Rubin's subjunctive variables). Suppose *Drug therapy* is manipulated to have the value 1 in all cases (i.e., everyone is given the drug.) Using the structural equation model, we can write:

Blood pressure$_{Man(Drug\ Therapy\ =\ 1)}$ = $a + b \times$ *Arterial disease* $+ 100 + \varepsilon_{bp}$

setting *Drug therapy* to the constant 1. (This is the approach to manipulation taken in Strotz and Wold 1960.) All of the subjunctive variables are now simple functions of occurent variables, so the joint distribution over the subjunctive and occurent variables follows from the distribution over the exogenous occurrent variables and the structural equations.

Prequels and Sequels 311

```
        Arterial Disease        Drug therapy_Unman
              ╱    ╲                   ╱
             ╱      ╲                 ╱
            ╱        ╲               ╱
           ▼          ▼             ▼
Blood pressure_Man(Drug Therapy = 1)  ◄─────►  Blood pressure_Unman
```

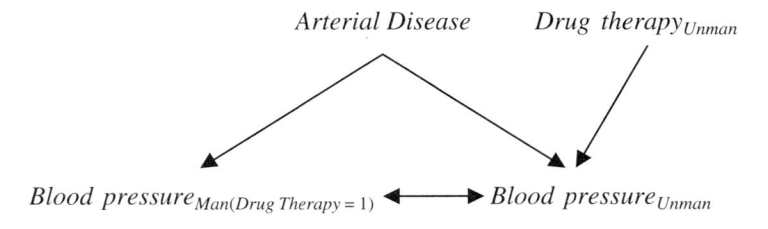

Figure 12.8

More generally if one wants a joint probability distribution on the values of B, C, and D, and E when A is manipulated to 0, and the unmanipulated variables A, B, C, and D, and E, then split each of the descendants of A into the unmanipulated and manipulated versions (in the case of figure 12.9, B, C, and D on the one hand, and $B_{Man(A = 0)}$, $C_{Man(A = 0)}$, and $D_{Man(A = 0)}$ on the other hand), add double-headed edges between each new variable and its counterpart, and an edge between two of the new variables if and only if there is an edge between their counterparts. Then apply m-separation. (See figure 12.9.) ($A_{Man(A = 0)}$ just has the constant value 0, so we do not include it in the MAG.)

Part of the price that is paid for the structural equation interpretation of subjunctive variables is that it posits the existence of a deterministic world with independent error terms. Dawid (1997) questions the existence of such independent error terms, and at the microscopic level, determinism is not compatible with the standard interpretation of quantum mechanics. The method of representation described here does not allow for arbitrary causal DAGs or MAGs among subjunctive and occurrent variables.

```
         E                                    E
         ↘                                    ↘
A ──► B ──► C ──► D    │    A ──► B ─────► C ─────► D
                       │         ↕         ↕         ↕
                       │         B_Man(A=0) ──► C_Man(A=0) ──► D_Man(A=0)

         G₁                              G₂
```

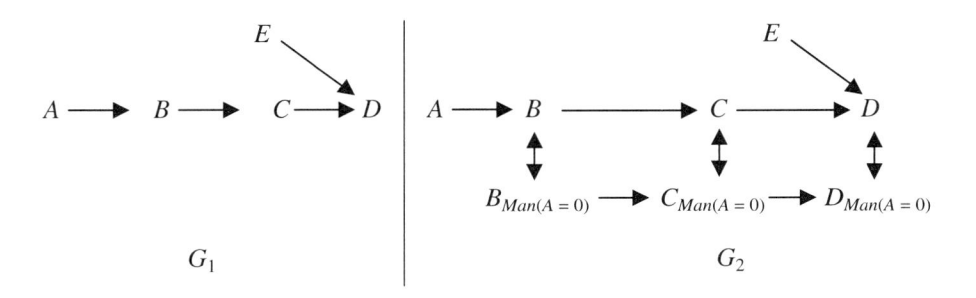

Figure 12.9

Joint distributions over subjunctive and occurent variables play an important role in Pearl's (2000) analysis of several different notions of causation. In Pearl's notation, $Y_x(u)$ is the response of variable Y to manipulating X to value x, when the exogenous variables take on value u (in Pearl's structural equation semantics of causal DAGs, the response of variable Y to manipulating X is a function of U). Let X and Y be binary variables, where x is the proposition that X takes the value *true* and x' is the proposition that X takes the value *false*. y_x is the proposition that Y takes the value *true* if X were manipulated to *true*, and y'_x is the proposition that Y takes the value *false* if X were manipulated to *true*. PS (the probability of sufficiency) is equal to $P(y_x|x', y')$, PN (the probability of necessity) is $P(y'_{x'}|x,y)$, and PNS (the probability of necessary and sufficient causation) is $P(y_x, y'_{x'})$. (In the notation of chapter 7 $P(y_x)$ is $P_{Man(X=true)}(Y=true)$, $P(y'_x)$ is $P_{Man(X=true)}(Y=false)$, $P(y_{x'})$ is $P_{Man(X=false)}(Y=true)$, and $P(y'_{x'})$ is $P_{Man(X=false)}(Y=false)$. However, there is no way in that notation to express $P(y_x, y'_{x'})$, $P(y'_{x'}|x, y)$, or $P(y_x|x', y')$, which mix occurrent and subjunctive variables, or subjunctive variables corresponding to different manipulations.)

There are several assumptions relevant to the conditions under which PN, PS, and PNS are identifiable. X is **exogenous** with respect to Y when there is no common cause of X and Y. X is **stochastically monotonic** with respect to Y when the probability of $Y = true$ given X is manipulated to *true* is greater than the probability of $Y = true$ given X is manipulated to *false*. X is **monotonic** with respect to Y when $y'_x \wedge y_{x'}$ is false. Robins and Greenland (1989) showed that even under the assumptions of exogeneity of X with respect to Y and stochastic monotonicity of X with respect to Y, PN is not identifiable; however they do calculate bounds for PN. Pearl (2000) showed that under the stronger assumptions of exogeneity of X with respect to Y and monotonicity of X with respect to Y, then PN, PS, and PNS are all identifiable. Pearl also showed that under the assumption of monotonicity, PN, PS, and PNS are all identifiable whenever $P(y_x)$ is identifiable.

In chapter 3, we discussed the relationship between causal DAGs and the Rubin (1978) subjunctive variable approach. Robins (1986, 1987) extended Rubin's theory to deal with time-varying treatments, outcomes, and covariates. Robins (1995) showed that causal DAGs can always be interpreted as subjunctive variable models. Galles and Pearl (1998) showed that for acyclic graphs, all of the conjunctive subjunctives derivable in structural equation semantics are entailed by the following two features of structural equation semantics:

- Composition: For any two singleton variables Y and W, and any set of variables **X** in a causal model, if $W_\mathbf{x}(u) = w$ then $Y_{\mathbf{x}w}(u) = Y_\mathbf{x}(u)$.
- Effectiveness: For all variables X and W, $X_{xw}(u) = x$.

(According to Galles and Pearl [1998] Robins suggested composition to Pearl in a personal communication.) Halpern (1997) found a complete set of axioms for the case of structural equation models represented by cyclic directed graphs.

12.3.2 Calculating the Effects of Interventions

Strotz and Wold (1960) pointed out that the effects of manipulating a variable X in structural equation models could be calculated by replacing the equation for X by an equation that set X to its manipulated value; this is the basic idea behind the Manipulation Theorem and Pearl's structural equation semantics. Robins (1986) derived the G-computation formula, which is equivalent to the Manipulation Theorem, though not formulated graphically.

An important special case of calculability (synonymous with "identifiability") is the case of sequential randomized trials, where the covariates may be affected by earlier treatments, and each treatment is a function of all of the earlier treatments. This has been studied since Robins (1986) under the name "G-computation algorithm formula." The theory has been translated into graphical terms in Pearl and Robins 1995. The formula expresses the probability of an outcome under a sequential randomized manipulation in terms of a sum and product of probabilities involving only the observed occurrent variables and the values the treatments were manipulated to. This formula can also be extended to the case in which there is a vector of outcomes included in the covariates, and the assumption that each treatment is a function of all earlier treatments can be relaxed. Robins (1986, 1987) also considers the extension to the case where the value that a treatment is manipulated to is a function of the preceding covariates.

One problem with the direct application of the G-computation formula is that standard parametric models of the conditional distributions that appear in it lead to a parameterization that will make the direct effect and total effect null hypotheses be rejected even when the null hypothesis of no direct effect is true. Robins (1993, 1994, 1997, 1998) develops the theory of structural nested models, which do not suffer from this defect. The only parametric models needed to test the no-direct effect hypothesis or estimate the size of the effect are parametric models of the probabilities of treatment.

Pearl (1995) proposes three rules, which he calls the "Calculus of Interventions." For disjoint sets $\mathbf{X}, \mathbf{Y}, \mathbf{Z}, \mathbf{W}$ of variables, it states rules for when various conditional probabilities containing manipulated quantities are equal to conditional probabilities that have fewer manipulated quantities. The rules are sound and all follow from Theorem 7.1. Theorem 7.1 and the Calculus of interventions are both equivalent to $P(\mathbf{Y}|\mathbf{Z})$ being invariant under manipulation when the policy variables are d-separated from $\mathbf{Y}$ given $\mathbf{Z}$.

In chapter 7 we defined a conditional manipulated probability as *calculable* if it was a function of the unmanipulated distribution, and of the manipulation. In chapter 7, the Prediction Algorithm (i) takes data as input, (ii) constructs a POIPG from the data, and (iii) uses consequences of theorem 7.1 to search for a way to express the manipulated quantity of interest in terms of other quantities, involving only observed variables, that, given the POIPG, are known to be invariant under manipulation. Pearl (1995) takes this method a step further, and shows how to use the Calculus of Interventions to write a manipulated conditional probability in terms of quantities which are not themselves invariant, but which are calculable, and hence ultimately functions of unmanipulated distributions. In contrast to our procedure, Pearl (1995) starts not with data but with a DAG that may contain latent variables, and searches for a way to express the manipulated quantity of interest in terms of other quantities involving only observed variables that, given the DAG, are known to be identifiable under manipulation. Galles and Pearl (1995) describe a set of rules for determining when a manipulated quantity is identifiable from applications of the Calculus of Interventions, and show that the identification of the causal effect between two variables (and a formula for calculating the quantity) can be established in a time polynomial in the number of variables in the graph.

The extension of predictions from interventions to circumstances where causal relations are reversible has also been investigated. Consider a bicycle with gears, so arranged that changing the speed the pedals rotate and the value of the gear setting influences the speed with which the rear wheel rotates, while changing the speed with which the rear wheel rotates (e.g., by pushing it by hand) changes the speed with which the pedals rotate but does not change the gear setting. One might try to represent the system by the cyclic graph of figure 12.10. Alternatively, one can introduce the kind of graph shown in figure 12.3 (iii). The predictions for each kind of intervention can be analyzed via the Manipulation Theorem. See also Richardson 1996a and Shafer 1996.

Gear setting

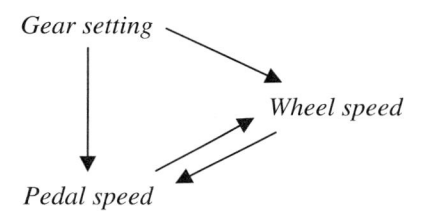

Wheel speed

Pedal speed

Figure 12.10

Further research is needed in this area, because neither procedure fully captures the dependencies in a simple dynamical system; for example, they do not tell us the speed of either the wheel or the pedal when countervailing forces are applied, although we have no difficulty making that calculation from elementary physical principles.

12.4 Consistency[2]

What assumptions guarantee the existence of "reliable" procedure for drawing causal conclusions from observational data, by any agent that has unlimited resources for search and computation? In this section, we will answer this question for several increasingly strong senses of "reliable." First we will consider what assumptions are needed to guarantee Bayes consistency, then what (stronger) assumptions are needed to guarantee the stronger condition of pointwise consistency, and finally what (still stronger) assumptions are needed to guarantee the still stronger condition of uniform consistency. (In every case we will assume the Causal Markov Condition and that causal relations for a population can be represented by a DAG.) We will then discuss the plausibility of the assumptions. We emphasize that the negative results described in this section apply to *any* method, not just the methods described in this book. We will consider what conclusions should be drawn by someone unwilling to make the assumptions required for the existence of reliable procedures for causal inference. The notation in this section, the negative results about the existence of uniformly consistent tests under some sets of assumptions, and some of the implications of the negative results, are based on Robins, Scheines, Spirtes, and Wasserman 1999.

As an illustration, consider the linear structural equation models in figure 12.11. We assume background knowledge gives a time order (B precedes C) and rules out selection bias, but does not rule out the possibility of latent common causes. In all three models ε_A, ε_B, and ε_C are independent Gaussians, A, B, and C are standard Gaussians, A is a latent variable, and B and C are observed. $\rho(B,C)$ is the correlation between B and C. In those cases in which several different population probability distributions are being discussed, $\rho_P(B,C)$ represents the correlation between B and C in the population with distribution P. Because the variables are standardized, x in Model M and in Model Q is a real valued variable that represents a linear coefficient in the structural equations which has values that range between -1 and 1. In Model N and Model Q, z is fixed at 0. (In Model M there is one other independent constraint on x, y, and z, namely that $\text{var}(C) = \text{var}(\varepsilon_C) + y^2 + z^2 + 2x \times y \times z = 1$, and hence $y^2 + z^2 + 2x \times y \times z \le 1$.) In Model M, $\rho(B,C) = (x \times y) + z$, in Model N $\rho(B,C) = 0$, and in Model Q $\rho(B,C) = x \times y$. In order to make Models M, N, and Q disjoint, $z = 0$ is not a legitimate parameter value in Model M, and $x = y = 0$ are not legitimate parameter values of Model Q.

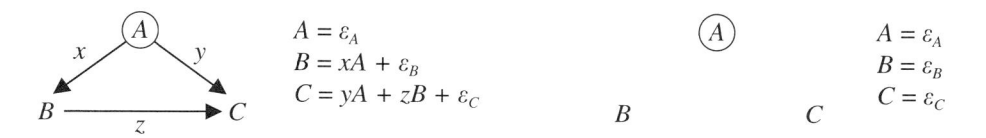

$A = \varepsilon_A$
$B = xA + \varepsilon_B$
$C = yA + zB + \varepsilon_C$

Model M: Graph G_M

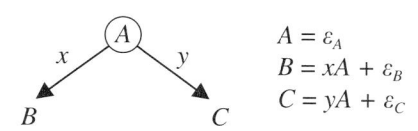

$A = \varepsilon_A$
$B = \varepsilon_B$
$C = \varepsilon_C$

Model N: Graph G_N

$A = \varepsilon_A$
$B = xA + \varepsilon_B$
$C = yA + \varepsilon_C$

Model Q: Graph G_Q

Figure 12.11. Model M, Model N, and Model Q

Model M and Model N entail the same observable population distribution ($\rho(B,C) = 0$) whenever $(x \times y) + z = 0$ in Model M. Model N and Model Q never entail the same population distribution. Call the ratio of a consequent change in C to a manipulated change in B the **treatment effect** of B on C. In Model N and Model Q, the treatment effect of B on

C is 0, while in Model M it is equal to z. Hence, Model M disagrees with Model N and Model Q on the treatment effect of B on C. In Model M, all of the legitimate values of x, y, and z that produce $\rho(B,C) = 0$ are unfaithful to DAG G_M. We will call these "unfaithful" parameter values for Model M and say that a distribution corresponding to an unfaithful parameter value is unfaithful to G_M.

Suppose that the sample estimate of $\rho(B,C)$ is zero. In that case, many methods for drawing causal conclusions from observational data would conclude that there is no treatment effect of B on C. For example, in many studies, a variable B is eliminated from consideration when the regression coefficient of B on the outcome variable is not significant. In this example, the regression coefficient of B for C is zero when $\rho(B,C)$ is zero. In addition, in the large sample limit, with probability 1, the BIC score of Model N is infinitely larger than the BIC score of Model M or Model Q. When $\rho(B,C) = 0$, for any prior which places a non-zero probability on Model N, and for which the distribution over the parameters is absolutely continuous with Lebesgue measure, in the large sample limit, the ratio of the posterior of Model N to the posterior of Model M or to Model Q approaches infinity. Also, the FCI algorithm (and the PC algorithm as well), concludes that the treatment effect of B on C is zero. Hence, in the large sample limit with probability 1, both the constraint based algorithms and various Bayesian scores prefer Model N to Model M or Model Q when $\rho(B,C) = 0$. If the true model is Model M with unfaithful parameter values so $z \neq 0$ then even in the large sample limit all of these search models will prefer Model N, and be incorrect; otherwise in the large sample limit they are all correct.

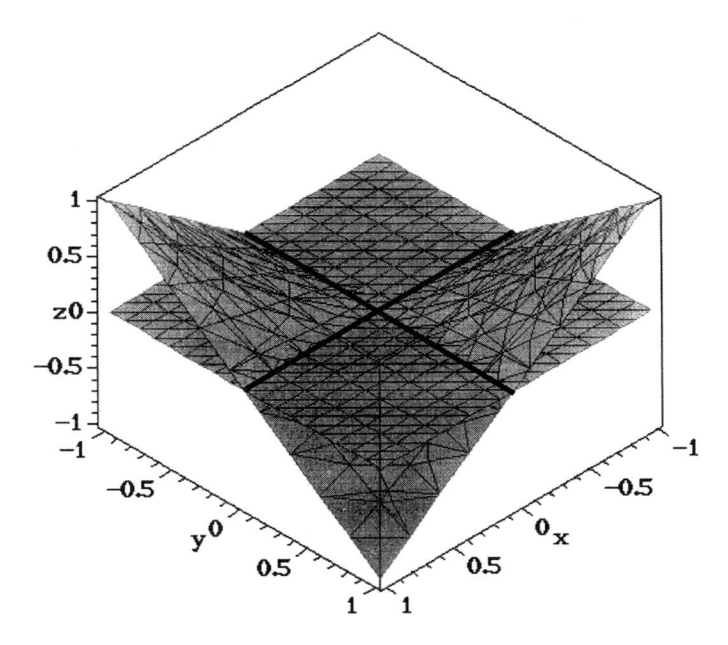

Figure 12.12. The set of unfaithful parameters for model M

Prequels and Sequels

Figure 12.12 shows the $z = 0$ plane and part of the surface of parameters for which in Model M $\rho(B,C) = (x \times y) + z = 0$. The two lines $x = 0$ and $y = 0$ in the $z = 0$ plane are also shown in figure 12.12. Henceforth, we will refer to the $\rho(B,C) = 0$ surface, excluding the non-legitimate parameter values $z = 0$, as the surface of unfaithful parameter values. (There are other unfaithful parameter values in the model, but only those shown lead to distributions that are unfaithful in the observed margin.) There are three important features of the surface of unfaithful parameter values. The first feature is that the surface is 2 dimensional, while the parameter space of Model M is higher dimensional. Hence the Lebesgue measure of the surface of unfaithful parameter values is 0.

The second feature is that in Model M *any* legitimate value of z is compatible with $\rho(B,C) = 0$ (because each value of z occurs somewhere on the surface of unfaithful parameter values.) For example, the four (x, y, z) points $(1, 1,-1)$, $(-1, -1,-1)$, $(1, -1,1)$ and $(-1,1,1)$ all occur on the surface of unfaithful parameters values. (The point $(-1, -1,-1)$ is hidden by the $z = 0$ plane in figure 12.12.) So in Model M the treatment effects (1 or -1) of B on C are both compatible with $\rho(B,C) = 0$, as well as with every other value.

The third feature is that for every value of z, there are points that are not on the surface of unfaithful parameter values that are arbitrarily close to the surface of unfaithfulness parameter values.

These three features of the surface of unfaithful parameter values are behind all of the various results about the possibility or impossibility of "reliably" discovering causal relations from observational data, in various senses of "reliability," under a variety of different assumptions.

12.4.1 Bayes Consistency

Let the set of vertices associated with DAG G be $\mathbf{V}_G$. Let Γ be a set of DAGs, such that for each $G \in \Gamma$, for a set of "observed" variables $\mathbf{O}$, $\mathbf{O} \subseteq \mathbf{V}_G$. Let $\mathbf{B}_G$ be the set of legitimate parameter values for the parameters of G. Let Π_G be the set of distributions over $\mathbf{V}_G$ that satisfy the Markov condition for G. Let γ be a function that maps $(\mathbf{B}_G, G)$ onto Π_G. In the examples of Model M, N, and Q of figure 12.11, γ is the usual function mapping linear structural equation model parameters (x, y, z) into Gaussian distributions. In the case of Model N, γ maps the parameters into a Gaussian distribution with a correlation matrix that is the identity matrix. In the case of Model M, γ maps (x, y, z) into a Gaussian distribution with correlation matrix

$$\begin{array}{c} & A & B & C \\ A \\ B \\ C \end{array} \left[\begin{array}{ccc} 1 & x & y + (x \times z) \\ x & 1 & z + (x \times y) \\ y + (x \times z) & z + (x \times y) & 1 \end{array} \right]$$

In the case of Model Q, γ maps (x, y) into a Gaussian distribution with correlation matrix

$$\begin{array}{c} & A & B & C \\ A & \begin{bmatrix} 1 & x & y \\ B & x & 1 & x \times y \\ C & y & x \times y & 1 \end{bmatrix} \end{array}$$

Let $\Pi_\Gamma = \bigcup_{G \in \Gamma} \Pi_G$ Let $O^n = O \times \ldots \times O$ where O is the range of the random variables in **O**. Assume we have a random sample $\mathbf{O}^n = (\mathbf{O}_1, \ldots, \mathbf{O}_n)$ from some $P(\mathbf{O}) \in \Pi_\Gamma(\mathbf{O})$. P^n is the n-fold product measure of P on O_n Let θ map $\mathrm{B}\Gamma = \bigcup_{G \in \Gamma} \bigcup_{\beta \in \mathrm{B}_G} (\beta, G)$ into the reals, i.e. θ is a parameter that for the moment we leave unspecified (e.g. the treatment effect of

$$\Pi_{\Gamma 0} = \bigcup_{G \in \Gamma} \{P \in \Pi_G : \exists \beta \in \mathrm{B}_G, \theta = \theta_0 \ \& \ \gamma(\beta, G) = P\}$$
$$\Pi_{\Gamma 1} = \bigcup_{G \in \Gamma} \{P \in \Pi_G : \exists \beta \in \mathrm{B}_G, \theta \neq \theta_0 \ \& \ \gamma(\beta, G) = P\}$$

B on C in Model M in figure 12.11). Let

Intuitively $\Pi_{\Gamma 0}$ is the set of distributions compatible with $\theta = \theta_0$, and $\Pi_{\Gamma 1}$ is the set of distributions compatible with $\theta \neq \theta_0$. Note that there may be a $P_1 \in \Pi_{\Gamma 0}$ and a $P_2 \in \Pi_{\Gamma 1}$ such that $P_1(\mathbf{O}) = P_2(\mathbf{O})$.

Suppose that there is a prior density $Pr(\mathrm{B}\Gamma)$, such that for $(\beta, G) \in \mathrm{B}\Gamma$, $Pr(\beta, G) = Pr(G)Pr(\beta|G)$. This prior, together with γ induces a prior Pr over $(\mathrm{B}\Gamma, \mathbf{O})$. Suppose that we test H_0: $\theta = \theta_0$ versus H_1: $\theta \neq \theta_0$. For our purposes, a test is a function φ_n: $\mathbf{O}^n \to \{0,1,2\}$, where $\phi_n(\mathbf{O}^n) = 0$ means "choose H_0", $\phi_n(\mathbf{O}^n) = 1$ means "choose H_1", and $\phi_n(\mathbf{O}^n) = 2$ means "don't know". We specify a test ϕ_n for each sample size n. In what follows all limits refer to the sample size n tending to ∞. Let $Pr^n(\mathbf{O}^n|\mathrm{B}\Gamma)$ be the n-fold product measure of $Pr(\mathbf{O}|\mathrm{B}\Gamma))$. A test that always returns "don't know" is trivially always correct, so we will eliminate such tests from consideration. A test is **non-trivial** if either

(i) for some $P \in \Pi_\Gamma$ $\lim_{n \to \infty} P^n(\varphi^n(\mathbf{O}^n) = 0) = 1$, or

(ii) for some $P \in \Pi_\Gamma$ $\lim_{n \to \infty} P^n(\varphi^n(\mathbf{O}^n) = 1) = 1$.

We will henceforth consider only non-trivial tests.

Definition 12.1: A test ϕ is Bayes consistent with respect to a prior $Pr(\mathrm{B}\Gamma)$ and a mapping γ which induces a prior Pr over $(\mathrm{B}\Gamma, \mathbf{O})$ if

$$\lim_{n \to \infty} Pr(H_0)Pr^n(\varphi_n(\mathbf{O}^n) = 1 | H_0) + Pr(H_1)Pr^n(\varphi_n(\mathbf{O}^n) = 0 | H_1) = 0$$

Intuitively, a test is Bayes consistent with respect to a prior when in the large sample limit the test is incorrect on a set of measure 0 under the prior. One trivial way to guarantee Bayes consistency is to have a prior that places all of its mass on a single point. The results in this section are more interesting, however, because we will consider diffuse priors. In the following theorem, G_M, G_N, and G_Q refer to the models in figure 12.11. Although it is non-standard to allow a test to return "don't know", we have allowed this for the following reason. An algorithm such as the FCI algorithm performs a statistical test of zero correlation, and returns 0 when the correlation is judged to be zero, and returns 2 ("don't know") when the correlation is judged to be non-zero. This is because a zero correlation entails a zero treatment effect except when there is a violation of faithfulness (which is of Lebesgue measure 0), but a non-zero correlation is compatible with either a direct effect of B on C (Model M), or with no direct effect and a common cause of of B and C (Model Q). (Although for the sake of simplicity, in the following discussion we do not consider all of the alternative models when B and C are the only variables measured, including the other models would not substantially change any of the arguments or conclusions.)

Theorem 12.1: If $\Gamma = \{G_M, G_N, G_Q\}$, $\theta = z$, and $\theta_0 = 0$, then there is a Bayes consistent test of $\theta = \theta_0$ against $\theta \neq \theta_0$ with respect to any prior Pr such that

$Pr(B_{G_M} | G_M)$, $Pr(B_{G_N} | G_N)$, and $Pr(B_{G_Q} | G_Q)$

are absolutely continuous with respect to Lebesgue measure.

Proof. There is a pointwise consistent test η (see section 12.4.2) of zero correlations against non-zero correlations. Let ϕ return 0 when η returns 0, and return 2 otherwise. Because ϕ never returns 1,

$$\lim_{n \to \infty} Pr^n(\phi_n(\mathbf{O}^n) = 1 | H_0) = 0.$$

Because η is pointwise consistent, for every P for which $\rho_P(B,C) \neq 0$,

$$\lim_{n \to \infty} Pr^n(\eta_n(\mathbf{O}^n) = 0) = 0$$

Hence

$$\lim_{n \to \infty} Pr^n(\phi_n(\mathbf{O}^n) = 0 | \rho(B,C) \neq 0) = 0$$

$\rho_P(B,C) = 0$ is incompatible with Model Q, and in Model M, $\rho_P(B,C) = 0$ only when $z = -x \times y \neq 0$. Because $Pr(B_{G_M} | G_M)$ is absolutely continuous with respect to Lebesgue measure, $Pr(z = -x \times y = 0 | G_m) = 0$. If $Pr(H_1) \neq 0$, then

$$\lim_{n \to \infty} Pr^n(\phi_n(\mathbf{O}^n) = 0 | H_1) = 0$$

Otherwise, if $Pr(H_1) = 0$, $\lim_{n \to \infty} Pr(H_1)Pr^n(\phi_n(\mathbf{O}^n) = 0 | H_1) = 0$ Q.E.D

In addition to Bayesian statistical tests, there are Bayesian versions of confidence intervals and estimators for a zero treatment effect of B on C.

The prior plays an important role in determining the existence of Bayesian consistent tests of $\theta = \theta_0$ versus $\theta \neq \theta_0$. Whenever $\rho(B,C) = 0$ there are two different kinds of theories that explain this: either $z = 0$ (Model N), or $z = -x \times y \neq 0$ (Model M). Because both of these theories make exactly the same prediction about the marginal population distribution over B and C, no sample from the marginal population distribution can ever distinguish between them. Whatever the ratio of the probability of $z = 0$ to the probability of $z = -x \times y \neq 0$ was prior to seeing the sample, it remains exactly the same after seeing the sample. So the choice between the faithful explanation and the unfaithful explanation is entirely based on the prior, and not on the evidence. The prior in this example assigned a zero probability to $z = -x \times y \neq 0$, so Bayes consistent tests exist for the example. For a different prior which assigned a non-zero prior probability to $z = -x \times y \neq 0$, there is no Bayes consistent test of $\theta = \theta_0$ versus $\theta \neq \theta_0$ with respect to that prior.

More generally, with a prior over the parameters for each DAG that assigns zero probability to unfaithful distributions, there are Bayes consistent tests of whether or not the DAG that generated a given sample is a member of a given **O**-Markov equivalence class. Theorem 12.2 is a slight variation of results proved in Robins and Wasserman (1999).

Theorem 12.2: Let Γ be a countable set of DAGs each of which contains at least the variables in **O**, and F an **O**-Markov equivalence class of DAGs that intersects Γ. Let H_0 be "G is a member of F", H_1 be "G is not a member of F", and $\mathrm{B}_{G,U}$ be the set of parameters β such that $\gamma(\beta,G)$ is unfaithful to G. If in Π_Γ, there are pointwise consistent tests of each conditional independence relation among the variables in **O**, and for each $G \in \Gamma$, $Pr(\mathrm{B}_{G,U}|G) = 0$, then there is a test ϕ of H_0 against H_1 that is Bayes consistent with respect to Pr.

Proof. Suppose there are pointwise consistent tests (see section 12.4.2) of conditional independence relations among the observed variables. Then there is a pointwise consistent test of a finite set of conditional independence relations, and hence a pointwise consistent test ϕ of membership in F. (Each **O**-Markov equivalence class of DAGs entails a finite unique set of conditional independence relations among the variables in **O**.) By reasoning analogous to the proof of Theorem 12.4.1, in the large sample limit, the output of ϕ is wrong in the large sample limit about membership in F only when the distribution generated by the true DAG is unfaithful to that DAG. But this has probability 0 by hypothesis. Q.E.D.

For both multinomial distributions, and Gaussian distributions, the Lebesgue measure of the usual parameterizations that produce unfaithful distributions conditional on a given G is 0. Hence for these distribution families and the usual priors (described in section 12.5.3) there is a Bayes consistent test. However, for a stronger sense of Bayes consistency which requires stronger assumptions for success, see Robins and Wasserman 1999.

12.4.2 Pointwise Consistency

Definition 12.2: A test ϕ is pointwise consistent over a set of distributions $\Pi_{\Gamma 0}$, $\Pi_{\Gamma 1}$ if
(i) for every $P \in \Pi_{\Gamma 0}$, $\lim_{n \to \infty} P^n(\phi_n(\mathbf{O}^n) = 1) = 0$, and

(ii) for every $P \in \Pi_{\Gamma 1}$, $\lim_{n \to \infty} P^n(\phi_n(\mathbf{O}^n) = 0) = 0$.

In constrast to Bayes consistency, this definition requires that with probability 1 the test does not fail in the large sample limit over *all* pairs $(\beta, G) \in B\Gamma$. Failing on a non-trivial set of measure 0 of pairs $(\beta, G) \in B\Gamma$ is enough to rule out pointwise consistency. Suppose now that $\Gamma = \{G_M, G_N, G_Q\}$ from figure 12.11, $\theta = z$, $\theta_0 = 0$, and we test $\theta = \theta_0$ against $\theta \neq \theta_0$.

Theorem 12.3: If $\Gamma = \{G_M, G_N, G_Q\}$ from figure 12.11, $\theta = z$, and $\theta_0 = 0$, then there is no pointwise consistent test of $\theta = \theta_0$ against $\theta \neq \theta_0$ with respect to $\Pi_{\Gamma 0}$ and $\Pi_{\Gamma 1}$.

Proof. For every $P \in \Pi_{\Gamma 0}$ with margin $P(\mathbf{O})$ (from Model N or Model Q) there is a $P' \in \Pi_{\Gamma 1}$ (from Model M) such that $P(\mathbf{O}) = P'(\mathbf{O})$, and vice-versa. Because any test ϕ depend only on the marginal distribution, it follows that there is no pointwise consistent test of $\theta = \theta_0$ against $\theta \neq \theta_0$. Q.E.D.

However, if the intersection of $\Pi_{\Gamma 0}$ and $\Pi_{\Gamma 1}$ in the observed margin are distributions where $\rho(B,C) = 0$, there is a pointwise consistent test of $\theta = \theta_0$ against $\theta \neq \theta_0$. Since the distributions in $\Pi_{\Gamma 1}$ with $\rho(B,C) = 0$ in the observed margin are just those corresponding to the surface of unfaithful parameter values in Model M, if those distributions are removed, there is a pointwise consistent test of test of $\theta = \theta_0$ against $\theta \neq \theta_0$. Let Ω_G be the set of distributions that satisfy the Markov condition for G *and* are faithful to G. Let $\Omega_\Gamma = \bigcup_{G \in \Gamma} \Omega_G$. Let

$$\Omega_{\Gamma 0} = \bigcup_{G \in \Gamma} \{P \in \Omega_G : \exists \beta \in B_G, \theta = \theta_0 \ \& \ \gamma(\beta, G) = P\}$$

$$\Omega_{\Gamma 1} = \bigcup_{G \in \Gamma} \{P \in \Omega_G : \exists \beta \in B_G, \theta \neq \theta_0 \ \& \ \gamma(\beta, G) = P\}.$$

Theorem 12.4: If $\Gamma = \{G_M, G_N, G_Q\}$, there is a pointwise consistent test of $\theta = \theta_0$ against $\theta \neq \theta_0$ with respect to $\Omega_{\Gamma 0}$, $\Omega_{\Gamma 1}$.

Proof. There is a pointwise consistent test η of zero correlation against non-zero correlations. Let ϕ return 0 when η returns 0, and otherwise ϕ returns 2. Since ϕ never returns 1, for every $P \in \Omega_{\Gamma 0}$, $P^n(\phi_n(\mathbf{O}^n) = 1) = 0$. Under the assumption of faithfulness, $\Omega_{\Gamma 1}$ contains only distributions for which $\rho_P(B,C) \neq 0$. Since for every $P \in \Omega_{\Gamma 1}$, $\lim_{n\to\infty} P^n(\eta_n(\mathbf{O}^n) = 0) = 0$, it follows that $\lim_{n\to\infty} P^n(\phi_n(\mathbf{O}^n) = 0) = 0$. Q.E.D.

Theorem 12.5: Let Γ be a countable set of DAGs each of which contains at least the variables in $\mathbf{O}$, and F an $\mathbf{O}$-Markov equivalence class of DAGs that intersects Γ. Let H_0 be "G is a member of F", and H_1 be "G is not a member of F". If in Ω_Γ, there are pointwise consistent tests of each conditional independence relation among the variables in $\mathbf{O}$, there is a pointwise consistent test ϕ of H_0 against H_1 with respect to a set of distributions $\Omega_{\Gamma 0}$, $\Omega_{\Gamma 1}$.

Proof. Under the assumption of faithfulness, a distribution P is compatible with a DAG G in $\mathbf{O}$-Markov equivalence class F if and only if it satisfies a certain finite set of conditional independence relations in the margin. No distribution from a DAG that is not in F satisfies the same set of conditional independence relations in the margin. If there are pointwise consistent tests of each conditional independence relation among the variables in $\mathbf{O}$, then there is a pointwise consistent test of the set of conditional independence relations that F entails, and hence a pointwise consistent test of membership in F. Q.E.D.

In the case of both multivariate Gaussian and multinomial distributions, there are pointwise consistent tests of conditional independence, and hence pointwise consistent tests of membership in an $\mathbf{O}$-Markov equivalence class.

12.4.3 Uniform Consistency

Let $\Pi_{\Gamma\delta 0} = \bigcup_{G \in \Gamma} \{P \in \Pi_G : \exists \beta \in B_G, |\theta - \theta_0| > \delta \ \& \ \gamma(\beta, G) = P\}$,

that is the set of distributions in Π_Γ compatible with being more than δ away from θ_0. (The '0' in the subscript of $\Pi_{\Gamma\delta 0}$ refers to θ_0, and the 'δ' refers to the distance from θ_0.)

Definition 12.3: A test ϕ of $\theta = \theta_0$ against $\theta \neq \theta_0$ is uniformly consistent over a set of distributions $\Pi_{\Gamma 0}$ and $\Pi_{\Gamma\delta 0}$ if

(i) $\lim_{n\to\infty} \sup_{P \in \Pi_{\Gamma 0}} P^n(\phi_n(\mathbf{O}^n) = 1) = 0$

(ii) $\forall \delta > 0, \lim_{n\to\infty} \sup_{P \in \Pi_{\Gamma\delta 0}} P^n(\phi_n(\mathbf{O}^n) = 0) = 0$

Suppose for the moment that $\Gamma = \{G_M, G_N\}$ from figure 12.11, θ_0 is $z = 0$, and we test $\theta = \theta_0$ against $\theta \neq \theta_0$. Consider for the moment a ϕ that either returns 0 or 1. Since ϕ is a function of the observed data, at each sample size it divides samples into those judged to come from H_0, those judged to come from H_1. For a test of a null hypothesis of independence, those samples judged to come from H_1 are in the rejection region, and those judged to come from H_0 are in the acceptance region. If ϕ is pointwise consistent, then for any $\delta > 0$, for any $P \in \Omega_{\Gamma\delta 0}$ it is possible to find an n such that a sample of size n drawn from P is very likely to fall into the rejection region for ϕ_n where n depends upon P. However, uniform consistency is stronger than pointwise consistency because the definition requires that for every $\delta > 0$, it is possible to find a single minimal n such that a sample of size n drawn from *any* $P \in \Omega_{\Gamma\delta 0}$ is very likely to fall into the rejection region for ϕ_n. The same idea generalizes to tests which allow "don't know" as an answer.

If no uniformly consistent test of $\theta = \theta_0$ exists, then there are no uniformly consistent non-trivial confidence intervals around θ, and no uniformly consistent estimators of θ. Uniform consistency is required in order to bound the error on θ (in the worst case over all models.)

Robins, Scheines, Spirtes, and Wasserman (1999) show that even when the unfaithful distributions are removed, for parameterizations of $\Gamma = \{G_M, G_N\}$ in which A, B and C are discrete, there is no non-trivial uniformly consistent test of $\theta = \theta_0$ against $\theta \neq \theta_0$. The original proof in Robins, Scheines, Spirtes, and Wasserman (1999) assumed that a test is non-trivial in the stronger sense that in the limit it does not return "don't know" for all of the distributions in the null hypothesis, or all of the distributions in the alternative. The proof has since been extended to cover the weaker sense of non-triviality proposed here.

Even if the unfaithful distributions are ruled out by assumption, for $\Gamma = \{G_M, G_N, G_Q\}$, there is no uniformly consistent test of $\theta = \theta_0$ against $\theta \neq \theta_0$. Informally, there is no uniformly consistent test because even after the surface of unfaithful parameter values have been removed from $\Pi_{\Gamma 0}$, for any $\delta > 0$, it is still possible to find a $P \in \Omega_{\Gamma\delta 0}$ for which $\rho_P(B,C)$ is arbitrarily close to 0. Consider the sequence of rejection regions for a test ϕ that is pointwise consistent with respect to $\Omega_{\Gamma 0} \cup \Omega_{\Gamma 1}$. For any given $P \in \Omega_{\Gamma\delta 0}$, no matter how close $\rho_P(B,C)$ is to zero, as long as it is not equal to 0, it is possible to find an n such that it is likely that a sample of size n falls into the rejection regions for ϕ_n. But there is always some other $P' \in \Omega_{\Gamma\delta 0}$ with $\rho_{P'}(B,C)$ even closer to zero, such that it is not likely that a sample of size n from P' will fall into the rejection region for ϕ_n. Let

$$\Omega_{\Gamma\delta 0} = \bigcup_{G \in \Gamma} \{P \in \Omega_G : \exists \beta \in B_G, |\theta - \theta_0| > \delta \ \& \ \gamma(\beta, G) = P\}.$$

Theorem 12.6: If $\Gamma = \{G_M, G_N, G_Q\}$, $\theta = z$, and $\theta_0 = 0$, there is no uniformly consistent test of $\theta = \theta_0$ against $\theta \neq \theta_0$ with respect to $\Omega_{\Gamma 0}$ and $\Omega_{\Gamma\delta 0}$.

Proof. Suppose that on the contrary there is a uniformly consistent test ϕ of $\theta = \theta_0$ against $\theta \neq \theta_0$. Because ϕ is non-trivial, either

(i) for some $P \in \Omega_\Gamma \lim_{n\to\infty} P^n(\varphi^n(\mathbf{O}^n) = 0) = 1$, or

(ii) for some $P \in \Omega_\Gamma \lim_{n\to\infty} P^n(\varphi^n(\mathbf{O}^n) = 1) = 1$.

Suppose that (ii) is the case. If P is in $\Omega_{\Gamma 0}$, φ is not uniformly consistent. Suppose then that P is in $\Omega_{\Gamma \delta 0}$. For every distribution P in $\Omega_{\Gamma \delta 0}$ (from Model M) there is a distribution D in $\Omega_{\Gamma 0}$ (from Model Q) with the same marginal over $\mathbf{O}$. Because ϕ is a function of just the margin over $\mathbf{O}$, $P^n(\phi_n(\mathbf{O}^n) = 1) = D^n(\phi_n(\mathbf{O}^n) = 1)$. Hence in the large sample limit there is a $D \in \Omega_{\Gamma 0}$ such that $D^n(\phi_n(\mathbf{O}^n) = 1) = 1$, and ϕ is not uniformly consistent.

Suppose now that (i) is the case. If P is in $\Omega_{\Gamma \delta 0}$, φ is not uniformly consistent. Suppose then that P is in $\Omega_{\Gamma 0}$. It follows that P is compatible with $z = 0$. Consider first the case where $\rho_P(B,C) = r \neq 0$ (i.e. if $z = 0$, P is compatible with Model Q but not Model N). There is a $\delta > 0$ and some distribution $D \in \Omega_{\Gamma \delta 0}$, such that $\rho_D(B,C) = r$, but D is compatible with $|z| > \delta$, (i.e. D is compatible with Model M, and has the same margin over B and C as P.) Because ϕ is a function of just the margin over B and C, $P^n(\phi_n(\mathbf{O}^n) = 0) = D^n(\phi_n(\mathbf{O}^n) = 0)$. Hence there is a $D \in \Omega_{\Gamma \delta 0}$ such that in the large sample limit $D^n(\phi_n(\mathbf{O}^n) = 0) = 1$, and hence ϕ is not uniformly consistent.

Consider finally the case where $z = 0$ and $\rho_P(B,C) = 0$ (i.e. P is from Model N.) There is a $\delta > 0$, and a distribution $D \in \Pi_{\Gamma \delta 0}$ (compatible with Model M with a z value of z_1, where $|z_1| > \delta$) and the same marginal as P over B and C. However, D is not faithful to Model M, and hence not a member of $\Omega_{\Gamma \delta 0}$. But there is an interval around zero such that for every value r in the interval, except for $r = 0$, there is some $D_n \in \Omega_{\Gamma \delta 0}$ compatible with Model M and $z = z_1$ such that $\rho_{Dn}(B,C) = r$. The Kullback-Liebler distance $I(\tilde{D}; \tilde{D}_n)$ equals $-1/2 \log(1 - r^2)$, which is a continuous function of r (where $\tilde{D}$ is the marginal of D over B and C). Hence $I(\tilde{D}^n; \tilde{D}_n^n)$ equals $-n/2 \log(1 - r^2)$. For every event A in the sample space,

$$\sup_A |\tilde{D}^n(A) - \tilde{D}_n^n(A)| \leq \frac{1}{2}\{I(\tilde{D}^n; \tilde{D}_n^n)\}^{1/2}$$

By choosing r small enough, there are distributions in $\Omega_{\Gamma \delta 0}$ with marginals that are arbitrarily close to $\tilde{D}$ and compatible with Model M and $z = z_1$. Hence for all n, and all $\varepsilon/2$, there is a distribution $D_n \in \Omega_{\Gamma \delta 0}$ (and hence faithful to Model M with $z = z_1$) such that $|\tilde{D}^n(\phi_n(\mathbf{O}^n) = 0) - \tilde{D}_n^n(\phi_n(\mathbf{O}^n) = 0)| < \varepsilon/2$.

Because ϕ is a function of just the margin over B and C, $P^n(\phi_n(\mathbf{O}^n) = 0) = \tilde{P}^n(\phi_n(\mathbf{O}^n) = 0) = \tilde{D}^n(\phi_n(\mathbf{O}^n) = 0) \leq \tilde{D}_n^n(\phi_n(\mathbf{O}^n) = 0) + \varepsilon/2 = D_n^n(\phi_n(\mathbf{O}^n) = 0) + \varepsilon/2$. Because $P^n(\phi_n(\mathbf{O}^n) = 0)$ converges to $1, (\forall \varepsilon/2 > 0)(\exists N)(\forall n > N)(P^n(\phi_n(\mathbf{O}^n) = 0) > 1 - \varepsilon/2$. It follows then that

$(\forall \varepsilon > 0)(\exists N)(\forall n > N)(D_n^n(\phi_n(\mathbf{O}^n) = 0) > 1 - \varepsilon/2 - \varepsilon/2) = 1 - \varepsilon$.

Since each $D_n \in \Pi_{\Omega\delta 0}$, it follows that

$$\lim_{n\to\infty} \sup_{P\in\Omega_{\Gamma\delta 0}} P^n(\phi_n(\mathbf{O}^n) = 0) = 1$$

and hence ϕ is not uniformly consistent. Q.E.D.

However if instead of assuming only that there are no unfaithful parameter values of Model M, suppose we assume that there are no "close to unfaithful" parameter values of M (and hence that there are no "close to unfaithful" distributions to M.) For example, in Model M, for any given fixed $\kappa > 0$, one could allow only those parameters such that $|z + (x \times y)| > \kappa|z|$, i.e. those parameter values for which the correlation is greater than a fixed percentage of the size of the treatment effect of B on C. If κ were 0.001, the assumption means that the correlation is a least 1/1000 the size of the treatment effect of B on C. For a fixed κ, call the set of parameter values such that $|z + (x \times y)| < \kappa|z|$ "close to unfaithful". The assumption that parameter values are not close to unfaithful is the assumption that small population correlations guarantee small treatment effects.

Let H_G be the set of distributions that satisfy the Markov condition for G and are not close to unfaithful to G, for some fixed κ, and

$$H_\Gamma = \bigcup_{G\in\Gamma} H_G, \ \theta = z, \ \theta_0 = 0$$

and

$$H_{\Gamma\delta 0} = \bigcup_{G\in\Gamma} \{P \in H_G : \exists \beta \in B_G, |\theta - \theta_0| > \delta \ \& \ \gamma(\beta, G) = P\}$$

(i.e. $H_{\Gamma\delta 0}$ is the set of distributions that satisfy the Markov Condition, are not close to unfaithful to G, and for which θ is more than δ from θ_0.

Theorem 12.7: If $\Gamma = \{G_M, G_N, G_Q\}$, there is a uniformly consistent test of $\theta = \theta_0$ against $\theta \neq \theta_0$ with respect to $H_{\Gamma 0}$ and $H_{\Gamma\delta 0}$.

Proof. There is a uniformly consistent test η of $\rho(B,C) = 0$ against $\rho(B,C) \neq 0$. Let ϕ return 0 when η returns 0, and let ϕ return 2 otherwise. Because ϕ never returns 1, for all $P \in \Gamma_{\Gamma 0}$, $P^n(\varphi_n(\mathbf{O}^n) = 1) = 0$. Let $T_{\Gamma\delta 0} = \bigcup_{G\in\Gamma} \{P \in \Pi_G : |\rho_P(B,C)| > \delta\}$.

By the assumption of no "close to unfaithful" parameter values, if the absolute value of the treatment effect of B on C is greater than δ then the absolute value of the correlation of B and C is greater than $\kappa\delta$. For every distribution P in $H_{\Gamma\delta}$, P is in $T_{\Gamma(\kappa\delta)0}$. Because $\phi_n(\mathbf{O}^n) = 0$ if and only if $\eta_n(\mathbf{O}^n) = 0$, it follows that

$$\forall \kappa\delta > 0, \lim_{n\to\infty} \sup_{P\in T_{\Gamma(\kappa\delta)0}} P^n(\eta_n(\mathbf{O}^n) = 0) = 0 \Rightarrow$$

$$\forall \delta > 0, \lim_{n\to\infty} \sup_{P\in H_{\Gamma\delta 0}} P^n(\phi_n(\mathbf{O}^n) = 0) = 0$$

The antecedent is true because η is a uniformly consistent test of zero correlation. Hence ϕ is a uniformly consistent test of $\theta = \theta_0$ against $\theta \neq \theta_0$ with respect to $H_{\Gamma 0}$ and $H_{\Gamma\delta}$. Q.E.D.

A zero treatment effect of B on C is a special case of a treatment effect that can in some cases be calculated from the Prediction Algorithm. Extending results about the existence of uniformly consistent tests of the size of treatment effects to all treatment effects that can be calculated from the Prediction Algorithm would require generalizing the concept of "close to unfaithful parameters" and generalizing the distance metric used in theorem 12.4.2. We conjecture that there are natural generalizations such that there is a uniformly consistent test of every treatment effect that can be calculated from the Prediction Algorithm, under the assumption of no "close to unfaithful" distributions. Similarly, extending results about uniform consistency to tests of membership in a given **O**-Markov equivalence class requires generalizing the concept of "close to unfaithful" to conditional independencies, and a metric measuring the distance between a pair (β, G) and an **O**-Markov equivalence class. We conjecture that there is a natural generalization of "close to unfaithful" and a natural metric under which there is a uniformly consistent test of membership in a given **O**-Markov equivalence class.

12.4.3 Interval Testing

Returning to Model M of figure 12.11, for a fixed $\varepsilon > 0$ let H_0 be "$|z| \leq \varepsilon$", and H_1 be "$|z| > \varepsilon$". If $Pr(B\Gamma)$ is a prior that assigns measure 0 to the close to unfaithful parameter values, then there is a test of H_0 against H_1 that is Bayes consistent with respect to Pr. Similarly, there are tests that are pointwise consistent tests with respect to $H_{\Gamma 0}$, $H_{\Gamma 1}$ (where $H_{\Gamma 0}$ is the set of distributions in H_Γ compatible with H_0, and $H_{\Gamma 1}$ is the set of distributions in H_Γ compatible with H_1), and uniformly consistent with repect to $H_{\Gamma 0}$ and $H_{\Gamma\delta 0}$ (where $H_{\Gamma\delta 0}$ is the set of distributions in H_Γ at leasts a distance δ from the null hypothesis.) The proof of the existence of the uniformly consistent test is analogous to the proof of Theorem 12.7 and the existence of the pointwise consistent test and the Bayes consistent test follow from the existence of the uniformly consistent tests.

12.4.5 Other Kinds of Background Knowledge

In the examples of figure 12.11, the background knowledge fixed a time order, but there was a possibility of unmeasured common causes. Consistency questions analogous to the ones considered in the previous sections can be raised for other kinds of background knowledge. For example, one kind of background knowledge is that there is no given time order, and also no unmeasured common causes; another is that there is a given time order but there are no unmeasured common causes. Assuming just faithfulness and Markov, we conjecture that there is in general no non-trivial uniformly consistent test of membership in a Markov equivalence class, given that there are no latents, but not given a time order. Given a time order, no latent variables, and no determinism, there is a non-trivial uniformly consistent test of membership in a Markov equivalence class.

12.4.6 Conclusions to Be Drawn from the Negative Results

We emphasize once more that the negative results described in this section apply to *any* method, not just the methods described in this book. Even given time order, without assuming faithfulness, or additional background knowledge such as that available for randomized clinical trials, there are no pointwise or uniformly consistent tests of zero treatment effect of B on C. It follows that there is no (non-trivial) uniformly consistent confidence interval for the size of the treatment effect of B on C, and no uniformly consistent estimator of the size of the treatment effect. No kind of search (constraint-based, greedy, Monte Carlo, simulated annealing, genetic, etc.), no kind of model selection based on any kind of score (posterior probabilities, BIC, AIC, MDL, etc.), no kind of model averaging, and no kind of test (χ^2 test, Fisher's exact test, t-tests, z-transformations), can get around these basic limitations. Nor can any informal method (using human judgements or "insight") escape these basic limitations.

What conclusions should we draw from the negative results? There are typically four strategies to follow when it is shown that no method can solve a given problem in a given sense of reliability (Kelly 1996): (i) strengthen the evidence; (ii) strengthen the background assumptions; (iii) weaken the sense of success required; or (iv) give up. We will discuss each of these strategies in turn.

One way of adding to the evidence is to provide the results of randomized trials. Certainly this is preferable when possible, but in most human studies and in psychology, randomized trials are not possible for practical, ethical, or theoretical reasons.

We have already seen several ways in which adding background assumptions can lead to success. Thus, if one adds the background assumption that there are no almost unfaithful distributions there are uniformly consistent tests of zero treatment effects.

We have also seen several ways of weakening the sense of success required, e.g. settling for Bayes consistency instead of pointwise consistency, or pointwise consistency instead of uniform consistency.

Another way of weakening the sense of success is to provide tests that are conditional on the strength of the association owing to unmeasured common causes. (See, e.g.,

Rosenbaum 1995. This method also applies to many cases where some of the algorithms proposed in this book simply say "don't know.") There are good reasons to carry out such sensitivity analyses; for example, the analysis clearly separates what are the assumptions and what role the data is playing in the analysis. But while this method makes clear what assumptions about the strength of confounding are needed for particular conclusions about the size of the treatment effect, if one is not willing to make these assumptions, then conclusions about the size of the treatment effect cannot be drawn. Without endorsing one or another of these additional background assumptions, a decision maker cannot make decisions that are uniformly consistent on the basis of such sensitivity analyses.

A third way of weakening the sense of success required is to calculate bounds on the size of the treatment effect (see Manski 1995). We have already seen however, that in the example of figure 12.11, if $\rho(B,C) = 0$, then there are no (non-trivial) bounds on the size of the treatment effect z. Even if $\rho(B,C) = a$ where a is positive, there are no (non-trivial) bounds on how negative the treatment effect might be. Unfortunately, without further assumptions, regardless of how large $\rho(B,C)$ might be, the bounds always include zero as a possibility, and the bounds will always include some negative treatment effects. Although there are interesting and useful bounds that can be obtained under a variety of assumptions, without these further assumptions, the bounds are generally not useful in practical decision making.

If someone insists on uniformly consistent tests as the minimally acceptable sense of success, is unwilling to accept the assumption that there are no nearly unfaithful distributions, and is unable to provide appropriate randomized trials, then that person should give up on the enterprise of inferring causal relations from observational data. "Giving up" does not mean substituting informal methods, or "human judgement" for automated techniques. Informal methods and "human judgement" are equally as subject to the limitations of the negative results as are formal or automated methods. "Giving up" means one should simply stop collecting data for such purposes, and stop looking at data in an attempt to make such inferences. This would involve halting most causal studies in epidemiology, sociology, psychology, and economics.

Is "giving up" the right policy? We still have to make decisions regarding health policy, social policy, economic policy, and so on. The question is whether, because we cannot obtain uniformly consistent tests without making an assumption such as "no nearly unfaithful distributions", we are better off giving up collecting evidence all together, or instead applying methods that do not satisfy strong consistencey requirements, but do satisfy weaker consistency requirements. We believe the latter.

This argument is not intended to show that any of the automated search algorithms that we have described are ultimately going to turn out to be useful tools. That question depends upon their performance on real data sets at real sample sizes, where the assumptions made are not going to hold exactly. It is intended to suggest however, that the non-existence of algorithms that satisfy strong consistency requirements making only weak assumptions about background knowledge is not by itself good reason to give up all attempts to draw causal inferences from observational data.

12.5 Search

Sections 12.5.1 through 12.5.6 originally appeared in a slightly altered form in Heckerman, Meek, and Cooper 1999, which contains some additional details.[1] Sections 12.5.1 and 12.5.2 review the Bayesian approach to model averaging and model selection and its applications to the discovery of causal DAG models. Section 12.5.3 discusses methods for assigning priors to model structures and their parameters. Section 12.5.4 compares the Bayesian and constraint-based methods for causal discovery with complete data, highlighting some of the advantages of the Bayesian approach. Section 12.5.5 notes computational difficulties associated with the Bayesian approach when data sets are incomplete—for example, when some variables are hidden—and discusses more efficient approximation methods including Monte-Carlo and asymptotic approximations. Section 12.5.6 discusses open problems in searching over models with latent variables, section 12.5.7 discusses search over equivalence classes of latent variable models, and section 12.5.8 discusses search over cyclic directed graphs. Section 12.5.9 describes some other recent approaches to search, and section 12.5.10 discusses what attitude should be adopted toward the output of a causal search algorithm. Other overviews of learning in Bayesian networks include Heckerman 1998, Buntine 1996, and Jordan 1998.

12.5.1 The Bayesian Approach

In a constraint-based approach to the discovery of causal DAG models, we[4] use data to make *categorical* decisions about whether or not particular conditional-independence constraints hold. We then piece these decisions together by looking for those sets of causal structures that are consistent with the constraints. To do so, we use the Causal Markov condition (discussed in chapter 3) to link lack of cause with conditional independence.

In the Bayesian approach, we also use the Causal Markov condition to look for structures that fit conditional-independence constraints. In contrast to constraint-based methods, however, we use data to make *probabilistic* inferences about conditional-independence constraints. For example, rather than conclude categorically that, given data, variables X and Y are independent, we conclude that these variables are independent with some probability. This probability encodes our uncertainty about the presence or absence of independence. Furthermore, because the Bayesian approach uses a probabilistic framework, we no longer need to make decisions about individual independence facts. Rather, we compute the probability that the independencies associated with an entire causal structure are true. Then, using such probabilities, we can average a particular hypothesis of interest—such as, "Does X cause Y?"—over all possible causal structures.

Let us examine the Bayesian approach in some detail. Suppose our problem domain consists of variables $\mathbf{X} = \{X_1,\ldots,X_n\}$. In addition, suppose that we have some data $D = \{\mathbf{x}_1,\ldots,\mathbf{x}_N\}$, which is a random sample from some unknown probability distribution for $\mathbf{X}$. For the moment, we assume that each case $\mathbf{x}$ in D consists of an observation of all the variables in $\mathbf{X}$. We assume that the unknown probability distribution can be encoded by some causal model with structure $\mathbf{m}$. We assume that the structure of this causal model is a DAG that encodes conditional independencies via the Causal Markov condition. We are uncertain about the structure and parameters of the model; and—using the Bayesian approach—we encode this uncertainty using probability. In particular, we define a discrete variable $\mathbf{M}$ whose states $\mathbf{m}$ correspond to the possible true models, and encode our uncertainty about $\mathbf{M}$ with the probability distribution $p(\mathbf{m})$. In addition, for each model structure $\mathbf{m}$, we define a continuous vector-valued variable Θ_m, whose values θ_m correspond to the possible true parameters. We encode our uncertainty about Θ_m using the (smooth) probability density function $p(\theta_m|\mathbf{m})$. The assumption that $p(\theta_m|\mathbf{m})$ is a smooth probability density function entails (measure 1) the assumption of faithfulness employed in constraint-based methods for causal discovery (Meek 1995).

Given random sample D, we compute the posterior distributions for each $\mathbf{m}$ and θ_m using Bayes's rule:

$$p(\mathbf{m}|D) = \frac{p(\mathbf{m})p(D|\mathbf{m})}{\sum_{m'} p(\mathbf{m}')p(D|\mathbf{m}')} \tag{12.1}$$

$$p(\theta_m|D,\mathbf{m}) = \frac{p(\theta_m|\mathbf{m})p(D|\theta_m,\mathbf{m})}{p(D|\mathbf{m})} \tag{12.2}$$

where

$$p(D|\mathbf{m}) = \int p(D|\theta_m,\mathbf{m})p(\theta_m|\mathbf{m})d\theta_m \tag{12.3}$$

is called the *marginal likelihood*. Given some hypothesis of interest, h, we determine the probability that h is true given data D by averaging over all possible models and their parameters:

$$p(h|D) = \sum_m p(\mathbf{m}|D)p(h|D,\mathbf{m}) \tag{12.4}$$

$$p(h|D,\mathbf{m}) = \int p(h|\theta_m,\mathbf{m})p(\theta_m|D,\mathbf{m})d\theta_m \tag{12.5}$$

For example, h may be the event that the next case $\mathbf{X}_{N+1}$ is observed in configuration $\mathbf{x}_{N+1}$. In this situation, we obtain

Prequels and Sequels

$$p(\mathbf{x}_{N+1} | D) = \sum_m p(\mathbf{m} | D) \int p(\mathbf{x}_{N+1} | \theta_m, \mathbf{m}) p(\theta_m | D, \mathbf{m}) d\theta_m \qquad (12.6)$$

where $p(\mathbf{x}_{N+1} | \theta_m, \mathbf{m})$ is the likelihood for the model. As another example, h may be the hypothesis that "X causes Y." We consider such a situation in detail in section 12.5.4.

Under certain assumptions, these computations can be done efficiently and in closed form. One assumption is that the likelihood term $p(\mathbf{x} | \theta_m, \mathbf{m})$ factors as follows:

$$p(\mathbf{x} | \theta_m, \mathbf{m}) = \prod_{i=1}^{n} p(x_i | \mathbf{pa}_i, \theta_i, \mathbf{m}) \qquad (12.7)$$

where each *local likelihood* $p(x_i | \mathbf{pa}_i, \theta_i, \mathbf{m})$ is in the exponential family. In this expression, $\mathbf{pa}_i$ denotes the configuration of the variables corresponding to parents of node x_i, and θ_i denotes the set of parameters associated with the local likelihood for variable x_i. One example of such a factorization occurs when each variable $X_i \in \mathbf{X}$ is discrete, having r_i possible values $x_i^1, \ldots, x_i^{r_i}$ and each local likelihood is a collection of multinomial distributions, one distribution for each configuration of $\mathbf{Pa}_i$—that is,

$$p(x_i^k | \mathbf{pa}_i^j, \theta_i, \mathbf{m}) = \theta_{ijk} > 0 \qquad (12.8)$$

where $\mathbf{pa}_i^1, \ldots, \mathbf{pa}_i^{q_i}$ $(q_i = \prod_{x_i \in \mathbf{Pa}_i} r_i)$ denote the configurations of $\mathbf{Pa}_i$ and

$$\theta_i = \left(\left(\theta_{ijk} \right)_{k=2}^{r_i} \right)_{j=1}^{q_i}$$

are the parameters. The parameter θ_{ij1} is given by

$$1 - \sum_{k=2}^{r_i} \theta_{ijk}$$

We shall use this example to illustrate many of the concepts in this paper. For convenience, we define the vector of parameters $\theta_{ij} = (\theta_{ij2}, \ldots, \theta_{ijr_i})$ for all i and j. A second assumption for efficient computation is that the parameters are mutually independent. For example, given the discrete-multinomial likelihoods, we assume that the parameter vectors θ_{ij} are mutually independent.

Let us examine the consequences of these assumptions for our multinomial example. Given a random sample D that contains no missing observations, the parameters remain independent:

$$p(\theta_m | D, \mathbf{m}) = \prod_{i=1}^{n} \prod_{j=1}^{q_i} p(\theta_{ij} | D, \mathbf{m}) \qquad (12.9)$$

Thus, we can update each vector of parameters θ_{ij} independently. Assuming each vector θ_{ij} has a conjugate prior[5]—namely, a Dirichlet distribution $\text{Dir}(\theta_{ij}|a_{ij1},\ldots,a_{ijr_i})$—we obtain the posterior distribution for the parameters

$$p(\theta_{ij} \mid D, \mathbf{m}) = \text{Dir}(\theta_{ij} \mid a_{ij1} + N_{ij1}, \ldots, a_{ijr_i} + N_{ijr_i}) \tag{12.10}$$

where N_{ijk} is the number of cases in D in which $X_i = x_i^k$ and $\mathbf{Pa}_i = \mathbf{pa}_i^j$. Note that the collection of counts N_{ijk} are sufficient statistics of the data for the model $\mathbf{m}$. In addition, we obtain the marginal likelihood (derived in Cooper and Herskovits 1992):

$$p(D \mid \mathbf{m}) = \prod_{i=1}^{n} \prod_{j=1}^{q_i} \frac{\Gamma(a_{ij})}{\Gamma(a_{ij} + N_{ij})} \prod_{k=1}^{r_i} \frac{\Gamma(a_{ijk} + N_{ijk})}{\Gamma(a_{ijk})} \tag{12.11}$$

where

$$a_{ij} = \sum_{k=1}^{r_i} a_{ijk} \text{ and } N_{ij} = \sum_{k=1}^{r_i} N_{ijk}$$

We then use equation (12.1) and equation (12.11) to compute the posterior probabilities $p(\mathbf{m}|D)$. Cooper and Yoo (1999) show that equation (12.11) also applies to a mixture of experimental and observational data, if N_{ijk} counts only those cases where X_i has not been experimentally manipulated.

As a simple illustration of these ideas, suppose our hypothesis of interest is the outcome of $\mathbf{X}_{N+1}$, the next case to be seen after D. Also suppose that, for each possible outcome $\mathbf{x}_{N+1}$ of $\mathbf{X}_{N+1}$, the value of X_i is x_i^k and the configuration of $\mathbf{Pa}_i$ is $\mathbf{pa}_i^j$, where k and j depend on i. To compute $p(\mathbf{x}_{N+1}|D)$, we first average over our uncertainty about the parameters. Using equations (12.2), (12.7.), and (12.8), we obtain

$$p(\mathbf{x}_{N+1} \mid D, \mathbf{m}) = \int \left(\prod_{i=1}^{n} \theta_{ijk} \right) p(\theta_m \mid D, \mathbf{m}) d\theta_m$$

Because parameters remain independent given D, we get

$$p(\mathbf{x}_{N+1} \mid D, \mathbf{m}) = \prod_{i=1}^{n} \int \theta_{ijk} p(\theta_{ij} \mid D, \mathbf{m}) d\theta_{ij}$$

Because each integral in this product is the expectation of a Dirichlet distribution, we have

$$p(\mathbf{x}_{N+1} \mid D, \mathbf{m}) = \prod_{i=1}^{n} \frac{a_{ijk} + N_{ijk}}{a_{ij} + N_{ij}} \tag{12.12}$$

Finally, we average this expression for $p(\mathbf{x}_{N+1}|D,\mathbf{m})$ over the possible models using equation (12.5) to obtain $p(\mathbf{x}_{N+1}|D)$.

12.5.2 Model Selection and Search

The full Bayesian approach is often impractical, even under the simplifying assumptions that we have described. One computation bottleneck in the full Bayesian approach is averaging over all models in equation (12.4). If we consider causal models with n variables, the number of possible structure hypotheses is at least exponential in n. Consequently, in situations where we can not exclude almost all of these hypotheses, the approach is intractable. Statisticians, who have been confronted by this problem for decades in the context of other types of models, use two approaches to address this problem: *model selection* and *selective model averaging*. The former approach is to select a "good" model (i.e., structure hypothesis) from among all possible models, and use that model as if it were the correct model. The latter approach is to select a manageable number of good models from among all possible models and pretend that these models are exhaustive. These related approaches raise several important questions. In particular, do these approaches yield accurate results when applied to causal structures? If so, how do we search for good models?

The question of accuracy is difficult to answer in theory. Nonetheless, several researchers have shown experimentally that the selection of a single model that is likely a posteriori often yields accurate predictions (Cooper and Herskovits 1992; Aliferis and Cooper 1994; Heckerman et al. 1995) and that selective model averaging using Monte-Carlo methods can sometimes be efficient and yield even better predictions (Herskovits 1991; Madigan et al. 1996).

Chickering (1996a) has shown that for certain classes of prior distributions the problem of finding the model with the highest posterior is NP-complete. However, a number of researchers have demonstrated that greedy search methods over a search space of DAGs work well. Also, constraint-based methods have been used as a first-step heuristic search for the most likely causal model (Singh and Valtorta 1993; Spirtes and Meek 1995). In addition, performing greedy searches in a space where Markov equivalent models (see definition below) are represented by a single model has improved performance (Spirtes and Meek 1995; Chickering 1996).

12.5.3 Priors

To compute the relative posterior probability of a model structure, we must assess the structure prior $p(\mathbf{m})$ and the parameter priors $p(\theta_m|m)$. Unfortunately, when many model structures are possible, these assessments will be intractable. Nonetheless, under certain assumptions, we can derive the structure and parameter priors for many model structures from a manageable number of direct assessments.

12.5.3.1 Priors for Model Parameters

First, let us consider the assessment of priors for the parameters of model structures. We consider the approach of Heckerman et al. (1995) who address the case where the local likelihoods are multinomial distributions and the assumption of parameter independence holds.

Their approach is based on two key concepts: Markov equivalence and distribution equivalence. Recall that two model structures for **X** are *Markov equivalent* (synonymous with faithfully indistinguishable) if they can represent the same set of conditional-independence assertions for **X** (Verma and Pearl 1990). For example, given **X** = {X,Y,Z}, the model structures $X \to Y \to Z$, $X \leftarrow Y \to Z$, and $X \leftarrow Y \leftarrow Z$, represent only the independence assertion that X and Z are conditionally independent given Y. Consequently, these model structures are equivalent. Another example of Markov equivalence is the set of *complete model structures* on **X**; a complete model is one that has no missing edge and which encodes no assertion of conditional independence. When **X** contains n variables, there are $n!$ possible complete model structures; one model structure for each possible ordering of the variables. All complete model structures for $p(\mathbf{x})$ are Markov equivalent. In general, two model structures are Markov equivalent if and only if they have the same structure ignoring arc directions and the same unshielded colliders (Verma and Pearl 1990; also see chapter 4).

The concept of distribution equivalence is closely related to that of Markov equivalence. Suppose that all causal models for **X** under consideration have local likelihoods in the family $\mathcal{F}$. This is not a restriction, per se, because $\mathcal{F}$ can be a large family. We say that two model structures $\mathbf{m}_1$ and $\mathbf{m}_2$ for **X** are *distribution equivalent with respect to (wrt)* $\mathcal{F}$ if they represent the same joint probability distributions for **X**—that is, if, for every θ_{m1}, there exists a θ_{m2} such that $p(\mathbf{x}|\theta_{m1},\mathbf{m}_1) = p(\mathbf{x}|\theta_{m2},\mathbf{m}_2)$, and vice versa. (This is a special case of **O**-distribution equivalence defined in section 12.2, where **O** is the entire set of variables in the DAG.)

Distribution equivalence wrt some $\mathcal{F}$ implies Markov equivalence, but the converse does not hold. For example, when $\mathcal{F}$ is the family of generalized linear-regression models, the complete model structures for $n \geq 3$ variables do not represent the same sets of distributions. Nonetheless, there are families $\mathcal{F}$—for example, multinomial distributions and linear-regression models with Gaussian noise—where Markov equivalence implies distribution equivalence wrt $\mathcal{F}$ (Heckerman and Geiger 1996). The notion of distribution equivalence is important, because if two model structures $\mathbf{m}_1$ and $\mathbf{m}_2$ are distribution equivalent wrt to a given $\mathcal{F}$, then it is often reasonable to expect that data can not help to discriminate them. That is, we expect $p(D|\mathbf{m}_1) = p(D|\mathbf{m}_2)$ for any data set D. Heckerman et al. (1995) call this property *likelihood equivalence*. Note that the constraint-based approach also does not discriminate among Markov equivalent structures.

Now let us return to the main issue of this section: the derivation of priors from a manageable number of assessments. Geiger and Heckerman (1995) show that the assumptions of parameter independence and likelihood equivalence imply that the parameters for any *complete* model structure $\mathbf{m}_c$ must have a Dirichlet distribution with constraints on the hyperparameters given by

$$a_{ijk} = a\, p(x_i^k, \mathbf{pa}_i^j \mid \mathbf{m}_c) \qquad (12.13)$$

where α is the user's equivalent sample size,[6] and

$$p(x_i^k, \mathbf{pa}_i^j \mid \mathbf{m}_c)$$

is computed from the user's joint probability distribution $p(\mathbf{x} \mid \mathbf{m}_c)$. This result is rather remarkable, as the two assumptions leading to the constrained Dirichlet solution are qualitative.

To determine the priors for parameters of *incomplete* model structures, Heckerman et al. (1995) use the assumption of *parameter modularity*, which says that if X_i has the same parents in model structures $\mathbf{m}_1$ and $\mathbf{m}_2$, then

$$p(\theta_{ij} \mid \mathbf{m}_1) = p(\theta_{ij} \mid \mathbf{m}_2)$$

for $j = 1,\ldots,q_i$. They call this property parameter modularity, because it says that the distributions for parameters θ_{ij} depend only on the structure of the model that is local to variable X_i—namely, X_i and its parents.

Given the assumptions of parameter modularity and parameter independence, it is a simple matter to construct priors for the parameters of an arbitrary model structure given the priors on complete model structures. In particular, given parameter independence, we construct the priors for the parameters of each node separately. Furthermore, if node X_i has parents $\mathbf{Pa}_i$ in the given model structure, we identify a complete model structure where X_i has these parents, and use equation (12.13) and parameter modularity to determine the priors for this node. The result is that all terms a_{ijk} for all model structures are determined by equation (12.13). Thus, from the assessments a and $p(\mathbf{x} \mid \mathbf{m}_c)$, we can derive the parameter priors for all possible model structures. We can assess $p(\mathbf{x} \mid \mathbf{m}_c)$ by constructing a causal model called a *prior model*, that encodes this joint distribution. Heckerman et al. (1995) discuss the construction of this model.

12.5.3.2 Priors for Model Structures

Now, let us consider the assessment of priors on model structures. The simplest approach for assigning priors to model structures is to assume that every structure is equally likely. Of course, this assumption is typically inaccurate and used only for the sake of convenience. A simple refinement of this approach is to ask the user to exclude various structures (perhaps based on judgments of cause and effect), and then impose a uniform prior on the remaining structures.

Buntine (1991) describes a set of assumptions that leads to a richer yet efficient approach for assigning priors. The first assumption is that the variables can be ordered (e.g., through a knowledge of time precedence). The second assumption is that the presence or absence of possible arcs are mutually independent. Given these assumptions, $n(n-1)/2$ probability assessments (one for each possible arc in an ordering) determines the prior probability of every possible model structures. One extension to this approach is to allow for multiple possible orderings. One simplification is to assume that the probability that an arc is absent or present is independent of the specific arc in question. In this case, only one probability assessment is required.

An alternative approach, described by Heckerman et al. (1995) uses a prior model. The basic idea is to penalize the prior probability of any structure according to some measure of deviation between that structure and the prior model. Heckerman et al. (1995) suggest one reasonable measure of deviation.

Madigan et al. (1995) give yet another approach that makes use of imaginary data from a domain expert. In their approach, a computer program helps the user create a hypothetical set of complete data. Then, using techniques such as those in section 12.5.1, they compute the posterior probabilities of model structures given this data, assuming the prior probabilities of structures are uniform. Finally, they use these posterior probabilities as priors for the analysis of the real data.

12.5.4 Example

In this section, we provide a simple example that applies Bayesian model averaging and Bayesian model selection to the problem of causal discovery. In addition, we compare these methods with a constraint-based approach.

Let us consider a simple domain containing three binary variables X, Y, and Z. Let h denote the hypothesis that variable X causally influences variable Z. For brevity, we will sometimes state h as "X causes Z."

First, let us consider Bayesian model averaging. In this approach, we use equation (12.4) to compute the probability that h is true given data D. Because our models are causal, the expression $p(D|\mathbf{m})$ reduces to an index function that is true when $\mathbf{m}$ contains an arc from node X to node Z. Thus, the right-hand-side of equation 12.4 reduces to

$$\sum_{m''} p(\mathbf{m}''|D)$$

where the sum is taken over all causal models $\mathbf{m}''$ that contain an arc from X to Z. For our three-variable domain, there are 25 possible causal models and, of these, there are eight models containing an arc from X to Z.

To compute $p(\mathbf{m}|D)$, we apply equation (12.1), where the sum over $\mathbf{m}'$ is taken over the 25 models just mentioned. We assume a uniform prior distribution over the 25

Prequels and Sequels							337

possible models, so that $p(\mathbf{m'}) = 1/25$ for every $\mathbf{m'}$. We use equation (12.11) to compute the marginal likelihood $p(D \mid \mathbf{m})$. In applying equation (12.11), we use the prior given by $a_{ijk} = 1/r_i q_i$, which we obtain from equation (12.13) using a uniform distribution for $p(\mathbf{x} \mid \mathbf{m}_c)$ and an equivalent sample $a = 1$. Because this equivalent sample size is small, the data strongly influences the posterior probabilities for h that we derive.

$$p(X = \text{true}) = 0.34$$
$$p(Y = \text{true}) = 0.57$$
$$p(Z = \text{true} \mid X = \text{true}, Y = \text{true}) = 0.36$$
$$p(Z = \text{true} \mid X = \text{true}, Y = \text{false}) = 0.64$$
$$p(Z = \text{true} \mid X = \text{false}, Y = \text{true}) = 0.42$$
$$p(Z = \text{true} \mid X = \text{false}, Y = \text{false}) = 0.81$$

Figure 12.13. A causal model used to generate data

To generate data, we first selected the model structure $X \rightarrow Z \leftarrow Y$ and randomly sampled its probabilities from a uniform distribution. The resulting model is shown in figure 12.13. Next, we sampled data from the model according to its joint distribution. As we sampled the data, we kept a running total of the number cases seen in each possible configuration of $\{X,Y,Z\}$. These counts are sufficient statistics of the data for any causal model $\mathbf{m}$. These statistics are shown in table 12.1 for the first 150, 250, 500, 1000, and 2000 cases in the data set.

Number of cases	Sufficient Statistics							
	$\bar{x}\bar{y}\bar{z}$	$\bar{x}\bar{y}z$	$\bar{x}y\bar{z}$	$\bar{x}yz$	$x\bar{y}\bar{z}$	$x\bar{y}z$	$xy\bar{z}$	xyz
150	5	36	38	15	7	16	23	10
250	10	60	51	27	15	25	41	21
500	23	121	103	67	19	44	79	44
1000	44	242	222	152	51	80	134	75
2000	88	476	431	311	105	180	264	145

Table 12.1

number of cases	p("X causes Z"|D)	output of Bayesian model selection	output of PC algorithm
150	0.036	X and Z unrelated	X and Z unrelated
250	0.123	X and Z unrelated	X causes Z
500	0.141	X causes Z or Z causes X	X and Z unrelated (with inconsistency)
1000	0.593	X causes Z	X causes Z
2000	0.926	X causes Z	X causes Z

Table 12.2

The second column in table 12.2 shows the results of applying equation (12.4) under the assumptions stated above for the first N cases in the data set. When $N = 0$, the data set is empty, in which case probability of hypothesis h is just the prior probability of "X causes Z": 8/25=0.32. Table 12.2 shows that as the number of cases in the database increases, the probability that "X causes Z" increases monotonically as the number of cases increases. Although not shown, the probability increases toward 1 as the number of cases increases beyond 2000. Column 3 in table 12.2 shows the results of applying Bayesian model selection. Here, we list the causal relationship(s) between X and Z found in the model or models with the highest posterior probability $p(\mathbf{m}|D)$. For example, when $N = 500$, there are three models that have the highest posterior probability. Two of the models have Z as a cause of X; and one has X as a cause of Z.

Column 4 in table 12.2 shows the results of applying the PC constraint-based causal discovery algorithm, which is part of the Tetrad II system (Scheines et al. 1994). PC is designed to discover causal relationships that are expressed using DAGs.[7] We applied PC using its default settings, which include a statistical significance level of 0.05. Note that, for $N = 500$, the PC algorithm detected an inconsistency. In particular, the independence tests yielded (1) X and Z are dependent, (2) Y and Z are dependent, (3) X and Y are independent given Z, and (4) X and Z are independent given Y. These relationships are not consistent with the assumption underlying the PC algorithm that the only independence facts found to hold in the sample are those entailed by the Causal Markov condition applied to the generating model. In general, inconsistencies may arise due to the use of thresholds in the independence tests.

There are several weaknesses of the Bayesian-model-selection and constraint-based approaches illustrated by our results. One is that the output is categorical—there is no indication of the strength of the conclusion. Another is that the conclusions may be incorrect in that they disagree with the generative model. Model averaging (column 2)

Prequels and Sequels

does not suffer from these weaknesses, because it indicates the strength of a causal hypothesis.

Although not illustrated here, another weakness of constraint-based approaches is that their output depends on the threshold used in independence tests. For causal conclusions to be correct asymptotically, the threshold must be adjusted as a function of sample size (N). In practice, however, it is unclear what this function should be.

Finally, we note that there are practical problems with model averaging. In particular, the domain can be so large that there are too many models over which to average. In such situations, the exact probabilities of causal hypotheses can not be calculated. However, we can use selective model averaging to derive approximate posterior probabilities, and consequently give some indication of the strength of causal hypotheses.

12.5.5 Methods for Incomplete Data and Hidden Variables

Among the assumptions that we described in section 12.5.1, the one that is most often violated is the assumption that all variables are observed in every case. In this section, we examine Bayesian methods for relaxing this assumption.

An important distinction for this discussion is that of hidden versus observable variable. A *hidden variable* is one that is unknown in all cases. An *observable variable* is one that is known in some (but not necessarily all) of the cases. We note that constraint-based and Bayesian methods differ significantly in the way that they represent missing data. Whereas constraint-based methods typically throw out cases that contain an observable variable with a missing value, Bayesian methods do not.

Another important distinction concerning missing data is whether or not the absence of an observation is dependent on the actual states of the variables. For example, a missing datum in a drug study may indicate that a patient became too sick—perhaps due to the side effects of the drug—to continue in the study. In contrast, if a variable is hidden, then the absence of this data is independent of state. Although Bayesian methods and graphical models are suited to the analysis of both situations, methods for handling missing data where absence is independent of state are simpler than those where absence and state are dependent. Here, we concentrate on the simpler situation. Readers interested in the more complicated case should see Rubin 1978, Robins 1986, Cooper 1995, and Spirtes et al. 1995, 1999.

Continuing with our example using discrete-multinomial likelihoods, suppose we observe a single incomplete case. Let $Y \subset X$ and $Z = X \setminus Y$ denote the observed and unobserved variables in the case, respectively. Under the assumption of parameter independence, we can compute the posterior distribution of θ_{ij} for model structure **m** as follows:

$$p(\theta_{ij} \mid \mathbf{y}, \mathbf{m}) = \sum_z p(\mathbf{z} \mid \mathbf{y}, \mathbf{m}) p(\theta_{ij} \mid \mathbf{y}, \mathbf{z}, \mathbf{m})$$

$$= (1 - p(\mathbf{pa}_i^j \mid \mathbf{y}, \mathbf{m}))\{p(\theta_{ij} \mid m)\} + \sum_{k=1}^{r_i} p(x_i^k, \mathbf{pa}_i^j \mid \mathbf{y}, \mathbf{m}) p(\theta_{ij} \mid x_i^k, \mathbf{pa}_i^j, \mathbf{m})$$

(12.14)

(See Spiegelhalter and Lauritzen 1990, for a derivation.) Each term

$$p(\theta_{ij} \mid x_i^k, \mathbf{pa}_i^j, \mathbf{m})$$

in equation (12.14) is a Dirichlet distribution. Thus, unless both X_i and all the variables in $\mathbf{Pa}_i$ are observed in case $\mathbf{Y}$, the posterior distribution of θ_{ij} will be a linear combination of Dirichlet distributions—that is, a Dirichlet mixture with mixing coefficients

$$(1 - p(\mathbf{pa}_i^j \mid \mathbf{y}, \mathbf{m})) \text{ and } p(x_i^k, \mathbf{pa}_i^j \mid \mathbf{y}, \mathbf{m}), k = 1, \dots, r_i$$

When we observe a second incomplete case, some or all of the Dirichlet components in equation (12.14) will again split into Dirichlet mixtures. That is, the posterior distribution for θ_{ij} will become a mixture of Dirichlet mixtures. As we continue to observe incomplete cases, each missing values for $\mathbf{Z}$, the posterior distribution for θ_{ij} will contain a number of components that is exponential in the number of cases. In general, for any interesting set of local likelihoods and priors, the exact computation of the posterior distribution for θ_m will be intractable. Thus, we require an approximation for incomplete data.

12.5.5.1 Monte-Carlo Methods

One class of approximations is based on Monte-Carlo or sampling methods. These approximations can be extremely accurate, provided one is willing to wait long enough for the computations to converge.

In this section, we discuss one of many Monte-Carlo methods known as *Gibbs sampling*, introduced by Geman and Geman (1984). Given variables $\mathbf{X} = \{X_1, \dots, X_n\}$ with some joint distribution $p(\mathbf{x})$, we can use a Gibbs sampler to approximate the expectation of a function $f(\mathbf{x})$, with respect to $p(\mathbf{x})$, as follows. First, we choose an initial state for each of the variables in $\mathbf{X}$ somehow (e.g., at random). Next, we pick some variable X_i, unassign its current state, and compute its probability distribution given the states of the other $n-1$ variables. Then, we sample a state for X_i based on this probability distribution, and compute $f(\mathbf{x})$. Finally, we iterate the previous two steps, keeping track of the average value of $f(\mathbf{x})$. In the limit, as the number of cases approach infinity, this average is equal to $E_{p(\mathbf{x})}(f(\mathbf{x}))$ provided two conditions are met. First, the Gibbs sampler must be *irreducible*. That is, the probability distribution $p(\mathbf{x})$ must be such that we can eventually sample any possible configuration of $\mathbf{X}$ given any possible initial configuration of $\mathbf{X}$. For example, if $p(\mathbf{x})$ contains no zero probabilities, then the Gibbs sampler will be irreducible. Second, each X_i must be chosen infinitely often. In practice, an algorithm for deterministically rotating through the variables is typically used. Introductions to Gibbs sampling and other Monte-Carlo methods—including methods for initialization and a discussion of convergence—are given by Neal (1993) and Madigan and York (1995).

To illustrate Gibbs sampling, let us approximate the probability density $p(\theta_m|D,\mathbf{m})$ for some particular configuration of θ_m, given an incomplete data set $D = \{\mathbf{y}_1,...,\mathbf{y}_N\}$ and a causal model for discrete variables with independent Dirichlet priors. To approximate $p(\theta_m|D,\mathbf{m})$, we first initialize the states of the unobserved variables in each case somehow. As a result, we have a complete random sample D_c. Second, we choose some variable X_{il} (variable X_i in case l) that is not observed in the original random sample D, and reassign its state according to the probability distribution

$$p(x'_{il} | D_c \setminus x_{il}, \mathbf{m}) = \frac{p(x'_{il}, D_c \setminus x_{il} | \mathbf{m})}{\sum_{x''_{il}} p(x''_{il}, D_c \setminus x_{il} | \mathbf{m})}$$

where $D_c \setminus x_{il}$ denotes the data set D_c with observation x_{il} removed, and the sum in the denominator runs over all states of variable X_{il}. As we have seen, the terms in the numerator and denominator can be computed efficiently (see equation (12.11)). Third, we repeat this reassignment for all unobserved variables in D, producing a new complete random sample D'_c. Fourth, we compute the posterior density $p(\theta_m|D'_c,\mathbf{m})$ as described in equations (12.9) and (12.10). Finally, we iterate the previous three steps, and use the average of $p(\theta_m|D'_c,\mathbf{m})$ as our approximation.

Monte-Carlo approximations are also useful for computing the marginal likelihood given incomplete data. One Monte-Carlo approach, described by Chib (1995) and Raftery (1996), uses Bayes's theorem:

$$p(D|\mathbf{m}) = \frac{p(\theta_m|\mathbf{m})p(D|\theta_m,\mathbf{m})}{p(\theta_m|D,\mathbf{m})} \qquad (12.15)$$

For any configuration of θ_m, the prior term in the numerator can be evaluated directly. In addition, the likelihood term in the numerator can be computed using causal-model inference (Jensen et al. 1990). Finally, the posterior term in the denominator can be computed using Gibbs sampling, as we have just described. Other, more sophisticated Monte-Carlo methods are described by DiCiccio et al. (1995).

12.5.5.2 The Gaussian Approximation
Monte-Carlo methods yield accurate results, but they are often intractable—for example, when the sample size is large. Another approximation that is more efficient than Monte-Carlo methods and often accurate for relatively large samples is the *Gaussian approximation* (e.g., Kass et al. 1988; Kass and Raftery 1995).

The idea behind this approximation is that, for large amounts of data, $p(\theta_m|D,\mathbf{m}) \propto p(D|\theta_m,\mathbf{m}) \times p(\theta_m|\mathbf{m})$ can often be approximated as a multivariate-Gaussian distribution. In particular, let

$$g(\theta_m) \equiv \log(p(D|\theta_m,\mathbf{m}) \times p(\theta_m|\mathbf{m})) \qquad (12.16)$$

Also, define $\overline{\theta}_m$ to be the configuration of θ_m that maximizes $g(\theta_m)$. This configuration also maximizes $p(\theta_m|D,\mathbf{m})$, and is known as the *maximum a posteriori* (MAP) configuration of θ_m. Using a second degree Taylor polynomial of $g(\theta_m)$ about $\overline{\theta}_m$ to approximate $g(\theta_m)$, we obtain

$$g(\theta_m) \approx g(\overline{\theta}_m) - \frac{1}{2}(\theta_m - \overline{\theta}_m)A(\theta_m - \overline{\theta}_m)^t \qquad (12.17)$$

where $(\theta_m - \overline{\theta}_m)^t$ is the transpose of row vector $(\theta_m - \overline{\theta}_m)$, and A is the negative Hessian of $g(\theta_m)$ evaluated at $\overline{\theta}_m$. Raising $g(\theta_m)$ to the power of e and using equation (12.16), we obtain

$$\begin{aligned}p(\theta_m|\mathbf{m},D) &\propto p(D|\theta_m,\mathbf{m})p(\theta_m|\mathbf{m}) \\ &\approx p(D|\overline{\theta}_m,\mathbf{m})p(\overline{\theta}_m|\mathbf{m})\exp\left\{-\frac{1}{2}(\theta_m - \overline{\theta}_m)A(\theta_m - \overline{\theta}_m)^t\right\}\end{aligned} \qquad (12.18)$$

Hence, the approximation for $p(\theta_m|D,\mathbf{m})$ is Gaussian.

To compute the Gaussian approximation, we must compute $\overline{\theta}_m$ as well as the negative Hessian of $g(\theta_m)$ evaluated at $\overline{\theta}_m$. In the following section, we discuss methods for finding $\overline{\theta}_m$. Meng and Rubin (1991) describe a numerical technique for computing the second derivatives. Raftery (1995) shows how to approximate the Hessian using likelihood-ratio tests that are available in many statistical packages. Thiesson (1995) demonstrates that, for multinomial distributions, the second derivatives can be computed using causal-model inference.

Using the Gaussian approximation, we can also approximate the marginal likelihood. Substituting equation (12.18) into equation (12.3), integrating, and taking the logarithm of the result, we obtain the approximation:

$$\log p(D|\mathbf{m}) \approx \log p(D|\overline{\theta}_m,\mathbf{m}) + \log p(\overline{\theta}_m|\mathbf{m}) + \frac{d}{2}\log(2\pi) - \frac{1}{2}\log|A| \qquad (12.19)$$

where d is the dimension of $g(\theta_m)$. For a causal model with multinomial distributions, this dimension is typically given by

$$\prod_{i=1}^{n} q_i(r_i - 1)$$

Sometimes, when there are hidden variables, this dimension is lower. See Geiger et al. (1996) for a discussion of this point. This approximation technique for integration is known as *Laplace's method*, and we refer to equation (12.19) as the *Laplace approximation*. Kass et al. (1988) have shown that, under certain regularity conditions, the relative error of this approximation is $O_p(1/N)$, where N is the number of cases in D. Thus, the Laplace approximation can be extremely accurate. For more detailed discussions of this approximation, see, for example, Kass et al. (1988) and Kass and Raftery (1995).

Although Laplace's approximation is efficient relative to Monte-Carlo approaches, the computation of $|A|$ is nevertheless intensive for large-dimension models. One simplification is to approximate $|A|$ using only the diagonal elements of the Hessian A. Although in so doing, we incorrectly impose independencies among the parameters, researchers have shown that the approximation can be accurate in some circumstances (see, e.g., Becker and Le Cun 1989, and Chickering and Heckerman 1997). Another efficient variant of Laplace's approximation is described by Cheeseman and Stutz (1995) and Chickering and Heckerman (1997).

We obtain a very efficient (but less accurate) approximation by retaining only those terms in equation (12.19) that increase with N; $\log p(D|\bar{\theta}_m, \mathbf{m})$, which increases linearly with N, and $\log |A|$, which increases as $d \log N$. Also, for large N, $\bar{\theta}_m$ can be approximated by $\hat{a}$, the maximum likelihood configuration of θ_m (see the following section). Thus, we obtain

$$\log p(D|\mathbf{m}) \approx \log p(D|\hat{\theta}_m, \mathbf{m}) - \frac{d}{2}\log(N) \qquad (12.20)$$

This approximation is called the *Bayesian information criterion* (BIC). Schwarz (1978) has shown that the relative error of this approximation is $O_p(1)$ for a limited class of models. Haughton (1988) has extended this result to curved exponential models.

The BIC approximation is interesting in several respects. First, roughly speaking, it does not depend on the prior. Consequently, we can use the approximation without assessing a prior.[8] Second, the approximation is quite intuitive. Namely, it contains a term measuring how well the parameterized model predicts the data $\log p(D|\hat{\theta}_m, \mathbf{m})$ and a term that punishes the complexity of the model ($d/2 \log (N)$). Third, the BIC approximation is exactly minus the Minimum Description Length (MDL) criterion described by Rissanen (1987).

12.5.5.3 The MAP and ML Approximations and the Algorithm

As the sample size of the data increases, the Gaussian peak will become sharper, tending to a delta function at the MAP configuration $\bar{\theta}_m$. In this limit, we can replace the integral over θ_m in equation (12.5) with $p(h|\bar{\theta}_m, \mathbf{m})$. A further approximation is based on the observation that, as the sample size increases, the effect of the prior $p(\theta_m|\mathbf{m})$ diminishes. Thus, we can approximate θ_m by the *maximum likelihood* (ML) configuration of θ_m:

$$\hat{\theta}_m = \arg\max_{\theta_m} p(D|\theta_m, \mathbf{m})$$

One class of techniques for finding a ML or MAP is gradient-based optimization. For example, we can use gradient ascent, where we follow the derivatives of $g(\theta_m)$ or the likelihood $p(D|\theta_m, \mathbf{m})$ to a local maximum. Russell et al. (1995) and Thiesson (1995) show how to compute the derivatives of the likelihood for a causal model with multinomial distributions. Buntine (1994) discusses the more general case where the likelihood comes from the exponential family. Of course, these gradient-based methods find only local maxima.

Another technique for finding a local ML or MAP is the expectation—maximization (EM) algorithm (Dempster et al. 1977). To find a local MAP or ML, we begin by assigning a configuration to θ_m somehow (e.g., at random). Next, we compute the *expected* sufficient statistics for a complete data set, where expectation is taken with respect to the joint distribution for $\mathbf{X}$ conditioned on the assigned configuration of θ_m and the known data D. In our discrete example, we compute

$$E_{p(\mathbf{x}|D,\theta_s,\mathbf{m})}(N_{ijk}) = \sum_{l=1}^{N} p(x_i^k, \mathbf{pa}_i^j | \mathbf{y}_l, \theta_m, \mathbf{m}) \qquad (12.21)$$

where $\mathbf{y}_l$ is the possibly incomplete l^{th} case in D. When X_i and all the variables in $\mathbf{Pa}_i$ are observed in case x_l, the term for this case requires a trivial computation: it is either zero or one. Otherwise, we can use any causal-model inference algorithm to evaluate the term. This computation is called the *expectation step* of the EM algorithm.

Next, we use the expected sufficient statistics as if they were actual sufficient statistics from a complete random sample D_c. If we are doing an ML calculation, then we determine the configuration of θ_m that maximizes $p(D_c|\theta_m,\mathbf{m})$. In our discrete example[9] we have

$$\theta_{ijk} = \frac{E_{p(\mathbf{x}|D,\theta_s,\mathbf{m})}(N_{ijk})}{\sum_{k=1}^{r_i} E_{p(\mathbf{x}|D,\theta_s,\mathbf{m})}(N_{ijk})}$$

If we are doing a MAP calculation, then we determine the configuration of θ_m that maximizes $p(\theta_m|D_c,\mathbf{m})$. In our discrete example, we have[7]

$$\theta_{ijk} = \frac{\alpha_{ijk} + E_{p(\mathbf{x}|D,\theta_s,\mathbf{m})}(N_{ijk})}{\sum_{k=1}^{r_i}\left(\alpha_{ijk} + E_{p(\mathbf{x}|D,\theta_s,\mathbf{m})}(N_{ijk})\right)}$$

This assignment is called the *maximization step* of the EM algorithm. Under certain regularity conditions, iteration of the expectation and maximization steps will converge to a local maximum. The EM algorithm is typically applied when sufficient statistics exist (i.e., when local likelihoods are in the exponential family), although generalizations of the EM algorithm have been used for more complicated local distributions (see, e.g., McLachlan and Krishnan 1997).

12.5.6 Open Problems in Latent Variable Search

The Bayesian framework gives us a conceptually simple framework for learning causal models. Nonetheless, the Bayesian solution often comes with a high computational cost. For example, when we learn causal models containing hidden variables, both the exact computation of marginal likelihood and model averaging/selection can be intractable. Although the approximations described in section 12.5.5 can be applied to address the difficulties associated with the computation of the marginal likelihood, model averaging and model selection remain difficult. The number of possible models with hidden variables is significantly larger than the number of possible DAGs over a fixed set of variables. Without constraining the set of possible models with hidden variables—for instance, by restricting the number of hidden variables—the number of possible models is infinite. On a positive note, the FCI algorithm has shown that constraint-based methods under suitable assumptions can sometimes indicate the existence of a hidden common cause between two variables. Thus, it may be possible to use the constraint-based methods to suggest an initial set of plausible models containing hidden variables that can then be subjected to a Bayesian analysis.

Another problem associated with learning causal models containing hidden variables is the assessment of parameter priors. The approach in section 12.5.5 can be applied in such situations, although the assessment of a joint distribution $p(\mathbf{x}|\mathbf{m}_c)$ in which $\mathbf{x}$ includes hidden variables can be difficult. Another approach may be to employ a property called *strong likelihood equivalence* (Heckerman 1995). According to this property, data should not help to discriminate among two models that are distribution equivalent with respect to the nonhidden variables. Heckerman (1995) showed that any method that uses this property will yield priors that differ from those obtained using a prior network.[10]

One possibility for avoiding this problem with hidden-variable models, when the sample size is sufficiently large, is to use BIC-like approximations. Such approximations are commonly used (Crawford 1994; Raftery 1995). Nonetheless, the regularity conditions that guarantee $O_p(1)$ or better accuracy do not typically hold when choosing among causal models with hidden variables. Additional work is needed to obtain accurate approximations for the marginal likelihood of these models.

Even in models without hidden variables there are many interesting issues to be addressed. In this section we discussed only discrete variables having one type of local likelihood: the multinomial. Thiesson (1995) discusses a class of local likelihoods for discrete variables that use fewer parameters. Geiger and Heckerman (1994) and Buntine (1994) discuss simple linear local likelihoods for continuous nodes that have continuous and discrete variables. Buntine (1994) also discusses a general class of local likelihoods from the exponential family for nodes having no parents. Nonetheless, alternative likelihoods for discrete and continuous variables are desired. Local likelihoods with fewer parameters might allow for the selection of correct models with less data. In addition, local likelihoods that express more accurately the data generating process would allow for easier interpretation of the resulting models.

12.5.7 MAG Search and PAG Search

Searching over latent variable DAG models faces several important computational and theoretical difficulties. Structuring the search can be difficult, because in addition to introducing, removing, or orienting edges, it requires deciding when to introduce latent variables. The exact calculation of a posterior distribution is typically computationally intractable. In the Gaussian and discrete cases, it is not known whether the BIC score is an $O_p(1)$ approximation to the posterior in the case of latent variable models (Geiger et al. 1999). In addition, the calculation of the dimension of a latent variable model is computationally expensive (Geiger et al. 1996) And none of this even begins to consider the problem of selection bias.

Some of the search difficulties can be overcome by searching the space of MAGs, rather than the space of latent variable DAGs. First, because every variable in a MAG is observed, a search over MAGs never requires the introduction of latent variables. Second, a MAG represents the conditional independence relations entailed by a DAG with both latent variables and selection bias. Third, in the Gaussian case, it is known how to parameterize MAGs (indeed each Gaussian MAG is a special case of a linear structural equation model—see Richardson and Spirtes 1999) in such a way that the only constraints imposed on the distributions are the conditional independence relations entailed by m-separation. In addition, in the case of a Gaussian MAG model it is known that the BIC score is an $O_p(1)$ approximation to the posterior. Assuming the Causal Markov Condition and a prior over the parameters which assigns zero probability to unfaithful parameter values, in the large sample limit, with probability 1, one of the MAGs with the highest BIC score (there may be several **O**-Markov equivalent MAGs with the same score) represents the true causal DAG with latent variables and selection bias. Moreover, calculating the dimension of a Gaussian MAG model is trivial (Spirtes et al. 1997). Standard structural equation model estimation techniques, available in such programs as EQS (Bentler 1985) and LISREL (Joreskog and Sorbom 1984) can be used to perform maximum likelihood estimates of the parameters. Examples of MAG search applied to actual data are given in Richardson and Spirtes 1999 and Richardson et al. 1999.

It is not currently known how to parameterize a MAG with discrete variables in such a way that the only constraints it imposes (other than the distributional family) are the conditional independence relations entailed by m-separation. However, Richardson (1999) has worked out a local Markov property of MAGs that is equivalent to m-separation, which may provide some guidance in devising a parameterization.

The limitations of searching over the space of MAGs, rather than the space of latent variable DAGs, is that a MAG gives only partial information about the DAGs it represents. Hence, even given the correct MAG, it may not be possible to predict the effects of some manipulations. Furthermore, latent variable DAGs that have very different posterior distributions might be represented by the same MAG; hence a latent variable DAG search, if it were feasible to carry out, could be more informative than a MAG search. And at small samples sizes, the MAG selected as the best may not represent the latent variable DAG that would be selected as the best, if the search over latent variable DAGs were feasible. However, the output of a MAG search could be used as the starting point of a DAG search.

PAGs were introduced as a representation of **O**-Markov equivalence class of DAGs. They can also be interpreted as a representation of an **O**-Markov equivalence class of MAGs. And just as searching over the space of patterns has some advantages over searching over the space of DAGs, searching over the space of PAGs has some advantages over the space of MAGs. However, a BIC (AIC, MDL) score based search over the space of PAGs is still difficult, because the different DAGs represented by a given PAG impose different nonindependence constraints on the margin, and hence receive different BIC (AIC, MDL) scores on the same data. In contrast, every MAG represented by a given PAG has the same BIC score for a given data set (because MAGs impose no nonindependence constraints on the margin.) Hence one can score a PAG by turning it into an arbitrary MAG that is represented by the PAG, scoring the MAG, and assigning that score to the PAG. The PAG score is not necessarily the highest BIC score among all of the DAGs represented by the PAG, but assuming the Causal Markov Condition and a prior over the parameters which assigns zero probability to unfaithful parameter values, in the large sample limit with probability 1, the PAG representing the true causal graph will have the highest score. Given a score for PAGs, it is possible to do a hill-climbing score-based search over the space of PAGs. A score-based PAG search of this kind is described in more detail in Spirtes et al. 1996.

12.5.8 Search over Cyclic Directed Graphs

Richardson (1996a, 1996b) describes constraint based methods of search over cyclic directed graphs, where it is assumed that the natural extension of d-separation to cyclic directed graphs characterizes the conditional independence constraints entailed by the graph. The input to Richardson's algorithm is a data set generated by an unknown cyclic directed graph G, which is used to test d-separation relations in G by performing tests of conditional independence. The output is a PAG with respect to a Markov equivalence class of directed graphs. The algorithm is polynomial in the number of variables if the maximum number of adjacencies of a vertex in all graphs is constant. The algorithm is correct with probability 1 in the large sample limit, assuming the Causal Markov and Faithfulness Conditions. For example, if the data is generated by the directed cyclic graph in figure 12.1, in the large sample limit with probability 1 the output of the algorithm is the PAG with respect to the Markov equivalence class of figure 12.1, which is shown in figure 12.4.

Score-based searches over cyclic directed graphs face some of the same problems that score-based searches over latent variable models face. Linear models represented by cyclic directed graphs entail nonconditional independence constraints. It is not known whether cyclic directed graphs represent curved exponential families, or whether the conditions under which BIC is an $O_p(1)$ approximation to the posterior distribution obtain.

12.5.9 Other Approaches to Search

One of the major obstacles to searching the space of DAGs is the problem of local maxima. There are several kinds of algorithms which can be used to overcome the problem of local maxima. For example, a Bayesian network search tecnhique which uses both genetic algorithms and simulated annealing is given in De Campos and Huete (1999).

Genetic algorithms are intended to mimic natural selection. Each individual is a potential solution to a problem, the set of individuals is a population, and there is a function which measures the fitness of each individual. An initial population is created, and then the most fit individuals are combined to get new individuals (crossover). Individuals can also spontaneously change (mutation) in order to get out of local maxima. The new individuals are added to the population, and the process is repeated for a fixed number of generations. The most fit individuals are then selected.

In a simulated annealing algorithm, there is a system of N variables and an "energy" E which is a function of the configuration c_i of the N variables, and which is to be minimized. In one step of the algorithm, a new configuration is generated by randomly perturbing the previous configuration. If the perturbation decreases the energy, the change is accepted. If the perturbation increases the energy it is accepted with probability $\exp(\Delta E/T)$, where T is a "temperature" parameter that is systematically decreased as the number of iterations increases. This allows the algorithm to get out of local maxima.

Stopping criteria can be functions of the energy, the temperature, or the number of iterations.

Wedelin (1996) describes a search based on MDL (Minimum Description Length). The search proceeds in two steps. First, the algorithm searches for an undirected graph (representing a random Markov field), and then the undirected graph is oriented, if possible. The original set of variables is transformed, and then the search starts by looking for first order interactions among the transformed variables. Search for higher order interactions is based on the heuristic that if there are $k-1$ order interactions among $\mathbf{Z}_1$ and $\mathbf{Z}_2$, and $\mathbf{Z} = \mathbf{Z}_1 \cup \mathbf{Z}_2$, then a k order interaction among the variables in $\mathbf{Z}$ is tested. If the k order interaction is found, then $\mathbf{Z}$ is made in to an undirected clique. Once the undirected graph is found, the algorithm takes all cliques of size greater than or equal to 3, and tests each possible ordering of the variables. If the test eliminates all but one of the directions, those orientations are added to the indirected graph.

Wallace et al. (1996) and Dai et al. (1997) describe a search over linear structural equation DAG models that is based on a minimum message length score, where message length is a joint encoding of the sample data and the causal model. The total message length can be expressed as the sum of the message length for data given the causal model plus the message length of the causal model; the latter in turn can be broken into the message length encoding the DAG, and the message length encoding the DAG parameters. (For larger models, they are not able to calculate the exact score.) In their encoding, Markov equivalent DAGs can receive different minimum message length scores. They report that when the edges in a DAG are weak, that at small sample sizes an MML based search out-performs the PC algorithm when the signficance level is set to 0.05, although they do not test whether the difference is statistically significant. (We generally have found that PC works better at small sample sizes when the significance level is set to higher than 0.05.)

Friedman (1997) considers the case where data are missing or there are hidden variables, and bases a search on a modification of the EM algorithm described in section 12.5.5.3. The structural EM algorithm maintains a current Bayesian network candidate, and at each iteration of the EM algorithm it estimates the sufficient statistics that are needed to evaluate alternative networks. Since the evaluation is done from complete data, Bayesian networks search techniques designed for no missing data can be used can be used at this point to look for improved structures. Thus the search for structure is interleaved into steps of the EM algorithm. In Boyen et al. 1999 the structural EM algorithm is applied to learning Bayesian networks representing dynamic systems. Ramoni (1996) also describes a Bayesian network search when there is missing data.

Friedman et al. (1999c) propose a class of algorithms called "sparse candidate" searches for searches over Bayesian networks without hidden variables. First, the set of possible parents of each vertex is restricted to a small number of candidates. Then, the procedure searches for the best Bayesian network that satisfies the candidate constraints. The best Bayesian network found is then used to generate a new set of possible candidates for each vertex. For example, if X and Y are selected as the initial candidate parents of Z, but X is not a parent of Z in the best Bayesian network with this restriction, at the next stage, another variable with a weaker connection to Z can replace X as a candidate parent.

Discretization of continuous variables can be considered a kind of nonparametric estimation technique. One problem with discretization is that continuous variables which are conditionally independent may have discretized counterparts that are not conditionally independent; preserving at least approximate conditional independence is important if the discretized variables are to be used to construct a Bayesian network that approximates the Bayesian network among the underlying continuous variables. So choosing a discretization policy that takes into account the interactions of the variables is important for Bayesian network search. Friedman and Goldszmidt (1996) propose a discretization policy that is based on MDL. Monti and Cooper (1998) represent discretization as a process that is itself represented in a Bayesian network B_D that is a modification of a Bayesian network B among the underlying continuous variables. Hence different discretization policies corresponding to different parameterization of B_D can be evaluated by the posterior probability of the network. However, this also implies that during search, when an alternative Bayesian network B' among the underlying continuous variables is considered, the Bayesian network B'_D representing the discretization process also changes, and the discretization policy has to be re-evaluated.

12.5.10 Attitude toward the Output of Search Algorithms

Some of the algorithms we have described are known (assuming the Causal Markov and Faithfulness Conditions) to pointwise converge to the truth in the large sample limit, while others are not. In either case, in practice, some of the assumptions made by the search algorithm applied will typically be only approximately true, and the sample size will not be infinite. What attitude should one have toward the output of causal search algorithms in these circumstances? First we will consider constraint based searches, and then we will consider Bayesian searches.

12.5.10.1 Constraint Based Search Algorithms

The power of the constraint based search algorithms against alternative models is an unknown and extremely complex function of the power of the statistical tests that the algorithm employs, and the distribution over the models tested. For that reason, the best answer that we can give about the reliability of these algorithm is based upon simulation studies, and actual cases (chapter 5, chapter 8, section 12.8). We, and others, have provided the results of a variety of simulation tests. The simulation studies should be interpreted as an upper bound on the reliability of the particular algorithm used, because in general the distributional assumptions are exactly satisfied in the simulations, and if a causal connection exists between variables in a study, we have limited how weak that causal connection can be. These studies suggest that one should be skeptical about the output at very small sample sizes, or when in the output there are variables with a large number of parents.

In general, the correctness of the output of a constraint based search depends upon nine factors:

1. The correctness of the background knowledge input to the algorithm (e.g., an initial starting model or no feedback.)
2. How closely the Causal Markov Condition holds (e.g., no interunit causation, no mixtures of subpopulations in which causal connections are in opposite directions).
3. How closely the Faithfulness Condition holds (e.g., no deterministic relations, no attempt to detect very small causal effects).
4. Whether the distributional assumptions made by the statistical tests hold (e.g., joint normality.)
5. The power of the statistical tests against alternatives.
6. The significance level used in the statistical tests.
7. The sample size.
8. The sampling method.
9. The sparseness of the true graphical model.

We do not have a formal mechanism for combining these factors into a score for the reliability of the output. However, there are some steps that can be taken to evaluate the output of the searches we have discussed.

Some of the factors that affect the reliability of the results can be judged from background knowledge. For example, the output may contain an edge which is known on substantive grounds not to exist (because e.g., it points from an earlier event to a later event.) Or the output may indicate that a distributional assumption has been violated. For example, in the education and fertility example of chapter 5 (Rindfuss et al. 1980), the variables of interest (education and age at which first child is born) can both be treated as continuous), but other variables, such as race and whether or not one lived on a farm, can not. The PC algorithm was run under the assumption of linearity. The edges of interest in that case point into education and age at first child, and are compatible with the assumption of linearity. However, other edges in the output pointed from continuous variables to binary variables, and hence are problematic because they indicate a violation of the assumption of linearity that the algorithm was run under.

In addition, the output may be very sensitive to the significance level chosen. Thus in the Spatina biomass example of chapter 8 (Rawlings 1988), the $pH \rightarrow BIO$ (where BIO represents the biomass of the grass) edge was quite robust over different significance levels, but the edges that appeared among the other variables changed at different significance levels.

It is also possible to test the output by various kinds of cross-validation. In chapter 8 we recommended performing a kind of parametric bootstrapping in which the search algorithm is run on a sample, the output of the search algorithm is turned into a DAG, the parameters of the DAG model are estimated, and Monte Carlo simulation techniques are used on the resulting parameterized DAG models to generate further samples. The search algorithm is then run on the additional samples, and the percentage of the time the search algorithm finds some feature of interest is calculated. We performed such a parametric bootstrap on the Weisberg (1985) rat liver data. In nonparametric bootstrapping, repeated subsamples of size N are drawn with replacement from the original sample, the search algorithm is run on each of the subsamples of size N, and the percentage of the time the search algorithm finds some feature of interest is calculated. Shipley (1997) applied nonparametric bootstrapping to search algorithms on small sample sizes. Friedman et al. (1999a and 1999b) also discuss the application of parametric and nonparametric bootstrapping to Bayesian network search.

In some cases the output of our search methods can be turned into a model on which a statistical test can be performed (as in the case of linear models.) In such cases, if there is a particular feature of interest, such as the existence of an edge from X to Y, a search can be run twice, once with the feature required, and once with the feature forbidden, and the two results compared; for example, one might pass a statistical test, while the other might fail a test. Alternatively the p-values can be used as a kind of informal score for the two models.

12.5.10.2 Score based Search Algorithms

For a Bayesian who can calculate the posterior of each causal model from a prior that represents his degrees of belief before seeing the data, it is clear how much confidence to put in each causal model. However, in practice, Bayesian searches cannot calculate posterior probabilities of causal models, they can only calculate the ratios of posteriors of different causal models, and priors are heavily influenced by mathematical convenience rather than conviction. This still leaves the question of how much confidence one should put in the output of a Bayesian (or other score-based) search algorithms.

Most of the same considerations that were used to judge the output of constraint based searches can also be used to judge the output of score-based searches. However, score-based searches have the major advantage that any two models in the space searched can be compared, and the investigator can get a sense of whether one model is overwhelmingly preferred to every other model visited during the search, or only slightly better than some alternatives.

12.6 Finite Samples

The question we consider here is the following: given a choice between the models of figure 12.11, what are the qualitative features of a prior distribution that, conditional on a small sample correlation between B and C, has a resulting posterior that places a high probability on the treatment effect of B on C being small? Since the FCI algorithm draws the conclusion that the treatment effect of B on C is zero when the sample correlation is small enough, this is related to the question of what qualitative features of prior distributions would make the output of the FCI algorithm (as typically employed)[11] a good approximation of Bayesian updating.

Note that for "approximate agreement" between the FCI algorithm we do not demand that the posterior place a high probability on the Markov equivalence class output by the FCI algorithm being exactly true. This is because in many cases, concluding that a treatment effect is zero when it is actually very small is of no practical significance. (However, there may be cases, especially in the medical domain where very small effects are important.) In addition, note that "approximate agreement" is defined here only for the kind of simple cases considered in figure 12.11. We leave it as an open problem to generalize this concept to more complex cases.

The prior over BΓ (defined in Section 12.4) has two distinct parts, the prior over the parameters given a DAG and the prior over the DAGs. We will discuss each of these in turn. Because the plausibility of a prior depends upon both the prior over the parameters and the prior over the DAGs, we will comment on the plausibility of various *combinations* of DAG priors and DAG parameter priors after we have pointed out the properties of the priors over the parameters.

There are three basic qualitative results that will be described in the following sections. First, the geometry of the parameter space favors small values of $|z|$ conditional on $\rho(B,C) = 0$ (that is even given a uniform distribution over the parameters, conditioning on $\rho(B,C) = 0$ increases the probability of small values of $|z|$.) Second, while there is one superficially plausible kind of prior probability P that leads to a high prior on "close to unfaithful" distributions, this prior also has the unintuitive consequence that there is almost certainly no significant confounding due to hidden variables. And finally, an obvious modification of P which avoids the unintuitive consequence that there is almost certainly no significant confounding due to hidden variables, is also a prior which gives a high posterior probability of a small value of $|z|$ conditional on a small value of $\rho(B,C)$.

12.6.1 The Prior over the Parameters[12]

In Model M, when $\rho(B,C)$ is zero, z ranges anywhere between $-\infty$ and ∞. However, when $z = 0$ and $\rho(B,C) = 0$, x can take on any value between -1 and 1. In contrast, as $|z|$ approaches infinity and $\rho(B,C) = 0$, the only legitimate values of $|x|$ approach 1. This suggests that even with a uniform prior over the legitimate parameter values of x, y, and z, $\rho(B,C) = 0$ favors small values of $|z|$. In order to calculate $f(z|\rho(B,C) = 0)$, where f is a uniform density over the legitimate values of x, y, z, the variables x, y, z can be transformed to $r_1, r_2,$ and r_3 in the following way:

$$r_1 = z + x \times y \qquad x = (r_1 - r_3)/r_2$$
$$r_2 = y \qquad y = r_2$$
$$r_3 = z \qquad z = r_3$$

r_1 is equal to $\rho(B,C)$. Let $|J|$ be the absolute value of the Jacobian of the transformation.

$$|J| = \det \begin{pmatrix} \frac{1}{r_2} & \frac{r_3 - r_1}{r_2^2} & -\frac{1}{r_2} \\ 0 & 1 & 0 \\ 0 & 0 & 1 \end{pmatrix} = \left|\frac{1}{r_2}\right|$$

When $\rho(B,C) = 0$, $z = -x \times y$. Because of the constraints on the variances, x varies from -1 to 1, and it follows that for a given value of z, y varies from $|z|$ to $\sqrt{z^2 + 1}$, and from $-|z|$ to $-\sqrt{z^2 + 1}$. Also, when $\rho(B,C) = 0$, z varies from $-\infty$ to ∞. Hence, when $r_1 = 0$, for a given value of r_3, r_2 varies from $|r_3|$ to $\sqrt{r_3^2 + 1}$, and from $-|r_3|$ to $-\sqrt{r_3^2 + 1}$; and when $r_1 = 0$, r_3 varies from $-\infty$ to ∞. For a uniform density, $f(x, y, z)$ is a constant c. In the transformed variables, $f(r_1, r_2, r_3) = |c/r_2|$. Hence one natural version of the conditional density is

$$f(z \mid \rho(B,C) = 0) = f(r_3 \mid r_1 = 0) =$$

$$\frac{f(r_1 = 0, r_3)}{f(r_1 = 0)} = \frac{c \left(\int_{|r_3|}^{+\sqrt{r_3^2+1}} \frac{dr_2}{r_2} + \int_{-|r_3|}^{-\sqrt{r_3^2+1}} \frac{-dr_2}{r_2} \right)}{c \left(\int_{-\infty}^{\infty} \int_{|r_3|}^{+\sqrt{r_3^2+1}} \frac{dr_2}{r_2} dr_3 + \int_{-\infty}^{\infty} \int_{-|r_3|}^{-\sqrt{r_3^2+1}} \frac{-dr_2}{r_2} dr_3 \right)} = 0.3183298862 \times \log\left|\frac{\sqrt{r_3^2+1}}{r_3}\right|$$

The uniform cumulative distribution of $|z|$ conditional on $\rho(B,C) = 0$ is shown in figure 12.14. Notice that conditional on $\rho(B,C) = 0$, the uniform measure tends to favor smaller values of $|z|$. For example, the probability of $|z| < 0.2$ is approximately 0.33. In figure 12.15 we compare the marginal distribution of $|z|$ and the distribution of $|z|$ conditional on $\rho(B,C) = 0$ for the uniform prior.

Note that the probability of $|z|$ being larger than 1 is approximately .28, and that while conditioning on $\rho(B,C) = 0$ substantially increases the probability of $|z|$ being less than 0.5, it does not substantially change the probability of $|z|$ being less than 1. Priors that put less mass on large values of x, y, z where $\rho(B,C) = 0$ than a uniform measure does, tend to increase the concentration of the posterior around $|z| = 0$ when $\rho(B,C) = 0$. When $|z|$ is large, the constraints on the variances of the observed variables also imply that $|y|$ is large, and the $|x|$ is near 1. Hence any prior which makes a large value of $|y|$, or a value of $|x|$ near 1 unlikely also makes the probability of a large value of $|z|$ unlikely. (Note that this analysis assumes Model M is true; placing positive probability on Model N being true greatly increases the probability of $|z|$ being small, as one would expect.)

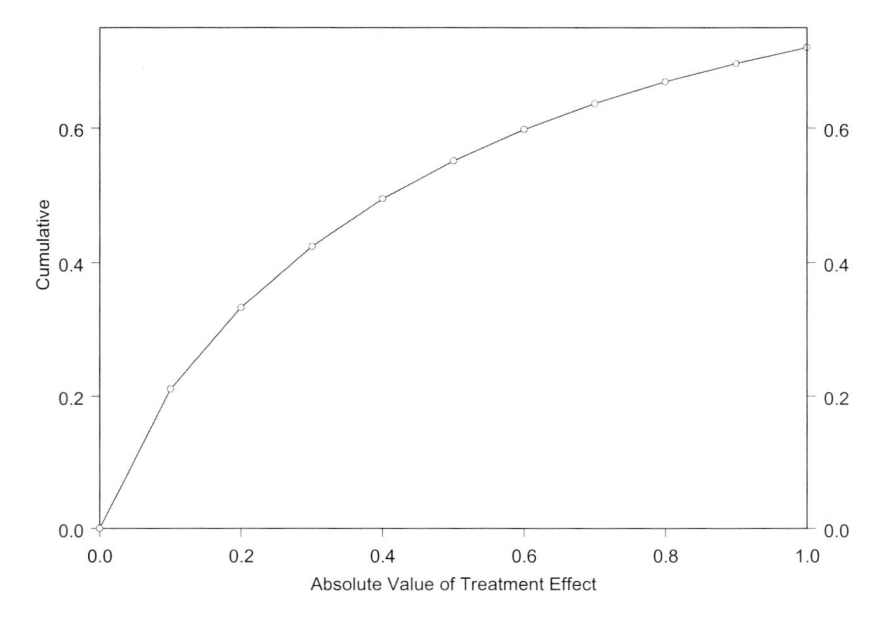

Figure 12.14. Cumulative distribution over $|z|$ conditional on $\rho(B,C) = 0$

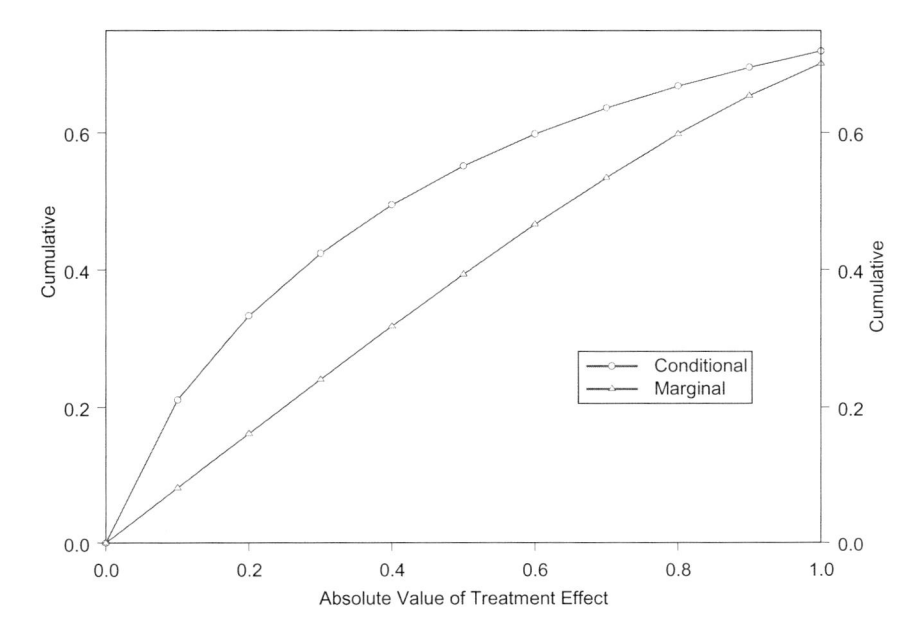

Figure 12.15. Comparing marginals and conditionals

12.6.2 The Prior over the Parameters of a Variable with Many Parents

Suppose that B and C are two time-ordered measured variables, and that B and C have k exogenous common causes U_1 through U_k, where each U_i and ε_B and ε_C has an independent standard Gaussian distribution with

$$B = \sum_{i=1}^{k} \beta_i U_i + \beta_0 \varepsilon_B \quad C = \sum_{i=1}^{k} \delta_i U_i + \delta_0 \varepsilon_C$$

It follows that if B and C have mean 0,

$$\text{var}(B) = E(B^2) = \sum_{i=0}^{k} \beta_i^2 \quad \text{var}(C) = E(C^2) = \sum_{i=0}^{k} \delta_i^2 \quad \text{cov}(B,C) = \sum_{i=1}^{k} \beta_i \delta_i$$

We will examine the consequences of several different kinds of prior distributions over the linear coefficients.

1. Independent Standard Gaussians

If the prior over the β and δ parameters are independent standard Gaussian distributions, then the prior distributions over var(B) and var(C) are χ^2 distributions with $k + 1$ degrees of freedom. It follows that in the prior over var(B) and var(C) the mean of var(B) and var(C) is $k + 1$, and the variance of var(B) and var(C) is $2(k + 1)$. Hence, both the mean and the variance of var(B) and var(C) approach ∞ as k approaches ∞. Also, it follows that the mean of cov(B,C) is zero, and the variance of cov(B,C) approaches ∞ as k approaches ∞. However, simulations (see figure 12.16) show that while the mean of $\rho(B,C)$ is zero, the variance of $\rho(B,C) \approx 1/k$. Thus, the distribution over the correlation is quite different than the distribution over the covariances, because the variance of $\rho(B,C)$ approaches zero as k approaches ∞. This means that the prior probability of significant confounding conditional on large k is small. The consequences of this kind of prior in combination with a prior over DAGs will be discussed in section 12.6.3.

2. Independent Gaussians with Variance $1/(k + 1)^2$

Suppose that the prior distributions over β and δ are independent Gaussians with mean 0 and variance $1/(k+1)^2$, where k is the number of latent variables. (This is equivalent to drawing each β and δ from a standard Gaussian, and multiplying the value drawn by $1/(k+1)$. This multiplication decreases the sample mean by a factor of $1/(k+1)$, and the sample variance by $1/(k+1)^2$.) Hence the mean of var(B) and var(C) is 1, regardless of k. However, the variance of var(B) and var(C) approaches zero as k approaches ∞. In addition, the mean of $\rho(B,C)$ is zero, and the variance of $\rho(B,C) \approx 1/k$, so the variance of $\rho(B,C)$ approaches zero as k approaches ∞. These facts are summarized in table 12.3. This implies that the prior probability of significant confounding conditional on large k

(i.e., many parents) is small. The consequences of this kind of prior in combination with a prior over DAGs will be discussed in section 12.6.3.

3. Place Prior Directly Over Mean and Variance

If the parents of a pair of observed variable are unobserved, it is possible to directly specify a prior over the variances and correlations of the observed variables, instead of deriving such a prior from a prior distribution over DAGs and DAG parameters. This represents the combined effect of all of the different latents as a single latent. However, if the parents are themselves observed, it is necessary to have a prior distribution over the DAGs and the DAG parameters.

4. Correlated Standard Gaussians

If the linear coefficients are correlated, there exist induced prior distributions over the variances of variables with many parents that have neither very high mean nor very low variance, unlike the priors discussed in *1* and *2*. For example, a prior could require that if 5 linear coefficients are large, then all of the others are almost certainly very much smaller. (If instead of placing a prior over linear coefficients, the prior is placed over standardized linear coefficients, then the prior necessarily correlates some coefficients being large with other coefficients being small, since the variance of each measured variable is 1.) If an edge coefficient in a model M_1 is very close to zero, M_1 can be approximated by a model M_2 in which the edge coefficient is zero (i.e., in which the edge is actually removed from the corresponding DAG.) Hence, a prior in which the linear coefficients are correlated in such a way that the probability is very large that the vast majority of edges from confounding latents have almost zero coefficients is approximately the same as a prior distribution in which probability is very large that the vast majority of edges from confounding latents have exactly zero coefficients. (This assumes that the coefficients are small enough that even the combined effect of a large number of them is negligible.) But the latter prior is a prior which places a high probability on there not being many confounders. So there are priors in which the linear coefficients are correlated in such a way that the prior approximates a prior over DAGs which places a high probability on there being few confounders; we will call these "approximately simple correlated priors."

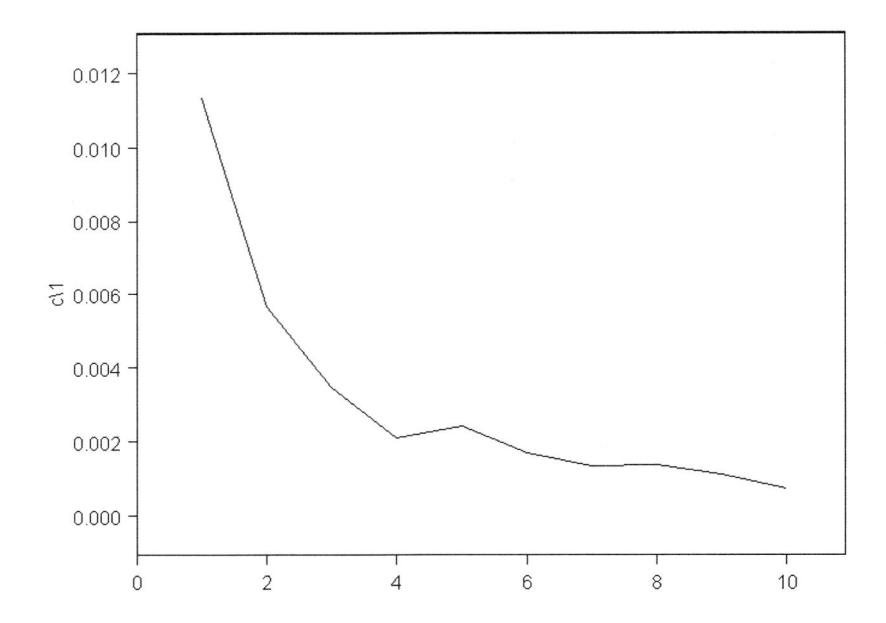

Figure 12.16. Variance of $\rho(B,C)$

Prior over Linear Coefficients				
	N(0,1)		N(0,1/(k+1)^2)	
	mean	variance	mean	variance
var(B)	∞	∞	1	0
var(C)	∞	∞	1	0
cov(B,C)	0	∞	0	0
$\rho(B,C)$	0	0	0	0

Table 12.3

A prior with correlated coefficients can also place a lower probability on some combinations of parameters that are almost unfaithful than does a corresponding prior that does not correlate the values of x, y, and z. For example in Model M, an almost unfaithful set of parameters occurs when $|z|$ is large and $\rho(B,C)$ is small; this occurs when $|x|$ and $|y|$ are also large. Hence a prior that correlates low values of $|x|$ and $|y|$ with large values of $|z|$, has a smaller probability of almost unfaithful sets of parameters than does a corresponding prior that does not correlate x, y, and z.

12.6.3 Prior Over DAGs

In this section, we will examine how the different priors over the DAG parameters described in section 12.6.1 interact with different priors over the DAGs.

1. Equal Probabilities For DAGs

The FCI algorithm outputs a Markov equivalence class of DAGs, rather than a single DAG. Let *FM* be the Markov equivalence class of Model *M*, and *FN* be the Markov equivalence class of Model *N* of figure 12.11. Any prior *P* such that the posterior probability of *FN* (conditioned on a small sample correlation) is extremely small compared to the posterior probability of *FM* (conditional on a small sample correlation), will not approximate the behavior of the FCI algorithm. However, as Robins and Wasserman (1999) point out, for a fixed number of possible unmeasured common causes, there are many more DAGs in the Markov equivalence class of Model *M* than there are in the Markov equivalence class of Model *N*.[13] Consider the following simplified extension of Model *M*. Suppose there are k exogenous standardized latent variables $U_1,...,U_n$, as well as the observed variables B and C. (Because the U_i are exogenous, there are no edges between them, simplifying the calculations.) Then for each latent variable U_i, there are four possible cases: (i) there are edges from U_i to B and C, or (ii) there is no edge from U_i to B but an edge from U_i to C, or (iii) there is no edge from U_i to C but an edge from U_i to B, or (iv) there are no edges out of U_i. In order to belong to the Markov equivalence class of Model *N*, there is no edge from B to C, and for each U_i, one of cases (ii), (iii), or (iv) holds. So there are 3^k DAGs in the Markov equivalence class of Model *N*. There are 2×4^k DAGs total (because each combination of latents can either have the edge from B to C or not). Hence a prior that puts equal weight on each DAG, places a prior on a DAG being in the Markov equivalence class of Model *N* of $1/2 \times (3/4)^k$. With this prior, even though an observed small correlation might boost the probability of the Markov equivalence class of Model *N* a great deal, it will not make it more probable than the Markov equivalence class of Model *M*, except at very large sample sizes. In other words, given a prior that places approximately equal weight on each DAG, the sample size not only has to be large in order for this prior to approximate the results of the FCI algorithm, it has to be large relative to the number of possible confounders.

One problem with placing equal probability over DAGs is that the prior places a high probability on the true DAG being complex (i.e., with many edges.) Thus the marginal prior (over all of the DAGs) approximates the prior conditional on a complex DAG. But if the DAG is complex, and there are independent standard Gaussians over the coefficients of parents of B, then the variance and the mean of var(B) both approach ∞.

On the other hand, if there are independent Gaussians with variances $1(k+1)^2$ over the coefficients of parents of B, then the variance of var(B) approaches 0. Neither of these alternatives seems plausible. Approximately simple correlated priors avoids both these problems, but then has the consequence that the prior probability that the actual distribution can be closely approximated by a simple DAG is high. While such a prior places very low probability on the output of the FCI algorithm being exactly correct, it can also place a high probability on the output of the FCI algorithm (in terms of the treatment effect of B on C) being approximately correct.

2. Equal Probabilities for Structural Classes of DAGs

In some cases, it makes sense to consider the number of distinct alternative causal structures to be less than the number of distinct DAGs. Suppose there are two unmeasured common causes U_1 and U_2. In DAG G_1 there are edges from U_1 to both B and C while there are no edges out of U_2. In DAG G_2 there are edges from U_2 to both B and C while there are no edges out of U_1. Are these two graphs really describing different facts, or should G_2 simply be considered a relabeling of G_1? If the list of possible unmeasured common causes is a list of actual variables such as Intelligence or Socio-Economic Status then clearly G_1 and G_2 are describing different possible facts. If however, someone has no *particular* unmeasured common causes in mind, then G_2 is simply a relabeling of G_1 and they should not count as two distinct DAGs. So we should consider alternatives to the priors that put equal probability on each DAG.

Given a set of k exogenous unmeasured variables, and two time ordered measured variables B and C, say that two DAGs are in the same **structural class** if they have the same number of unmeasured variables which are parents of both B and C, the same number of unmeasured variables which are parents of B but not C, the same number of unmeasured variables which are parents of C and not B, the same number of unmeasured variables which are parents of neither B nor C, and the same number of edges (0 or 1) from B to C. The total number of different structural classes with no latent confounding and no edge from B to C (i.e., in the same Markov equivalence class as Model N) is equal to

$$\sum_{r=0}^{k} k-r+1 = \binom{k+2}{2}$$

This is because if there is no latent confounding, each latent variable falls into one of three classes (is a parent of B but not C, a parent of C but not B, or is a parent of neither). If there are r latent variables in the first class, then the remaining $k - r$ latents can be divided among the two remaining classes in $k - r + 1$ different ways.

The total number of structural classes is

$$2\sum_{s\neq 0}^{k}\sum_{r\neq 0}^{k-s} k-r-s+1 = 2\binom{k+3}{3}$$

The reasoning is similar to the previous case. The factor of 2 occurs because each possible structural class of latent variables may be combined either with an edge from B to C or no edge from B to C.

The ratio of the number of structural classes in the Markov equivalence class of Model N to the total number of structural classed is

$$\binom{k+2}{2} \Big/ 2\binom{k+3}{3} = \frac{3}{2(k+3)}$$

For a given k, the prior that places equal probability on each structural class puts a much higher probability on the Markov equivalence class of Model N than does the prior which places equal probability on each DAG. Nevertheless, for very large k, the prior that places equal probability on each structural class still places a relatively low probability on the Markov equivalence class of Model N.

3. Higher probability on Simple DAGs

A prior that places higher probability on simpler DAGs than complex DAGs (i.e., with many edges) more closely approximates the behavior of the FCI algorithm, because it makes up for the greater number of DAGs that occur in a Markov equivalence class with many edges, by making those greater number of DAGs less probable. Such a prior also implies that while the induced prior distribution over the variances and correlations of observed variables *conditional* on a DAG G in which the observed variables have many parents has very low variance (assuming the linear coefficients are uncorrelated). However, the marginal induced prior (over all of the DAGs) over the variances and correlations of observed variables does not necessarily have either very low variance or very high variance.

12.7 Structural Equation Models

There have been many developments in SEM theory since 1993, most of which we cannot cover here. We will focus on work that extends the ideas in chapter 10. The MIMbuild procedure described in chapter 10 uses vanishing tetrad differences to test 0 and 1st order independence among the latent variables of a SEM with a pure measurement model. Spirtes (1996) generalized MIMbuild so that it can now test independence relations of any order among the latent variables of a SEM with a pure measurement model. This allows, in effect, the PC or FCI algorithm to be applied to the latent variables in a SEM. The procedure has been tested on simulated data, and performs well at large sample sizes with data generated by models that satisfy the assumptions required by the algorithm (Spirtes 1996). In another development, Scheines, Boomsma, and Hoijtink (1999) applied Markov Chain Monte Carlo methods in order to do Bayesian estimation of SEMs, and the technique has been used to make inferences about the effect of Lead exposure on IQ in children (Scheines 1997).

12.7.1 Generalizing MIMbuild

In a "pure" measurement model, each indicator variable measures exactly one latent variable, and is d-separated from every other variable in the model by its associated latent. This corresponds to the Local Independence Assumption in IRT models, Latent Class models, and other Factor Analytic models. Anderson and Gerbing (1982) recommend a two-step model search in which the first step detects whether a measurement model is "uni-dimensional" (or in the terminology of chapter 10 "pure"), and then if the measurement model is pure, conducts a search over connections between latent variables. They stated necessary but not sufficient conditions for purity. In chapter 10, and in Scheines 1993, we describe (assuming faithfulness) necessary and sufficient conditions for there being at least three pure indicators for each latent variable in a linear SEM model, and describe a search for finding a pure measurement model that is a sub-model of the original measurement model, if one exists.

For example, figure 12.17 (A) shows a pure measurement model and (B) shows an impure one. The novelty of the Purify procedure is that in the multivariate Gaussian case it allows an initially specified measurement model to be modified until it can be confirmed by the data to be pure without making any assumptions about the causal structure among the latent variables. The General MIMbuild procedure begins with a pure measurement model, and constructs test models to investigate independence of *any* order among the latents.

Prequels and Sequels

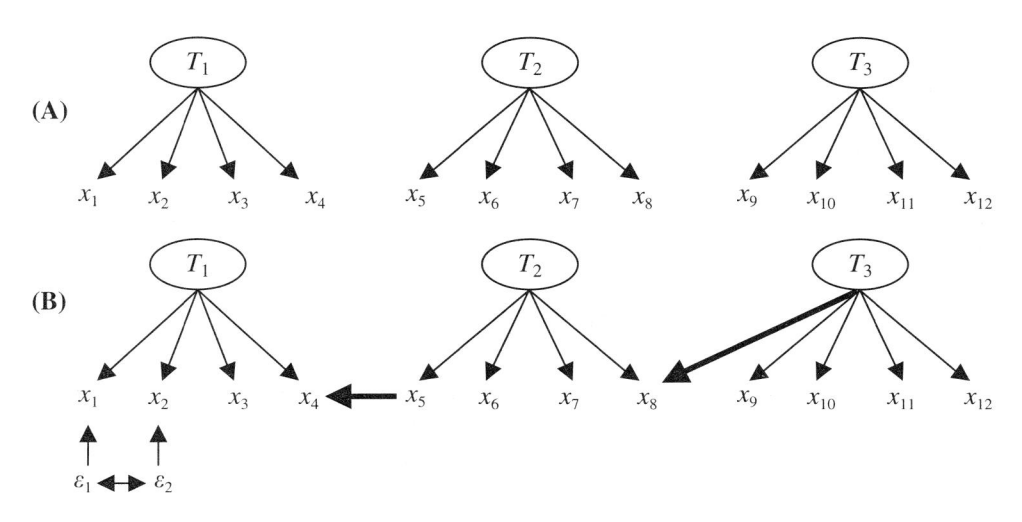

Figure 12.17. Pure and impure measurement models

Suppose that we have a pure measurement model with latents $\mathbf{L} = \{L_1 \ldots L_k\}$. This means that for each latent L_i, there is a a set of pure indicators $I(L_i) = \{X_{i1}\ldots X_{im}\}$. Suppose that we wish to test $L_i \perp\!\!\!\perp L_j | \mathbf{Q}$, where $\mathbf{Q} \subseteq \mathbf{L}$ and contains neither L_i nor L_j. The strategy is to construct two nested SEMs containing L_i, L_j, $\mathbf{Q}$, and their measurement models, such that testing one model against the other is a test of the constraint $L_i \perp\!\!\!\perp L_j | \mathbf{Q}$.

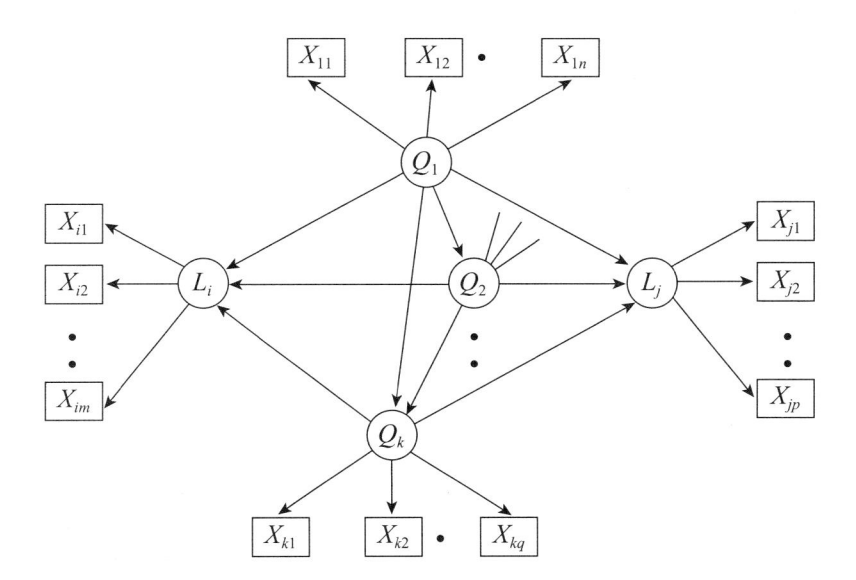

Figure 12.18. Model M_0 for testing $L_i \perp\!\!\!\perp L_j | \mathbf{Q}$.

The simpler model M_0 is constructed so that there is a complete graph among the variables in **Q** (it does not matter which complete graph) and there is an edge from each variable in **Q** both to L_i and to L_j, but no edge from L_i to L_j. (See figure 12.18.) The model M_1 is the same as M_0, except that it also includes the edge $L_i \rightarrow L_j$. The models can be compared by the difference in their χ^2 score, which itself is distributed as χ^2 distribution with one degree of freedom (Bollen 1989). Alternatively, one can simply estimate model M_1 and perform a significance test on the parameter associated with the edge $L_i \rightarrow L_j$.

12.7.2 Bayesian Estimation of SEM

Maximum Likelihood (ML) estimation for Structural Equation Models has been available since the 1970s, and is now standard with statistical programs like LISREL, EQS, AMOS, and SAS Proc-Calis. Programs like LISREL (Jöreskog and Sörbom 1993) calculate the ML estimate θ_{ML} as well as estimates of the asymptotic standard errors of each parameter estimate. Because it relies on asymptotic theory, appropriate statistical inferences for the ML estimates require a large sample size. Several robustness studies show that SEM estimators behave badly at small n; see for instance Bearden, Sharma, and Teel 1982; Boomsma 1982, 1983; Baldwin 1986; Chou, Bentler, and Satorra 1991; Hu, Bentler, and Kano 1992; Yung and Bentler 1994; and Hoogland and Boomsma 1998. Further, the distribution of likelihood-ratio fit statistics is not known for small N. These problems hold for other estimation methods as well, like generalized least squares (GLS) and weighted least squares (WLS).

Given a prior distribution over the parameters of a SEM, $p(\theta)$, if the likelihood function is known then joint and marginal posterior distributions, $p(\theta)$ and $p(\theta|\mathbf{S})$ (where **S** is the sample covariance matrix) can be numerically approximated to arbitrary precision, for any finite sample size n, with Markov Chain Monte Carlo (MCMC) methods, and in particular with a single-component Metropolis-Hastings algorithm, a specific case of which is the Gibbs sampler (Geman and Geman 1984; Chib and Greenberg 1995). Given a sample covariance matrix **S**, and the assumption that the variables are distributed as multivariate normal, the log-likelihood for a SEM is:

$$\log L(\theta|\mathbf{S}) = -(n-1)/2 \; \{\log|\Sigma(\theta)| + \text{tr}[\mathbf{S}\Sigma^{-1}(\theta)]\} \; ,$$

where $\Sigma(\theta)$ is the covariance matrix implied by the model as a function of its parameters θ.

The Gibbs sampler (section 12.5.5.1) is an iterative procedure that, after it has converged, renders a dependent sample from the posterior $p(\theta|\mathbf{S})$. In each iteration $m = 1,...,M$, each parameter is sampled from its posterior conditional on the current values of the other parameters, any constraints appropriate for the parameter at hand, and the sample covariance matrix **S**. An accessible but detailed introduction to the Gibbs sampler can be found in Casella and George 1992, and more elaborate discussions are in Gelfand and Smith 1990, Tierney 1994, and Smith and Roberts 1993. BUGS is a general purpose

Gibbs sampling program developed by Spiegelhalter, Thomas, Best, and Gilks that can be applied to graphical models and can be obtained from <http://www.iph.cam.ac.uk/bugs/mainpage.html>.

Scheines, Hoijtink, and Boomsma (1999) implemented a Gibbs Sampler for linear SEM in TETRAD III, and used it to estimate the effect of low levels of Lead Exposure on the cognitive capacities (IQ) of children (Scheines 1997), and to show that the likelihood surface for SEMs with latent variables is not only nonnormal at small N, but actually multimodal (Scheines, Hoijtink, and Boomsma 1997). Here we briefly describe the Lead-IQ case and the problem of multimodality in the likelihood surface.

12.7.3 Lead and IQ

The description of this case is based on Scheines, Hoijtink, and Boomsma (1999), which contains additional details. In a 1985 article in *Science*, Needleman, Geiger, and Frank reanalyzed data they had previously collected on the effect of lead exposure on the verbal IQ score of 221 suburban children. After eliminating approximately 35 potential confounders with backward stepwise regression, they settled on regressing child's IQ on measured lead exposure, controlling for five measures of genetic factors, environmental stimulation, and physical factors that might compromise the child's cognitive endowment. Using the Build Module in TETRAD II (Scheines et al. 1994), Scheines, Hoijtink, and Boomsma were able to eliminate all the physical factor variables with almost no predictive loss.[14] The final set of variables they used are as follows:

ciq   the child's verbal IQ score
lead the measured concentration of lead in the child's baby teeth
med   the mother's level of education, in years
piq   the parent's IQ scores

Standardizing all the measured variables (which we do throughout this analysis), the regression solution is as follows, with t-statistics in parentheses:

$\hat{ciq}$ = − .177 *lead* + .251 *med* + .253 *piq* .
       (2.89)   (3.50)   (3.59)

All coefficients are significant at 0.05, R^2 = .243, and the estimates are very close to those obtained by including the physical factor variables (see Scheines 1997).

As Klepper (1988) points out, however, the measured regressor variables are really proxies that almost surely contain substantial measurement error. Although an errors-in-all-variables SEM explicitly modeling the regressor variables as latents as in figure 12.19 seems a more reasonable specification, unless the amount of measurement error for each regressor is known precisely, this model is underidentified.

Several strategies have been discussed for handling models of this type and underidentified models in general. One is instrumental variable estimation (Bollen 1989), another is a sensitivity analysis (Greene and Ernhart 1993) and still another is to bound parameters rather than produce a point estimate for them (Klepper and Leamer 1984). An additional strategy, made possible by the Gibbs sampler, is Bayesian estimation.

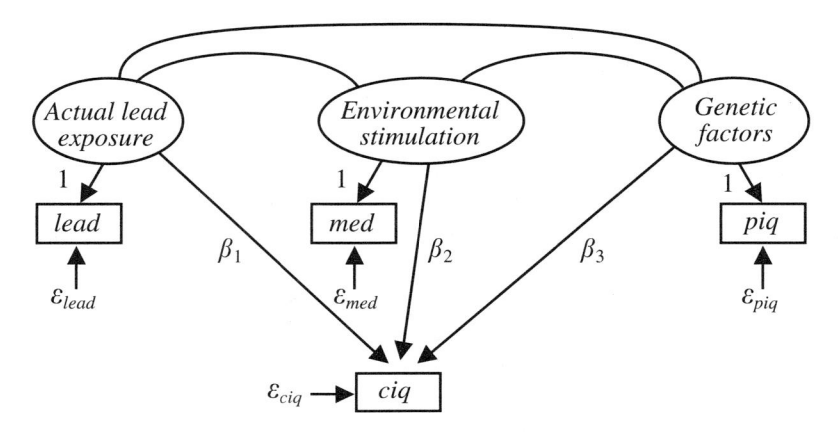

Figure 12.19 An errors-in-variables model for the lead exposure and IQ

If we standardize the measured variables in the model shown in figure 12.19, then the amount of measurement error for *lead*, which measures *Actual Lead Exposure*, and for *med*, which measures *Environmental Stimulation*, and for *piq*, which measures *Genetic factors*, is parameterized by var(ε_{lead}), var(ε_{med}), and var(ε_{piq}), respectively. Since the model implies that var(*lead*) = var(*Actual lead exposure*) + var(ε_{lead}), for example, and we are constraining var(*lead*) to unity, then if we were to set var(ε_{lead}) = 0.25, we would be asserting that 25% of the variance of measured lead comes from measurement error, while 75% comes from *Actual lead exposure*. In this case, and many others like it, there is reasonable prior information about the amount of measurement error present, but it is not specific enough to assign a unique value to the parameters associated with measurement error. Needleman pioneered a technique of inferring cumulative lead exposure from measures of the accumulated lead in a child's baby teeth. In Needleman's view,[15] between 0% and 40% of the variance in Needleman's proxy is probably from measurement error, with 20% a conservative best guess. For the measures of environmental stimulation and genetic factors, he is less confident, so guesses that between 0% and 60% of the variance in *med* and *piq* is from measurement error, with 30% his best guess.

Using a normal prior distribution truncated by removing below 0 values for the measurement error parameters, and flat prior elsewhere, Scheines, Hoijtink, and Boomsma produced 50,000 iterations with the Gibbs sampler in TETRAD III as a sample

Prequels and Sequels 367

from the posterior. The histogram in figure 12.20 shows the shape of the marginal posterior over β_1, the crucial coefficient representiong the influence of *Actual lead exposure* on children's *IQ*.

The results support Needleman's original conclusion, but do not require the unrealistic assumption of zero measurement error. The Bayesian point estimate of the effect of *Actual lead exposure* on *IQ*, $\hat{\beta}_{1.EAP}$ is –0.215, and since the central 95% region of its marginal posterior lies between –0.420 and –0.038, we conclude that exposure to environmental lead is indeed deleterious conditional on this model and our prior uncertainty as specified.

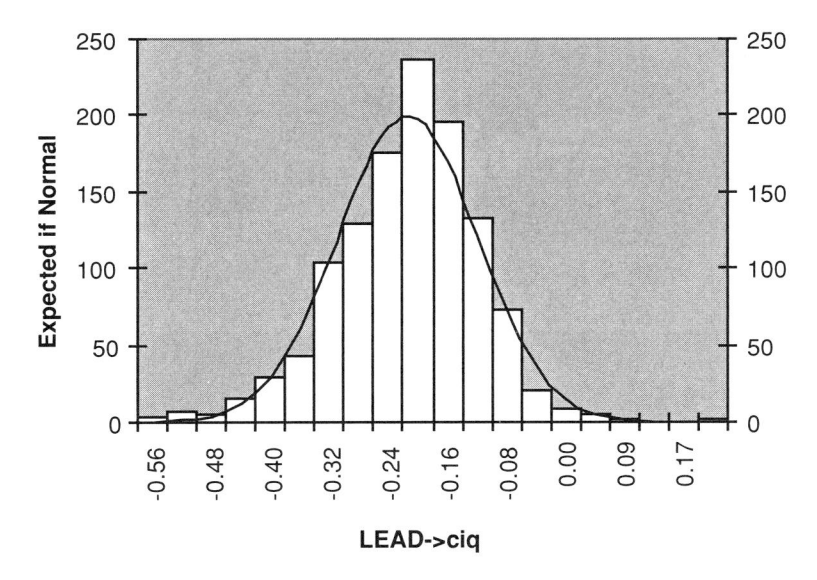

Figure 12.20. Marginal posterior of the causal effect of *Actual lead exposure* on IQ.

12.8 Applications

The practical value of the methods of search and prediction we have described comes from their use in applied sciences for classification, for forecasting, for predicting the effects of interventions, and for reconstructing causal relations independently known by other means. Chapter 5, chapter 8, and section 12.7.3 give some examples, and in this final section we will review a number of other studies conducted since 1993. We do not consider applications of Bayesian networks which are not generated by search, nor do we consider any nonconstraint-based search applications.

12.8.1 College Dropouts

Druzdzel and Glymour (1999) used the U.S. News and World Report database on American colleges and universities for 1992 and 1993 to investigate policies for lowering dropout rates. Using the TETRAD II program, they found that the average percentile score of the freshman class on ACT or SAT examinations is a "controlling" variable, analogous to the role of pH in the study of Spartina grass in chapter 8. That is, other variables in the database are independent of dropout rate conditional on average test scores of the entering class. The independence held quite closely in 1992, and less closely in 1993. (Regression predicted that other variables in the database directly influence dropout rate in both years.) Of course, this relation is not causal—SAT scores are a proxy for whatever background, resources and skills enable students to find their first year of college to be satisfactory.

The study was conducted at the request of the provost of Carnegie Mellon University, an institution with a history of high dropout rates in its freshman classes in the 1980s and early 1990s. Glymour and Druzdzel reported that the university could reduce its dropout rate by increasing the average SAT scores of the freshman class, but proposed no mechanism to do so. Beginning with the class of 1994, the university changed its formula for awarding scholarships, and received a larger number of applicants allowing for more selectivity, and there was a resulting increase in mean SAT scores of the entering class in that year and every succeeding one. In every year except one (1997) the dropout rate of the freshman class declined from the rate in the preceding year. The direction of the change is in accordance with the predictions of the Glymour and Druzdzel model, but they did not compare the quantitative prediction of the model with subsequent events at Carnegie Mellon University. Other unknown factors may also have affected the dropout rate.

12.8.2 In Flight Recalibration of a Mass Spectrometer Aboard an Earth Satellite

The Swedish Freja satellite carried a number of instruments to study the composition of the lower magnetosphere and upper ionosphere. One of these instruments, a three dimensional ion composition spectrometer (TICS) is essentially a mass spectrometer designed to measure hydrogen and oxygen ions and the two ions of helium. The instrument had 32 distinct detection channels, and calibration required matching signals at a particular channel with particular ion species. The correct matching depends on the incident energy of the ions, which varies within and between orbits. Unfortunately, the instrument was miscalibrated before launch, and two kinds of errors resulted: TICS values for the relative frequencies of the various species differed widely from their relative frequencies calculated theoretically from data from another instrument (a plasma detector); and the densities of ions according to TICS were a quarter to a fifth of the densities calculated from the plasma detector. Working at the University of Umea and the Swedish Institute for Space Physics, Waldemark and Norqvist (1999) recalibrated the instrument after launch using TETRAD II, principal components, and neural networks with backpropagation.

Ideally, different ions would be recorded at different channels, and there would be no leakage of signal from one channel to others spatially close to it. The correct causal description would then have four latent variables, one for each species, with directed edges from each latent into a set of channels for that species. If the sources are uncorrelated, in that ideal case, an analysis of the correlations of the 32 channel signals should yield four cliques, one for each distinct ion source. A TETRAD II analysis instead found two clusters of channels, with a few channels connected to both clusters. Principal components also gave a two factor model. The physical significance is that, at most orbits, the instrument cannot distinguish between helium and hydrogen ions (although for data from special orbits TETRAD II found a distinct cluster of channels for helium) because there is leakage between channels and because of instrument errors in determining physical locations on the detector. The clusterings differed with different energy levels.

Waldemark and Norqvist then used backpropagation in a neural network to find the channels that worked best for hydrogen and helium ions as against oxygen ions over a range of energies. The differences between the recalibrated TICS relative frequencies and those theoretically calculated from the plasma detector were reduced by half and the sensitivity of the instrument was increased considerably.

12.8.3 Economic Analysis and Forecasting

Bessler and his collaborators (Guven and Bessler, 1997; Akleman and Bessler, 1998; Akleman et al., 1998) have applied the PC and FCI algorithms and modifications of them to a number of econometric data sets. In a study of the dependency of corn exports on exchange rates, Akleman et al. found that graphical methods produced better forecasts than did a search procedure (Hsiao search) widely used in econometric forecasting. They have also used the techniques to study the relation between farm and retail meat prices, and, most recently, Loper and Bessler have used the methods on international data on GNP increases and the size of the agricultural sector in developing nations.

12.8.4 Comparing Machine and Expert Causal Judgment in Medicine

Ideal tests of the usefulness of search algorithms in domains such as medicine and epidemiology would compare predictions obtained by applying the algorithms to well designed observational databases with the outcomes of randomized clinical trials. Unfortunately, because of the rarity of adequate observational data sets paired with appropriate randomized clinical trials, and the inaccessibility of data, to our knowledge no such comparisons have been made. A second best alternative is to compare predictions from observational data with human expert judgment. Cooper and Spirtes (1998) compared predictions from a simplified (but correct) algorithm applied to a database on hospitalized pneumonia patients with the judgments of physicians. There studies shows some of the difficulties of this sort of test, not least because of the considerable variation in expert medical judgment of causal relations, and because of the difficulty of appropriate controls.

Recall that a measured variable *V* is exogenous in a causal DAG if there is no arrow directed into it. Assume that there is no causal relation between the sampling mechanism and the measured variables (i.e., there is no selection bias). Then the following theorem follows simply from Cooper (1997) and Spirtes et al. (1995).

THEOREM 12.8.1: Assuming the Causal Markov Condition, if

- *E* is exogenous, and
- each causal DAG containing the variables <*E,A,B*> in which *E* is exogenous has a nonzero prior probability,
- the prior probability of the parameters of each DAG is absolutely continuous with the BDe metric (Heckerman et al. 1994),
- $E \rightarrow A \rightarrow B$ has the highest posterior probability among all DAGs containing the variables <*E,A,B*> in which *E* is exogenous,

then with probability 1 in the large sample limit, in the true causal DAG, *A* is an ancestor of *B* (i.e., *A* is a cause of *B*) and there are no latent causes (i.e., unmeasured confounders) of *A* and *B*.

This result justifies a simple algorithm for causal inference with background knowledge, the Instrumental Variable (IV) algorithm. The IV algorithm takes as input background knowledge about which variables are exogenous, and a database consisting of patient records. An exogenous variable is also called an *instrumental variable*. The algorithm outputs a list of causal conclusions of the form "*A* causes *B*." The algorithm consists of the following steps:

1. Select a subset of variables **E** that are known to be exogenous. In the case of the pneumonia data (see below), the exogenous variables we used were race, age, and gender.
2. For each vertex *E* in **E**, search for measured variables *A* and *B* such that *A* is highly dependent on *E*, *B* is highly dependent on *A*, and *E* is independent of *B* given *A*. In the case of the data, we defined "highly dependent" to mean that the *p* value of the g^2 statistic measuring the dependence of discrete variables was less than 0.01, and "*E* is independent of *B* given *A*" means that the *p* value of the g^2 statistic measuring the conditional dependence of *E* and *B* given *A* is greater than 0.5.
3. For each triple of vertices <*E,A,B*> selected in step 2, for each DAG *G* that can be constructed out of the triple in which *E* is exogenous, calculate the posterior probability of *G*. If no DAG has a higher posterior probability than the DAG $E \rightarrow A \rightarrow B$ then output "*A* causes *B*."

Cooper and Spirtes assume each DAG compatible with the exogeneity of *E* has an equal prior probability. For each DAG, the prior probability over the parameters is the BDe prior described in Heckerman et al. 1994. The IV algorithm was tested on a pneumonia database of community acquired pneumonia patients (see Fine 1997 for details), which is called the pneumonia PORT database. Based on chart review, hundreds of data items were collected for each of the 2287 patients in the database. The causal conclusions of the IV algorithm applied to the database are shown in table 12.4.

A physician familiar with the pneumonia database but not with the algorithm was presented with a set of pairs of variables, some output by the algorithm as bearing a cause-effect relation to each other, and some chosen at random; the order of the pairs of variables was listed randomly. The physician was asked to classify each pair of variables into one of three classes: "Confident that A does cause B," "Don't know whether A causes B," or "Confident that A does not cause B." The results were that for all 10 pairs of variables suggested by the IV algorithm, the physician judge was confident that the relationship was cause and effect. (One pair of variables suggested by the IV algorithm which are definitionally related is not shown in Table 12.4) For the randomly chosen pairs of variables, he was confident that 10 were not cause and effect; and in 7 cases he was not sure. The hypothesis that the algorithm's decision that a relationship is causal is independent of the physician's is rejected by Fisher's exact test ($p = .0002$).

A second test used five physicians who regularly see pneumonia patients as part of their practice. Given a series of variable pairs and asked to judge whether the pairs were causally related, the physicians showed poor agreement across raters. To control as much as possible for the fact that the pairs selected by the IV algorithm were very highly correlated, the variable pairs selected by the IV algorithm were interspersed with other variable pairs that were also highly correlated. When the pooled judgments of the physicians were used in a test similar to the first, the hypothesis of independence (of the algorithm's causal claims and the pooled physician's claims) could not be rejected.

However, the results obtained did suggest some obvious improvements to the IV algorithm. Among the pairs selected by the IV algorithm, the 5 pairs that the physicians were most dubious about all involved current employment status as a cause. There are a number of obviously relevant features that the more dubious pairs output by the IV algorithm have in common.

Instrument	Cause	Effect	Score
age	coronary artery disease	myocardial infarction	18.41
age	current employment status	intravenous drug use (non-prescribed)	14.52
age	nausea	vomiting	9.28
gender	# of comorbid conditions	dire outcome (i.e., mortality or serious complications)	8.47
gender	sputum	cough	7.99
age	current employment status	chronic obstructive pulmonary disease	7.55
age	current employment status	prior hospitalization within 30 days	4.87
age	current employment status	a history of chronic obstructive pulmonary disease requiring prior ICU admission	4.42
age	current employment status	days since last hospital discharge	0.56

Table 12.4

• 4 of the 5 dubious causal relations have the 4 lowest scores.

• If the Bayes Information Criterion were used to score the models rather than the posterior probability, 2 of the dubious causal relations (the 2 with the lowest scores) would not have been suggested by the algorithm at all.

• All of the dubious effects contained categories with relatively few members, in contrast with the effects chosen by the IV algorithm that the doctors agreed with.

• When conducting statistical tests of the association of the cause with the effect, on four of the five dubious effects the statistical program we used issued a warning that the chi-squared test of independence may not be appropriate because the expected value of some cells was less than 5. It did not issue this warning on any of the 4 nondubious effects.

These features suggest that the performance of the IV algorithm could be improved by eliminating pairs of variables for which the test of independence is dubious because some expected cell sizes are less than 5, and/or by raising the score threshold of what is considered a positive result for the algorithm.

12.8.5 Infant Mortality

Mani and Cooper (1999) used an algorithm related to the IV algorithm to look for causal relations in a random sample of size 41,155 form the U.S. Linked Birth/Infant Death database. They selected a set of 85 clinically interesting, nonredundant variables to examine. The LCD2 algorithm searches for triples of variables with causal relations $W \to X \to Y$, where W is known from background knowledge to be exogenous. Given a set of exogenous variables **W**, the algorithm outputs "X causes Y" if there is an exogenous variable W such that W and Y are dependent, W and X are dependent, W and X are dependent given Y, X and Y are dependent, X and Y are dependent given W, and W and Y are independent given X. Assuming Causal Markov, Causal Faithfulness, the correctness of the independence tests, and the exogeneity of W, it can be shown that the algorithm is correct. It is not complete because there are cases where, using higher order conditional independence tests, it may be possible to determine that X causes Y, but the $X \to Y$ pair will not be in the algorithm's output. However, it has advantages both in terms of reliability at small sample sizes and speed over more complete searches.

The exogenous variables were race of the mother and child gender. The algorithm found 9 causal relations: *Maternal education* $\to$ *Delivery conductor*, *Maternal education* $\to$ *Maternal age*, *Marital status mother* $\to$ *Delivery conductor*, *Marital status mother* $\to$ *Maternal age*, *Prenatal care start* $\to$ *Delivery facility*, *Prenatal care start* $\to$ *Delivery conductor*, *Prenatal care adequacy* $\to$ *Prenatal care start*, *Birth weight* $\to$ *Infant outcome one year*, *Birth weight* $\to$ *Delivery conductor*. In all 9 cases, the exogenous variable was *Maternal race*. The meanings of the variables are described in table 12.5.

The relationship between *Prenatal care adequacy* and *Prenatal care start* is actually definitional, because *Prenatal care adequacy* is defined (in part) in terms of *Prenatal care start*. The other 8 causal relations all appear plausible. *Maternal education* $\to$ *Delivery conductor* is plausible because education can have an important effect on access to health care. *Birth weight* $\to$ *Infant outcome one year* is a well-documented causal relationship. The authors plan to ask OB/GYN clinicians to judge the plausibility of each member of a list of causal relations, including the 9 suggested by the algorithm intermixed with randomly generated variable pairs.

Variable Name	Variable meaning
Maternal education	Years of education of the mother
Delivery conductor	Care giver conducting delivery
Maternal age	Age of mother at delivery
Marital status mother	Marital status of the mother
Prenatal care start	Trimester prental care began
Delivery facility	Place or facility of delivery
Prenatal care adequacy	Adequacy of care
Birth weight	Weight of infant at birth
Infant outcome one year	If the child was alive on first birthday

Table 12.5

12.8.6 Biological Applications

Experimental research is difficult in ecology, and explanations founded on observational data are common, although sample sizes are often quite small. Shipley has applied directed graph search techniques, with a number of innovations, to ecological studies and to plant physiology. Shipley (1995) and his collaborators (Pyankov et al. 1999) have applied the techniques to study the causes of variation in leaf mass and area among related species, and causes of variation in relative growth between species (McKenna and Shipley 1999) He has developed a number of new search methods, including a bootstrapping technique for small samples (Shipley 1997) that generalizes the bootstrapping idea in the Weisberg example of chapter 8, and is discussed in section 12.5.10 (see also Friedman 1999b), and performs much better on small samples than the PC algorithm. Shipley (1999) has also provided an algorithm for obtaining, from any directed acyclic graph without latent variables, a set of independent partial correlation constraints; the output of the procedure can be used to test entire models by chi-square. He is preparing a monograph on structural equation models and search methods for causal explanations in biology.

12.8.7 Automated Mineral Identification from Near Infra-red Spectra

For many reasons, including power demands and limits on available antenna time, it would be valuable to have extra-planetary robots do some scientific analysis autonomously, on-board, rather than transmit all data to Earth for analysis. Visible and near infrared spectrometry has long been a standard tool in the identification of chemical species and minerals, and very light weight instruments have recently become available. An issue is whether fast computational procedures can be found that can identify minerals from rock and soil targets *in situ* from reflectance spectra, with a reliability comparable to that of expert human geophysical spectroscopists. The identification of water, hydrates and carbonates is of particular interest. In recent work for NASA on carbonate recognition, DeFazio et al. (1999) compared a simplified version of the PC algorithm with regression, with an expert system, and with a human expert.

Samples of spectra from rocks and soils *in situ* near Silver Lake, California were obtained from NASA field trials of a robot in the winter of 1999. Paul Gazis of NASA Ames Research Center provided an automated test for excess noise (owing to instrument error or atmospheric effects), and after that test was applied, 21 samples suitable for analysis were obtained. Each sample was examined in the field by expert geologists, and many of the samples were tested chemically and by analysis of transmitted light through thin slices. 13 of the samples were judged to be carbonates and eight were judged to be non-carbonates

The spectra were then given to a simplified version of the PC algorithm (essentially the PC algorithm in this book, but ignoring associations among causes), a regression algorithm from MiniTab, and an expert system modeled on a human expert spectroscopist. The PC algorithm and regression used a reference library of spectra from the Jet Propulsion Laboratory. Each procedure was tuned to give the best possible separation of carbonates from noncarbonates. Thirteen of the samples actually contained carbonates, according to the field geologists. The PC algorithm identified 12 of the 13 carbonates, and misidentified no non-carbonates. Regression identified 11 of the carbonate samples, and misidentified 4 noncarbonates. The expert system identified 9 of the carbonates and misidentified no noncarbonates.

As a further test, the PC algorithm, regression and the human expert (rather than the program simulating him) attempted to identify samples with carbonate composiiton from a library of spectra from Johns Hopkins University of 192 rock and soil samples, 91 of which actually contained some carbonate minerals. In addition, a commercial program, Model 1, was given the same task. The tuning of the PC algorithm and regression was the same as that used in the previous experiment. The PC algorithm identified 38 samples with carbonates, and misclassified 3 non-carbonate samples; the human expert correctly identified 24 carbonate samples and misidentified 1; regression claimed 154 of the samples contained carbonate, including 75 of the samples actually with carbonate and 79 of the samples without carbonate. The Model 1 program found 27 actual carbonate samples of 41 samples it claimed were carbonates.

Properly tuned, the simplified PC algorithm performs considerably better at this task than does regression, a human expert, and a commercial program, and requires minimal computational resources.

12.9 Foundational Issues and Relations to Other Disciplines

There is a voluminous literature on such questions as whether counterfactual conditionals have truth values (or merely acceptability conditions), what the truth conditions are, whether they can be meaningfully nested, etc. (e.g., Lewis 1973). Various representative attempts at definitions of causality, and the relationship between causation and counterfactuals, are explored in Sosa and Tooley (1993). Heckerman and Shachter (1995) attempt to define causal relations in terms of decision theory. Shafer (1996) explicates various related concepts of causality in terms of event trees.

There have been several attempts to find models of belief change which, like deductive logic are qualitative and deductively closed, but like probability can be held with varying degrees of firmness and can be retracted. Alchourrón et al. (1985) propose a set of axioms appropriate for revising a data-base in the face of new evidence (belief revision), while Katsumo and Mendelson (1991) propose a system for revising a data-base in the face of an external intervention (belief update). Goldszmidt and Pearl (1992) propose a system Z^+ for both belief revision and belief update that incorporates a qualitative version of the Causal Markov Condition. Formal learning theory also studies learning without probabilities. Kelly (1996) considers the problem of learning causes in the long run without using probabilities.

Iwasaki and Simon (1994) describe graphical representations of dynamic equations that are expressed as differential equations, and hence often involve both a variable and its differential. They do not relate the graphical representation to any conditional independence relations or statistical model.

Matuš and Studený have shown that there are 18300 sets of conditional independence relations among four variables that can be realized by some probability distribution, which is far larger than the number of different subsets of conditional independence relations that can be represented by graphical models. Matuš and Studený (1995) and Matuš (1995) investigate properties common to all of the realizeable sets of conditional independence relations among four variables. Studený (1992) shows that there is no finite complete characterization of probabilistic conditional independence.

13 Proofs of Theorems

We will adopt the following notational conventions. "w.l.g." abbreviates "without loss of generality," "r.h.s." abbreviates "right hand side," and "l.h.s." abbreviates "left hand side." Any sum over the empty set is equal to 0 and any product over the empty set is 1. $R(I,J)$ represents a directed path from I to J. If U is an undirected path from A to B, and X and Y occur on U, then we will denote the subpath of U between X and Y as $U(X,Y)$. $T(I,J)$ represents a trek in $\mathbf{T}(I,J)$. The definitions of all technical terms in this chapter that have not been defined in chapters 2 or 3 have been placed in a glossary following the chapter.

13.1 Theorem 2.1

THEOREM 2.1: If $P(\mathbf{V})$ is a positive distribution, then for any ordering of the variables in $\mathbf{V}$, P satisfies the Markov and Minimality conditions for the directed independence graph of $P(\mathbf{V})$ for that ordering.

Proof. See Pearl 1988.

13.2 Theorem 3.1

THEOREM 3.1: If S is an LCT, and S' is a random coefficient LCT with the same directed acyclic graph, the same set of noncoefficient random variables, the same variances for each noncoefficient exogenous variable, and for each random coefficient a'_{IJ} in S', $E(a'_{IJ}) = a_{IJ}$ in S, then a partial correlation is equal to 0 in S if and only if it is equal to 0 in S'

Let a **linear causal theory** be (**LCT**) be <<**R,M,E**>, (Ω,f,P), **EQ,L,Err**> where

(i) (Ω,f,P) is a probability space, where Ω is the sample space, f is a sigma-field over Ω, and P is a probability distribution over f.
(ii) <**R,M,E**> is a directed acyclic graph. **R** is a set of random variables over (Ω,f,P).
(iii) The variables in **R** have a joint distribution. Every variable in **R** has a nonzero variance. **E** is a set of directed edges between variables in **R**. (**M** is the set of marks that occur in a directed graph, that is, {EM, >}.
(iv) **EQ** is a consistent set of independent homogeneous linear equations in random variables in **R**. For each X_i in **R** of positive indegree there is an equation in **EQ** of the form

$$X_i = \sum_{X_j \in \mathbf{Parents}(X_i)} a_{ij} X_j$$

where each a_{ij} is a nonzero real number and each X_i is in **R**. This implies that each vertex X_i in **R** of positive indegree can be expressed as a linear function of all and only its parents. There are no other equations in **EQ**. A nonzero value of a_{ij} is the **equation coefficient** of X_j in the equation for X_i.

(v) If vertices (random variables) X_i and X_j are exogenous, then X_i and X_j are pairwise statistically independent.

(vi) L is a function with domain E such that for each e in E, $L(e) = a_{ij}$ iff **head**$(e) = X_j$ and **tail**$(e) = X_j$. $L(e)$ will be called the **label** of e. By extension, the product of labels of edges in any acyclic undirected path U will be denoted by $L(U)$, and $L(U)$ will be called the **label** of U. The label of an empty path is fixed at 1.

(vii) There is a subset of **S** of **R** called the **error variables**, each of indegree 0 and outdegree 1. For every X_i in **R** of indegree $\neq 0$ there is exactly one error variable with an edge into X_i. We assume that the partial correlations of all orders involving only non-error variables are defined.

Note that the variance of any endogenous variable I conditional on any set of variables that does not contain the error variable of I is not equal to zero.

The definition of a **random coefficient linear causal theory** is the same as that of a linear causal theory except that each linear coefficient is a random variable independent of the set of all other random variables in the model.

A **linear causal form** is an unestimated LCT in which the linear coefficients and the variances of the exogenous variables are real variables instead of constants. This entails that an edge label in an LCF is a real variable instead of a constant (except that the label of an edge from an error variable is fixed at one.) More formally, let a linear causal form (**LCF**) be <<**R,M,E**>, **C**, **V**, **EQ,L,Err**> where

(i) <**R,M,E**> is a directed acyclic graph. **Err** is a subset of **R** called the **error variables**. Each error variable is of indegree 0 and outdegree 1. For every X_i in **R** of indegree $\neq 0$ there is exactly one error variable with an edge into X_i.

(ii) c_{ij} is a unique real variable associated with an edge from X_j to X_i, and **C** is the set of c_{ij}. **V** is the set of variables σ_i^2, where X_i is an exogenous variable in <**R,M,E**> and σ_i^2 is a variable that ranges over the positive real numbers.

(iii) L is a function with domain E such that for each e in E, $L(e) = c_{ij}$ iff $head(e) = X_j$ and $tail(e) = X_i$. $L(e)$ will be called the **label** of e. By extension, the product of labels of edges in any acyclic undirected path U will be denoted by $L(U)$, and $L(U)$ will be called the **label** of U. The label of an empty path is fixed at 1.

iv. **EQ** is a consistent set of independent homogeneous linear equationals in variables in **R**. For each X_i in **R** of positive indegree there is an equation in **EQ** of the form

$$X_i = \sum_{X_j \in \mathbf{Parents}(X_i)} c_{ij} X_j$$

where each c_{ij} is a real variable in **C** and each X_i is in **R**. There are no other equations in **EQ**. c_{ij} is the **equation coefficient** of X_j in the equation for X_i.

An LCT S is an **instance** of an LCF F if and only if the directed acyclic graph of S is isomorphic to the directed acyclic graph of F. In an LCF, a quantity (e.g., a covariance) X **is equivalent to a polynomial in the coefficients and variances of exogenous variables** if and only if for each LCF $F = $ <<**R,M,E**>, **C**, **V**, **EQ,L,Err**> and in every LCT $S = $ <<**R',M',E'**>, (Ω,f,P), **EQ',L',Err'**> that is an instance of F, there is a polynomial in the variables in **C** and **V** such that X is equal to the result of substituting the linear coefficients of S in as values for the corresponding variables in **C**, and the variances of the exogenous variables in S as values for the corresponding variables in **V**.

In an LCT or LCF S, a variable X_i is **independent** iff X_i has zero indegree (i.e., there are no edges directed into it); otherwise it is **dependent**. Note that the *property* of independence is completely distinct from the *relation* of statistical independence. The context will make clear in which of these senses the term is used. For a directed acyclic graph G, **Ind** is the set of independent variables in G. Given a directed acyclic graph G, $\mathbf{D}(X_i, X_j)$ is the set of all directed paths from X_i to X_j. In an LCF <<**R,M,E**>, **C**, **V**, **EQ,L,S**> an equation is an **independent equational for a dependent variable** X_j if and only if it is implied by **EQ** and the variables in **R** which appear on the r.h.s. are independent and occur at most once. $^{Ind}a_{IJ}$ is the coefficient of J in the independent equational for I.

LEMMA 3.1.1: In an LCF S, if J is an independent variable, then

$$^{Ind}a_{IJ} = \sum_{U \in \mathbf{D}(J,I)} L(U)$$

Proof. This is a special case of Mason's rule for calculating the "total effect" of a variable J on a variable I. See Glymour et al. 1987. ∴

The following two lemmas show how to calculate the variance of random variables and covariances between random variables in terms of the covariances between other random variables. The proofs of these lemmas can be found in Freund and Walpole 1980. We denote the covariance of I and J by γ_{IJ}, the variance of I by σ^2_I, the correlation of I and J by ρ_{IJ}, the partial correlation of I and J given the set **H** by $\gamma_{IJ.\mathbf{H}}$, and the partial covariance of I and J given **H** by $\rho_{IJ.\mathbf{H}}$. The correlation of two subscripted variables such as X_i and X_j we will write as ρ_{ij} for legibility, and similarly for partial correlations, etc.

LEMMA 3.1.2: If **Q** is a set of random variables with a joint probability distribution and

$$Y = \sum_{I \in \mathbf{Q}} a_{YI} I$$

and

$$Z = \sum_{J \in \mathbf{Q}} a_{ZJ} J$$

then

$$\gamma_{YZ} = \sum_{I \in \mathbf{Q}} \sum_{J \in \mathbf{Q}} a_{YI} a_{ZJ} \gamma_{IJ}$$

Lemmas 3.1.3, 3.1.5, and 3.1.7 are not used in the proof of theorem 3.1, but they are used in later theorems, and we include them here because they follow easily from the other lemmas in this section.

LEMMA 3.1.3: If $\mathbf{Q}$ is a set of random variables with a joint probability distribution and

$$Y = \sum_{I \in \mathbf{Q}} a_{YI} I$$

then

$$\sigma_Y^2 = \sum_{I \in \mathbf{Q}} \sum_{J \in \mathbf{Q}} a_{YI} a_{YJ} \gamma_{IJ}$$

In an LCF S, $\mathbf{U}_X$ is the set of all independent variables that are the source of a directed path to X. (Note that if X is independent then $X \in \mathbf{U}_X$ since there is an empty path from every vertex to itself.) In an LCF S, $\mathbf{U}_{XY}$ is $\mathbf{U}_X \cap \mathbf{U}_Y$.

LEMMA 3.1.4: If S is an LCF,

$$Y = \sum_{I \in \mathbf{Ind}}^{Ind} a_{YI} I$$

and

$$Z = \sum_{I \in \mathbf{Ind}}^{Ind} a_{ZI} I$$

then

$$\gamma_{YZ} = \sum_{I \in \mathbf{U}_{YZ}} {}^{Ind}a_{YI} \, {}^{Ind}a_{ZI} \sigma_I^2$$

Proof. **Ind** is a set of independent variables. It follows that γ_{IJ} is equal to 0 if $I \neq J$, and γ_{IJ} is equal to σ_I^2 if $I = J$. Substituting these values for γ_{IJ} into the r.h.s. of the equation for γ_{YZ} in lemma 3.1.2 shows that

$$\gamma_{YZ} = \sum_{I \in \mathbf{Ind}} {}^{Ind}a_{YI} \, {}^{Ind}a_{ZI} \sigma_I^2 \tag{13.1}$$

If I is in **Ind**, but I is not in $\mathbf{U}_{YZ}$ then there is no pair of directed acyclic paths from I to Y and Z. By lemma 3.1.1, if there is no pair of directed acyclic paths from I to Y and Z, then the coefficient of I in the independent equation for either Y or Z is zero. So, the only nonzero terms in equation 1 are for $I \in \mathbf{U}_{YZ}$. ∴

LEMMA 3.1.5: If S is an LCF,

$$Y = \sum_{I \in \mathbf{Ind}} {}^{Ind}a_{YI} I$$

then

$$\sigma_Y^2 = \sum_{I \in \mathbf{U}_Y} {}^{Ind}a_{YI}^2 \sigma_I^2$$

Proof. **Ind** is a set of independent random variables. It follows that γ_{IJ} is equal to 0 if $I \neq J$, and γ_{IJ} is equal to σ_I^2 if $I = J$. Substituting these values for γ_{IJ} into the r.h.s. of the equation for σ_Y^2 in lemma 3.1.1 proves that

$$\sigma_Y^2 = \sum_{I \in \mathbf{Ind}} {}^{Ind}a_{YI}^2 \sigma_I^2 \tag{13.2}$$

If I is in **Ind**, but I is not in $\mathbf{U}_Y$, then there is no directed path from I to Y. It follows from lemma 3.1.1 that a_{YI} is zero. Hence the only nonzero terms in equation 2 come from $I \in \mathbf{U}_Y$. ∴

LEMMA 3.1.6: If S is an LCF,

$$\gamma_{IJ} = \sum_{K \in \mathbf{U}_{IJ}} \sum_{R \in \mathbf{D}(K,I)} \sum_{R' \in \mathbf{D}(K,J)} L(R)L(R')\sigma_K^2$$

Proof. This follows immediately from lemmas 3.1.2 and 3.1.4. ∴

LEMMA 3.1.7: If S is an LCF,

$$\sigma_I^2 = \sum_{K \in \mathbf{U}_I} \left(\left(\sum_{R \in \mathbf{D}(K,L)} L(R) \right)^2 \sigma_K^2 \right)$$

Proof. This follows immediately from lemmas 3.1.1 and 3.1.5. ∴

THEOREM 3.1: If S is an LCT, and S' is a random coefficient LCT with the same directed acyclic graph, the same set of noncoefficient random variables, the same variances for each exogenous variable, and for each random coefficient a'_{IJ} in S', $E(a'_{IJ}) = a_{IJ}$ in S, then a partial correlation is equal to 0 in S if and only if it is equal to 0 in S'.

$$\gamma_{IJ} = \sum_{K \in \mathbf{U}_{IJ}} \sum_{R \in \mathbf{D}(K,I)} \sum_{R' \in \mathbf{D}(K,J)} L(R)L(R')\sigma_K^2$$

The label of a path is equal to the product of the labels of the edges and because the random coefficients are independent of each other and all the random variables that are not coefficients, it follows that

$$E\left(\prod_{edge \in U} L(edge) \right) = \prod_{edge \in U} E(L(edge))$$

Transform all of the variables so that they have mean 0; this does not affect the value of any of the covariances. In T, $\gamma_{IJ} = E(IJ)$ and

$$E(IJ) = E\left(\sum_{H \in \mathbf{U}_I} \sum_{U \in \mathbf{D}(H,X)} \sum_{F \in \mathbf{U}_J} \sum_{V \in \mathbf{D}(F,Y)} L(U)L(V)HF\right) =$$

$$\sum_{H \in \mathbf{U}_{IJ}} \sum_{U \in \mathbf{D}(H,X)} \sum_{V \in \mathbf{D}(H,Y)} E(L(U)L(V)H^2) =$$

$$\sum_{H \in \mathbf{U}_{IJ}} \sum_{U \in \mathbf{D}(H,X)} \sum_{V \in \mathbf{D}(H,Y)} E\left(\prod_{edge \in U} L(edge) \prod_{edge \in V} L(edge)H^2\right) =$$

$$\sum_{H \in \mathbf{U}_{IJ}} \sum_{U \in \mathbf{D}(H,X)} \sum_{V \in \mathbf{D}(H,Y)} \prod_{edge \in U} E(L(edge)) \prod_{edge \in V} E(L(edge))E(H^2)$$

because for exogenous variables $E(HF) = 0$ unless $H = F$.

By hypothesis, $E(L(edge))$ in $S' = L(edge)$ in S. Hence the expression γ_{IJ} is the same for both random and constant coefficients. The partial correlations are a function of the covariance matrix so the partial correlations are the same in S and S'. ∴

13.3 Theorem 3.2

THEOREM 3.2: Let M be an LCF with n free linear coefficients $a_1,\ldots, a_n$ and k positive variances $v_1,\ldots,v_k$. Let $M(<u_1,\ldots,u_n,u_{n+1},\ldots,u_{n+k}>)$ be the distributions consistent with specifying values $<u_1,\ldots,u_n, u_{n+1},\ldots,u_{n+k}>$ for $a_1,\ldots, a_n$ and $v_1,\ldots,v_k$. Let Π be the set of probability measures P on the space $\Re_{n+k}$ of values of the parameters of M such that for every subset $\mathbf{V}$ of $\Re_{n+k}$ having Lebesgue measure zero, $P(\mathbf{V}) = 0$. Let $\mathbf{Q}$ be the set of vectors of coefficient and variance values such that for all q in $\mathbf{Q}$ every probability distribution consistent with $M(q)$ has a vanishing partial correlation that is not linearly implied by M. Then for all P in Π $P(\mathbf{Q}) = 0$.

LEMMA 3.2.1: In an LCF S, $\rho_{ij.\mathbf{X}} = 0$ is equivalent to a polynomial equation in the linear coefficients and variances of the independent variables.

Proof. We will prove more generally that a polynomial equation in partial covariances is equivalent to a polynomial equation in the linear coefficients and variances of the independent variables. If $\mathbf{X}$ contains n distinct variables, then say $\rho_{ij.\mathbf{X}}$ is a partial correlation of order n. Let the pc-order (partial covariance order) of a polynomial in partial covariances be the highest order of any partial covariance appearing in the polynomial. The proof is by induction on the pc-order of the polynomials.

Base Case: If polynomial Q is of pc-order 0, then by lemma 3.1.2, Q is equivalent to a polynomial equation in the linear coefficients and variances of the independent variables.

Induction Case: Suppose that the lemma is true for polynomials of pc-order $n-1$, and let Q be a polynomial of pc-order n. The recursion formula for partial covariances is

$$\gamma_{ij.\mathbf{Y} \cup r} = \gamma_{ij.\mathbf{Y}} - \frac{\gamma_{ir.\mathbf{Y}} \gamma_{jr.\mathbf{Y}}}{\gamma_{rr.\mathbf{Y}}}$$

Form Q' by using this recursion formula to replace each covariance of pc-order n appearing in Q by an algebraic combination of covariances of pc-order $n-1$. Form Q'' by multiplying Q' by the lowest common denominator of all of the terms in Q', producing a polynomial of pc-order n–1. By the induction hypothesis, Q'' is equivalent to a polynomial equation in the linear coefficients and variances of the independent variables. Hence, a polynomial equation in partial covariances is equivalent to a polynomial equation in the linear coefficients and variances of the independent variables.

By definition,

$$\rho_{ij.\mathbf{X}} = \frac{\gamma_{ij.\mathbf{X}}}{\sqrt{\gamma_{ii.\mathbf{X}}} \sqrt{\gamma_{jj.\mathbf{X}}}}$$

so $\rho_{ij.\mathbf{X}} = 0$ iff $\gamma_{ij.\mathbf{X}} = 0$. Since the latter is a polynomial equation in partial covariances, it is equivalent to a polynomial equation in the linear coefficients and variances of the independent variables. It follows that the former is also equivalent to a polynomial equation in the linear coefficients and variances of the independent variables. ∴.

THEOREM 3.2: Let M be a linear model with directed acyclic graph G and n linear coefficients $a_1,..., a_n$ and k positive variances of exogenous variables $v_1,..., v_k$. Let $M(<u_1,...,u_n, u_{n+1},...,u_{n+k}>)$ be the distributions consistent with specifying values $<u_1,...,u_n, u_{n+1},...,u_{n+k}>$ for $a_1,..., a_n$ and $v_1,...v_k$. Let Π be the set of probability measures P on the space $\Re^{n+k}$ of values of the parameters of M such that for every subset $\mathbf{V}$ of $\Re^{n+k}$ having Lebesgue measure zero, $P(\mathbf{V}) = 0$. Let $\mathbf{Q}$ be the set of vectors of coefficient and variance values such that for all q in $\mathbf{Q}$ every probability distribution in with $M(q)$ has a vanishing partial correlation that is not linearly implied by G. Then for all P in Π $P(\mathbf{Q}) = 0$.

Proof. For any LCF, each partial correlation is equivalent to a polynomial in the linear coefficients and the variances of the exogenous variables: the rest of the features of the distribution have no bearing on the partial correlation. Hence for a vanishing partial correlation to be linearly implied by the directed acyclic graph of the theory, it is necessary and sufficient that the corresponding polynomial in the linear coefficient and variance parameters vanish identically. Thus any vanishing partial correlation not linearly implied by an LCF represents a polynomial P in variables consisting of the linear coefficients and variances of that theory, and the polynomial does not vanish identically.

So the set of linear coefficient and variance values satisfying P is an algebraic variety in $\Re^{n+k}$. Any connected component of such a variety has Lebesgue measure zero. But an algebraic variety has at most a finite number of connected components (Whitney 1957). ∴

13.4 Theorem 3.3

THEOREM 3.3: $P(\mathbf{V})$ is faithful to directed acyclic graph G with vertex set $\mathbf{V}$ if and only if for all disjoint sets of vertices $\mathbf{X}$, $\mathbf{Y}$, and $\mathbf{Z}$, $\mathbf{X}$, and $\mathbf{Y}$ are independent conditional on $\mathbf{Z}$ if and only if $\mathbf{X}$ and $\mathbf{Y}$ are d-separated given $\mathbf{Z}$.

The "if" portion of the theorem was first proved in Verma 1986 and the "only if" portion of the theorem was first proved in Geiger and Pearl 1989a. The proof produced here is considerably different, but since the bulk of it is a series of lemmas that we also need to prove other theorems, we state it here.

G' is an **inducing path graph over O for directed acyclic graph** G if and only if $\mathbf{O}$ is a subset of the vertices in G, there is an edge between variables A and B with an arrowhead at A if and only if A and B are in $\mathbf{O}$, and there is an inducing path in G between A and B relative to $\mathbf{O}$ that is into A. (Using the notation of chapter 2, the set of marks in an inducing path graph is {>, EM}.) We will refer to the variables in $\mathbf{O}$ as **observed** variables. Unlike a directed acyclic graph, an inducing path graph can contain double-headed arrows. However, it does not contain any edges with no arrowheads. If there is an inducing path between A and B in G that is into A, then the edge between A and B in G' is into A. However, if there is an inducing path between A and B in G that is out of A, it does not follow that the edge in G' between A and B is out of A. Only if *no* inducing path between A and B in G is into A is the edge between A and B in G' out of A. The definitions of directed path, d-separability, inducing path, collider, ancestor, and descendant are the same as those for directed graphs, that is, a directed path in an inducing path graph, as in an acyclic directed graph, contains only directed edges (e.g., $A \rightarrow B$). However, an undirected path in an inducing path graph can contain either directed edges, or bi-directed edges (e.g., $C \leftrightarrow D$.) Also, if $A \leftrightarrow B$ in an inducing path graph, A is not a parent of B. Note that if G is a directed acyclic graph, and G' the inducing path graph for G over $\mathbf{O}$, then there are no directed cycles in G'.

Lemma 3.3.1 states a method for constructing a path between X and Y that d-connects X and Y given $\mathbf{Z}$ out of a sequence of paths.

LEMMA 3.3.1: In a directed acyclic graph G (or an inducing path graph G) over $\mathbf{V}$, if X and Y are not in $\mathbf{Z}$, there is a sequence S of distinct vertices in $\mathbf{V}$ from X to Y, and there is a set T of undirected paths such that

(i). for each pair of adjacent vertices V and W in S there is a unique undirected path in T that d-connects V and W given $\mathbf{Z}\backslash\{V,W\}$, and (ii). if a vertex Q in S is in $\mathbf{Z}$, then the paths in T that contain Q as an endpoint collide at Q, and

(iii). if for three vertices V, W, Q occurring in that order in S the d-connecting paths in T between V and W, and W and Q collide at W then W has a descendant in $\mathbf{Z}$,

then there is a path U in G that d-connects X and Y given $\mathbf{Z}$. In addition, if all of the edges in all of the paths in T that contain X are into (out of) X then U is into (out of) X, and similarly for Y.

Proof. Let U' be the concatenation of all of the paths in T in the order of the sequence S. U' may not be an acyclic undirected path, because it may contain some vertices more than once. Let U be the result of removing all of the cycles from U'. If each edge in U' that contains X is into (out of) X, then U is into (out of) X, because each edge in U is an edge in U'. Similarly, if each edge in U' that contains Y is into (out of) Y, then U is into (out of) Y, because each edge in U is an edge in U'. We will prove that U d-connects X and Y given $\mathbf{Z}$.

We will call an edge in U containing a given vertex V an endpoint edge if V is in the sequence S, and the edge containing V occurs on the path in T between V and its predecessor or successor in S; otherwise the edge is an internal edge.

First we prove that every member R of $\mathbf{Z}$ that is on U is a collider on U. If there is an endpoint edge containing R on U then it is into R because by assumption the paths in T containing R collide at R. If an edge on U is an internal edge with endpoint R then it is into R because it is an edge on a path that d-connects two variables A and B not equal to R given $\mathbf{Z}\backslash\{A,B\}$, and R is in $\mathbf{Z}$. All of the edges on paths in T are into R, and hence the subset of those edges that occur on U are into R.

Next we show that every collider R on U has a descendant in $\mathbf{Z}$. R is not equal to either of the endpoints X or Y, because the endpoints of a path are not colliders along the path. If R is a collider on any of the paths in T then R has a descendant in $\mathbf{Z}$ because it is an edge on a path that d-connects two variables A and B not equal to R given $\mathbf{Z}\backslash\{A,B\}$. If R is a collider on two endpoint edges then it has a descendant in $\mathbf{Z}$ by hypothesis. Suppose then that R is not a collider on the path in T between A and B, and not a collider on the path in T between C and D, but after cycles have been removed from U', R is a collider on U. In that case U' contains an undirected cycle containing R. Because G is acyclic, the undirected cycle contains a collider. Hence R has a descendant that is a collider on U'. Each collider on U' has a descendant in $\mathbf{Z}$. Hence R has a descendant in $\mathbf{Z}$. ∴

LEMMA 3.3.2: If G is a directed acyclic graph (or an inducing path graph), R is d-connected to Y given $\mathbf{Z}$ by undirected path U, and W and X are distinct vertices on U not in $\mathbf{Z}$, then $U(W,X)$ d-connects W and X given $\mathbf{Z} = \mathbf{Z}\backslash\{W,X\}$.

Proof. Suppose G is a directed acyclic graph, R is d-connected to Y given $\mathbf{Z}$ by undirected path U, and W and X are distinct vertices on U not in $\mathbf{Z}$. Each noncollider on $U(W,X)$

except for the endpoints is a noncollider on U, and hence not in **Z**. Every collider on $U(W,X)$ has a descendant in **Z** because each collider on $U(W,X)$ is a collider on U, which d-connects R and Y given **Z**. It follows that $U(W,X)$ d-connects W and X given **Z** = **Z**\{W,X}. ∴

LEMMA 3.3.3: If G is a directed acyclic graph (or an inducing path graph), R is d-connected to Y given **Z** by undirected path U, there is a directed path D from R to X that does not contain any member of **Z**, and X is not on U, then X is d-connected to Y given **Z** by a path U' that is into X. If D does not contain Y, then U' is into Y if and only if U is.

Proof. Let D be a directed path from R to X that does not contain any member of **Z**, and U an undirected path that d-connects R and Y given **Z** and does not contain X. Let Q be the point of intersection of D and U that is closest to Y on U. Q is not in **Z** because it is on D.

If D does contain Y, then $Y = Q$, and $D(Y,X)$ is a path into X that d-connects X and Y given **Z** because it contains no colliders and no members of **Z**.

If D does not contain Y then $Q \neq Y$. $X \neq Q$ because X is not on U and Q is. By lemma 3.3.2 $U(Q,Y)$ d-connects Q and Y given **Z**\{Q,Y} = **Z**. Also, $D(Q,X)$ d-connects Q and X given **Z**\{Q,X} = **Z**. $D(Q,X)$ is out of Q, and Q is not in **Z**. By lemma 3.3.1, there is a path U' that d-connects X and Y given **Z** that is into X. If Y is not on D, then all of the edges containing Y in U' are in $U(Q,Y)$, and hence by lemma 3.3.1 U' is into Y if and only if U is. ∴

In a directed acyclic graph G, **ND(Y)** is the set of all vertices that do not have a descendant in **Y**

LEMMA 3.3.4: If $P(\mathbf{V})$ satisfies the Markov condition for directed acyclic graph G over **V**, **S** is a subset of **V**, and **ND(Y)** is included in **S**, then

$$\overrightarrow{\sum_{\mathbf{S}}} \left(\prod_{V \in \mathbf{V}} P(V|\mathbf{Parents}(V)) \right) = \overrightarrow{\sum_{\mathbf{S} \setminus \mathbf{ND(Y)}}} \left(\prod_{V \in \mathbf{V} \setminus \mathbf{ND(Y)}} P(V|\mathbf{Parents}(V)) \right)$$

Proof. **S** can be partitioned into **S\ND(Y)** and **S** ∩ **ND(Y)** = **ND(Y)**. If V is in **V\ND(Y)** then no variable occurring in the term $P(V|\mathbf{Parents}(V))$ occurs in **ND(Y)**; hence for each V in **V\ND(Y)**, $P(V|\mathbf{Parents}(V))$ can be removed from the scope of the summation over the values of variables in **ND(Y)**.

$$\overrightarrow{\sum_{\mathbf{S}}} \left(\prod_{V \in \mathbf{V}} P(V|\mathbf{Parents}(V)) \right) = \quad (1)$$

$$\overrightarrow{\sum_{\mathbf{S} \setminus \mathbf{ND}(\mathbf{Y})}} \left(\prod_{V \in \mathbf{V} \setminus \mathbf{ND}(\mathbf{Y})} P(V|\mathbf{Parents}(V)) \times \left(\overrightarrow{\sum_{\mathbf{ND}(\mathbf{Y})}} \left(\prod_{V \in \mathbf{ND}(\mathbf{Y})} P(V|\mathbf{Parents}(V)) \right) \right) \right)$$

We will now show that

$$\overrightarrow{\sum_{\mathbf{ND}(\mathbf{Y})}} \left(\prod_{V \in \mathbf{ND}(\mathbf{Y})} P(V|\mathbf{Parents}(V)) \right) = 1$$

unless for some value of $\mathbf{S} \setminus \mathbf{ND}(\mathbf{Y})$ the set of values of $\mathbf{ND}(\mathbf{Y})$ such that $P(V|\mathbf{Parents}(V))$ is defined for each V in $\mathbf{ND}(\mathbf{Y})$ is empty, in which case on the l.h.s of (1) no term containing that value of $\mathbf{S} \setminus \mathbf{ND}(\mathbf{Y})$ appears in the sum, and on the r.h.s. of (1) every term in the scope of the summation over $\mathbf{S} \setminus \mathbf{ND}(\mathbf{Y})$ that contains that value of $\mathbf{S} \setminus \mathbf{ND}(\mathbf{Y})$ is zero.

Let $P(W|\mathbf{Parents}(W))$ be a term in the factorization such that W does not occur in any other term, that is, W is not the parent of any other variable. If $\mathbf{ND}(\mathbf{Y})$ is not empty W is in $\mathbf{ND}(\mathbf{Y})$.

$$\overrightarrow{\sum_{\mathbf{ND}(\mathbf{Y})}} \left(\prod_{V \in \mathbf{ND}(\mathbf{Y})} P(V|\mathbf{Parents}(V)) \right) =$$

$$\overrightarrow{\sum_{\mathbf{ND}(\mathbf{Y}) \setminus \{W\}}} \left(\prod_{V \in \mathbf{ND}(\mathbf{Y}) \setminus \{W\}} P(V|\mathbf{Parents}(V)) \right) \times \left(\overrightarrow{\sum_{W}} P(W|\mathbf{Parents}(W)) \right)$$

The latter expression can now be written as

$$\overrightarrow{\sum_{\mathbf{ND}(\mathbf{Y}) \setminus \{W\}}} \left(\prod_{V \in \mathbf{ND}(\mathbf{Y}) \setminus \{W\}} P(V|\mathbf{Parents}(V)) \right)$$

because $\overrightarrow{\sum_{W}} P(W|\mathbf{Parents}(W))$ equal to one. Now some element in $\mathbf{ND}(\mathbf{Y}) \setminus \{W\}$ is not a parent of any other member of $\mathbf{ND}(\mathbf{Y}) \setminus \{W\}$, and the process can be repeated until each element is removed from $\mathbf{ND}(\mathbf{Y})$. ∴

In a directed acyclic graph G, if $\mathbf{Y} \cap \mathbf{Z} = \emptyset$, then V is in $\mathbf{IV}(\mathbf{Y},\mathbf{Z})$ (informative variables for $\mathbf{Y}$ given $\mathbf{Z}$) if and only if V is d-connected to $\mathbf{Y}$ given $\mathbf{Z}$, and V is not in $\mathbf{ND}(\mathbf{YZ})$. (This entails that V is not in $\mathbf{Y} \cup \mathbf{Z}$ by definition of d-connection.) In a directed acyclic graph G, if $\mathbf{Y} \cap \mathbf{Z} = \emptyset$, W is in $\mathbf{IP}(\mathbf{Y},\mathbf{Z})$ (W has a parent that is an informative variable for $\mathbf{Y}$ given $\mathbf{Z}$) if and only if W is a member of $\mathbf{Z}$, and W has a parent in $\mathbf{IV}(\mathbf{Y},\mathbf{Z}) \cup \mathbf{Y}$.

Proofs of Theorems 389

LEMMA 3.3.5: If P satisfies the Markov condition for directed acyclic graph G over $\mathbf{V}$, then

$$P(\mathbf{Y}|\mathbf{Z}) = \frac{\overset{\rightarrow}{\sum}_{\mathbf{IV}(\mathbf{Y},\mathbf{Z})} \prod_{W \in \mathbf{IV}(\mathbf{Y},\mathbf{Z}) \cup \mathbf{IP}(\mathbf{Y},\mathbf{Z}) \cup \mathbf{Y}} P(W|\mathbf{Parents}(W))}{\overset{\rightarrow}{\sum}_{\mathbf{IV}(\mathbf{Y},\mathbf{Z}) \cup \mathbf{Y}} \prod_{W \in \mathbf{IV}(\mathbf{Y},\mathbf{Z}) \cup \mathbf{IP}(\mathbf{Y},\mathbf{Z}) \cup \mathbf{Y}} P(W|\mathbf{Parents}(W))}$$

for all values of $\mathbf{V}$ for which the conditional distributions in the factorization are defined, and for which $P(\mathbf{z}) \neq 0$.

Proof. Let $\mathbf{V}' = \mathbf{V} \backslash \mathbf{ND}(\mathbf{YZ})$, that is, the subset of $\mathbf{V}$ with descendants in $\mathbf{YZ}$. It follows from the definition of conditional probability that

$$P(\mathbf{Y}|\mathbf{Z}) = \frac{P(\mathbf{YZ})}{P(\mathbf{Z})} = \frac{\overset{\rightarrow}{\sum}_{\mathbf{V} \backslash \mathbf{YZ}} \prod_{W \in \mathbf{V}} P(W|\mathbf{Parents}(W))}{\overset{\rightarrow}{\sum}_{\mathbf{V} \backslash \mathbf{Z}} \prod_{W \in \mathbf{V}} P(W|\mathbf{Parents}(W))}$$

By lemma 3.3.4,

$$\frac{\overset{\rightarrow}{\sum}_{\mathbf{V} \backslash \mathbf{YZ}} \prod_{W \in \mathbf{V}} P(W|\mathbf{Parents}(W))}{\overset{\rightarrow}{\sum}_{\mathbf{V} \backslash \mathbf{Z}} \prod_{W \in \mathbf{V}} P(W|\mathbf{Parents}(W))} = \frac{\overset{\rightarrow}{\sum}_{\mathbf{V}' \backslash \mathbf{YZ}} \prod_{W \in \mathbf{V}'} P(W|\mathbf{Parents}(W))}{\overset{\rightarrow}{\sum}_{\mathbf{V}' \backslash \mathbf{Z}} \prod_{W \in \mathbf{V}'} P(W|\mathbf{Parents}(W))}$$

First we will show that we can factor the numerator and the denominator into a product of two sums. The second term in both the numerator and the denominator is the same, so it cancels. In the case of the denominator, we show that

$$\overset{\rightarrow}{\sum}_{\mathbf{V}' \backslash \mathbf{Z}} \prod_{W \in \mathbf{V}'} P(W|\mathbf{Parents}(W)) =$$

$$\overset{\rightarrow}{\sum}_{\mathbf{IV}(\mathbf{Y},\mathbf{Z}) \cup \mathbf{Y}} \prod_{W \in \mathbf{IV}(\mathbf{Y},\mathbf{Z}) \cup \mathbf{IP}(\mathbf{Y},\mathbf{Z}) \cup \mathbf{Y}} P(W|\mathbf{Parents}(W)) \times \overset{\rightarrow}{\sum}_{\mathbf{V}' \backslash (\mathbf{IV}(\mathbf{Y},\mathbf{Z}) \cup \mathbf{YZ})} \prod_{W \in \mathbf{V}' \backslash (\mathbf{IV}(\mathbf{Y},\mathbf{Z}) \cup \mathbf{IP}(\mathbf{Y},\mathbf{Z}) \cup \mathbf{Y})} P(W|\mathbf{Parents}(W))$$

by demonstrating that if W is in $\mathbf{IV}(\mathbf{Y},\mathbf{Z}) \cup \mathbf{IP}(\mathbf{Y},\mathbf{Z}) \cup \mathbf{Y}$, then neither W nor any parent of W occurs in the scope of the summation over $\mathbf{V}' \backslash (\mathbf{IV}(\mathbf{Y},\mathbf{Z}) \cup \mathbf{YZ})$, and also that if W is in $\mathbf{V}\backslash(\mathbf{IV(Y,Z)} \cup \mathbf{IP(Y,Z)} \cup \mathbf{Y})$ then neither W nor any parent of W is in the scope of the summation over $\mathbf{IV(Y,Z)} \cup \mathbf{Y}$.

First we demonstrate that if W is in $\mathbf{IV(Y,Z)} \cup \mathbf{IP(Y,Z)} \cup \mathbf{Y}$ then W is not in $\mathbf{V}\backslash(\mathbf{IV(Y,Z)} \cup \mathbf{YZ})$. If W is in $\mathbf{IV(Y,Z)} \cup \mathbf{Y}$ then trivially it is not in $\mathbf{V}\backslash(\mathbf{IV(Y,Z)} \cup \mathbf{YZ})$. If W is in $\mathbf{IP(Y,Z)}$ then W is in $\mathbf{Z}$, so W is not in $\mathbf{V}\backslash(\mathbf{IV(Y,Z)} \cup \mathbf{YZ})$.

Now we will demonstrate that if W is in $\mathbf{IV(Y,Z)} \cup \mathbf{IP(Y,Z)} \cup \mathbf{Y}$ then no parent of W is in $\mathbf{V}\backslash(\mathbf{IV(Y,Z)} \cup \mathbf{YZ})$. Suppose first that W is in $\mathbf{IV(Y,Z)}$ and T is a parent of W. If T is in $\mathbf{IV(Y,Z)} \cup \mathbf{IP(Y,Z)} \cup \mathbf{Y}$ this reduces to the previous case. Assume then that T is not in $\mathbf{IV(Y,Z)} \cup \mathbf{IP(Y,Z)} \cup \mathbf{Y}$. We will show that T is in $\mathbf{YZ}$. T is not d-connected to $\mathbf{Y}$ given $\mathbf{Z}$. However, W, a child of T, is d-connected to $\mathbf{Y}$ given $\mathbf{Z}$ by some path U. If T is on U then T is d-connected to $\mathbf{Y}$ given $\mathbf{Z}$, contrary to our assumption, unless T is in $\mathbf{YZ}$. If T is not on U, and U is not into W, then the concatenation of the edge between T and W with U d-connects T and $\mathbf{Y}$ given $\mathbf{Z}$, contrary to our assumption, unless T is in $\mathbf{YZ}$. If T is not on U, but U is into W, then because W is in $\mathbf{IV(Y,Z)}$ it has a descendant in $\mathbf{YZ}$. If W has a descendant in $\mathbf{Z}$, then W is a collider on the concatentation of the edge between T and W with U, and has a descendant in $\mathbf{Z}$; hence T is d-connected to $\mathbf{Y}$ given $\mathbf{Z}$, contrary to our assumption, unless T is in $\mathbf{YZ}$. If W does not have a descendant in $\mathbf{Z}$, then there is a directed path D from W to $\mathbf{Y}$ that does not contain any member of $\mathbf{Z}$. The concatenation of the edge from T to W and D d-connects T and $\mathbf{Y}$ given $\mathbf{Z}$, contrary to our assumption, unless T is in $\mathbf{YZ}$. In any case, T is in $\mathbf{YZ}$, and not in $\mathbf{V}\backslash(\mathbf{IV(Y,Z)} \cup \mathbf{YZ})$.

Suppose next that W is in $\mathbf{IP(Y,Z)}$ and T is a parent of W. It follows that some parent R of W is in $\mathbf{IV(Y,Z)}$ or in $\mathbf{Y}$, and W is in $\mathbf{Z}$. If T is in $\mathbf{IV(Y,Z)} \cup \mathbf{IP(Y,Z)} \cup \mathbf{Y}$ this reduces to the previous case. Assume then that T is not in $\mathbf{IV(Y,Z)} \cup \mathbf{IP(Y,Z)} \cup \mathbf{Y}$. If R is in $\mathbf{Y}$, then T is d-connected to $\mathbf{Y}$ given $\mathbf{Z}$ by the concatenation of the edge from R to W and the edge from W to T, contrary to our assumption, unless T is in $\mathbf{YZ}$. Hence T is in $\mathbf{YZ}$, and not in $\mathbf{V}\backslash(\mathbf{IV(Y,Z)} \cup \mathbf{YZ})$. Assume then that R is in $\mathbf{IV(Y,Z)}$. R is d-connected to $\mathbf{Y}$ given $\mathbf{Z}$ by some path U. If T is on U then T is d-connected to $\mathbf{Y}$ given $\mathbf{Z}$ unless T is in $\mathbf{YZ}$. If W is on U, but T is not, then W is a collider on U, because W is in $\mathbf{Z}$. W is also a collider on the concatenation of the edge from T to W with the subpath of U from W to $\mathbf{Y}$; hence this path d-connects T and $\mathbf{Y}$ given $\mathbf{Z}$ unless T is in $\mathbf{YZ}$. If neither T nor W is on U, then the concatentation of the edge between T and W, the edge between W and R, and U, is a path on which W is a collider and R is not a collider (because R is a parent of W); hence this path d-connects T and $\mathbf{Y}$ given $\mathbf{Z}$, unless W is in $\mathbf{YZ}$. By hypothesis, T is not d-connected to $\mathbf{Y}$ given $\mathbf{Z}$ because T is not in $\mathbf{IV(Y,Z)}$; it follows that T is in $\mathbf{YZ}$. Hence T is not in $\mathbf{V}\backslash(\mathbf{IV(Y,Z)} \cup \mathbf{YZ})$.

Suppose finally that W is in $\mathbf{Y}$ and T is a parent of W. It follows that T is d-connected to $\mathbf{Y}$ given $\mathbf{Z}$ unless T is in $\mathbf{YZ}$. By hypothesis, T is not d-connected to $\mathbf{Y}$ given $\mathbf{Z}$ because T is not in $\mathbf{IV(Y,Z)}$ so T is in $\mathbf{YZ}$. Hence T is not in $\mathbf{V}\backslash(\mathbf{IV(Y,Z)} \cup \mathbf{YZ})$.

Now we will demonstrate by contraposition that if W is in $\mathbf{V}\backslash(\mathbf{IV(Y,Z)} \cup \mathbf{IP(Y,Z)} \cup \mathbf{Y})$ then neither W nor any parent of W is in the scope of the summation over $\mathbf{IV(Y,Z)} \cup \mathbf{Y}$. Suppose W or some parent T of W is in $\mathbf{IV(Y,Z)} \cup \mathbf{Y}$. If W is in $\mathbf{IV(Y,Z)} \cup \mathbf{Y}$ it follows trivially that W is not in $\mathbf{V'}\backslash(\mathbf{IV(Y,Z)} \cup \mathbf{IP(Y,Z)} \cup \mathbf{Y})$. Suppose T is in $\mathbf{IV(Y,Z)} \cup \mathbf{Y}$ but W is not. We will show that W is in $\mathbf{YZ}$. If T is in $\mathbf{Y}$, then W is d-connected to $\mathbf{Y}$ given $\mathbf{Z}$, contrary to our assumption, unless T is in $\mathbf{YZ}$. If T is in $\mathbf{IV(Y,Z)}$ it follows that there is a path U d-connecting T and $\mathbf{Y}$ given $\mathbf{Z}$. If W is on U, then W is d-connected to $\mathbf{Y}$ given $\mathbf{Z}$, contrary to our hypothesis, unless W is in $\mathbf{YZ}$. If W is not on U, then the concatenation of the edge between W and T with U d-connects W and $\mathbf{Y}$ given $\mathbf{Z}$ (because T is not a collider and not in $\mathbf{Z}$), contrary to our hypothesis, unless W is in $\mathbf{YZ}$. It follows that W is in $\mathbf{YZ}$. If W is in $\mathbf{Z}$, then W is in $\mathbf{IP(Y,Z)}$, and hence not in $\mathbf{V'}\backslash(\mathbf{IV(Y,Z)} \cup \mathbf{IP(Y,Z)} \cup \mathbf{Y})$. If W is in $\mathbf{Y}$, then W is not in $\mathbf{V'}\backslash(\mathbf{IV(Y,Z)} \cup \mathbf{IP(Y,Z)} \cup \mathbf{Y})$. Hence by contraposition, if W is in $\mathbf{V'}\backslash(\mathbf{IV(Y,Z)} \cup \mathbf{IP(Y,Z)} \cup \mathbf{Y})$ then neither W nor any parent of W is in the scope of the summation over $\mathbf{IV(Y,Z)} \cup \mathbf{Y}$.

The proof for the numerator is essentially the same. Hence,

$$\frac{\overset{\rightarrow}{\sum_{\mathbf{V'}\backslash\mathbf{YZ}}} \prod_{W \in \mathbf{V'}} P(W|\mathbf{Parents}(W))}{\overset{\rightarrow}{\sum_{\mathbf{V'}\backslash\mathbf{Z}}} \prod_{W \in \mathbf{V'}} P(W|\mathbf{Parents}(W))} =$$

$$\frac{\overset{\rightarrow}{\sum_{\mathbf{IV(Y,Z)}}} \prod_{W \in \mathbf{IV(Y,Z)} \cup \mathbf{IP(Y,Z)} \cup \mathbf{Y}} P(W|\mathbf{Parents}(W))}{\overset{\rightarrow}{\sum_{\mathbf{IV(Y,Z)} \cup \mathbf{Y}}} \prod_{W \in \mathbf{IV(Y,Z)} \cup \mathbf{IP(Y,Z)} \cup \mathbf{Y}} P(W|\mathbf{Parents}(W))} \times$$

$$\frac{\overset{\rightarrow}{\sum_{\mathbf{V'}\backslash(\mathbf{IV(Y,Z)} \cup \mathbf{YZ})}} \prod_{W \in \mathbf{V'}\backslash(\mathbf{IV(Y,Z)} \cup \mathbf{IP(Y,Z)} \cup \mathbf{Y})} P(W|\mathbf{Parents}(W))}{\overset{\rightarrow}{\sum_{\mathbf{V'}\backslash(\mathbf{IV(Y,Z)} \cup \mathbf{YZ})}} \prod_{W \in \mathbf{V'}\backslash(\mathbf{IV(Y,Z)} \cup \mathbf{IP(Y,Z)} \cup \mathbf{Y})} P(W|\mathbf{Parents}(W))} =$$

$$\frac{\overset{\rightarrow}{\sum_{\mathbf{IV(Y,Z)}}} \prod_{W \in \mathbf{IV(Y,Z)} \cup \mathbf{IP(Y,Z)} \cup \mathbf{Y}} P(W|\mathbf{Parents}(W))}{\overset{\rightarrow}{\sum_{\mathbf{IV(Y,Z)} \cup \mathbf{Y}}} \prod_{W \in \mathbf{IV(Y,Z)} \cup \mathbf{IP(Y,Z)} \cup \mathbf{Y}} P(W|\mathbf{Parents}(W))}$$

∴

LEMMA 3.3.6: In a directed acyclic graph G, if V is d-connected to $\mathbf{Y}$ given $\mathbf{Z}$, and $\mathbf{X}$ is d-separated from $\mathbf{Y}$ given $\mathbf{Z}$, then V is d-connected to $\mathbf{Y}$ given $\mathbf{XZ}$.

Proof. Suppose $\mathbf{X}$ is d-separated from $\mathbf{Y}$ given $\mathbf{Z}$. If V is d-separated from $\mathbf{Y}$ given $\mathbf{XZ}$, but d-connected to $\mathbf{Y}$ given $\mathbf{Z}$, then there is a path U that d-connects V and some Y in $\mathbf{Y}$

given **Z**, but not given **XZ**. It follows that some noncollider X on U is in **X**. Hence $U(X,Y)$ d-connects **X** and **Y** given **Z**. ∴

LEMMA 3.3.7: In a directed acyclic graph G, if V is d-connected to **Y** given **XZ**, and **X** is d-separated from **Y** given **Z**, then V is d-connected to **Y** given **Z**.

Proof. Suppose **X** is d-separated from **Y** given **Z**. If V is d-separated from **Y** given **Z**, but d-connected to **Y** given **XZ**, then there is a path U that d-connects V and **Y** given **XZ**, but not given **Z**. Some vertex on U is a collider with a descendant in **X**, but not in **Z**. Let C be the vertex on U closest to Y that is the source of a directed path to some X in **X** that contains no member of **Z**. C is d-connected to Y given **Z**. If X is on U then $U(X,Y)$ d-connects X and Y given **Z**. If X is not on U, then there is a directed path from C to X that does not contain any member of **Z**, and hence X is d-connected to Y given **Z**, contrary to our assumption. ∴

LEMMA 3.3.8: In a directed acyclic graph G, if **X** is d-separated from **Y** given **Z**, and P satisfies the Markov condition for G, then **X** is independent of **Y** given **Z**.

Proof. We will show if **X** is d-separated from **Y** given **Z** that $P(\mathbf{Y}|\mathbf{XZ}) = P(\mathbf{Y}|\mathbf{Z})$ by showing that **IV(Y,XZ)** = **IV(Y,Z)** and **IP(Y,XZ)** = **IP(Y,Z)** and applying lemma 3.3.5.

Suppose that V is in **IV(Y,Z)**. V is d-connected to **Y** given **Z** and has a descendant in **YZ**. Hence V has a descendant in **XYZ**. It follows by lemma 3.3.6 that V is d-connected to **Y** given **XZ**. Hence V is in **IV(Y,XZ)**.

Suppose then that V is in **IV(Y,XZ)**; we will show that V is also in **IV(Y,Z)**. Because V is in **IV(Y,XZ)**, V is not in **XYZ**, V has a descendant in **XYZ** and is d-connected to **Y** given **XZ**. Because V is not in **XYZ** it is not in **XZ**. By lemma 3.3.7 V is d-connected to **Y** given **Z**. If V has a member X of **X** as a descendant, but no member of **YZ** as a descendant then there is a directed path from V to X that contains no member of **Y** or **Z**. It follows by lemma 3.3.3 that X is d-connected to **Y** given **Z**, contrary to our hypothesis. Hence V has a member of **YZ** as a descendant, and is in **IV(Y,Z)**.

Suppose that V is in **IP(Y,Z)**. If V has a parent in **Y**, then **V** is in **IP(Y,XZ)**. If V has a parent T in **IV(Y,Z)** then T is in **IV(Y,XZ)** because **IV(Y,Z)** = **IV(Y,XZ)**. Hence V is in **IP(Y,XZ)**.

Suppose that V is in **IP(Y,XZ)**. Because V is in **IP(Y,XZ)** V is in **XZ** and has a parent in **IV(Y,XZ)** ∪ **Y**. We have already shown that **IV(Y,XZ)** ∪ **Y** = **IV(Y,Z)** ∪ **Y**. We will now show that V is not in **X**. If V is in **X** and has a member of **Y** as a parent, then **X** is d-connected to **Y** given **Z**, contrary to our hypothesis. If V is in **X** and has some W in **IV(Y,XZ)** as a parent, then W is in **IV(Y,Z)**. It follows that X is d-connected to **Y** given **Z**, contrary to our hypothesis. Hence V is not in **X**, and **IP(Y,XZ)** = **IP(Y,Z)**.

By lemma 3.3.5, $P(\mathbf{Y}|\mathbf{XZ}) = P(\mathbf{Y}|\mathbf{Z})$, and hence **X** is independent of **Y** given **Z**. ∴

Proofs of Theorems

LEMMA 3.3.9: In a directed acyclic graph G, if X is not a descendant of Y, and X and Y are not adjacent, then X and Y are d-separated by **Parents**(Y).

Proof. (A slight variant of this is stated in Pearl 1989.) Suppose on the contrary that some undirected path U d-connects X and Y given **Parents**(X). If U is into Y then it contains some member of **Parents**(Y) not equal to X as a noncollider. Hence it does not d-connect X and Y given **Parents**(Y), contrary to our assumption. If U is out of Y, then because X is not a descendant of Y, U contains a collider. Let C be the collider on U closest to Y. If U d-connects X and Y given **Parents**(Y) then C has a descendant in **Parents**(Y). But then C is an ancestor of Y, and Y is an ancestor of C, so G is cyclic, contrary to our assumption. Hence no undirected path between X and Y d-connects X and Y given **Parents**(Y). ∴

THEOREM 3.3: $P(\mathbf{V})$ is faithful to directed acyclic graph G with vertex set $\mathbf{V}$ if and only if for all disjoint sets of vertices $\mathbf{X}$, $\mathbf{Y}$, and $\mathbf{Z}$, $\mathbf{X}$, and $\mathbf{Y}$ are independent conditional on $\mathbf{Z}$ if and only if $\mathbf{X}$ and $\mathbf{Y}$ are d-separated given $\mathbf{Z}$.

Proof. ⇒ Suppose that P is faithful to G. It follows that P satisfies the Markov condition for G. By lemma 3.3.8 if $\mathbf{X}$ and $\mathbf{Y}$ are d-separated given $\mathbf{Z}$ then $\mathbf{X}$ and $\mathbf{Y}$ are independent conditional on $\mathbf{Z}$. By lemma 3.5.8 (proved below) there is a distribution P' that satisfies the Markov condition for G such that if $\mathbf{X}$ and $\mathbf{Y}$ are not d-separated given $\mathbf{Z}$ then $\mathbf{X}$ and $\mathbf{Y}$ are not independent conditional on $\mathbf{Z}$. It follows that if $\mathbf{X}$ and $\mathbf{Y}$ are not d-separated given $\mathbf{Z}$ then the Markov condition does not entail that $\mathbf{X}$ and $\mathbf{Y}$ independent conditional on $\mathbf{Z}$.

⇐ Suppose that $\mathbf{X}$ and $\mathbf{Y}$ are independent conditional on $\mathbf{Z}$ in P if and only if $\mathbf{X}$ and $\mathbf{Y}$ are d-separated given $\mathbf{Z}$. It follows from lemma 3.3.9 that that P satisfies the Markov condition for G because **Parents**(V) d-separates V from V\\(**Descendants**(V) ∪ **Parents**(V)). Hence all of the conditional independence relations entailed by the Markov condition are true of P. If the independence of $\mathbf{X}$ and $\mathbf{Y}$ conditional on $\mathbf{Z}$ is not entailed by the Markov condition for G then by lemma 3.5.8 $\mathbf{X}$ and $\mathbf{Y}$ are not d-separated in G, and $\mathbf{X}$ and $\mathbf{Y}$ are not independent conditional on $\mathbf{Z}$. It follows that P is faithful to G. ∴

13.5 Theorem 3.4

THEOREM 3.4: If $P(\mathbf{V})$ is faithful to some directed acyclic graph, then $P(\mathbf{V})$ is faithful to directed acyclic graph G with vertex set $\mathbf{V}$ if and only if

(i) for all vertices X, Y of G, X, and Y are adjacent if and only if X and Y are dependent conditional on every set of vertices of G that does not include X or Y; and

(ii) for all vertices X, Y, Z such that X is adjacent to Y and Y is adjacent to Z and X and Z are not adjacent, $X \rightarrow Y \leftarrow Z$ is a subgraph of G if and only if X, Z are dependent conditional on every set containing Y but not X or Z.

Proof. The theorem follows from a theorem first proved in Verma and Pearl 1990b. ∴

13.6 Theorem 3.5

THEOREM 3.5: Let S be an LCT with directed acyclic graph G over the set of non-error variables **V**. Then for any two non-error vertices A, B in **V** and any subset **H** of **V**\{A,B}, G linearly implies that $\rho_{AB.H} = 0$ if and only if A, B are d-separated given **H**.

The **distributed form** of an expression or equation E is the result of carrying out every multiplication, but no additions, subtractions, or divisions in E. If there are no divisions in an equation then its distributed form is a sum of terms. For example, the distributed form of the equation $u = (a + b)(c + d)v$ is $u = acv + adv + bcv + bdv$. In an LCF or LCT T, if an expression is equal to ce, where c is a nonzero constant, and e is a product of equation coefficients raised to positive integral powers, then e is the **equation coefficient factor**(e.c.f.) of ce, and c is the **constant factor** (c.f.) of ce.

An acyclic directed graph G over **V** is an **I-map** of probability distribution $P(\mathbf{V})$ iff for every **X**, **Y**, and **Z** that are disjoint sets of random variables in **V**, if **X** is d-separated from **Y** given **Z** in G then **X** is independent of **Y** given **Z** in $P(\mathbf{V})$. An acyclic graph G is a **minimal I-map** of probability distribution P iff G is an I-map of P, and no proper subgraph of G is an I-map of P. An acyclic graph G over **V** is a **D-map** of probability distribution $P(\mathbf{V})$ iff for every **X**, **Y**, and **Z** that are disjoint sets of random variables in **V**, if **X** is not d-separated from **Y** given **Z** in G then **X** is not independent of **Y** given **Z** in $P(\mathbf{V})$. However, when minimal I-map, I-map, or D-map is applied to the graph in an LCT or LCF, the quantifiers in the definitions apply only to sets of *non-error* variables.

A **trek** $T(I,J)$ between two distinct vertices I and J is an unordered pair of acyclic directed paths from some vertex K to I and J respectively that intersect only at K. The source of the paths in the trek is called the **source** of the trek. I and J are called the **termini of the trek**. Given a trek $T(I,J)$ between I and J, $I(T(I,J))$ will denote the path in $T(I,J)$ from the source of $T(I,J)$ to I and $J(T(I,J))$ will denote the path in $T(I,J)$ from the source of $T(I,J)$ to J. One of the paths in a trek may be an empty path. However, since the termini of a trek are distinct, only one path in a trek can be empty. $\mathbf{T}(I,J)$ is the set of all treks between I and J. $T(I,J)$ will represent a trek in $\mathbf{T}(I,J)$. $S(T(I,J))$ represents the source of the trek $T(I,J)$.

The proofs of the following two lemmas are trivial.

LEMMA 3.5.1: In a directed acyclic graph G, every undirected path $V = <V_1,V_2,...V_{n-1},V_n>$ without colliders contains a vertex V_k such that $<V_k,...,V_1>$ and $<V_k,...,V_n>$ are directed subpaths of V that intersect only at V_k.

Hence, corresponding to each undirected path $V = <V_1,V_2,...V_{n-1},V_n>$ without colliders is a trek $T = (<V_k,...,V_1>,<V_k,...,V_n>)$. When V is a directed path, one of the paths is empty; for example, $V_k = V_1$.

LEMMA 3.5.2: In a directed acyclic graph G, for every trek $(<V_1,...,V_n>,<V_1,...,V_m>)$, the concatenation of $<V_n,...,V_1>$ with $<V_1,...,V_m>$ is an undirected path from V_n to V_m without colliders.

We will say that a directed acyclic graph has error variables if every vertex of indegree not equal to 0 has an edge into it from a vertex of indegree 0 and outdegree 1. If each independent random variable in an LCT S is normally distributed, then the joint distribution of the set of all random variables in the LCT is multivariate normal. We will say the random variables in such an LCT have a linear multivariate normal distribution. The next series of lemmas demonstrate that every directed acyclic graph with error variables is faithful to some LCT S in which the joint distribution Q of the random variables in S is linear multivariate normal.

LEMMA 3.5.3: If S is an acyclic multivariate normal LCT with directed acyclic graph G' and distribution P, **V** is the set of non-error terms in S, G is the subgraph of G' over **V**, and the exogenous variables are jointly independent, then G is a minimal I-map of $P(\mathbf{V})$.

Proof. Let **V** be the set of non-error terms in S, and G be the subgraph of G' over **V**. First we will show that if A and B are distinct variables in **V**, and B is not a descendant of A or a parent of A in G, then A is independent of B given **Parents**(G,A). ε_A is normally distributed and uncorrelated with any of the parents of A or B. B is not a linear function of **Parents**(G,A) because the distribution is positive. Hence, if we write A as a linear function of **Parents**(G,A), B, and ε_A, this is a regression model of A. The coefficient of B in such an equation is zero. The coefficient of B in such a linear equation for A is zero if and only if A and B are independent conditional on **Parents**(G,A). (See Whittaker 1990.) Hence B is independent of A given **Parents**(G,A). Because the joint distribution is normal, it follows that A is independent of the set of its nonparental nondescendants given its parents. Hence G is an I-map of $P(\mathbf{V})$.

We will now show that $P(\mathbf{V})$ satisfies the Minimality Condition for G. Suppose, on the contrary, that G is not a minimal I-map of $P(\mathbf{V})$. It follows that some some subgraph of G is an I-map of $P(\mathbf{V})$. Let G_{Sub} be a subgraph of G that is an I-map of $P(\mathbf{V})$, and in which the only difference between G and G_{Sub} is that X is a parent of Y in G, but not in G_{Sub}. Because **Parents**$(G_{Sub},Y) \cup \{X\} = $ **Parents**(G,Y), when Y is written as a linear function of **Parents**(G_{Sub},Y), X, and ε_Y, the coefficient of X is not zero. But because X is not a parent of Y in G_{Sub}, and not a descendant of Y in G_{Sub}, it follows that X and Y are d-separated given **Parents**(G_{Sub},Y). Because G_{Sub} is an I-map of P(**V**), X and Y are independent given **Parents**(G_{Sub},Y). But this entails that the coefficient of X in the linear equation for Y in terms of **Parents**(G,Y) and ε_Y is zero, which is a contradiction. ∴

LEMMA 3.5.4: If a polynomial equation Q in real variables <X_1,...,X_n> is not an identity, then for every solution a of Q, and for every $\varepsilon > 0$ there is a nonsolution b of Q such that |b - a| < ε, where |b - a| is the Euclidean distance between a and b.

Proof. The proof is by induction on the number n of variables in Q.

Base case: If n = 1, then there are only a finite number of solutions of Q. It follows that for every solution a of Q, and for every $\varepsilon > 0$ there is a nonsolution b of Q such that |b - a| < ε.

Induction case: Suppose that Q is a polynomial equation in <X_1,...,X_n>, Q is not an identity, and the lemma is true for n–1. Take an arbitrary solution <a_1,...,a_n> of Q. Transform Q into a polynomial equation Q' in X_n by fixing the variables <X_1,...,X_{n-1}> at the value <a_1,...,a_{n-1}>. There are two cases.

In the first case, Q' is not an identity. Hence, by the induction hypothesis, there is a nonsolution of Q' whose distance from a_n is < ε. Let a'_n be this nonsolution of Q'. Then a' = <a_1,...a_{n-1},a'_n> is a nonsolution of Q, and |a - a'| < ε.

In the second case, Q' is an identity. Rewrite Q so that it is of the form

$$\sum_m Q_m X_n^m$$

where each Q_m is a polynomial in at most X_1,...,X_{n-1}.

For each m, the equation $Q_m = 0$ is a polynomial equation in less than n variables. If Q' is an identity, then when terms of the same power of X_n are added together, the coefficient of each power of X_n is zero. This implies that <a_1,...,a_{n-1}> is a solution to $Q_m = 0$ for each m. If, for each m, $Q_m = 0$ is an identity, then so is Q; hence for some m, $Q_m = 0$ is not an identity. For this value of m, by the induction hypothesis, there is a nonsolution <a'_1,...,a'_{n-1}> to $Q_m = 0$ that is less than distance ε from <a_1,...,a_{n-1}>. If <a'_1,...,a'_{n-1}> is substituted for <X_1,...,X_{n-1}> in Q, the resulting polynomial equation in X_n is not an identity. This reduces to the first case. ∴

LEMMA 3.5.5: If G' is a subgraph of G, and there is some LCT S' with directed acyclic graph G' and distribution P' such that $\rho_{IJ.Z} \neq 0$ in P', then there is some LCT S containing G and distribution P such that $\rho_{IJ.Z} \neq 0$ in P.

Proof. By lemma 3.2.1, in S' $\rho_{IJ.\mathbf{Z}} = 0$ is equivalent to a polynomial equation in the linear coefficients and variances of independent variables in S'. Since there is some LCT S' containing G' such that $\rho_{IJ.\mathbf{Z}} \neq 0$ in S', the polynomial equation is not an identity.

Let S be an LCT with directed acyclic graph G such that for all variables J, I, if the coefficient c' of J in the equation for I in S' is not equal to zero, then the coefficient of J in the equation for I in S is equal to c'. In S, $\rho_{IJ.\mathbf{Z}} = 0$ is equivalent to a polynomial equation E in the linear coefficients and variances of independent variables in S. When labels of the edges in G but not in G' are set to zero, the polynomial in E equals the polynomial in E'. No label of an edge in G but not in G' occurs in E'. Hence when the labels of the edges in G but not in G' are set to nonzero values, the polynomial in E contains all of the terms that are in E' and possibly some extra terms. Let us say that two terms in a polynomial equation are **like terms** if they contain the same variables raised to the same powers. Each of the terms that are in E but not E' contain some linear coefficient that does not appear in any term in E'; hence each of the additional terms in E is not like any term in E'.

If E were an identity, then the sum of the coefficients of like terms in E would be equal to zero. Since E' is not an identity, there are like terms in E' such that the sum of their coefficients is not zero. These same like terms appear in E. Furthermore, since the only additional terms in E that are not in E' are not like any term in E', it follows that if the sum of the coefficients of like terms in E' is not zero, then the sum of the coefficients of the same like terms in E is not identically zero. Hence E is not identically zero, and there is some LCT S containing G such that $\rho_{IJ.\mathbf{Z}} \neq 0$ in S. $\therefore$

The next lemma states that given a set $\mathbf{Z}$ of partial correlations and a directed acyclic graph G, if it is possible to construct a set $\mathbf{S}$ of LCTs with directed acyclic graph G such that each Z in $\mathbf{Z}$ fails to vanish for some one of the LCTs in $\mathbf{S}$, then it is possible to construct a single LCT with directed acyclic graph G such that all of the Z in $\mathbf{Z}$ fail to vanish.

LEMMA 3.5.6: Given a set of partial correlations $\mathbf{Z}$ and a directed acyclic graph G, if for all Z in $\mathbf{Z}$ there exists an LCT S' with directed acyclic graph G and distribution P' such that $Z \neq 0$ in P', then there exists a single LCT S with directed acyclic graph G and distribution P such that for all Z in $\mathbf{Z}$, $Z \neq 0$ in P.

Proof. The proof is by induction on the cardinality of $\mathbf{Z}$.

Base Case: If the only member of $\mathbf{Z}$ is Z, then by assumption there is an LCT S containing G such that $Z \neq 0$.

Induction Case: Suppose that the lemma is true for each set of cardinality $n-1$, $\mathbf{Z}$ is of cardinality n, and for each Z_i in $\mathbf{Z}$, there is an LCT S' with directed acyclic graph G and distribution P' such that $Z_i \neq 0$ in P'. By the induction hypothesis, there is an LCT S with directed acyclic graph G and distribution P such that $Z_i \neq 0$, $i \leq 1 \leq n-1$. Let **V** be a set of values for the linear coefficients and variances of independent variables such that $Z_i \neq 0$, $i \leq 1 \leq n-1$. The valuation **V** either makes Z_n equal to zero or it doesn't. If it doesn't, then the proof is done. If it does, we will show how to perturb V by a small amount to make $Z_n \neq 0$, while keeping each $Z_i \neq 0$, $i \leq 1 \leq n-1$.

By lemma 3.2.1, each of the partial correlations in Z_i in **Z** is equivalent to a polynomial Q_i in the linear coefficients and the variances of independent variables in G. Suppose that the smallest nonzero value for any of the Q_i under the valuation **V** is δ. By lemma 3.5.4, for arbitrarily small ε there is a nonsolution V' to $Z_n = 0$ within distance ε of **V**. Choose an ε small enough so that the largest possible change in any of the Q_i is less than δ. For the valuation **V'** then $Z_i \neq 0$, $i \leq 1 \leq n$. ∴

Recall that if a graph with error variables is a D-map of some distribution P, then we consider only dependencies among the non-error variables.

LEMMA 3.5.7: For every directed acyclic graph G with error variables, there is an LCT S with directed acyclic graph G and joint linear multivariate normal distribution Q, such that G is a D-map of Q.

Proof. In order to show that G is a D-map of Q, we must show that for all disjoint sets of variables **X**, **Y**, and **Z**, if **X** and **Y** are not d-separated in G, then **X** is not independent of **Y** given **Z** in Q. In a linear multivariate normal distribution, if **X**, **Y**, and **Z** are disjoint sets of variables, then $\mathbf{X} \perp\!\!\!\perp \mathbf{Y}|\mathbf{Z}$ iff $X \perp\!\!\!\perp Y|\mathbf{Z}$ for each X in **X** and Y in **Y**; similarly if **X**, **Y**, and **Z** are disjoint sets of variables then **X** and **Y** are d-separated given **Z** iff for all X in **X** and Y in **Y**, X and Y are d-separated given **Z**. Hence, we need consider only dependency statements of the form X and Y are not independent given **Z**, where X and Y are individual variables. Also in a linear multivariate normal distribution, $\rho_{XY\cdot \mathbf{Z}} = 0$ iff $X \perp\!\!\!\perp Y|\mathbf{Z}$. So it suffices to prove that there is an LCT S with directed acyclic graph G and distribution P such that for each X, Y, and **Z** in G such that X and Y are not d-separated given **Z** in G, $\rho_{XY\cdot \mathbf{Z}} \neq 0$ in P. The proof is by induction. We assume that in all of the LCTs constructed, the independent random variables are normally distributed.

Base Case: If **Z** is empty, then by lemma 3.5.1, X and Y are not d-separated given **Z** iff there is a trek connecting them. Form a subgraph G' and a sub-LCT S' with directed acyclic graph G' and distribution P', such that there is exactly one trek between X and Y. It was proved in Glymour et al. (1987) that in this case the covariance between X and Y is equal to the product of the labels of the edges in the trek (the linear coefficients) times the variance of the source of the trek. If each of these quantities is nonzero, so is the covariance, and also the correlation in P'. By lemma 3.5.5 if ρ_{XY} is not identically zero in S' it is also not identically zero in some LCT S with directed acyclic graph G. By lemma 3.5.6 there exists a LCT containing G in which for all X and Y, if X and Y are not d-separated by the empty set then the correlation between X and Y is not zero.

Induction Case: Suppose that there is an LCT S with directed acyclic graph G and distribution P such that for each X, Y, and for each $\mathbf{A}$ of cardinality less than n that does not contains X or Y, such that X and Y are *not* d-separated given $\mathbf{A}$ in G, $\rho_{XY.\mathbf{A}} \neq 0$ in P. Let $\mathbf{Z}$ be of cardinality n. Suppose that X and Y are not d-separated by $\mathbf{Z}$ in G. It follows that there is an undirected path U between X and Y such that every noncollider is not in $\mathbf{Z}$, and every vertex V_i on U that is a collider is the source of a directed path U_i from V_i to a variable in $\mathbf{Z}$. Form a subgraph G', such that G' contains only the undirected path U, one directed path U_i from each collider V_i on U, the vertices in those paths, and the vertices in $\mathbf{Z}$. Shorten each U_i so that it contains only one variable in $\mathbf{Z}$. Finally, if two variables V_n and V_m that are colliders on U are the sources of directed paths U_n and U_m that intersect, let F be the first point of intersection of U_n and U_m. Replace the subpath of U from V_n to V_m by the concatenation of the subpaths of $U_n(V_n,F)$ and $U_m(F,V_m)$, and replace U_n and U_m by $U_n(F,Z)$, where Z is in $\mathbf{Z}$. The new path has one fewer collider than the old path. Repeat this process until none of the U_i intersect each other or there are no colliders on U. There are two cases.

In the first case, U contains no vertices with a collider, and hence no vertices in $\mathbf{Z}$. By lemma 3.5.1 there is a trek between X and Y that contains no vertices in $\mathbf{Z}$. Let R be an arbitrary vertex in $\mathbf{Z}$, and $\mathbf{W} = \mathbf{Z}\setminus\{R\}$. There is a trek between X and Y that contains no vertices in $\mathbf{W}$. It follows that $\mathbf{W}$ does not d-separate X and Y, so by the induction hypothesis, there is an LCT with directed acyclic graph G' and distribution P' such that $\rho_{XY.\mathbf{W}} \neq 0$. It follows from lemma 3.5.3 that in P' that $\rho_{XR.\mathbf{W}} = 0$ and $\rho_{YR.\mathbf{W}} = 0$ because by construction there are no undirected paths from X to R or Y to R. By the recursion formula for partial correlation, $\rho_{XY.\mathbf{Z}} = 0$ iff $\rho_{XY.\mathbf{W}} = \rho_{XR.\mathbf{W}} \times \rho_{YR.\mathbf{W}}$. But $\rho_{XY.\mathbf{W}}$ is nonzero in P', and $\rho_{XR.\mathbf{W}} \times \rho_{YR.\mathbf{W}}$ is zero in P'. Hence $\rho_{XY.\mathbf{Z}} \neq 0$ in P'. By lemma 3.5.5, there is some LCT S'' with directed acyclic graph G and distribution P'' such that $\rho_{XY.\mathbf{Z}} \neq 0$ in P''.

In the second case, U contains vertices with colliders, but every vertex that is not a collider is not in $\mathbf{Z}$. (See figure 13.1.)

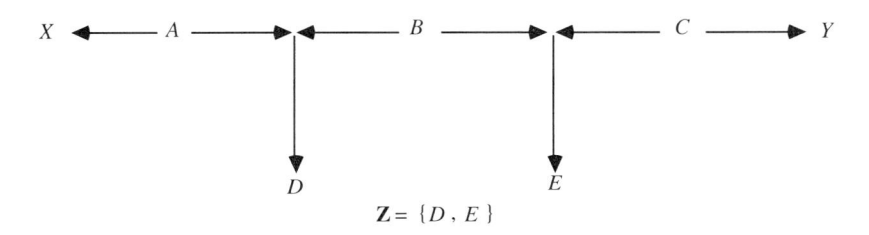

```
X ←──── A ────→←──── B ────→←──── C ────→ Y
                │              │
                ↓              ↓
                D              E
            Z = {D, E}
```

Figure 13.1

Let E be the vertex that is the sink of the directed path from the collider closest to Y on U, and $\mathbf{W} = \mathbf{Z}\setminus\{E\}$. Since by construction there is a trek between Y and E that does not contain any variables in $\mathbf{W}$, Y and E are not d-separated by $\mathbf{W}$. There is also an undirected path from X to E such that every vertex that is not a collider is not in $\mathbf{W}$, and every vertex that does contain a collider has a descendant in $\mathbf{W}$. Hence X and E are not d-separated by $\mathbf{W}$. By the induction hypothesis, there is an LCT S' with directed acyclic graph G' and distribution P' such that $\rho_{XE.\mathbf{W}} \neq 0$, and $\rho_{YE.\mathbf{W}} \neq 0$ in P'.

On the other hand, since path U was constructed so that each vertex that is a collider has only one descendant in $\mathbf{Z}$, and $\mathbf{W}$ does not contain E, X and Y are d-separated by $\mathbf{W}$. Hence by lemma 3.5.3 $\rho_{XY.\mathbf{W}} = 0$ in P'.

$\rho_{XY.\mathbf{W}} = 0$ iff $\rho_{XY.\mathbf{W}} = \rho_{XE.\mathbf{W}} \times \rho_{YE.\mathbf{W}}$. Since $\rho_{XY.\mathbf{W}} = 0$, while $\rho_{XE.\mathbf{W}} \times \rho_{YE.\mathbf{W}} \neq 0$, $\rho_{XY.\mathbf{Z}} \neq 0$ in P'. By lemma 3.5.5, there is an LCT S'' with directed acyclic graph G and distribution P'' such that $\rho_{XY.\mathbf{Z}} \neq 0$ in P''.

Since for each triple X, Y, $\mathbf{Z}$ such that X and Y are not d-separated given $\mathbf{Z}$ in G there is an LCT S' with directed acyclic graph G and distribution P' such that $\rho_{XY.\mathbf{Z}} \neq 0$ in P', by lemma 3.5.6 there is an LCT S'' with directed acyclic graph G and distribution P'' such that for each triple X, Y, $\mathbf{Z}$ for which X and Y are not d-separated given $\mathbf{Z}$ in G, $\rho_{XY.\mathbf{Z}} \neq 0$ in P''. Because the LCTs constructed in lemmas 3.5.5 and 3.5.6 don't change the normality of the independent variables, the joint distribution of the random variables in S is linear multivariate normal. Hence there is an LCT S such that Q is a linear multivariate normal distribution and G is a D-map of Q. ∴

LEMMA 3.5.8: For every directed acyclic graph G with error variables, there is an LCT S containing G with a linear multivariate normal distribution Q such that G is faithful to Q.

Proof. This follows immediately from lemmas 3.5.7 and 3.5.3. ∴

The next theorem states that the d-separability relations between sets of non-error variables can be determined from a subgraph that does not include error terms.

LEMMA 3.5.9: In an acyclic LCT S with directed acyclic graph G, let G' be the subgraph of G over the non-error variables. Given three disjoint sets $\mathbf{X}$, $\mathbf{Y}$, and $\mathbf{Z}$ of non-error variables, $\mathbf{X}$ is d-separated from $\mathbf{Y}$ given $\mathbf{Z}$ in G iff $\mathbf{X}$ is d-separated from $\mathbf{Y}$ given $\mathbf{Z}$ in G'.

Proof. If an error variable occurs on an undirected path, then that error variable is either the source or the sink of the undirected path. Hence, error variables do not occur on any undirected path between non-error variables. It follows that the undirected paths in G and G' between non-error variables are exactly the same. The lemma then follows from the definition of d-separability. ∴

A directed acyclic graph G **linearly implies** $\rho_{AB.\mathbf{H}} = 0$ if and only if $\rho_{AB.\mathbf{H}} = 0$ in all distributions linearly represented by G. (We assume all partial correlations exist for the distribution.) Kiiveri and Speed (1982) explicitly notes the connection between the Markov Condition and zero partial correlations.

LEMMA 3.5.10: In an LCT S with directed acyclic graph G over the set of non-error variables $\mathbf{V}$ and the distribution $P(\mathbf{V})$, if $\mathbf{Y}$ d-separates X and Z, then S linearly implies that $\rho_{XZ.\mathbf{Y}} = 0$.

Proof. Suppose $\mathbf{Y}$ d-separates X and Z in G. The values of the partial correlations in $P(\mathbf{V})$ are completely determined by the values of the linear coefficients and the variances of the independent variables. Consider a multivariate normal distribution $P'(\mathbf{V})$ in the LCT with the same linear coefficients and the same variances of independent variables as S, but in which the independent variables are normally distributed and jointly independent. By lemma 3.5.3, G is an I-map of $P'(\mathbf{V})$, and because $\mathbf{Y}$ d-separates X and Z, $X \perp\!\!\!\perp Z|\mathbf{Y}$ in $P'(\mathbf{V})$. Because $P'(\mathbf{V})$ is a multivariate normal distribution, $X \perp\!\!\!\perp Z|\mathbf{Y}$ if and only $\rho_{XZ.\mathbf{Y}} = 0$. It follows that $\rho_{XZ.\mathbf{Y}} = 0$ in $P'(\mathbf{V})$, and hence $\rho_{XZ.\mathbf{Y}} = 0$ in $P(\mathbf{V})$. ∴

THEOREM 3.5: Let S be an LCT with directed acyclic graph G over the set of non-error variables $\mathbf{V}$. Then for any two non-error vertices A, B in $\mathbf{V}$ and any subset $\mathbf{H}$ of $\mathbf{V}\setminus\{A,B\}$, G linearly implies that $\rho_{AB.\mathbf{H}} = 0$ if and only if A, B are d-separated given $\mathbf{H}$.

Proof. The if clause follows from lemma 3.5.10.
The only if clause follows from lemma 3.5.7. By lemma 3.5.7 there is an LCT S such that Q, the joint distribution of the random variables is linear multivariate normal, and G is a D-map of Q. In S, if A and B are not d-separated given $\mathbf{H}$, then A and B are not independent given $\mathbf{H}$, and $\rho_{AB.\mathbf{H}} \neq 0$. Hence if A and B are not d-separated given $\mathbf{H}$, G does not linearly imply that $\rho_{AB.\mathbf{H}} = 0$. ∴

COROLLARY 3.5.1: In an LCT $S = <G, (\Omega,f,P), \mathbf{EQ}, L>$ in which the exogenous variables are jointly independent, if X and Z are distinct non-error variables, and $\mathbf{Y}$ is a set of non-error variables not including X and Z, if $\rho_{XZ.\mathbf{Y}}$ is linearly implied to vanish then $X \perp\!\!\!\perp Z|\mathbf{Y}$.

COROLLARY 3.5.2: In an LCT $S = <G, (\Omega,f,P), \mathbf{EQ}, L>$, if P is faithful to G, X and Z are distinct non-error variables, and $\mathbf{Y}$ is a set of non-error variables not including X and Z, G linearly implies that $\rho_{XZ.\mathbf{Y}} = 0$ if and only if $X \perp\!\!\!\perp Z|\mathbf{Y}$.

13.7 Theorem 3.6 (Manipulation Theorem)

THEOREM 3.6: (Manipulation Theorem): Given directed acyclic graph G_{Comb} over vertex set $\mathbf{V} \cup \mathbf{W}$ and distribution $P(\mathbf{V} \cup \mathbf{W})$ that satisfies the Markov condition for G_{Comb}, if changing the value of $\mathbf{W}$ from $\mathbf{w1}$ to $\mathbf{w2}$ is a manipulation of G_{Comb} with respect to $\mathbf{V}$, G_{Unman} is the unmanipulated graph, G_{Man} is the manipulated graph, and

$$P_{Unman(\mathbf{W})}(\mathbf{V}) = \prod_{X \in \mathbf{V}} P_{Unman(\mathbf{W})}(X|\mathbf{Parents}(G_{Unman}, X))$$

for all values of **V** for which the conditional distributions are defined, then

$$P_{Man(\mathbf{W})}(\mathbf{V}) =$$
$$\prod_{X \in \mathbf{Manipulated(W)}} P_{Man(\mathbf{W})}(X|\mathbf{Parents}(G_{Man}, X)) \times$$
$$\prod_{X \in \mathbf{V} \setminus \mathbf{Manipulated(W)}} P_{Unman(\mathbf{W})}(X|\mathbf{Parents}(G_{Unman}, X))$$

for all values of **V** for which each of the conditional distributions is defined.

If G is a directed acyclic graph over a set of variables $\mathbf{V} \cup \mathbf{W}$, and $\mathbf{V} \cap \mathbf{W} = \emptyset$, then **W is exogenous with respect to V** in G if and only if there is no directed edge from any member of **V** to any member of **W**. If G_{Comb} is a directed acyclic graph over a set of variables $\mathbf{V} \cup \mathbf{W}$, and $P(\mathbf{V} \cup \mathbf{W})$ satisfies the Markov condition for G_{Comb}, then changing the value of **W** from **w1** to **w2** is a **manipulation** of G_{Comb} with respect to **V** if and only if **W** is exogenous with respect to **V**, and $P(\mathbf{V}|\mathbf{W} = \mathbf{w1}) \neq P(\mathbf{V}|\mathbf{W} = \mathbf{w2})$.

We define $P_{Unman(\mathbf{W})}(\mathbf{V}) = P(\mathbf{V}|\mathbf{W} = \mathbf{w1})$, and $P_{Man(\mathbf{W})}(\mathbf{V}) = P(\mathbf{V}|\mathbf{W} = \mathbf{w2})$, and similarly for various marginal and conditional distributions formed from $P(\mathbf{V})$.

We refer to G_{Comb} as the **combined graph**, and the subgraph of G_{Comb} over **V** as the **unmanipulated graph** G_{Unman}.

V is in **Manipulated(W)** (that is, V is a variable directly influenced by one of the manipulation variables) if and only if V is in **Children(W)** $\cap$ **V**; we will also say that the variables in **Manipulated(W)** have been **directly manipulated**. We will refer to the variables in **W** as **policy variables**.

The **manipulated graph**, G_{Man} is a subgraph of G_{Unman} for which $P_{Man(\mathbf{W})}(\mathbf{V})$ satisfies the Markov Condition and which differs from G_{Unman} in at most the parents of members of **Manipulated(W)**.

Lemmas 3.6.1 and 3.6.2 show that distributions satisfying the antecedent of theorem 3.6 exist.

In a directed acyclic graph G over **V**, X is in **Nondescendants**$(G,\mathbf{Y})$ if and only if X is in **V** and there is no directed path from any member of **Y** to X in G.

LEMMA 3.6.1: Given directed acyclic graph G_{Comb} over vertex set $\mathbf{V} \cup \mathbf{W}$ and distribution $P(\mathbf{V} \cup \mathbf{W})$ that satisfies the Markov condition for G, if changing the value of **W** from **w1** to **w2** is a manipulation of G_{Comb} with respect to **V**, and G_{Unman} is the unmanipulated graph, then $P_{Unman(\mathbf{W})}(\mathbf{V})$ satisfies the Markov Condition for G_{Unman}.

Proof. $P_{Unman(\mathbf{W})}(\mathbf{V})$ satisfies the Markov Condition for G_{Unman} if for each vertex V in **V**, V is independent of **Nondescendants**$(G_{Unman},V)\setminus$**Parents**(G_{Unman},V) conditional on **Parents**$(G_{Unman},V) \cup \mathbf{W}$. Suppose that on the contrary that for some V in **V**, V is dependent on **Nondescendants**(G_{Unman},V)**Parents**(G_{Unman},V) conditional on **Parents** (G_{Unman},V) $\cup$ **W**. It follows that there is some path U in G_{Comb} that d-connects V and some member X in **Nondescendants**(G_{Unman},V) given **Parents**(G_{Unman},V) $\cup$ **W**. Every member of **W** that occurs on U is a collider on U because U d-connects X and V given **Parents**(G_{Unman},V) $\cup$ **W**. Because **W** is exogenous to **V**, U contains no member of **W**. It follows that no collider on U has a descendant in **W**. Hence U d-connects V and X given **Parents**(G_{Unman},V) in G_{Comb}. The path corresponding to U in G_{Unman} also d-connects V and X given **Parents**(G_{Unman},V). But this contradicts lemma 3.3.9. ∴

LEMMA 3.6.2: Given directed acyclic graph G_{Comb} over vertex set **V** $\cup$ **W** and distribution $P(\mathbf{V} \cup \mathbf{W})$ that satisfies the Markov condition for G_{Comb}, if changing the value of **W** from **w1** to **w2** is a manipulation of G_{Comb} with respect to **V**, and G_{Unman} is the unmanipulated graph, then $P_{Man(\mathbf{W})}(\mathbf{V})$ satisfies the Markov Condition for some subgraph of G_{Unman}.

Proof. The proof that $P_{Man(\mathbf{W})}(\mathbf{V})$ satisfies the Markov Condition for G_{Unman} is essentially the same as that of lemma 3.6.1. Because G_{Unman} is an (improper) subgraph of itself, $P_{Man(\mathbf{W})}(\mathbf{V})$ satisfies the Markov Condition for some subgraph of G_{Unman}.

THEOREM 3.6: (Manipulation Theorem): Given directed acyclic graph G_{Comb} over vertex set **V** $\cup$ **W** and distribution $P(\mathbf{V} \cup \mathbf{W})$ that satisfies the Markov condition for G_{Comb}, if changing the value of **W** from **w1** to **w2** is a manipulation of G_{Comb} with respect to **V**, G_{Unman} is the unmanipulated graph, G_{Man} is the manipulated graph, and

$$P_{Unman(\mathbf{W})}(\mathbf{V}) = \prod_{X \in \mathbf{V}} P_{Unman(\mathbf{W})}(X | \mathbf{Parents}(G_{Unman}, X))$$

for all values of **V** for which the conditional distributions are defined, then

$$P_{Man(\mathbf{W})}(\mathbf{V}) = \prod_{X \in \mathbf{Manipulated}(\mathbf{W})} P_{Man(\mathbf{W})}(X | \mathbf{Parents}(G_{Man}, X)) \times \prod_{X \in \mathbf{V} \setminus \mathbf{Manipulated}(\mathbf{W})} P_{Unman(\mathbf{W})}(X | \mathbf{Parents}(G_{Unman}, X))$$

for all values of **V** for which each of the conditional distributions is defined.

Proof. By assumption, $P_{Man(\mathbf{W})}(\mathbf{V})$ satisfies the Markov Condition for G_{Man}. Hence

$$P_{Man(\mathbf{W})} = \prod_{X \in \mathbf{V}} P(X|\mathbf{Parents}(G_{Man}, X)) =$$

$$\prod_{X \in \mathbf{Manipulated}(\mathbf{W})} P(X|\mathbf{Parents}(G_{Man}, X)) \times \prod_{X \in \mathbf{V} \setminus \mathbf{Manipulated}(\mathbf{W})} P(X|\mathbf{Parents}(G_{Man}, X))$$

for all values of **V** for which the conditional distributions exist. No member of **W** is a descendant of any variable in **V** in G_{Comb}, so for each V in **V\Manipulated(W)**, **W** is d-separated from V given **Parents**(G_{Comb},V) in G_{Comb}. For any member X of **V\Manipulated(W)**, **Parents**(G_{Comb},X) = **Parents**(G_{Unman}, X) = **Parents**(G_{Man},X). It follows that $P(V|\mathbf{Parents}(GMan,X),\mathbf{W} = \mathbf{w2}) = P(V|\mathbf{Parents}(G_{Man},X)) = P(V|\mathbf{Parents}(G_{Man},X),\mathbf{W} = \mathbf{w1}) = P(V|\mathbf{Parents}(G_{Unman},X),\mathbf{W} = \mathbf{w1})$. Hence

$$P_{Man(\mathbf{W})}(\mathbf{V}) =$$

$$\prod_{X \in \mathbf{Manipulated}(\mathbf{W})} P_{Man(\mathbf{W})}(X|\mathbf{Parents}(G_{Man}, X)) \times \prod_{X \in \mathbf{V} \setminus \mathbf{Manipulated}(\mathbf{W})} P_{Unman(\mathbf{W})}(X|\mathbf{Parents}(G_{Unman}, X))$$

for all values of **V** for which the conditional distributions are defined. ∴

13.8 Theorem 3.7

THEOREM 3.7: If G is a directed acyclic graph over **V**, **X**, **Y**, and **Z** are disjoint subsets of **V**, and $P(\mathbf{V})$ satisfies the Markov condition for G and the deterministic relations in **Deterministic(V)** then if **X** and **Y** are D-separated given **Z** and **Deterministic(V)**, **X** and **Y** are independent given **Z** in P.

We will say that a set of variables **Z** determines the set of variables **A**, when every variable in **A** is a deterministic function of the variables in **Z**, and not every variable in **A** is a deterministic function of any proper subset of **Z**. Suppose G is a directed acyclic graph over **V**, and **Deterministic(V)** is a set of ordered tuples of variables in **V**, where for each tuple D in **Deterministic(V)**, if D is $<V_1,...,V_n>$ then V_n is a deterministic function of $V_1,..., V_{n-1}$ and is not a deterministic function of any subset of $V_1,..., V_{n-1}$; we also say $\{V_1,..., V_{n-1}\}$ **determines** V_n. For a given **Deterministic(V)**, if **Z** is included in **V**, then **Det(Z)** is the set of variables determined by any subset of **Z**. Note that **Z** is included in **Det(Z)**.

If G is a directed acyclic graph over **V**, and **Z** is included in **V**, then G' is in **Mod**(G) relative to **Deterministic(V)** and **Z** if and only if for each V in **V**

(i) if there exists a set of vertices included in **Z** that are nondescendants of V in G and that determine V, then **Parents**(G',V)= **X**, where **X** is some set of vertices included in **Z** that are nondescendants of V in G and that determine V;
(ii) if there is no set **X** of vertices included in **Z** that are nondescendants of V in G and that determine V, then **Parents**(G',V) = **Parents**(G,V).

Proofs of Theorems 405

If G is a directed acyclic graph with vertex set **V**, **Z** is a set of vertices not containing X or Y, and $X \neq Y$, then X and Y are **D-separated** given **Z** and **Deterministic**(**V**) if and only if there is no undirected path U in G between X and Y such that each collider on U has a descendant in **Z**, and no other vertex on U is in **Det**(**Z**); otherwise if $X \neq Y$ and X and Y are not in **Z**, then X and Y are **D-connected** given **Z** and **Deterministic**(**V**). Similarly, if **X**, **Y**, and **Z** are disjoint sets of variables, and **X** and **Y** are non-empty, then **X** and **Y** are D-separated given **Z** and **Deterministic**(**V**) if and only if each pair $<X,Y>$ in the Cartesian product of **X** and **Y** are **D-separated** given **Z** and **Deterministic**(**V**); otherwise if **X**, **Y**, and **Z** are disjoint, and **X** and **Y** are non-empty, then **X** and **Y** are **D-connected** given **Z** and **Deterministic**(**V**).

If G is a directed acyclic graph over **V**, **Z** is a subset of **V** that does not contain X or Y, and $X \neq Y$, then X and Y are **det-separated** given **Z** and **Deterministic**(**V**) if and only if either X and Y are d-separated given $\mathbf{Z} \cup \mathbf{Det}(\mathbf{Z})$ in some **Mod**(G) relative to **Deterministic**(**V**) and **Z**, or X or Y is in **Det**(**Z**); otherwise if $X \neq Y$ and X and Y are not in **Z**, then X and Y are **det-connected** given **Z** and **Deterministic**(**V**). If **X**, **Y** and **Z** are disjoint sets of variables in **V**, and **X** and **Y** are non-empty, then **X** and **Y** are **det-separated** given **Z** if and only if every member X of **X** and every member Y of **Y** are det-separated given **Z**; otherise if **X**, **Y** and **Z** are disjoint sets of variables in **V**, and **X** and **Y** are non-empty, then **X** and **Y** are **det-connected** given **Z** and **Deterministic**(**V**).

LEMMA 3.7.1: Let G be a directed acyclic graph with vertex set **V**, Ord an ordering of variables in **V** such that if A is before B in Ord then **A** is not a descendant of B in G, **Predecessors**(Ord,V) the set of all vertices before V in Ord, and $P(\mathbf{V})$ a distribution over **V**. $P(\mathbf{V})$ satisfies the Minimality and Markov Conditions for G if and only if for each V in **V**, V is independent of **Predecessors**(Ord,V)**Parents**(G,V) given **Parents**(G,V) and for no proper subset $\mathbf{X}(V)$ of **Parents**(G,V), V is independent of **Predecessors**(Ord,V)\\$\mathbf{X}(V)$ given $\mathbf{X}(V)$.

Proof. See Pearl 1988. ∴

LEMMA 3.7.2: If G is a directed acyclic graph over **V**, and **X**, **Y**, and **Z** are disjoint subsets of **V**, and $P(\mathbf{V})$ satisfies the Markov condition for G and the deterministic relations in **Deterministic**(**V**), then if **X** and **Y** are det-separated given **Z** and **Deterministic**(**V**), **X** and **Y** are independent given **Z** in P.

Proof. First we will prove that $P(\mathbf{V})$ satisfies the Markov condition for each directed acylic graph G' in **Mod**(G). First form an acceptable ordering Ord of the variables in **V** for G. Let **Predecessors**(Ord,V) be the variables that precede V in Ord. From lemma 3.7.1 it follows that if G' is a directed acyclic graph in which for each V in **V**, V is independent of **Predecessors**(V)**Parents**(V) given **Parents**(V), then G' is an I-map of

$P(\mathbf{V})$. If $\mathbf{X}$ is a subset of $\mathbf{Parents}(V)$ that determines V, it follows that V is independent of $\mathbf{Predecessors}(V)\backslash\mathbf{X}$ given $\mathbf{X}$. Hence if in G', $\mathbf{Parents}(V) = \mathbf{X}$, G' is still an I-map of $P(\mathbf{V})$.

If either $\mathbf{X}$ or $\mathbf{Y}$ is included in $\mathbf{Det}(\mathbf{Z})$, it follows that $\mathbf{X}$ and $\mathbf{Y}$ are independent given $\mathbf{Z} \cup \mathbf{Det}(\mathbf{Z})$. Suppose then that neither $\mathbf{X}$ nor $\mathbf{Y}$ is included in $\mathbf{Det}(\mathbf{Z})$. By definition of det-separability, $\mathbf{X}\backslash\mathbf{Det}(\mathbf{Z})$ and $\mathbf{Y}\backslash\mathbf{Det}(\mathbf{Z})$ are d-separated given $\mathbf{Z} \cup \mathbf{Det}(\mathbf{Z})$. Hence

$$P((\mathbf{X} \cup \mathbf{Y}) \setminus \mathbf{Det}(\mathbf{Z})|\mathbf{Z} \cup \mathbf{Det}(\mathbf{Z})) = P(\mathbf{X} \setminus \mathbf{Det}(\mathbf{Z})|\mathbf{Z} \cup \mathbf{Det}(\mathbf{Z}))P(\mathbf{Y} \setminus \mathbf{Det}(\mathbf{Z})|\mathbf{Z} \cup \mathbf{Det}(\mathbf{Z}))$$

It now follows that $\mathbf{X}$ is independent of $\mathbf{Y}$ given $\mathbf{Z}$ because

$$\begin{aligned}P(\mathbf{X} \cup \mathbf{Y}|\mathbf{Z}) &= P(\mathbf{X} \cup \mathbf{Y}|\mathbf{Z} \cup \mathbf{Det}(\mathbf{Z})) = P((\mathbf{X} \cup \mathbf{Y}) \setminus \mathbf{Det}(\mathbf{Z})|\mathbf{Z} \cup \mathbf{Det}(\mathbf{Z})) = \\ &\quad P(\mathbf{X} \setminus \mathbf{Det}(\mathbf{Z})|\mathbf{Z} \cup \mathbf{Det}(\mathbf{Z}))P(\mathbf{Y} \setminus \mathbf{Det}(\mathbf{Z})|\mathbf{Z} \cup \mathbf{Det}(\mathbf{Z})) = \\ &\quad P(\mathbf{X}|\mathbf{Z} \cup \mathbf{Det}(\mathbf{Z}))P(\mathbf{Y}|\mathbf{Z} \cup \mathbf{Det}(\mathbf{Z})) = P(\mathbf{X}|\mathbf{Z})P(\mathbf{Y}|\mathbf{Z})\end{aligned}$$

∴

THEOREM 3.7: If G is a directed acyclic graph over $\mathbf{V}$, $\mathbf{X}$, $\mathbf{Y}$, and $\mathbf{Z}$ are disjoint subsets of $\mathbf{V}$, and $P(\mathbf{V})$ satisfies the Markov condition for G and the deterministic relations in $\mathbf{Deterministic}(G)$ then if $\mathbf{X}$ and $\mathbf{Y}$ are D-separated given $\mathbf{Z}$ and $\mathbf{Deterministic}(\mathbf{V})$, $\mathbf{X}$ and $\mathbf{Y}$ are independent given $\mathbf{Z}$ in P.

Proof. We will prove that if X and Y are det-connected given $\mathbf{Z}$ and $\mathbf{Deterministic}(\mathbf{V})$, then X and Y are D-connected given $\mathbf{Z}$ and $\mathbf{Deterministic}(\mathbf{V})$. It follows then that if X and Y are D-separated given $\mathbf{Z}$ and $\mathbf{Deterministic}(\mathbf{V})$, then X and Y are det-separated given $\mathbf{Z}$ and $\mathbf{Deterministic}(\mathbf{V})$, and by lemma 3.7.1, X and Y are independent given $\mathbf{Z}$ in P.

Suppose some X in $\mathbf{X}$ is det-connected to some Y in $\mathbf{Y}$ given $\mathbf{Z}$ and $\mathbf{Deterministic}(\mathbf{V})$. It follows by definition that X and Y are not in $\mathbf{Z}$ and not in $\mathbf{Det}(\mathbf{Z})$. Because X and Y are det-connected given $\mathbf{Z}$ there is an undirected path U' that d-connects X and Y given $\mathbf{Z}$ in some graph G' in $\mathbf{Mod}(G)$.

First, we will show that the path U corresponding to U' exists in G; then we will show that U D-connects X and Y given $\mathbf{Z}$ and $\mathbf{Deterministic}(\mathbf{V})$ in G.

No member of $\mathbf{Det}(\mathbf{Z})$ is a noncollider on U' because U' d-connects X and Y given $\mathbf{Z} \cup \mathbf{Det}(\mathbf{Z})$. Hence for each noncollider A on U', $\mathbf{Parents}(G',A)$ equals $\mathbf{Parents}(G,A)$. It follows that if there is an edge into A in G', there is a corresponding edge into A in G.

Suppose then that A is a collider on U'. If there is an edge into A in G' that does not exist in G, then every parent of A is in $\mathbf{Z}$. It follows that either the endpoints of U' are in $\mathbf{Z}$, or some noncollider on U' is in $\mathbf{Z}$. But then U' does not d-connect X and Y given $\mathbf{Z} \cup \mathbf{Det}(\mathbf{Z})$ in G'. Hence if there is an edge into A on U', then the corresponding edge exists in G.

It follows that the path U in G corresponding to U' in G' exists.

The endpoints of U are not in $\mathbf{Z} \cup \mathbf{Det}(\mathbf{Z})$, because they are equal to the endpoints of U', which are not in $\mathbf{Z} \cup \mathbf{Det}(\mathbf{Z})$.

Proofs of Theorems

No noncollider on U is in $\mathbf{Z} \cup \mathbf{Det(Z)}$, because each noncollider on U is a noncollider on U', and no noncollier on U' is in $\mathbf{Z} \cup \mathbf{Det(Z)}$.

Finally suppose that A is a collider on U'. It follows that A has a descendant in $\mathbf{Z} \cup \mathbf{Det(Z)}$ in G'. There are two cases.

If A has a descendant in $\mathbf{Z}$ in G', then it has a descendant in $\mathbf{Z}$ in G. Suppose that A has a descendant X in $\mathbf{Z}$ in G, and let $D(A,X)$ be a directed path from A to X in G'. Let Z be the member of $\mathbf{Z}$ closest to A on $D(A,X)$. Every edge that is in G' but not in G is out of a member of $\mathbf{Z}$. $D(A,Z)$ has no edges out of a member of $\mathbf{Z}$. Hence every edge in $D(A,Z)$ exists in G, and A has a descendant in $\mathbf{Z}$ in G.

Suppose A does not have a descendant in $\mathbf{Z}$ in G'. It follows that there is a directed path $D(A,X)$ from A to a member X of $\mathbf{Det(Z)}\backslash\mathbf{Z}$ in G'. If A itself is in $\mathbf{Det(Z)}$ then it has parents not in $\mathbf{Z}$, because U' d-connects X and Y given $\mathbf{Z} \cup \mathbf{Det(Z)}$. Because G' is in $\mathbf{Mod}(G)$, it follows from the fact that A has a parent not in $\mathbf{Z}$ that A has a descendant in $\mathbf{Z}$ in G. If A is not in $\mathbf{Det(Z)}$ then $D(A,X)$ is not an empty path, and it does not contain any member of $\mathbf{Z}$. Hence X has a parent that is not in $\mathbf{Z}$. Because G' is in $\mathbf{Mod}(G)$, it follows from the fact that X has a parent not in $\mathbf{Z}$ that X has a descendant in $\mathbf{Z}$ in G. $D(A,X)$ exists in G because every edge in G' but not in G is out of a member of $\mathbf{Z}$, and $D(A,X)$ contains no member of $\mathbf{Z}$. Hence A has a descendant in $\mathbf{Z}$ in G.

It follows that U D-connects X and Y given $\mathbf{Z}$ and $\mathbf{Deterministic(V)}$ in G.∴

13.9 Theorem 4.1

THEOREM 4.1: Two directed acyclic graphs G_1, G_2, are strongly statistically indistinguishable if and only if (i) they have the same vertex set $\mathbf{V}$, (ii) vertices V_1 and V_2 are adjacent in G_1 if and only if they are adjacent in G_2, and (iii) for every triple V_1, V_2, V_3 in $\mathbf{V}$, the graph $V_1 \rightarrow V_2 \leftarrow V_3$ is a subgraph of G_1 if and only if it is a subgraph of G_2.

Proof. $\Leftarrow$ Suppose two directed acyclic graphs G_1 and G_2 contain the same vertices, the same adjacencies and the same colliders, and G_1 is a minimal I-map of P. By theorem 3.4 the same distributions are faithful to G_1 and G_2 so they have the same d-separability relations, and hence G_2 is also an I-map of P.

G_2 is also minimal. Every subgraph of G_1 has the same d-separability relations as does the corresponding subgraph of G_2 because removing corresponding vertices and adjacencies from both graphs leaves subgraphs that contain the same vertices, adjacencies and colliders. Hence, if a subgraph of G_2 is an I-map of P, then the corresponding subgraph of G_1 is an I-map of P. But by supposition, no proper subgraph of G_1 is an I-map of P. Hence no proper subgraph of G_2 is an I-map of P. By definition, G_2 is a minimal I-map of P. It follows that G_1 and G_2 are s.s.i.

$\Rightarrow$ Now consider the case where G_1 and G_2 differ either in their sets of vertices, their adjacencies, or their colliders. We will show that there exists a distribution P such that G_1

is a minimal I-map of P, while G_2 is not. By definition, it follows that G_1 and G_2 are not s.s.i.

Case 1. Suppose first that G_1 and G_2 differ in their sets of vertices. By definition they are not s.s.i.

Case 2. Suppose that G_1 and G_2 differ in their adjacencies. Suppose without loss of generality that G_1 contains an adjacency not in G_2. Then there is a pair of vertices X and Y such that X and Y are d-separated given a subset **S** in G_2, while X and Y are not d-separated given **S** in G_1. There is a distribution P faithful to G_1. G_1 is also a minimal I-map of P. In G_1, X and Y are dependent conditional on **S**. But because X and Y are d-separated given a subset **S** in G_2, G_2 is not an I-map of P. Hence G_1 and G_2 are not s.s.i.

Case 3. Suppose that G_1 and G_2 differ in their unshielded colliders but not in any adjacencies. Let Y be an unshielded collider on the path $<X,Y,Z>$ in G_1, but not in G_2. Let P be a distribution faithful to G_1. It follows that G_1 is a minimal I-map of P. In G_2, X and Z are d-separated given a set **S** containing Y, while in G_1 X and Z are not d-separated given **S**. Since G_1 is faithful to P, X and Z are dependent conditional on **S**. Hence G_2 is not a minimal I-map of P, and G_1 and G_2 are not s.s.i.

Case 4. Finally, suppose that G_1 and G_2 differ in their shielded colliders but not in any adjacencies or unshielded colliders. Let Y be a shielded collider on the path $<X,Y,Z>$ in G_1, but not in G_2. Suppose G_2' is the subgraph of G_2 with the edge between X and Z removed. G_2' is faithful to some distribution P. G_2 is not a minimal I-map of P (because it contains a subgraph which is an I-map of P). We will now show that G_1 is a minimal I-map of P.

First, G_1 is an I-map of P. G_1 is f.i. to G_2. G_2 is a proper supergraph of G_2', and so the d-separation relations true of G_2 are included in the d-separation relations true of G_2'; hence the d-separation relations true of G_1 are included in the d-separation relations true of G_2'. It follows that G_1 is an I-map of P.

G_1 is also minimal. If G_1' is a subgraph obtained by deleting from G_1 any edge other than the X - Z edge, by Case 2, the subgraph is not an I-map of P. If G_1' is a subgraph obtained by deleting from G_1 just the X - Z edge, then G_1' contains an unshielded collider at Y that does not occur in G_2'. By Case 3, G_1' is not an I-map of P.

Because G_1 is a minimal I-map of P, and G_2 is not, G_1 and G_2 are not s.s.i. ∴

13.10 Theorem 4.2

THEOREM 4.2: Two directed acyclic graphs G and H are faithfully indistinguishable if and only if (i) they have the same vertex set, (ii) any two vertices are adjacent in G if and only if they are adjacent in H, and (iii) any three vertices, X, Y, Z, such that X is adjacent

Proofs of Theorems

to Y and Y is adjacent to Z but X is not adjacent to Z in G or H, are oriented as $X \rightarrow Y \leftarrow Z$ in G if and only if they are so oriented in H.

Proof. This was proved in Verma and Pearl 1990b. It also follows directly from theorem 3.4. ∴

13.11 Theorem 4.3

THEOREM 4.3: Two directed acyclic graphs are faithfully indistinguishable if and only if some distribution faithful to one is faithful to the other and conversely; that is, they are f.i. if and only if they are w.f.i.

Proof. Suppose G_1 and G_2 are f.i. By lemma 3.5.8 there is some distribution P faithful to G_1. Hence P is faithful to G_2, and G_1 and G_2 are w.f.i.
  Suppose that G_1 and G_2 are w.f.i. Then there is some distribution P faithful to G_1 and G_2. It follows that G_1 and G_2 have the same d-separation relations, so any distribution faithful to G_1 is also faithful to G_2 and vice-versa. ∴

13.12 Theorem 4.4

THEOREM 4.4: If probability distribution P satisfies the Markov Condition for directed acyclic graphs G and H, and P is faithful to H, then for all vertices X, Y, if X, Y are adjacent in H they are adjacent in G.

Proof. If P is faithful to H then X is adjacent to Y in H only if X, Y are dependent conditional on every set of vertices not containing X or Y. Suppose then that P satisfies the Markov condition for G but, contrary to the claim, X and Y are not adjacent in G. Then X is not a parent of Y and Y is not a parent of X. Either X is not a descendant of Y or Y is not a descendant of X; suppose without loss of generality that X is not a descendant of Y. Then by the Markov Condition, X and Y are independent in P conditional on the set of all parents of Y, which is a contradiction. ∴

13.13 Theorem 4.5

THEOREM 4.5: If probability distribution P satisfies the Markov and Minimality Conditions for directed acyclic graphs G, and P is faithful to graph H, then (i) for all X, Y, Z such that $X \rightarrow Y \leftarrow Z$ is in H and X is not adjacent to Z in H, either $X \rightarrow Y \leftarrow Z$ in G or X, Z are adjacent in G and (ii) for every triple X, Y, Z of vertices such that $X \rightarrow Y \leftarrow Z$ is in G and X is not adjacent to Z in G, if X is adjacent to Y in H and Y is adjacent to Z in H then $X \rightarrow Y \leftarrow Z$ in H.

Proof.
(i) Suppose that P satisfies the Markov and Minimality Conditions for directed acyclic graphs G, and P is faithful to graph H. Suppose $X \rightarrow Y \leftarrow Z$ is in H and X is not adjacent to Z in H. By theorem 4.4, X is adjacent to Y and Y is adjacent to Z in G. Suppose Y is not a collider on $<X, Y, Z>$ in G and X and Z are not adjacent in G. Then by the Markov Condition X and Z are independent conditional on some set containing Y; but since H is faithful, this is impossible.

(ii) Suppose Y is an unshielded collider on the path $<X,Y,Z>$ in G. Then X and Z are d-separated in G given some set of vertices, and hence d-separated given **Parents**(G,X) or **Parents**(G,Z). It follows that X and Z are independent given **Parents**(G,X) or **Parents**(G,Z) in P. Y is not a parent of X or Z in G; hence in P, X and Z are independent given some set not containing Y. But if X, Y and Y, Z are adjacent in H and Y is not a collider on $<X, Y, Z>$, then there is a trek between X and Z containing only X, Y, and Z; hence in H, X and Z are not d-separated given any set of variables not containing Y. Because P is faithful to H, X and Z are not independent given any set of variables containing Y. This is a contradiction. ∴

COROLLARY 4.1: If probability distribution P satisfies the Markov condition for directed acyclic graph G and P is faithful to directed acyclic graph H and G and H agree on an ordering of the variables (as, for example, by time) such that $X \rightarrow Y$ only if $X < Y$ in the order, then H is a subgraph of G.

Proof. An immediate consequence of theorem 4.4.

13.14 Theorem 4.6

THEOREM 4.6: No two distinct s.s.i. directed acyclic graphs with the same vertex set are rigidly statistically indistinguishable.

Proof. Suppose G_1 and G_2 are distinct s.s.i. directed acyclic graphs with vertex set **V**. Because they are s.s.i they have the same adjacencies; hence if they are distinct graphs there is some edge $A \rightarrow B$ in G_1 and $B \rightarrow A$ in G_2. Let U_1 and U_2 be variables not in **V**. Embed G_1 and G_2 in H_1 and H_2 respectively by adding edges from U_1 to A and U_2 to B. Then H_1 and H_2 are not s.s.i because they have different colliders. ∴

13.15 Theorem 5.1

THEOREM 5.1: If the input to the PC, SGS, PC–1, PC–2, PC* or IG algorithms is data faithful to directed acyclic graph G, the output is a pattern that represents the faithful indistinguishability class of G.

In a graph G, let V be in **Undirected**(X,Y) if and only if V lies on some undirected path between X and Y.

LEMMA 5.1.1: In a directed acyclic graph G, if X is not a descendant of Y, and Y and X are not adjacent in G, then X is d-separated from Y given **Parents**$(Y) \cap$ **Undirected**(X,Y).

Proof. Suppose on the contrary that some undirected path U d-connects X and Y given **Parents**$(X) \cap$ **Undirected**(X,Y). If U is into Y then it contains some member of **Parents**$(Y) \cap$ **Undirected**(X,Y) not equal to X as a noncollider. Hence it does not d-connect X and Y given **Parents**$(Y) \cap$ **Undirected**(X,Y), contrary to our assumption. If U is out of Y, then because X is not a descendant of Y, U contains a collider in **Undirected**(X,Y). Let C be the collider on U closest to Y. If U d-connects X and Y given **Parents**$(Y) \cap$ **Undirected**(X,Y) then C has a descendant in **Parents**$(Y) \cap$ **Undirected**(X,Y). But then C is an ancestor of Y, and Y is an ancestor of C, so G is cyclic, contrary to our assumption. Hence no undirected path between X and Y d-connects X and Y given **Parents**$(Y) \cap$ **Undirected**(X,Y). ∴

LEMMA 5.1.2: In a directed acyclic graph G, if X is adjacent to Y, and Y is adjacent to Z, and X is not adjacent to Z, then the edges are oriented as $X \rightarrow Y \leftarrow Z$ if and only for every subset **S** of **V**, X is d-connected to Z given $\{Y\} \cup$ **S**$\setminus\{X,Z\}$.

Proof. This follows from theorem 3.4. ∴

LEMMA 5.1.3 was suggested in Pearl(1990a).

LEMMA 5.1.3: In a directed acyclic graph G, if X is adjacent to Y, and Y is adjacent to Z, and X is not adjacent to Z, then either Y is in every set of variables that d-separates X and Z, or it is in no set of variables that d-separates X and Z.

Proof. Assume that in G, X and Z are not adjacent but X is adjacet to Y and Y is adjacent to Z. Since X, Z are not adjacent, they are d-separated given some subset **S**$\setminus\{X,Z\}$. In G, the X - Y and Y - Z edges collide at Y if and only if there is no set **S** containing Y and not X or Z such that X and Z are d-separated given **S**. If the X - Y and Y - Z edges do not collide at Y, then there is an undirected path U between X and Z that contains no colliders (including Y). Any set **S**$\setminus\{X,Z\}$ that does not contain Y will fail to d-separate X and Z because of this path. ∴

THEOREM 5.1: If the input to the PC, SGS, PC–1, PC–2, PC* or IG algorithms is data faithful to directed acyclic graph G, the output is a pattern that represents the faithful indistinguishability class of G.

Proof. The correctness of the SGS algorithm is evident from theorem 3.4 since the procedure simply verifies the conditions for faithfulness given in that theorem.

Let G' be the output of one any of the algorithms except SGS. Suppose that X and Y are not adjacent in G'. None of the algorithms removes an edge between X and Y unless X and Y are d-separated given some subset of $\mathbf{V} \backslash \{X,Y\}$. If X and Y are d-separated given some subset of $\mathbf{V} \backslash \{X,Y\}$, then they are not adjacent in G. Hence if X and Y are not adjacent in G', X and Y are not adjacent in G.

Suppose X and Y are adjacent in the output G' of any of the algorithms except PC*. It follows that in G, X and Y are not d-separated given any subset of the adjacencies of X or any of the adjacencies of Y in G'. From what we have just proved, the adjacencies of X in G' are a superset of **Parents**(G,X) and the adjacencies of Y in G' are a superset of **Parents**(G,Y). Hence X and Y are not d-separated given **Parents**(X,G) or **Parents**(Y,G) in G. It follows from lemma 3.5.9 that X and Y are adjacent in G.

Suppose X and Y are adjacent in the output G' of PC*. **Undirected**(X,Y) in G' is a superset of **Undirected**(X,Y) in G. This, together with lemmas 3.5.9 and 5.1.1 entails that X and Y are adjacent in G.

We will show by induction on the number of applications of orientation rules in the repeat loop of the algorithm that the orientations are correct in the output G'.

Base Case: Suppose that $X \rightarrow Y$ is oriented by the rule that if X is adjacent to Y, and Y is adjacent to Z, and X is not adjacent to Z, then the edges are oriented as $X \rightarrow Y \leftarrow Z$ if and only Y is not in **Sepset**(X,Z). This is a correct orientation by lemmas 5.1.2 and 5.1.3.

Induction Case: Suppose that the orientations of G' after n applications of orientation rules are correct. Suppose first that $X \rightarrow Y$ is oriented because there is a directed path from X to Y in G'. It follows from the induction hypothesis that there is a directed path from X to Y in G, and hence $X \rightarrow Y$ in G because G is acyclic. Suppose next that $X \rightarrow Y$ is oriented because there is an edge $Z \rightarrow X$ and the edge between X and Y in G' has no arrowhead at X. It follows that Y is in **Sepset**(X,Z), and hence Y is not a collider on the path $<X,Y,Z>$ in G. Also by the induction hypothesis $Z \rightarrow X$ in G, and hence $X \rightarrow Y$ in G. ∴

13.16 Theorem 6.1

THEOREM 6.1: (Verma and Pearl)**:** If **V** is a set of vertices, **O** is a subset of **V** containing A and B, and G is a directed acyclic graph over **V** (or an inducing path graph over **O**) then A and B are not d-separated by any subset of $\mathbf{O} \backslash \{A,B\}$ if and only if there is an inducing path over the subset **O** between A and B.

(Theorem 6.1 was first stated and proved in Verma and Pearl 1990 for directed acyclic graphs, but that paper did not include the parts of the lemmas relating the existence of an inducing path that is into (or out of) its endpoints to the existence of d-connecting paths that are into (or out of) their endpoints.)

If G is a directed acyclic graph over a set of variables **V**, **O** is a subset of **V** containing A and B, and $A \neq B$, then an undirected path U between A and B is an **inducing path relative to O** if and only if every member of **O** on U except for the endpoints is a collider on U, and every collider on U is an ancestor of either A or B. We will sometimes refer to members of **O** as **observed** variables. In a graph G, an edge between A and B is **into** A if and only if the mark at the A end of edge is an ">." If an undirected path U between A and B contains an edge into A we will say that U is **into** A. In a graph G, an edge between A and B is out of A if and only if the mark at the A endpoint is the empty mark. If an undirected path U between A and B contains an edge out of A we will say that U is **out of** A.

LEMMA 6.1.1: If **V** is a set of vertices, **O** is a subset of **V**, G is a directed acyclic graph over **V** (or an inducing path graph over **O**) if there is an inducing path relative to **O** between A and B that is out of A and into B, then for any subset **Z** of **O**\{A,B} there is an undirected path C that d-connects A and B given **Z** that is out of A and into B.

Proof. Let U be an inducing path over **O** between A and B that is out of A and into B. Every observed vertex on U except for the endpoints is a collider, and every collider is an ancestor of either A or B.

If every collider on U has a descendant in **Z**, then let $C = U$. C d-connects A and B given **Z** because every collider has a descendant in **Z**, and no noncollider is in **Z**. C is out of A and into B.

Suppose that not every collider on U has a descendant in **Z**. Let R be the collider on U closest to A that does not have a descendant in **Z**, and W be the collider on U closest to A. $R \neq A$ and $R \neq B$ because A and B are not colliders on U.

Suppose first that $R = W$. There is a directed path from R to B that does not contain A, because otherwise there is a cycle in G. R is not in **Z** because R has no descendant in **Z**. B is not on $U(A,R)$. $U(A,R)$ d-connects A and R given **Z**, and is out of A. By lemma 3.3.3 there is a d-connecting path C between A and B given **Z** that is out of A and into B.

Suppose then that $R \neq W$. Because U is out of A, W is a descendant of A. W has a descendant in **Z** by definition of R. It follows that every collider on U that is an ancestor of A has a descendant in **Z**. Hence R is an ancestor of B, and not of A. B is not on $U(A,R).U(A,R)$ d-connects A and R given **Z** and is out of A. By hypothesis, there is a directed path D from R to B that does not contain A or any member of **Z**. By lemma 3.3.3, there is a path that d-connects A and B given **Z** that is out of A and into B. ∴

LEMMA 6.1.2: If **V** is a set of vertices, **O** is a subset of **V**, G is a directed acyclic graph over **V** (or an inducing path graph over **O**), and there is an inducing path U over **O** between A and B that is into A and into B, then for every subset **Z** of **O**\{A,B} there is an undirected path C that d-connects A and B given **Z** that is into A and into B.

Proof. If every collider on U has a descendant in **Z**, then U is a d-connecting path between A and B given **Z** that is into A and into B. Suppose then that there is a collider that does not have a descendant in **Z**. Let W be the collider on U closest to A that does not have a descendant in **Z**. Suppose that W is the source of a directed path D to B that does not contain A. B is not on $U(A,W)$. $U(A,W)$ is a path that d-connects A and W given **Z**, and is into A. By lemma 3.3.3, there is an undirected path C that d-connects A and B given **Z** and is into A and into B. Similarly, if the first collider W on U after B that does not have a descendant in **Z** is the source of a directed path D to A that does not contain B, then by lemma 3.3.3, A and B are d-connected given **Z** by an undirected path into A and into B.

Suppose then that the collider W on U closest to A that does not have a descendant in **Z** is not the source of a directed path to B that does not contain A, and that the collider R on U closest to B that does not have a descendant in **Z** is not the source of a directed path to A that does not contain B. It follows that there exist two colliders E and F on U such that E is an ancestor of A, F is an ancestor of B, and every collider between E and F is an ancestor of a member of **Z**. $U(E,F)$ d-connects E and F given **Z**\{E,F} because no member of **O** is a noncollider on $U(E,F)$ except for the endpoints, and every collider on $U(E,F)$ has a descendant in **Z**. The directed path from E to A d-connects E and A given **Z**\{E,A} and the directed path from F to B d-connects F and B given **Z**\{F,B}. By lemma 3.3.3 there is an undirected path that d-connects A and B given **Z** that is into A and into B. ∴

In a graph G, Let **A**(A,B) be the union of the ancestors of A or B.

LEMMA 6.1.3: If **V** is a set of vertices, **O** is a subset of **V**, G is a directed acyclic graph over **V** (or an inducing path graph over **O**) and an undirected path U in G d-connects A and B given (**A**(A,B) ∩ **O**)\{A,B} then U is an inducing path between A and B over **O**.

Proof. If there is a path U that d-connects A and B given (**A**(A,B) ∩ **O**)\{A,B} then every collider on U is an ancestor of a member of (**A**(A,B) ∩ **O**)\{A,B}, and hence an ancestor of A or B. Every vertex on U is an ancestor of either A or B or a collider on U, and hence every vertex on U is an ancestor of A or B. If U d-connects A and B given (**A**(A,B) ∩ **O**)\{A,B}, then every member of (**A**(A,B) ∩ **O**)\{A,B} that is on U, except for the endpoints, is a collider. Since every vertex on U is in **A**(A,B), every member of **O** that is on U, except for the endpoints, is a collider. Hence U is an inducing path between A and B over **O**. ∴

The following pair of lemmas state some basic properties of inducing paths.

LEMMA 6.1.4: If G is a directed acyclic graph over **V**, **O** is a subset of **V** that contains A and B, and G contains an inducing path over **O** between A and B that is out of A, then there is a directed path from A to B in G.

Proofs of Theorems

Proof. Let U be an inducing path between A and B relative to **O** that is out of A. If U does not contain a collider, then U is a directed path from A to B. If U does contain a collider, let C be the first collider after A. By definition of inducing path, there is a directed path from C to B or C to A. There is no path from C to A because there is no cycle in G; hence there is a directed path from C to B. Because U is out of A, and C is the first collider after A, there is a directed path from A to C. Hence there is a directed path from A to B. ∴

LEMMA 6.1.5: If **V** is a set of vertices, **O** is a subset of **V**, G is a directed acyclic graph over **V** (or an inducing path graph over **O**) that contains an inducing path relative to **O** between A and B that is out of A, then every inducing path relative to **O** between A and B is into B.

Proof. By lemma 6.1.4, if there an inducing path out of A, and an inducing path out of B, there is a cycle in G. ∴

THEOREM 6.1: (Verma and Pearl)**:** If **V** is a set of vertices, **O** is a subset of **V** containing A and B, G is a directed acyclic graph over **V** (or an inducing path graph over **O**) A and B are not d-separated by any subset of **O**\{A,B} if and only if there is an inducing path over the subset **O** between A and B.

Proof. This follows from lemmas 6.1.1, 6.1.2, 6.1.3, and 6.1.5. ∴

13.17 Theorem 6.2

THEOREM 6.2: In an inducing path graph G' over **O**, where A and B are in **O**, if A is not an ancestor of B, and A and B are not adjacent then A and B are d-separated given a subset of **D-SEP**(A,B).

If G' is an inducing path graph over **O** and $A \neq B$, let $V \in$ **D-SEP**(A,B) if and only if $A \neq V$ and there is an undirected path U between A and V such that every vertex on U is an ancestor of A or B, and (except for the endpoints) is a collider on U.

LEMMA 6.2.1: If G' is the inducing path graph for G over **O** and there is a directed path from A to B in G', then there is a directed path from A to B in G.

Proof. Suppose there is a directed path D from A to B in G'. Let X and Y be any two vertices adjacent on the directed path and that occur in that order. There is a directed edge from X to Y in G'. By the definition of inducing path graph, there is an inducing path between X and Y in G that is out of X. Hence by lemma 6.1.4, there is a directed path from X to Y in G.

In G, the concatenation of the directed paths between vertices that are adjacent on D contains a subpath that is a directed path from A to B. ∴

LEMMA 6.2.2: If G' is the inducing path graph for G over $\mathbf{O}$, and there is a path U d-connecting A and B given $\mathbf{Z}$ in G' then there is a path d-connecting A and B given $\mathbf{Z}$ in G.

Proof. Suppose that U d-connects A and B in G'. If there are vertices R, S, and T on U such that R and S are adjacent on U, and S and T are adjacent on U, and S is in $\mathbf{Z}$, then S is a collider on U. By the definition of inducing path graph, in G there are inducing paths over $\mathbf{O}$ between R and S, and S and T, such that each of them is into S. By lemmas 6.1.1 and 6.1.2, in G there is a d-connecting path given $\mathbf{Z}\backslash\{R,S\}$ between R and S, and a d-connecting path given $\mathbf{Z}\backslash\{S,T\}$ between S and T, such that each of them is into S.

If there are vertices R, S, and T on U such that R and S are adjacent on U, and S and T are adjacent on U, and S is a collider on U, then S has a descendant in $\mathbf{Z}$ in G'. By the definition of inducing path graph, in G there are inducing paths between R and S, and S and T, that are both into S. By lemmas 6.1.1 and 6.1.2, in G there is a d-connecting path given $\mathbf{Z}\backslash\{R,S\}$ between R and S, and a d-connecting path given $\mathbf{Z}\backslash\{S,T\}$ between S and T, and both are into S. If S has a descendant in $\mathbf{Z}$ in G' then by lemma 6.2.1 it has a descendant in $\mathbf{Z}$ in G.

By lemma 3.3.1, there is a path in G that d-connects A and B given $\mathbf{Z}$. ∴

LEMMA 6.2.3: If G' is the inducing path graph for directed acyclic graph G over $\mathbf{O}$ and there is an inducing path U over $\mathbf{O}$ between A and C in G', then there is an edge between A and C in G'.

Proof. Suppose there is an inducing path over $\mathbf{O}$ between A and C in G'. By lemmas 6.1.1 and 6.1.2, in G' there is an undirected path d-connecting A and C given $\mathbf{A}(A,C) \cap \mathbf{O}\backslash\{A,C\}$. Hence by lemma 6.2.2 there is an undirected path in G such that A and C are d-connected given $\mathbf{A}(A,C) \cap \mathbf{O}\backslash\{A,C\}$ in G. By lemma 6.1.3 there is an inducing path over $\mathbf{O}$ between A and C in G. It follows by definition that there is an edge between A and C in G'. ∴

Let a total order *Ord* of variables in an inducing path graph or directed acyclic graph G' be **acceptable** if and only if whenever $A \neq B$ and there is a directed path from A to B in G', A precedes B in *Ord*. In a graph G, vertex X is **after** vertex Y if and only if there is a directed path from Y to X in G, and it is **before** vertex Y if and only if there is a directed path from X to Y in G. For inducing path graph G' and acceptable total ordering *Ord*, let **Predecessors**(*Ord*,V) equal the set of all variables that precede V (not including V) according to *Ord*. For inducing path graph G' and acceptable total ordering *Ord*, W is in **SP**(*Ord*,G',V) (separating predecessors of V in G' for ordering *Ord*) if and only if $W \neq V$ and there is an undirected path U between W and V such that each vertex on U except for V precedes V in *Ord* and every vertex on U except for the endpoints is a collider on U. Notice that by this definition each parent of V is in **SP**(*Ord*,G',V). For example in figure 13.2, if *Ord* = <X,S,T,R,M,Z,Q,Y>, then **SP**(*Ord*,G',Y) = $\{Q,T,S\}$ and if *Ord* = <X,S,T,R,M,Z,Y,Q> then **SP**(*Ord*,G',Y) = $\varnothing$.

LEMMA 6.2.4: If G' is an inducing path graph and Ord an acceptable total ordering then **Predecessors**(Ord,X)**SP**(Ord,G',X) is d-separated from X given **SP**(Ord,G',X).

Proof. Suppose on the contrary that there is a path U that d-connects some V in **Predecessors**(Ord,X)**SP**(Ord,G',X) to X given **SP**(Ord,G',X). There are three cases.

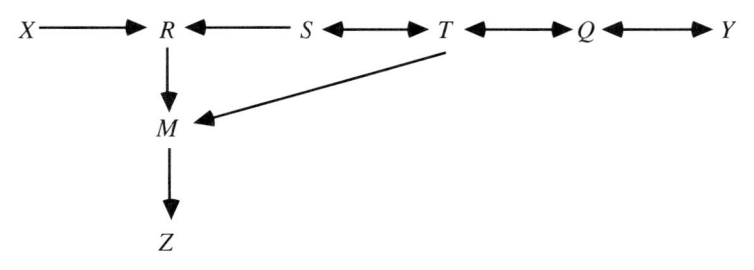

Figure 13.2

First suppose U has an edge into X that is not a double-headed arrow. (By a double-headed arrow we mean e.g., $A \leftrightarrow B$.) Then some parent R of X is on U, and is not a collider on U. R is in **SP**(Ord,G',X) and hence is not equal to V. Because R is not a collider on U, U does not d-connect V to X given **SP**(Ord,G',X), contrary to our assumption.

Next suppose U has an edge out of X. Since V is in **Predecessors**(Ord,X)**SP**(Ord,G',X) it precedes X in Ord; hence there is no directed path from X to V. It follows that U contains a collider. Let the first collider after X on U be R. R is a descendant of X, and the descendants of R are descendants of X. It follows that no descendant of R (including R itself) is in **SP**(Ord,G',X), and hence U does not d-connect V and X, contrary to our assumption.

Suppose finally that U contains a double-headed arrow into X. Because U d-connects X and V given **SP**(Ord, G', X), each collider along U has a descendant in $SP(Ord, G',X)$ and hence precedes X in Ord; it follows that every ancestor of a collider on U precedes X in Ord. Let W be the vertex on U closest to X not in **SP**(Ord,G,X), and R be the vertex adjacent to W on U and between W and X. If R is not a collider on U, then U does not d-connect V and X given **SP**(Ord,G',X). If R is a collider on U, then $W \mathbin{*\!\rightarrow} R$ on U. W is either an ancestor of V or of a collider on U, in which case it precedes X, and is a member of **SP**(Ord,G',X), contrary to our assumption. ∴

THEOREM 6.2: In an inducing path graph G' over **O**, where A and B are in **O**, if A is not an ancestor of B, and A and B are not adjacent then A and B are d-separated given a subset of **D-SEP**(A,B).

Proof. Suppose that A and B are not adjacent, and A is not an ancestor of B. Let the total order *Ord* on the variables in G' be such that all ancestors of A and all ancestors of B except for A are prior to A, and all other vertices are after A. Then **SP**(Ord,G',A) is a subset of **D-SEP**(A,B). Hence by lemma 6.2.4, if B is not in **D-SEP**(A,B) then **D-SEP**(A,B) d-separates A from B in G. B is in **D-SEP**(A,B) if and only if there is a path from A to B in which each vertex except the endpoints is a collider on the path, and each vertex on the path is an ancestor of A or B. But then there is an inducing path between A and B, and, by lemma, 6.2.3 A and B are adjacent, contrary to our assumption. ∴

13.18 Theorem 6.3

THEOREM 6.3: If the input to the CI algorithm is data over **O** that is faithful to G, the output is a partially oriented inducing path graph of G over **O**.

It is proved in lemma 7.3.2 that if G' is the inducing path graph for G over **O**, and there is a path U d-connecting A and B given **Z** in G then there is a path d-connecting A and B given **Z** in G'.

In an inducing path graph G', U is a **discriminating path** for B if and only if U is an undirected path between X and Y containing B, $B \neq X$, $B \neq Y$, and

(i) if V and V' are adjacent on U, and V' is between V and B on U, then $V \mathbin{*\!\!\rightarrow} V'$ on U,
(ii) if V is between X and B on U and V is a collider on U then $V \rightarrow Y$ in G', else $V \leftarrow\!\!* \, Y$ in G',
(iii) if V is between Y and B on U and V is a collider on U then $V \rightarrow X$ in G', else $V \leftarrow\!\!* \, X$ in G',
(iv) X and Y are not adjacent in G'.

B is a **definite noncollider** on undirected path U if and only if either B is an endpoint of U, or there exist vertices A and C such that U contains one of the subpaths $A \leftarrow B \mathbin{*\text{-}\!*} C$, $A \mathbin{*\text{-}\!*} B \rightarrow C$, or $A \mathbin{*\text{-}\underline{* \, B \, *}\text{-}*} C$.

In a partially oriented inducing path graph π, U is a **definite discriminating path** for B if and only if U is an undirected path between X and Y containing B, $B \neq X$, $B \neq Y$, every vertex on U except for B and the endpoints is a collider or a definite noncollider on U, and

(i) if V and V' are adjacent on U, and V' is between V and B on U, then $V \mathbin{*\!\!\rightarrow} V'$ on U,
(ii) if V is between X and B on U and V is a collider on U then $V \rightarrow Y$ in π, else $V \leftarrow\!\!* \, Y$ in π,
(iii) if V is between Y and B on U and V is a collider on U then $V \rightarrow X$ in π, else $V \leftarrow\!\!* \, X$ in π,
(iv) X and Y are not adjacent in π.

Proofs of Theorems

LEMMA 6.3.1: If G' is an inducing path graph, U is a discriminating path for B between X and Y, and X and Y are d-separated given $\mathbf{S}$, then for every vertex V on U not equal to B, V is in $\mathbf{S}$ if and only if V is a collider on U.

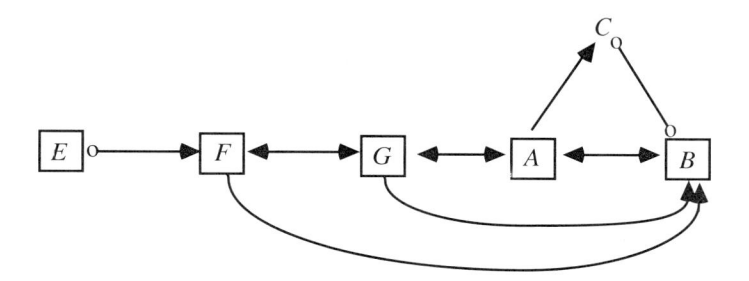

Figure 13.3. <E,F,G,A,C,B> is a definite discriminating path for C

Proof. First we will prove for each vertex V on U between X and B that V is in $\mathbf{S}$ if and only if V is a collider on U. The proof is by induction on the number of vertices between X and V on U.

Base Case: Let A be the first vertex on U after X. If $A = B$, then trivially for every vertex V between X and A, V is in $\mathbf{S}$ if and only if V is a collider on U. Suppose then that $A \ne B$. If A is a collider on U then there is an edge from A to Y. A is not a collider on the concatenation of $U(X,A)$ and the edge between A and Y, and hence that path d-connects X and Y given $\mathbf{S}$ unless A is in $\mathbf{S}$. If A is not a collider on U then there is an edge between Y and A that is into A. By definition of discriminating path, the edge between X and A is into A. Hence A is a collider on the concatenation of $U(X,A)$ and the edge between A and Y. Hence that path d-connects X and Y given $\mathbf{S}$ unless A is not in $\mathbf{S}$.

Induction Case: Suppose that if there are n or fewer vertices between X and V on U, then V is in $\mathbf{S}$ if and only if V is a collider on U. If there are only n vertices between X and B then we are done. Otherwise let A be the vertex such that there are $n+1$ vertices between X and A on U. Except for the endpoints, if V is on $U(X,A)$ then V is a collider on U if and only if U is in $\mathbf{S}$. If A is a collider on U, then there is a directed edge from A to Y. A is not a collider on the concatenation of $U(X,A)$ and the edge from A to Y, so that path d-connects X and Y given $\mathbf{S}$ unless A is in $\mathbf{S}$. If A is not a collider on U, then there is an edge between A and Y that is into A. Hence A is a collider on the concatenation of $U(X,A)$ and the edge from A to Y, so that path d-connects X and Y given $\mathbf{S}$ unless A is not in $\mathbf{S}$.

Similarly, if V is between Y and B, V is in $\mathbf{S}$ if and only if V is a collider on U. ∴

LEMMA 6.3.2: If G' is an inducing path graph, U is a discriminating path for B between X and Y, and X and Y are d-separated given $\mathbf{S}$, then B is in $\mathbf{S}$ if and only if B is not a collider on U.

Proof. By lemma 6.3.1, for every vertex V on U not equal to B, V is a collider on U if and only if V is in **S**. If B is a collider and in **S**, then U d-connects X and Y given **S**, contrary to our assumption. If B is not a collider and not in **S**, then U d-connects X and Y given **S**, contrary to our assumption. Hence B is in **S** if and only if B is not a collider on U. ∴

THEOREM 6.3: If the input to the CI algorithm is data over **O** that is faithful to G, the output is a partially oriented inducing path graph of G over **O**.

Proof. The proof is by induction on the number of applications of orientation rules in the repeat loop of the Causal Inference Algorithm. Let G' be the inducing path graph of G. Let the object constructed by the algorithm after the nth iteration of the repeat loop be π_n.

Base Case: Suppose that the only orientation rule that has been applied is that if A *-* B *-* C in F, but A and C are not adjacent in F, A *-* B *-* C is oriented as A *→ B ←* C if B is not a member of **Sepset**(A,C) and as A *-* $\underline{B}$ *-* C if B is a member of **Sepset**(A,C). Suppose A *→ B ←* C in π_0, but not in G'. It follows that B is not a member of **Sepset**(A,C), and either B is a parent of A or a parent of C in G'. If B is a parent of either A or C in G', then there is an undirected path between A and C that does not collide at B, and except for the endpoints contains only B. For any subset **S**, if that path in G' does not d-connect A and C given **S**, then **S** contains B. It follows that **Sepset**(A,C) contains B, which is a contradiction.

Suppose that A *-* $\underline{B}$ *-* C in π_0, but the edges between A and B, and B and C collide at B in G'. It follows that **Sepset**(A,C) does contain B but every set that d-separates A and C in G' does not contain B. Hence **Sepset**(A,C) does not contain B, which is a contradiction.

Induction Case: Suppose π_n is a partially oriented inducing path graph of G. We will now show that π_{n+1} is a partially oriented inducing path graph of G.

Case 1: There is a directed path from A to B and an edge A *-* B in π_n, so A *-* B is oriented as A *→ B. By the induction hypothesis if there is an edge $R \to S$ in π_n, then there is an edge $R \to S$ in G'. It follows that if there is a directed path from A to B in π_n, then there is a directed path from A to B in G'. Because G' is acyclic, A *→ B in G'.

Case 2: If B is a collider along <A,B,C> in π_n, B is adjacent to D, and D is in **Sepset**(A,C), then orient B *-* D as B ←* D. By the induction hypothesis, B is a collider along <A,B,C> and D is adjacent to B in G'. If in G' A and C are not d-connected given **Sepset**(A,C) (which contains D) by <A,B,C> then B has no descendant in $\{D\}$. Hence D *→ B in G'.

Cas 3: If U is a definite discriminationg path between A and B for M in π_n, and P and R are adjacent to M on U, and P-M-R is a trangle, then

if M is in **Sepset**(A,B) then mark M as a noncollider on subpath P *-* $\underline{M}$ *-* R
else orient P *-* M *-* R as P *$\rightarrow$ M $\leftarrow$ * R.

By the induction hypothesis, if U is a definite discriminating path for M in π_n, then it is a discriminating path for M in G'. By lemma 6.3.2, in G', if U is a discriminating path for M, then M is a collider on $<P,Q,R>$ if and only if M is not in **Sepset**(A,B).

Case 4: If P *$\rightarrow$ $\underline{M}$ *-* R then the orientation is changed to P *$\rightarrow$ M $\rightarrow$ R. By the induction hypothesis, if P *$\rightarrow$ $\underline{M}$ *-* R in π_n, then in G' the edge from P to M is into M, but M is not a collider on P *$\rightarrow$ M *-* R. It follows that P *$\rightarrow$ M $\rightarrow$ R in G'. ∴

13.19 Theorem 6.4

THEOREM 6.4: If the input to the FCI algorithm is data over **O** that is faithful to G, the output is a partially oriented inducing path graph of G over **O**.

If $A \neq B$ in partially oriented inducing path graph π, V is in **Possible-D-Sep**(A,B) in π if and only if $V \neq A$, and there is an undirected path U between A and V in π such that for every subpath $<X,Y,Z>$ of U either Y is a collider on the subpath, or Y is not a definite noncollider on U, and X, Y, and Z form a triangle in π.

LEMMA 6.4.1: If G' is the inducing path graph of directed acyclic graph G over **O**, and F' is the partially oriented graph constructed in step C) of Fast Causal Inference Algorithm for G over **O**, A and B are in **O**, and A is not an ancestor of B in G', then every vertex in **D-SEP**(A,B) in G' is in **Possible-D-SEP**(A,B) in F.

Proof. Suppose that A is not an ancestor of B. If V is in **D-SEP**(A,B) in G', then there is an undirected path U from A to V in which every vertex except the endpoints is a collider. It follows that in G' for every subpath $<X,Y,Z>$ of U, Y is a collider on the subpath. Hence in π, Y is either a collider, or X, Y, and Z form a triangle in π and Y is not a definite noncollider. ∴

THEOREM 6.4: If the input to the FCI algorithm is data over **O** that is faithful to G, the output is a partially oriented inducing path graph of G over **O**.

Proof. This follows immediately from theorem 6.3 and lemma 6.4.1. ∴

13.20 Theorem 6.5

THEOREM 6.5: If π is a partially oriented inducing path graph of directed acyclic graph G over $\mathbf{O}$, and there is a directed path U from A to B in π, then there is a directed path from A to B in G.

LEMMA 6.5.1: If π is a partially oriented inducing path graph of directed acyclic graph G over $\mathbf{O}$, and $A \rightarrow B$ in π, then there is a directed path from A to B in G.

Proof. Let G' be the inducing path graph of G. If $A \rightarrow B$ in π, then $A \rightarrow B$ in G'. If $A \rightarrow B$ in G', then in G there is an inducing path from A to B that is not into A. Hence by lemma 6.1.4 there is a directed path from A to B in G. $\therefore$

THEOREM 6.5: If π is a partially oriented inducing path graph of directed acyclic graph G over $\mathbf{O}$, and there is a directed path U from A to B in π, then there is a directed path from A to B in G.

Proof. By lemma 6.5.1, for each edge between R and S in U there is a directed path from R to S in G. The concatenation of the directed paths in G contains a subpath that is a directed path from A to B in G. $\therefore$

13.21 Theorem 6.6

THEOREM 6.6: If π is the CI partially oriented inducing path graph of directed acyclic graph G over $\mathbf{O}$, and there is no semidirected path from A to B in π, then there is no directed path from A to B in G.

LEMMA 6.6.1: Suppose that G is a directed acyclic graph, and in G there is a sequence of vertices M starting with A and ending with C, and a set of paths F such that for every pair of vertices I and J adjacent in M there is exactly one inducing path W over $\mathbf{O}$ between I and J in F. Suppose further that if $J \neq C$ then W is into J, and if $I \neq A$ then W is into I, and I and J are ancestors of either A or C. Then in G there is an inducing path T over $\mathbf{O}$ between A and C such that if the path in F between A and its successor in M is into A then U is into A, and if the path in F between C and its predecessor in M is into C then U is into C.

Proof. Suppose that in G there is a sequence M of vertices in $\mathbf{O}$ starting with A and ending with C, and a set of paths F such that for every pair of vertices I and J adjacent in M there is exactly one inducing path W over $\mathbf{O}$ between I and J, and if $J \neq C$ then W is into J, and if $I \neq A$ then W is into A, and I and J are ancestors of either A or C. Let T' be the concatenation of the paths in F. T' may not be an acyclic undirected path because it might contain undirected cycles. Let T be an acyclic undirected subpath of T' between A and C. We will now show that except for the endpoints, every vertex in **O** on T is a collider, and every collider on T is an ancestor of A or C.

If V is a vertex in **O** that is on T but that is not equal to A or C, every edge on every path in F is into V. Hence, every edge on T that contains V is into V because the edges on T are a subset of the edges on inducing paths in F.

Let R and S be the endpoints of W. We will now show that every vertex on W is either an ancestor of A or an ancestor of C. By hypothesis, R is an ancestor of either A or C, and S is an ancestor of either A or C. Because W is an inducing path over **O**, every collider on W is an ancestor of either R or S, and hence an ancestor of either A or C. Every noncollider on W is either an ancestor of R or S, or an ancestor of a collider on W. Hence every vertex on W is an ancestor of either A or C. It follows that every collider on T is an ancestor of A or C, because the vertices on T are a subset of the vertices on paths in F.

By definition, T is an inducing path between A and C over **O**. Suppose the path in F between A and its successor is into A. If the edge on T with endpoint A is on path in F on which A is an endpoint, then T is into A because by hypothesis that inducing path is into A. If the edge on T with endpoint A is on an inducing path over **O** in which A is not an endpoint of the path, then T is into A because A is in **O**, and hence a collider on every inducing path for which it is not an endpoint. Similarly, T is into C if in F the path between C and its predecessor is into A. ∴

In an inducing path or directed acyclic graph G that contains an undirected path U between X and Y, the the edge between V and W is **substitutable** for $U(V,W)$ in U if and only if V and W are on U, V is between X and W on U, G contains an edge between V and W, V is a collider on the concatenation of $U(X,V)$ and the edge between V and W if and only if it is a collider on U, and W is a collider on the concatenation of $U(Y,W)$ and the edge between V and W if and only if it is a collider on U.

LEMMA 6.6.2: If G' is an inducing path graph for directed acyclic graph G over **O**, C is a descendant of B in G, and U is an undirected path in G' between X and R containing subpath $A \mathrel{*\!\!\rightarrow} B \leftrightarrow C$ where A is between X and B, then in G' there is a vertex E on U between X and A inclusive and an edge between E and C that is substitutable for $U(E,C)$ in U. Furthermore the concatenation of $U(X,E)$ and the edge between E and C is into C, and if U is into X, then the concatenation of $U(X,E)$ and the edge between E and C is into X.

Proof. Suppose G' is an inducing path graph for directed acyclic graph G over **O**, C is a descendant of B in G, and U is an undirected path in G' between X and R containing subpath $A \mathrel{*\!\!\rightarrow} B \leftrightarrow C$ where A is between X and B. If E and F are on U, we will say that F is the successor of E on U if and only if there is an edge between E and F on U and E is between X and F or $E = X$. Let Y be the successor of X on U.

First we consider the case where there is no vertex V on U between X and A inclusive such that the edge from V to C is substitutable for $U(V,C)$ in U, but each vertex on U between Y and A inclusive is adjacent to C in G'. We will show that there is a directed path from Y to B.

Suppose that $U(Y,B)$ is not a directed path from Y to B. Let E be the vertex on U closest to B such that $U(E,B)$ is not a directed path from E to B. Let F be the successor of E on U. F is an ancestor of B in G', not a collider on U unless $F = B$, and by assumption F is adjacent to C. The edge between C and F is not out of C and into F, because G' is acyclic. Hence it is into C. If $F = B$, then $A \leftrightarrow B \leftrightarrow C$ in G'. It follows that in G there is an inducing path betweeen A and C that is into A and C, and hence $A \leftrightarrow C$ in G', and the edge between A and C is substitutable for the subpath of U between A and C. Suppose then that $F \neq B$. $U(F,B)$ is a directed path from F to B in G'. Because the edge between F and C is not substitutable for $U(F,C)$ in U it follows that F is a collider on the concatenation of $U(X,F)$ with the edge between F and C. Hence the edge between F and C is into F and into C, and the edge between E and F on U is into F. It follows that the edge between E and F is also into E because E is not an ancestor of B, and F is. Hence G' contains the path $E \leftrightarrow F \leftrightarrow C$. Because F is an ancestor of B in G', it is an ancestor of B in G. Since F is an ancestor of B in G, and B is an ancestor of C in G, F is an ancestor of B in G. It follows by lemma 6.6.1 that there is an inducing path between E and C in G relative to $\mathbf{O}$ that is into E and into C. But then in G' the edge between E and C is substitutable for $U(E,C)$ in U, which is a contradiction.

We have shown that $U(Y,B)$ is a directed path from Y to B. It follows that Y is an ancestor of B in G, and because B is an ancestor of C in G, Y is an ancestor of C in G. We have shown that the edge between Y and its successor on U is out of Y. Hence Y is not a collider on U. By assumption there is an edge between Y and C in G'. If the edge between Y and C is not substitutable for $U(Y,C)$ in U, then the edge between Y and C is into Y, and because G' is acyclic (i.e., there is no directed cycle in G'), the edge between Y and C is also into C. Because the edge between Y and C is not substitutable for $U(Y,C)$ in U, and the edge between Y and C is into Y, it follows that the edge between X and Y is into Y. Hence G' contains the path $X \mathbin{*}\!\!\rightarrow Y \leftrightarrow C$, and Y is an ancestor of C in G. It follows that there is an inducing path between X and C in G relative to $\mathbf{O}$ that is into C, and if U is into X, also into X. But then the edge between X and C is substitutable for $U(X,C)$ in U, which is a contradiction.

Next we consider the case where there is no vertex V on U between X and A inclusive such that the edge from V to C is substitutable for $U(V,C)$ in U, but some vertex on U between Y and A inclusive is not adjacent to C. Let E be the vertex on U closest to C and between X and C that is not adjacent to C, and let F be the successor of E on U. $E \neq A$, because by lemma 6.6.1 there is an inducing path between A and C in G, and hence A is adjacent to C in G'. From the previous case, it follows that either there is an edge between V on $U(E,C)$ and C that is substitutable for $U(V,C)$ in $U(E,C)$ or F is an ancestor of B in G'. Suppose first that there is an edge between V on $U(E,C)$ and C that is substitutable for

$U(V,C)$ in $U(E,C)$. E is not adjacent to C, so $V \neq E$, and V lies on $U(F,C)$. If the edge between V and C is substitutable for $U(V,C)$ in $U(E,C)$, then it is also substitutable for $U(V,C)$ in U, which is a contradiction. Hence F is an ancestor of B in G'. By the definition of E, F is adjacent to C in G'. The edge between F and C is not out of C and into F, because G' is acyclic. The edge between F and C is not out of F and into C because the edge between F and C is not substitutable for $U(F,C)$ in $U(E,C)$, and $U(F,B)$ is a directed path from F to B. Hence the edge between F and C is into F and C. If the edge $E \leftarrow F$ is on U, then the $F \leftrightarrow C$ edge is substitutable for $U(F,C)$ in U. If $E \ast\!\rightarrow F$ in G' then G' contains the path $E \ast\!\rightarrow F \leftrightarrow C$, and F is an ancestor of C in G' and hence in G; it follows that there is an inducing path between E and C relative to $\mathbf{O}$ in G, and E is adjacent to C in G'. This is a contradiction.

It follows that for some vertex E on U between X and A inclusive there is an edge from E to C that is substitutable for $U(E,C)$ in U and is into C. If $E = X$ then there is an inducing path between X and C that contains the edge on U with X as endpoint. If $E \neq X$ then there is some vertex $E \neq X$ on U such that there is an edge between E and C that is substitutable for $U(E,C)$ in U. In the first case, the inducing path is into X if U is into X and hence the edge between C and X is into X. In the second case the path consisting of the concatenation of $U(X,V)$ and the edge between V and C contains the edge on U with X as endpoint, and hence is into X if U is. ∴

LEMMA 6.6.3: If π is the CI partially oriented inducing path graph of graph G over $\mathbf{O}$, and $A \ast\!\rightarrow B$ in π, then every inducing path in G between A and B is into B.

Proof. We will prove that each orientation rule in the Causal Inference Algorithm is such that if the rule orients the edge between A and B as $A \ast\!\rightarrow B$, then every inducing path between A and B over $\mathbf{O}$ in G is into B. Let G' be the inducing path graph of G.

Case 1: By lemma 6.5.1 any of the rules that orients the edge between A and B as $A \rightarrow B$ in π entails that there is a directed path from A to B in G. If there is an inducing path over $\mathbf{O}$ between A and B in G that is out of B, and there is a directed path from B to A in G. But G is not cyclic, so there is no inducing path between A and B in G that is not into B.

Case 2: Suppose the edge between A and B is oriented as $A \ast\!\rightarrow B$ in order to avoid a cycle in π because there is a directed path from A to B in π. By theorem 6.5 there is a directed path from A to B in G. If there is an inducing path over $\mathbf{O}$ between A and B in G that is out of B, then there is a directed path from B to A in G. But G is not cyclic, so there is no inducing path over $\mathbf{O}$ between A and B in G that is out of B.

Case 3: Suppose that the edge between A and B is oriented as $A \ast\!\rightarrow B$ because there is a vertex C such that A and B are adjacent in π, B and C are adjacent in π, A and C are not adjacent in π, and B is not in **Sepset**(A,C). It follows that $A \ast\!\rightarrow B \leftarrow\!\ast C$ in G'. By the construction of G' it follows that in G there is an inducing path over **O** between A and B into B, and an inducing path over **O** between B and C into B. Suppose contrary to the theorem that there is another inducing path over **O** between A and B in G that is out of B. By lemma 6.1.4, A is a descendant of B in G. By lemma 6.6.1 there is an inducing path over **O** between A and C. But if there is an inducing path over **O** between A and C in G, then A and C are adjacent in π, contrary to our assumption.

Case 4: Suppose that the edge between A and B is oriented as $A \;*\!\!\rightarrow B$ because B is a collider along $<C,B,D>$ in π, B is adjacent to A, and A is not in Sepset (C,D). Suppose, contrary to the theorem, that in G there is an inducing path over **O** between A and B that is out of B. It follows that A is a descendant of B in G. Because there is an edge between C and B that is into B in π, there is an edge between C and B that is into B in G'. The edge between C and B in G' d-connects C and B given A and is into B. By lemmas 6.1.1 and 6.1.2 there is a path in G that d-connects C and B given A that is into B. Similarly, there is a path in G that d-connects D and B given A that is into B. By lemma 3.3.1, C and D are d-connected given A in G. By lemma 5.1.3, this is a contradiction.

Case 5: Suppose the edge between A and B in π is oriented as $A \;*\!\!\rightarrow B$ because in π U is a definite discriminating path for B between X and Y, B is in a triangle on U, and B is not in **Sepset**(X,Y). Let A and C be the vertices adjacent to B on U. If U is a definite discriminating path for B in π, then by the induction hypothesis, the corresponding path U' in G' is a discriminating path for B. In G', X and Y are d-separated given **Sepset**(X,Y) because by definition of definite discriminating path they are not adjacent. If X and Y are d-separated given **Sepset**(X,Y) in G', then by lemma 6.3.1 every collider on U' except for B is in **Sepset**(X,Y), and every noncollider on U' is not in **Sepset**(X,Y).

Suppose that there is an inducing path over **O** between B and A in G that is out of B. It follows that there is a directed path from B to A in G and that $A \leftrightarrow B$ in G'. By definition of discriminating path it follows that A is a collider on U' or $A = X$. By lemma 6.3.1 A is in **Sepset**(X,Y). Hence B is a collider on U' in G', and B has a descendant in **Sepset**(X,Y) in G.

If some vertex Z on U is in **Sepset**(X,Y) then Z is a collider on U. Let R and T be the vertices on U' that are adjacent to Z on U'. By the definition of inducing path graph, in G there are inducing paths over **O** between R and Z, and Z and T, such that each of them is into Z. By lemmas 6.1.1 and 6.1.2, in G there is a d-connecting path given $S\backslash\{R,Z\}$ between R and Z, and a d-connecting path given $S\backslash\{Z,T\}$ between Z and T, such that each of them is into Z.

If there are vertices R, Z, and T on U' such that R and Z are adjacent on U, and Z and T are adjacent on U', and Z is a collider on U', then either Z is in **Sepset**(X,Y) (if $Z \neq B$), or Z has a descendant in **Sepset**(X,Y) in G (if $Z = B$). In either case Z has a descendant in **Sepset**(X,Y) in G. By the definition of inducing path graph, in G there are inducing paths over **O** between R and Z, and Z and T, that are both into Z. By lemmas 6.1.1 and 6.1.2, in

G there is a d-connecting path given **Sepset**(X,Y)\{R,Z} between R and Z, and a d-connecting path given **Sepset**(X,Y)\{Z,T} between Z and T, that are both into Z. By lemma 3.3.1, there is a path in G that d-connects X and Y given **Sepset**(X,Y). But this contradicts the assumption that X and Y are d-separated given **Sepset**(X,Y). Hence there is no inducing path in G that is out of B. ∴

A **semidirected path from** A to B in partially oriented inducing path graph π is an undirected path U from A to B in which no edge contains an arrowhead pointing toward A, that is, there is no arrowhead at A on U, and if X and Y are adjacent on the path, and X is between A and Y on the path, then there is no arrowhead at the X end of the edge between X and Y.

THEOREM 6.6: If π is the CI partially oriented inducing path graph of directed acyclic graph G over **O**, and there is no semidirected path from A to B in π, then there is no directed path from A to B in G.

Proof. Suppose there is a directed path P from A to B in G. Let P' in π be the sequence of vertices in **O** along P in the order in which they occur. P' is an undirected path in π because for each pair of vertices X and Y adjacent in P' for which X is between A and Y or X = A there is an inducing path over **O** in G that is out of X. P' is a semidirected path from X to Y in π because by lemma 6.6.3, there is no arrowhead into X on P'. ∴

13.22 Theorem 6.7

THEOREM 6.7: If π is a partially oriented inducing path graph of directed acyclic graph G over **O**, A and B are adjacent in π, and there is no undirected path between A and B in π except for the edge between A and B, then in G there is a trek between A and B that contains no variables in **O** other than A or B.

Proof. Suppose that every trek between A and B in G contains some member of **O** other than A or B. Because there is an edge between A and B in π, there is an inducing path between A and B in G. Hence, A and B are d-connected given the empty set in G, and there is a trek T between A and B. Let U be the sequence of observed vertices on T. Each subpath of T between variables adjacent in U is an inducing path relative to **O**. Hence U is an undirected path in π that contains a member of **O** other than A or B. ∴

13.23 Theorem 6.8

THEOREM 6.8: If π is the CI partially oriented inducing path graph of directed acyclic graph G over **O**, and every semidirected path from A to B contains some member of **C** in π, then every directed path from A to B in G contains some member of **C**.

Proof. Suppose that U is a directed path in G from A to B that does not contain a member of **C**. Let the sequence of observed variables on U in G be U'. Let X and Y be two adjacent vertices in U', where X is between A and Y. $U(X,Y)$ is a directed subpath of U that contains no observed variables except for the endpoints. Hence $U(X,Y)$ is an inducing path between X and Y given **O** that is out of X. It follows that there is an edge between X and Y in π, and by lemma 6.6.3 the edge between X and Y is not into X. Hence U' is a semidirected path from A to B in π that does not contain any member of **C**. ∴

13.24 Theorem 6.9

THEOREM 6.9: If π is a partially oriented inducing path graph of directed acyclic graph G over **O**, and $A \leftrightarrow B$ in π, then there is a latent common cause of A and B in G.

Proof. By theorem 6.6, every inducing path over **O** in G between A and B is into B and into A. By lemma 6.1.2, there is in G a d-connecting path U between A and B given the empty set that is into A and into B in G. Because U d-connects A and B given the empty set in G it contains no colliders, and hence no members of **O** except A and B. Because U contains an edge into A and an edge into B, U is not a single edge between A and B. Hence there is some vertex C not in **O** on U that is a common cause of A and B. ∴

13.25 Theorem 6.10 (Tetrad Representation Theorem)

TETRAD REPRESENTATION THEOREM 6.10: In an acyclic LCF G, there exists an $LJ(T(I,J),T(K,L),T(I,L),T(J,K))$ choke point or an $IK(T(I,J),T(K,L),T(I,L),T(J,K))$ choke point iff G linearly implies $\rho_{IJ}\rho_{KL} - \rho_{IL}\rho_{JK} = 0$.

In a graph G, the **length** of a path equals the number of vertices in the path minus one. In a graph G, a path U of length n is an **initial segment** of path V of length m iff $m \geq n$, and for $1 \leq i \leq n+1$, the ith vertex of V equals the ith vertex of U. In a graph G, path U of length n is a **final segment** of path V of length m, iff $m \geq n$, and for $1 \leq i \leq n+1$, the ith vertex of U equals the $(m-n+i)^{th}$ vertex of V. A path U of length n is a **proper initial segment** of path V of length m iff U is an initial segment of V and $U \neq V$. A path U of length n is a **proper final segment** of path V of length m iff U is a final segment of V and $U \neq V$.

The proofs of the following lemma are obvious.

LEMMA 6.10.1: In a directed graph G, if $R(U,I)$ is an acyclic path, and X is a vertex on $R(U,I)$, then there is a unique initial segment of $R(U,I)$ from U to X.

Because the proofs refer to many different paths, we will usually designate a directed path by $R(X,Y)$ where X and Y are the endpoints of the path. When there is a path $R(U,I)$ in a proof, and a vertex X on $R(U,I)$, $R(U,X)$ will refer to the unique initial segment of $R(U,I)$ from U to I, and $R(X,I)$ will refer to the unique final segment of $R(U,I)$ from X to I.

In a directed acyclic graph G, the **last point of intersection** of directed path $R(U,I)$ with directed path $R(V,J)$ is the last vertex on $R(U,I)$ that is also on $R(V,J)$. Note that if G is a directed acyclic graph, the last point of intersection of directed path $R(U,I)$ with directed path $R(V,J)$ equals the last point of intersection of $R(V,J)$ with $R(U,I)$; this is not true of directed cyclic paths.

LEMMA 6.10.2: If G is a directed acyclic graph, for all variables Y and Z in G, if $Y \neq Z$ and R and R' are two intersecting directed paths with sinks Y and Z respectively then there is a trek between Y and Z that consists of subpaths of R and R'.

Proof. Since R and R' intersect, they have a last point of intersection X. Let the source of the trek to be constructed be X. $R(X,Y)$ and $R(X,Z)$ do not intersect anywhere except at X. Since $Y \neq Z$, one of $R(X,Y)$ and $R(X,Z)$ is not empty. Hence $\{R(X,Y),R(X,Z)\}$ is a trek. ∴

In a directed acyclic graph, directed paths $R(U,I)$ and $R(U,J)$ **contain trek** T iff $I(T(I,J))$ is a final segment of $R(U,I)$ and $J(T(I,J))$ is a final segment of $R(U,J)$.

LEMMA 6.10.3: In a directed acyclic graph, if $R(U,I)$ and $R(U,J)$ are directed paths that contain both $T(I,J)$ and $T'(I,J)$, then $T(I,J) = T'(I,J)$.

Proof. In a directed acyclic graph, there is a unique last point of intersection of $R(U,I)$ and $R(U,J)$, and unique final segments of R and R' whose source is the last point of intersection of $R(U,I)$ and $R(U,J)$. ∴

If G is a directed acyclic graph, let Let $\mathbf{P}_{XY}$ be the set of all directed paths in G from X to Y. In an LCF S, the **path form of a product of covariances** $\gamma_{IJ}\gamma_{KL}$ is the distributed form of

$$\left(\sum_{U \in \mathbf{U}_{IJ}} \left(\sum_{R \in \mathbf{P}_{UI}} \sum_{R' \in \mathbf{P}_{UJ}} L(R)L(R')\sigma_U^2 \right) \right) \left(\sum_{V \in \mathbf{U}_{KL}} \left(\sum_{R'' \in \mathbf{P}_{VK}} \sum_{R''' \in \mathbf{P}_{VL}} L(R'')L(R''')\sigma_V^2 \right) \right)$$

$\gamma_{IJ}\gamma_{KL} - \gamma_{IL}\gamma_{JK}$ is in **path form** iff both terms are in path form.

Henceforth, we will assume that all variances, covariances, products of covariances, and tetrad differences are in path form unless otherwise stated.

We will adopt the following terminology. Suppose that m is a term in the path form of a product of covariances $\gamma_{IJ}\gamma_{Kl}$. By definition, m is of the form $L(R(U,I))L(R(U,J))L(R(V,K))L(R(V,L))$ $\sigma_U^2 \sigma_V^2$. Let the paths associated with m be the ordered quadruple $<R(U,I),R(U,J),R(V,K),R(V,L)>$. There is a one-to-one correspondence between terms in the path form of a product of covariances, and such ordered quadruples. We will consider terms m and m' to be distinct terms if their associated paths are different (i.e., the terms may contain the same number of occurrences of the same edge labels, but in different orders.) Note that under this criterion of identity of terms, no term appears twice in the path form of a product of covariances or tetrad difference. Henceforth when we consider sets of terms appearing in some expression, we will do so under the assumption that each term occurs at most once in the expression (although distinct terms that have identically equal values may occur in the expression). We will say that a term m contains a path or trek X if its associated quadruple contains X.

LEMMA 6.10.4: A tetrad difference $\gamma_{IJ}\gamma_{KL} - \gamma_{IL}\gamma_{JK}$ is not linearly implied to vanish by an LCF S if there is a term m in the path form of $\gamma_{IJ}\gamma_{KL}$ such that every term m' in the path form of $\gamma_{IL}\gamma_{JK}$ contains an edge not in m.

Proof. Suppose that there is a term m in the path form of $\gamma_{IJ}\gamma_{KL}$ such that every term m' in the path form of $\gamma_{IL}\gamma_{JK}$ contains an edge not in m. Set every variable not in m to be zero. Then $\gamma_{IL}\gamma_{JK}$ is zero since every term in $\gamma_{IL}\gamma_{JK}$ contains a variable not in m. Set every variable in m to be positive. Then every nonzero term in the path form of $\gamma_{IJ}\gamma_{KL}$ is positive, since the e.c.f. of each nonzero term is positive, and the c.f. of each nonzero term is positive. $\gamma_{IJ}\gamma_{KL}$ is not zero since every term in it is either 0 or positive, and some are positive. Hence the tetrad difference is not linearly implied to vanish. ∴.

LEMMA 6.10.5: In an LCF S, if the paths in a term m in the path form of a tetrad difference have different sources than the paths in a term m', then m contains some variable not in m'.

Proof. Each of the sources of the paths in m and m' are independent random variables, and it is not the case that all of the paths in m or m' are empty. Let $\{I,J\}$ be the sources of the paths in m, and $\{K,Z\}$ be the sources of the paths in m' and suppose that $\{I,J\} \neq \{K,Z\}$. Suppose w.l.g. that $I \neq K$. Since I, K, and Z are independent I does not occur on any paths with source K or Z. m contains at least one edge X out of I. Since I does not occur on any path with source K or Z, X does not occur on any path with source K or Z. Hence m contains a variable (the label of X) that does not occur in m'. ∴.

In an LCF F, $e(S)$ is equal to S if S is an independent variable, and it is equal to the error variable into S if S is not an independent variable.

LEMMA 6.10.6: In an LCF S, if there exist $T(I,J) \in \mathbf{T}(I,J)$ and $T(K,L) \in \mathbf{T}(K,L)$ such that $I(T(I,J)) \cap K(T(K,L)) = \emptyset$, $J(T(I,J)) \cap L(T(K,L)) = \emptyset$, and $I(T(I,J)) \cap L(T(K,L)) = \emptyset$, then there exists a term m in $\gamma_{IJ}\gamma_{KL}$ such that every term m' in $\gamma_{IL}\gamma_{JK}$ contains an edge not in m.

Proof. Let S be the source of $T(I,J)$ and S' be the source of $T(K,L)$. (Note that since $I(T(I,J))$ does not intersect $L(T(K,L))$, the source of $T(I,J)$ does not equal the source of $T(K,L)$, and hence $e(S)$ does not equal $e(S')$. See figure 13.4.)

Proofs of Theorems 431

Let $m = L(R(e(S),I))L(R(e(S),J))L(R(e(S'),K))L(R(e(S'),L))$. m is the coefficient of a term in $\gamma_{IJ}\gamma_{KL}$ (the full term also contains a factor equal to the product of the variances of the sources of paths in m.)

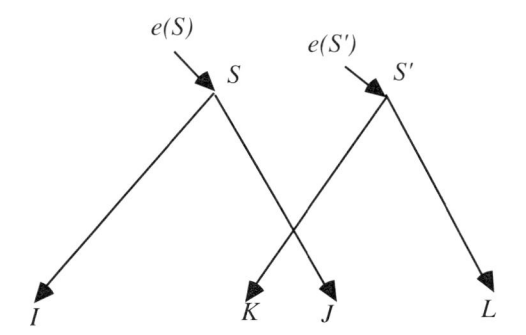

Figure 13.4

Suppose there is a term m' in $\gamma_{IL}\gamma_{JK}$ whose associated paths contain only edges in m. m' contains the product of the labels of edges in a trek $T(I,L)$. Let the source of $T(I,L)$ be S''. If $S'' \neq S$ and $S'' \neq S'$, then $e(S'') \neq e(S)$ and $e(S'') \neq e(S')$. Since $e(S'')$ is an independent variable, and the only independent variables in m are $e(S)$ and $e(S')$, if $e(S'') \neq e(S)$ and $e(S'') \neq e(S')$, then $T(I,L)$ contains an edge label not in m. Suppose then w.l.g. that $S'' = S$. There is a path $R(S,L)$ containing edge labels only in m. Since $J(T(I,J)) \cap L(T(K,L)) = \emptyset$, and $I(T(I,J)) \cap L(T(K,L)) = \emptyset$, the only path in m that contains L is $L(T(K,L))$. Hence $R(S,L)$ intersects $L(T(K,L))$ at some vertex. The only two paths in m with source S are $I(T(I,J))$ and $J(T(I,J))$, and neither of them intersects $L(T(K,L))$. Hence one of them intersects some other paths that in turn intersects $L(T(K,L))$. The only other path in m that intersects $L(T(K,L))$ is $K(T(K,L))$. So $R(S,L)$ intersects $K(T(K,L))$. Since the last point of intersection of $L(T(K,L))$ and $K(T(K,L))$ is S', $R(S,L)$ intersects $K(T(K,L))$ at or before S'. But the only paths with source S in m are $J(T(I,J))$ and $I(T(I,J))$, and neither of them intersects $K(T(K,L))$ at or before S'. Hence, there is no path from S to L containing only edge labels in m. Similarly it can be shown that there is no path from S' to I containing only edge labels in m. Hence m' contains an edge label not in m. ∴.

LEMMA 6.10.7: In an LCF S, if there exists a $T(I,J) \in \mathbf{T}(I,J)$ and $T(K,L) \in \mathbf{T}(K,L)$ such that $I(T(I,J) \cap K(T(K,L)) = \emptyset$, and $L(T(K,L)) \cap J(T(I,J)) = \emptyset$, or there exists a $T(I,L) \in \mathbf{T}(I,L)$ and $T(J,K) \in \mathbf{T}(J,K)$ such that $I(T(I,L)) \cap K(T(J,K)) = \emptyset$, and $L(T(I,L)) \cap J(T(J,K)) = \emptyset$, then S does not linearly imply that $\gamma_{IJ}\gamma_{Kl} - \gamma_{IL}\gamma_{JK}$ vanishes.

Proof. Suppose w.l.g. that $I(T(I,J)) \cap K(T(K,L)) = \emptyset$, and $L(T(K,L)) \cap J(T(I,J)) = \emptyset$. There are four cases: either (i) $I(T(I,J)) \cap L(T(K,L)) = \emptyset$ and $J(T(I,J)) \cap K(T(K,L)) = \emptyset$, or (ii) $I(T(I,J)) \cap L(T(K,L)) = \emptyset$ and $J(T(I,J)) \cap K(T(K,L)) \neq \emptyset$, or (iii) $I(T(I,J)) \cap L(T(K,L)) \neq \emptyset$ and $J(T(I,J)) \cap K(T(K,L)) = \emptyset$, or (iv) $I(T(I,J)) \cap L(T(K,L)) \neq \emptyset$ and $J(T(I,J)) \cap K(T(K,L)) \neq \emptyset$.

In the first three cases, by lemma 6.10.6 there exists a term m in $\gamma_{IJ}\gamma_{KL}$ such that every m' in $\gamma_{IL}\gamma_{JK}$ contains an edge label not in m.

In the fourth case, let X be the last point of intersection of $I(T(I,J))$ and $L(T(K,L))$, and Y be the last point of intersection of $J(T(I,J))$ and $K(T(K,L))$. X is not the source of either trek, since otherwise $I(T(I,J)) \cap K(T(K,L)) \neq \emptyset$ or $J(T(I,J)) \cap L(T(K,L)) \neq \emptyset$. Similarly, Y is not the source of either trek. $\{R(X,I), R(X,L)\}$ is a trek $T(I,L)$ between I and L, by lemma 6.10.2. Similarly, $\{R(Y,J), R(Y,K)\}$ form a trek $T(J,K)$. (See figure 13.5.)

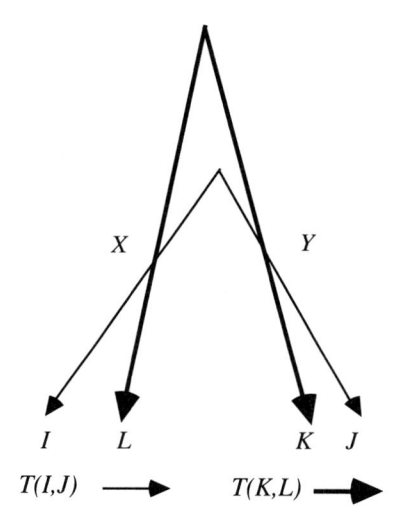

Figure 13.5

Now we will show that $T(I,L) \cap T(J,K) = \emptyset$. $I(T(I,L)) \cap J(T(J,K)) = \emptyset$ since $I(T(I,L))$ is a proper subpath of $I(T(I,J))$ and $J(T(J,K))$ is a proper subpath of $J(T(I,J))$, and the last point of intersection of $I(T(I,J))$ and $J(T(I,J))$ is the source of $T(I,J)$. $I(T(I,L)) \cap K(T(J,K)) = \emptyset$, since $I(T(I,L))$ is a subpath of $I(T(I,J))$ and $K(T(J,K))$ is a subpath of $K(T(K,L))$, and $I(T(I,J)) \cap K(T(K,L)) = \emptyset$ by hypothesis. For similar reasons, $L(T(I,L)) \cap J(T(J,K)) = \emptyset$, and $L(T(I,L)) \cap K(T(J,K)) = \emptyset$. It follows from lemma 6.10.6 there exists a term m in $\gamma_{IL}\gamma_{JK}$ such that every m' in $\gamma_{IJ}\gamma_{KL}$ contains an edge label not in m.

Since there exists a term m in $\gamma_{IL}\gamma_{JK}$ such that every m' in $\gamma_{IJ}\gamma_{KL}$ contains an edge not in m, by lemma 6.10.4 $\gamma_{IJ}\gamma_{KL} - \gamma_{IL}\gamma_{JK}$ is not linearly implied. ∴

A vanishing tetrad difference is a constraint upon the covariances of four pairs of variables: $\langle I,J \rangle$, $\langle K,L \rangle$, $\langle I,L \rangle$ and $\langle J,K \rangle$. Roughly speaking, a choke point for such a foursome of variable pairs is a point where all of the treks between I and J intersect all of the treks between K and L, and all of the treks between I and L intersect all of the treks between J and K. (A more precise definition is given later.) In this section, we will prove that in an LCF G, the existence of such a choke point is a necessary condition for the corresponding tetrad difference to vanish in distributions perfectly represented by G. We will prove this by showing that the existence of a choke point in G is equivalent to a condition that has already been proved to be a necessary condition for S to linearly imply a vanishing tetrad difference; namely, the trek intersection condition described in lemma 6.10.7. Unfortunately, this proof is long and tedious because there are many different ways in which a choke point can fail to exist, depending upon which treks are assumed to intersect and which treks are assumed not to intersect. In each case we show that the non-existence of a choke point implies the violation of the necessary condition described in lemma 6.10.7.

Two strategies are employed in the proofs. The first is to show that the assumptions about which treks intersect and don't intersect lead to contradictions. The second is to show that it is possible to construct a pair of treks $T'(I,J)$ and $T'(K,L)$ such that $I(T'(I,J))$ and $K(T'(K,L))$ don't intersect, and $J(T'(I,J))$ and $L(T'(K,L))$ don't intersect, or to construct a pair of treks $T'(I,L)$ and $T'(J,K)$ such that $I(T'(I,L))$ and $K(T'(J,K))$ don't intersect, and $J(T'(J,K))$ and $L(T'(I,L))$ don't intersect. In either case, by lemma 6.10.7, it follows that $\gamma_{IJ}\gamma_{KL} - \gamma_{IL}\gamma_{JK}$ is not linearly implied by G.

In general, when constructing a trek $T(I,J)$ we will speak as if it suffices to show how to construct a pair of (acyclic) directed paths R and R' from a common source S to sinks I and J respectively, without showing that the pair of directed paths constructed do not intersect. This is because even if R and R' do not form a trek because they intersect each other at some vertex other than S, we have shown in lemma 6.10.2 that directed subpaths of R and R' do form a trek, and the existence of the directed subpaths of R and R' is enough for our purposes. We are generally interested in showing that particular pairs of trek branches fail to intersect. If R_1 and R_2 fail to intersect, then directed subpaths of R_1 and R_2 also fail to intersect. Hence, if the goal is to show that trek branches T and T' fail to intersect, it suffices to show that R_1 and R_2 fail to intersect, even if T and T' are actually equal to directed subpaths of R_1 and R_2 respectively.

Let **S** be a set of vertices, and $\mathbf{R_K(S)}$ be the set of all directed paths with sink K and a source in **S**. Let $R(S,I)$ be a directed path from S in **S** to I. Let X_n be the nth vertex on $R(S,I)$ such that some directed path in $\mathbf{R_K(S)}$ intersects it. Let the set of sources of directed paths in $\mathbf{R_K(S)}$ whose first point of intersection with $R(S,I)$ is X_n be $\mathbf{S_n}$. Let the last vertex in $R(S,I)$ that is the first intersection of some directed path in $\mathbf{R_K(S)}$ with $R(S,I)$ be X_{max}. Note that X_{max} is not necessarily the last point of intersection of some directed path in $\mathbf{R_K(S)}$ with $R(S,I)$; it is merely the last of the first points of intersection. (See figure 13.6.)

LEMMA 6.10.8: In a directed acyclic graph G, if $R(M,I)$ is a directed path, and $\mathbf{R_K(S)}$ is the set of all directed paths to K from a given set of sources **S**, and there does not exist a vertex Z such that all of the directed paths in $\mathbf{R_K(S)}$ intersect $R(M,I)$ at Z, then there is a pair of directed paths, R and R', with the following properties: M is the source of R, R' has a source in **S**, either R has sink I and R' has sink K or R has sink K and R' has sink I, and R does not intersect R'.

Proof. If there is a path R' in $\mathbf{R_K(S)}$ that does not intersect $R(M,I)$ the proof is done. Assume then that every path in $\mathbf{R_K(S)}$ intersects $R(M,I)$. Let S'' be the source of a path in $\mathbf{S_{max}}$ (the set of all sources of paths in $\mathbf{R_K(S)}$ whose first intersection with $R(M,I)$ is X_{max}.) The proof is by induction on the number of distinct vertices in which the paths in $\mathbf{R_K(S)}$ intersect $R(M,I)$.

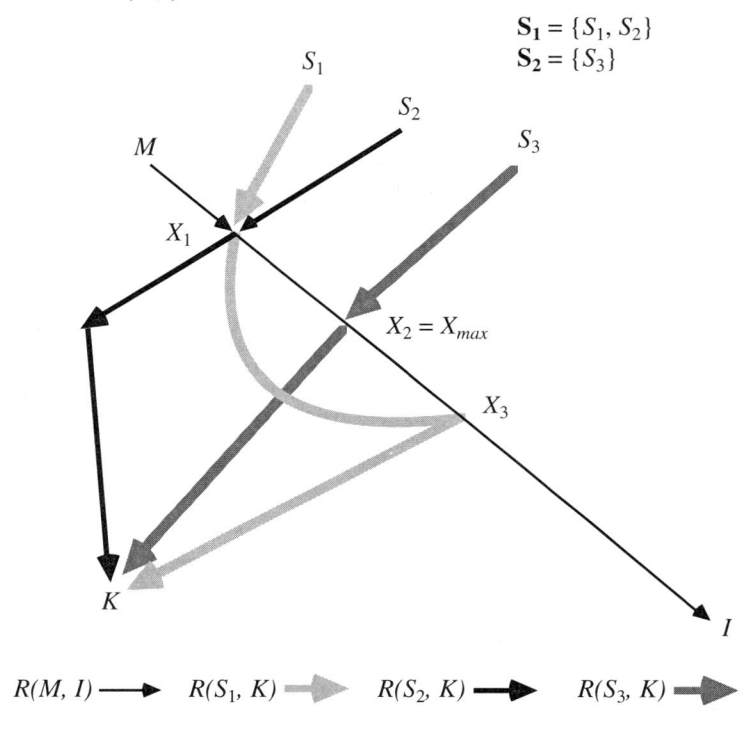

Figure 13.6

Base Case: Suppose the antecedent in the statement of the lemma is true. The paths in $\mathbf{R_K(S)}$ intersect $R(M,I)$ in two distinct vertices. There is a path $R(S',K)$ that does not intersect $R(M,I)$ at X_2 (= X_{max}), since otherwise all paths in $\mathbf{R_K(S)}$ would intersect X_2, contrary to our hypothesis. In addition, $R(S',K)$ does not intersect $R(M,I)$ at any vertex prior to X_1, since otherwise the paths in $\mathbf{R_K(S)}$ would intersect $R(M,I)$ at more than two distinct vertices, contrary to our hypothesis. Similarly, there is a path $R(S'',K)$ that intersects $R(M,I)$ only at X_2.

Let $R(X_1,K)$ be a final segment of $R(S',K)$ and $R(S'', X_2)$ an initial segment of $R(S'',K)$. There are two cases.

Proofs of Theorems 435

1. $R(X_1,K)$ does not intersect $R(S'', X_2)$. (See figure 13.7.) Let $R(M, X_1)$ be an initial segment of $R(M, I)$, $R(X_2, I)$ be a final segment of $R(M,I)$, $R = R(M, X_1)\&R(X_1, K)$ and $R' = R(S'', X_2)\&R(X_2, I)$. R and R' do not intersect for the following reasons.

$R(M,X_1)$ does not intersect $R(S'',X_2)$. $R(S'',X_2)$ is a subpath of $R(S'',K)$, which, by hypothesis intersects $R(M,I)$ only at X_2. Since X_2 occurs after X_1 on $R(M,I)$, X_2 does not occur on $R(M,X_1)$. $R(M,X_1)$ does not intersect $R(X_2,I)$. $R(M,X_1)$ and $R(X_2,I)$ are both subpaths of $R(M,I)$, G is acyclic, and by hypothesis X_1 occurs before X_2. $R(X_1,K)$ does not intersect $R(S'',X_2)$ by hypothesis. $R(X_1,K)$ does not intersect $R(X_2,I)$. $R(X_1,K)$ is a subpath of $R(S',K)$ and $R(X_2,I)$ is a subpath of $R(M,I)$; by hypothesis $R(S',K)$ intersects $R(M,I)$ only at X_1, which does not occur on $R(X_2,I)$.

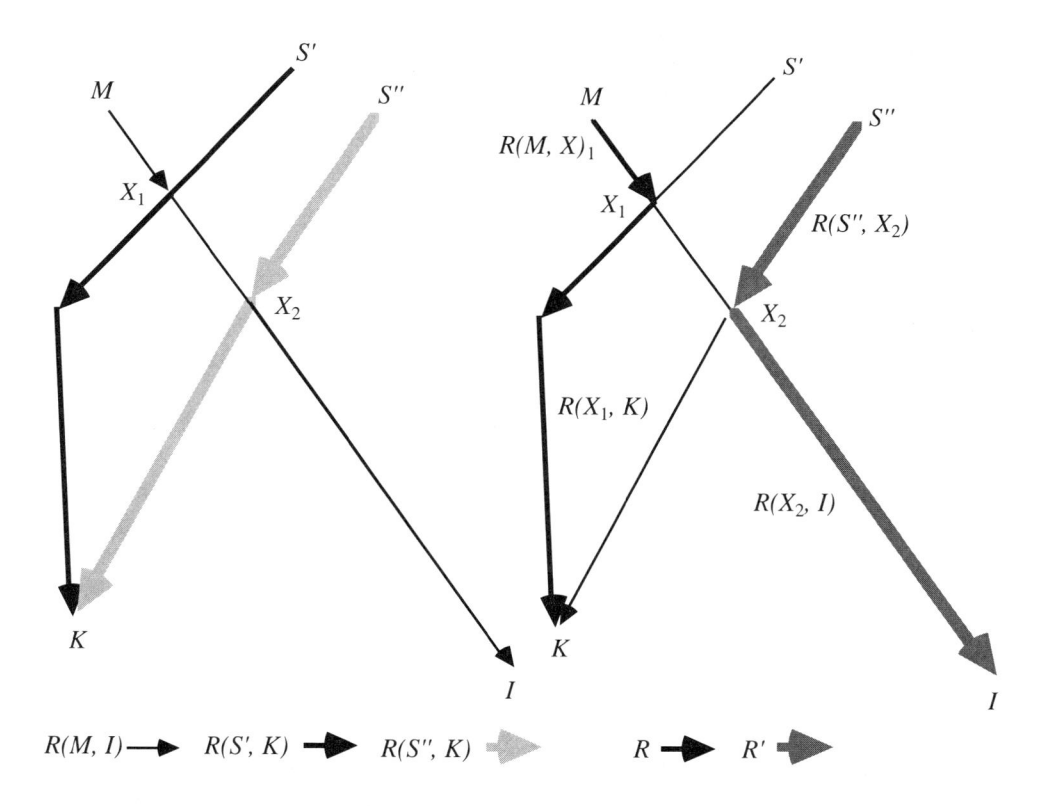

Figure 13.7

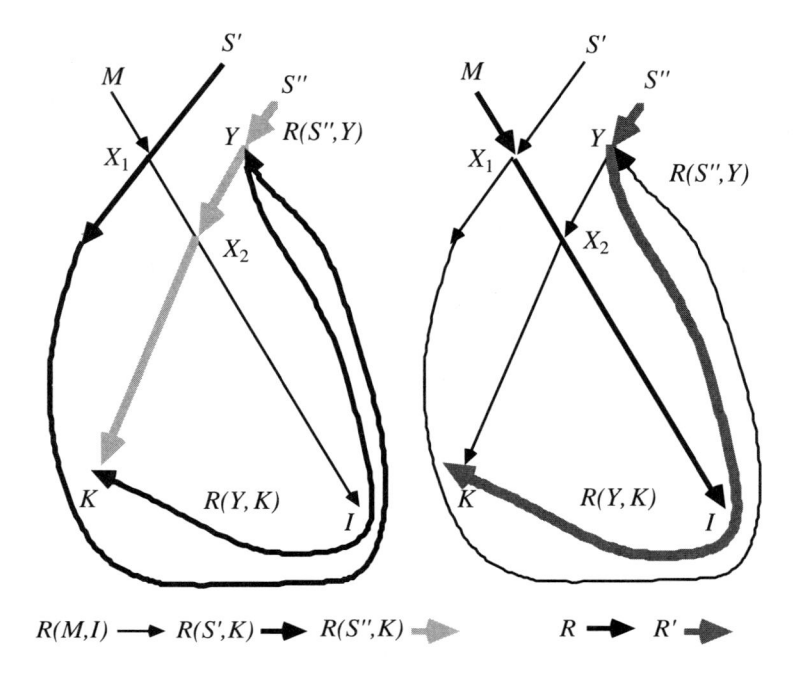

Figure 13.8

2. $R(X_1,K)$ does intersect $R(S'',X_2)$ at Y. (See figure 13.8.) Let $R(S'',Y)$ be an initial segment of $R(S'',K)$, $R(Y,K)$ be a final segment of $R(S',K)$, $R = R(M,I)$ and $R' = R(S'',Y)\&R(Y,K)$. R and R' do not intersect for the following reasons.

First we will show that $R(M,I)$ does not intersect $R(S'',Y)$. $Y \neq X_2$ since $R(X_1,K)$ intersect $R(M,I)$ only at X_1. Also, G is acyclic, Y is prior to X_2 on $R(S'',K)$, and X_2 is the first point of intersection of $R(S'',K)$ with $R(M,I)$. Next we will show that $R(M,I)$ does not intersect $R(Y,K)$. Y is on $R(S'',K)$ which does not contain X_1; hence Y is not equal to X_1. It follows that $R(Y,K)$ does not contain X_1, since Y occurs after X_1 on $R(S',K)$, and $R(S'',K)$. By hypothesis $R'(M,K)$ intersects $R(M,I)$ only at X_1, so that $R(Y,K)$ does not intersect $R(M,I)$ at all.

Induction Case: Assume that the antecedent is true, and that the theorem is true for all $m < n$. If there is a path in $\mathbf{R_K}(S)$ that does not intersect $R(M,I)$, the proof is done. Suppose then that every path in $\mathbf{R_K}(S)$ intersects $R(M,I)$ and that the set of paths in $\mathbf{R_K}(S)$ intersects $R(M,I)$ at exactly n distinct vertices. Let $R(X_{max},I)$ be a final segment of $R(M,I)$. Since not every path in $\mathbf{R_K}(S)$ intersects $R(M,I)$ at X_{max}, there is a point of intersection prior to X_{max} on $R(M,I)$. Hence the number of distinct points of intersection of the paths in $\mathbf{R_K}(S)$ with $R(X_{max},I)$ is less than n. By the induction hypothesis, there is a path R_1 with source X_{max} and a path R_1' with a source S' in the sources of $\mathbf{R_K}(S)$, such that one of R_1 and R_1' has a sink I, the other has sink K, and R_1 and R_1' do not intersect. Suppose w.l.g.

that R_1 has sink I and R_1' has sink K. Since R_1' does not contain X_{max}, its first point of intersection with $R(M,I)$ is some vertex X_r, which occurs on $R(M,I)$ before X_{max} (by definition of X_{max}.) Let $R_1'(X_r,K)$ be a final segment of R_1', $R(S'',K)$ be a path in $\mathbf{R_K}(S)$ whose first point of intersection with $R(M,I)$ is X_{max}, and $R(S'', X_{max})$ an initial segment of $R(S'',K)$. There are two cases.

1. Assume that $R(X,K)$ does not intersect $R(S'', X_{max})$. Let $R = R(M,X_r)\&R_1'(X_r,K)$ and $R' = R(S'', X_{max})$ & R_1. R and R' do not intersect for reasons analogous to those in case 1 of the base case (with X_r substituted for X_1, and X_{max} substituted for X_2; see figure 13.9.)

2. Assume that $R_1'(X_r,K)$ does intersect $R(S'',X_{max})$, and the last point of intersection is Y. $Y \neq X_{max}$ because it lies on $R_1'(X_r,K)$ and $R_1'(X_r,K)$ does not contain X_{max}. Let $R_1'(Y,K)$ be a final segment of $R_1'(X_r,K)$. There are two cases.

a. Assume that $R_1'(Y,K)$ intersects $R(M,X_{max})$ and the first point of intersection is Z. Let $R(S'',Y)$ be an initial segment of $R(S'',X_{max})$, $R(Y,Z)$ an initial segment of $R_1'(Y,K)$, and $R(M,Z)$ an initial segment of $R(M,I)$. $Z \neq X_{max}$ because $R_1'(Y,K)$ does not intersect X_{max}. (See figure 13.9.)

We will now prove Z is not after X_{max}. Consider the path $R(S'',Y)\&R(Y,Z)$. $R(S'',Y)$ does not intersect $R(M,I)$ because Y occurs before X_{max}, $R(S'',Y)$ is an initial segment of $R(S'',K)$ and the first point of intersection of $R(M,I)$ and $R(S'',K)$ is X_{max}. The first point of intersection of $R(Y,Z)\&R(M,I)$ is Z, since $R(Y,Z)$ is an initial segment of $R_1'(Y,K)$ and Z is the first point of intersection of $R_1'(Y,K)$ and $R(M,I)$. Hence the first point of intersection of $R(S'',Y)\&R(Y,Z)$ with $R(M,I)$ is Z. $R(S'',Y)\&R(Y,Z)$ is an initial segment of a path from S'' to K that is in $\mathbf{R_K}(S)$. It follows that there is a path in $\mathbf{R_K}(S)$ whose first point of intersection with $R(M,I)$ is Z. If Z is after X_{max}, then there is a path in $\mathbf{R_K}(S)$ whose first point of intersection with $R(M,I)$ is after X_{max}, contrary to the definition of X_{max}.

Let $R = R(M,Z)\&R_1'(Z,K)$ and $R' = R(S'',X_{max})\&R_1$. $R(M,Z)$ does not intersect $R(S'',X_{max})$ since $R(S'',X_{max})$ is an initial segment of $R(S'',K)$ and $R(M,Z)$ is an initial segment of $R(M,I)$ and the first point of intersection of $R(M,I)$ and $R(S'',K)$ is X_{max}. $R(M,Z)$ does not intersect R_1 (which has source X_{max}) since Z occurs before X_{max} and the directed graph is acyclic. $R_1'(Z,K)$ does not intersect R_1 since $R_1'(Z,K)$ is a subpath of R_1' that does not intersect R_1 by construction. $R_1'(Z,K)$ does not intersect $R(S'',X_{max})$ since $R_1'(Z,K)$ is a final segment of $R_1'(X_r,K)$, Z is after Y, and Y is the last point of intersection of $R_1'(X_r,K)$ and $R(S'',X_{max})$.

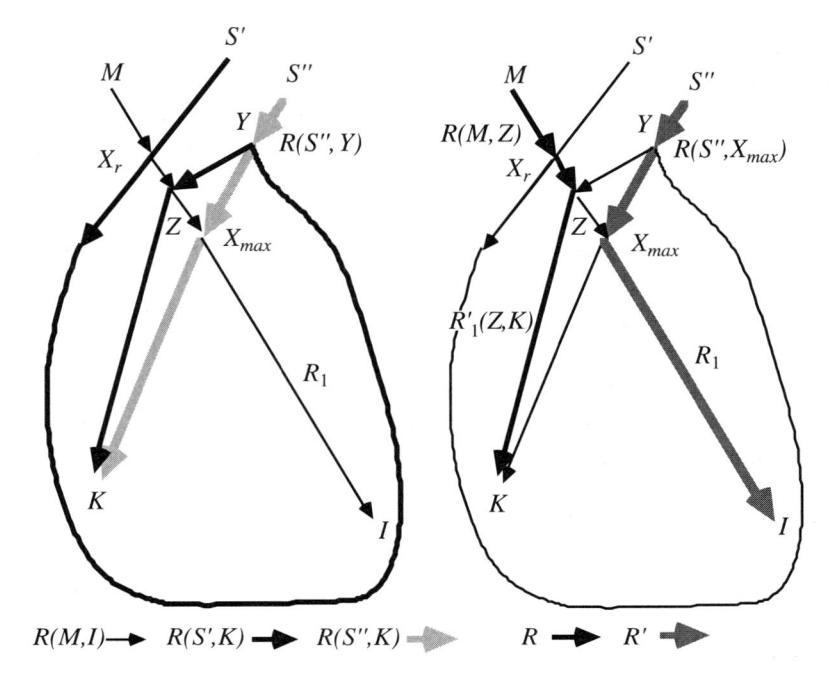

Figure 13.9

b. Assume that $R_1'(Y,K)$ does not intersect $R(M,X_{max})$. (This is similar to part 2 of the Base case, with X_{max} substituted for X_2. See figure 13.8.) Let $R' = R(S'',Y) \& R_1'(Y,K)$ and $R = R(M,X_{max}) \& R_1$. We have already shown that $R(S'',Y)$ does not intersect $R(M,I)$ and $R(M,X_{max})$ is an initial segment of $R(M,I)$. $R(S'',Y)$ does not intersect R_1 because Y is before X_{max}, and the directed graph is acyclic. $R_1'(Y,K)$ does not intersect $R(M,X_{max})$ by hypothesis, and $R_1'(Y,K)$ does not intersect R_1 because it is a subpath of R_1' that does not intersect R_1 by construction. ∴

In a directed acyclic graph G, if all $L(T(K,L))$ and all $J(T(I,J))$ intersect at a vertex Q, then Q is an $LJ(T(I,J),T(K,L))$ **choke point**. Similarly, if all $L(T(K,L))$ and all $J(T(I,J))$ intersect at a vertex Q, and all $L(T(I,L))$ and all $J(T(J,K))$ also intersect at Q, then Q is a $LJ(T(I,J),T(K,L),T(I,L),T(J,K))$ **choke point**.

LEMMA 6.10.9: In a directed acyclic graph G, if there is no $LJ(T(I,J),T(K,L))$ choke point, then either there is a trek $T'(K,L)$ such that there is no vertex V' that occurs in the intersection of all $J(T(I,J))$ with $L(T'(K,L))$, or there is a trek $T'(I,J)$ such that there is no vertex V' that occurs in the intersection of all $L(T(K,L))$ with $J(T'(I,J))$

Proof. Suppose that the lemma is false. Then, for each trek $T'(K,L)$ there is a non-empty set of points $\mathbf{P}(T'(K,L))$ such that every point in $\mathbf{P}(T'(K,L))$ is in the intersection of all

Proofs of Theorems

$J(T(I,J))$ with $L(T'(K,L))$. Similarly, for each trek $T'(I,J)$ there is a non-empty set of points $\mathbf{P}(T'(I,J))$ such that every point in $\mathbf{P}(T'(I,J))$ is in the intersection of all $L(T(K,L))$ with $J(T'(I,J))$. Every $J(T(I,J))$ contains every vertex in $\bigcup_{T(K,L) \in \mathbf{T}(K,L)} \mathbf{P}(T(K,L))$

(since every $J(T(I,J))$ intersects each $L(T'(K,L))$ at some vertex in $\mathbf{P}(T'(K,L))$), and every vertex in $\bigcup_{T(K,L) \in \mathbf{T}(K,L)} \mathbf{P}(T(K,L))$ occurs on some trek $L(T'(K,L))$. Similarly, every $L(T(K,L))$ contains every vertex in $\bigcup_{T(I,J) \in \mathbf{T}(I,J)} \mathbf{P}(T(I,J))$.

Furthermore, for every vertex in $\bigcup_{T(K,L) \in \mathbf{T}(K,L)} \mathbf{P}(T(K,L))$ there is some $L(T'(K,L))$ that does not contain it (else all $J(T(I,J))$ and all $L(T(K,L))$ intersect at a single vertex), and some $L(T''(K,L))$ that does contain it. Similarly, for every vertex in $\bigcup_{T(I,J) \in \mathbf{T}(I,J)} \mathbf{P}(T(I,J))$ there is some $J(T'(I,J))$ that does not contain it and some $J(T''(I,J))$ that does contain it.

Since every vertex in $\bigcup_{T(K,L) \in \mathbf{T}(K,L)} \mathbf{P}(T(K,L))$ occurs on every $J(T(I,J))$, they can be ordered by the order of their occurrence on some $J(T(I,J))$; similarly every vertex in $\bigcup_{T(I,J) \in \mathbf{T}(I,J)} \mathbf{P}(T(I,J))$ can be ordered. By the antecedent of the lemma, there are at least two vertices in each of $\bigcup_{T(K,L) \in \mathbf{T}(K,L)} \mathbf{P}(T(K,L))$ and $\bigcup_{T(I,J) \in \mathbf{T}(I,J)} \mathbf{P}(T(I,J))$.

(See figure 13.10.) Let A be the first vertex in $\bigcup_{T(I,J) \in \mathbf{T}(I,J)} \mathbf{P}(T(I,J))$ and B be the first vertex in $\bigcup_{T(K,L) \in \mathbf{T}(K,L)} \mathbf{P}(T(K,L))$. Suppose w.l.g. that A is before B. There exists an $L(T'(K,L))$ that contains A (since every $L(T(K,L))$ contains A), that does not contain B, but that does contain some vertex C ($\neq B$) in $\bigcup_{T(K,L) \in \mathbf{T}(K,L)} \mathbf{P}(T(K,L))$.

There is also a $J(T'(I,J))$ that contains A. Let S be the source of $T'(I,J)$, $R(S,A)$ an initial segment of $J(T'(I,J))$, $R(A,C)$ a segment of $L(T'(K,L))$, and $R(C,J)$ a final segment of $J(T'(I,J))$. Let $J(T''(I,J)) = R(S,A) \& R(A,C) \& R(C,J)$, and $I(T''(I,J)) = I(T'(I,J))$. $J(T''(I,J))$ does not contain B for the following reasons. $R(S,A)$ does not contain B because A occurs before B. $R(A,C)$ does not contain B because it is a segment of $L(T'(K,L))$ which does not contain B. $R(C,J)$ does not contain B because it is a segment of $J(T'(I,J))$, and since B is the first vertex in $\bigcup_{T(K,L) \in \mathbf{T}(K,L)} \mathbf{P}(T(K,L))$ it occurs before C on $J(T'(I,J))$.

But this contradicts the fact that for every $T(I,J)$, $J(T(I,J))$ contains B. ∴

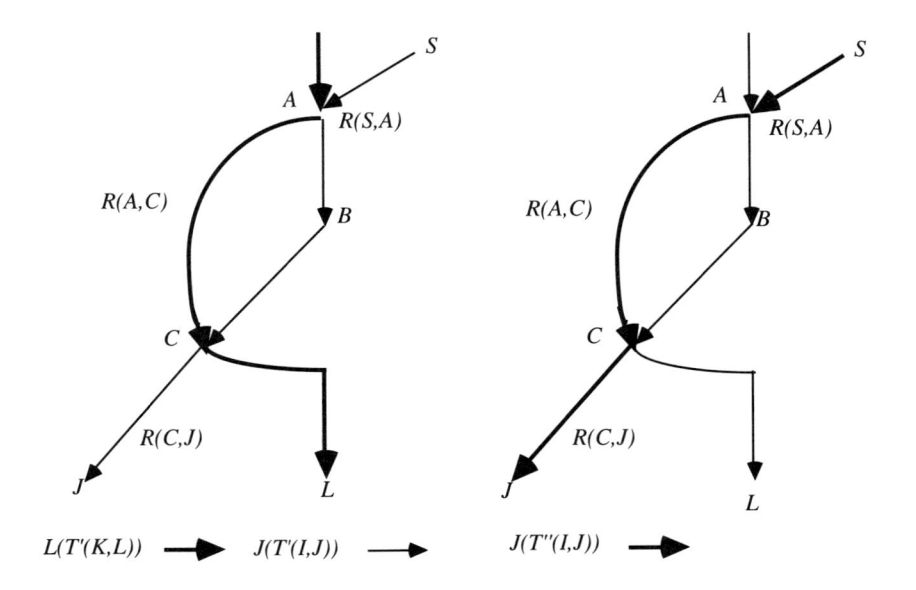

Figure 13.10

LEMMA 6.10.10: In a directed acyclic graph G, if there is no $IK(T(I,J),T(K,L))$ choke point, then either there is a trek $T'(K,L)$ such that there is no vertex V' that occurs in the intersection of all $I(T(I,J))$ with $K(T'(K,L))$, or there is a trek $T'(I,J)$ such that there is no vertex V' that occurs in the intersection of all $K(T(KL))$ with $I(T'(I,J))$

Proof. The proof of lemma 6.10.10 is the same as that of lemma 6.10.9 with I, J, K, L permuted. ∴

LEMMA 6.10.11: In an acyclic LCF G, if there is a trek $T'(K,L)$ such that there is no vertex V' that occurs in the intersection of all $J(T(I,J))$ with $L(T'(K,L))$, then either there are treks $T''(I,J)$ and $T''(K,L)$ such that $J(T''(I,J))$ does not intersect $L(T''(K,L))$ or $\rho_{IJ}\rho_{Kl} - \rho_{IL}\rho_{JK}$ is not linearly implied by G.

Proof. Let S be the source of $T'(K,L)$, and **S** be the set of sources of treks between I and J. By lemma 6.10.8 it is possible to construct a pair of paths R and R', with sources S and S' (in **S**), and sinks J and L, such that R and R' do not intersect. There are two cases.

1. If R is a path from S to L, and R' is a path from S' to J, then the following treks can be formed from subpaths of R and R'. (See figure 13.11.) $J(T''(I,J)) = R'$, $I(T''(I,J)) = I(T'(I,J))$, $K(T''(K,L)) = K(T'(K,L))$, and $L(T''(K,L)) = R$. By construction R does not intersect R'; hence $J(T''(I,J))$ does not intersect $L(T''(K,L))$.

Proofs of Theorems 441

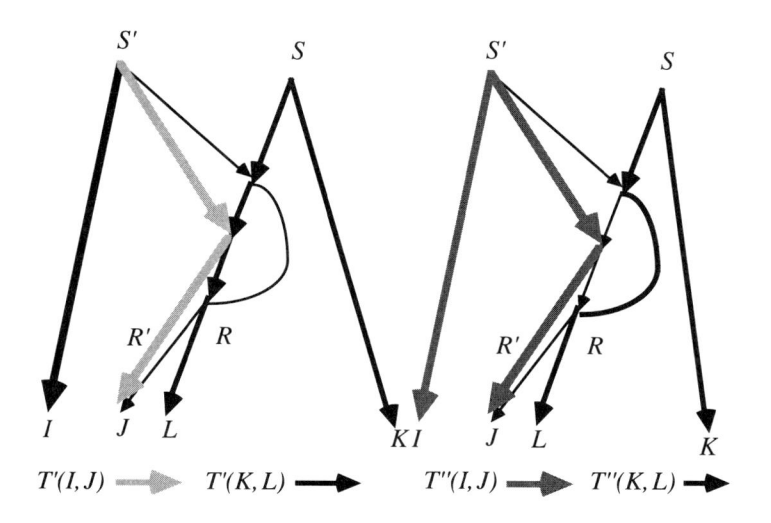

Figure 13.11

2. If R is a path from S to J, and R' is a path from S' to L, there are two cases.

a. $K(T'(K,L))$ intersects $I(T'(I,J))$, and the first vertex of intersection is Y. Let $R(S,Y)$ be an initial segment of $K(T'(K,L))$, $R(Y,K)$ a final segment of $K(T'(K,L))$, $R(S',Y)$ an initial segment of $I(T'(I,J))$, $R(Y,I)$ a final segment of $I(T'(I,J))$, $J(T''(I,J)) = R$, $I(T''(I,J)) = R(S,Y)\&R(Y,I)$, $K(T''(K,L)) = R(S',Y)\&R(Y,K)$, and $L(T''(K,L)) = R'$. (See figure 13.12.) By construction, $J(T''(I,J))$ and $L(T''(K,L))$ do not intersect.

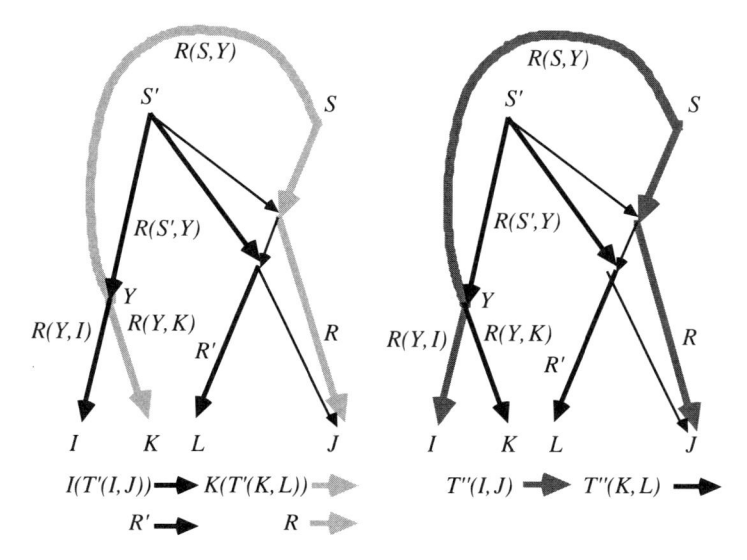

Figure 13.12

b. If $K(T'(K,L))$ does not intersect $I(T'(I,J))$, the following treks can be formed. (See figure 13.13.) $I(T'(I,L)) = I(T'(I,J))$, $L(T'(I,L)) = R'$, $J(T'(J,K)) = R$, and $K(T'(J,K)) = K(T'(K,L))$. By hypothesis, $K(T'(J,K))$ does not intersect $I(T'(I,L))$. By construction, $L(T'(I,L))$ does not intersect $J(T'(J,K))$. Hence by lemma 6.10.7, $\rho_{IJ}\rho_{KL} - \rho_{IL}\rho_{JK}$ is not linearly implied by G. ∴

LEMMA 6.10.12: In an acyclic LCF G, if there is a trek $T'(I,J)$ such that there is no vertex V' that occurs in the intersection of all $L(T(K,L))$ with $J(T'(I,J))$, then either there are treks $T''(I,J)$ and $T''(K,L)$ such that $J(T''(I,J))$ does not intersect $L(T''(K,L))$ or $\rho_{IJ}\rho_{KL} - \rho_{IL}\rho_{JK} = 0$ is not linearly implied by G.

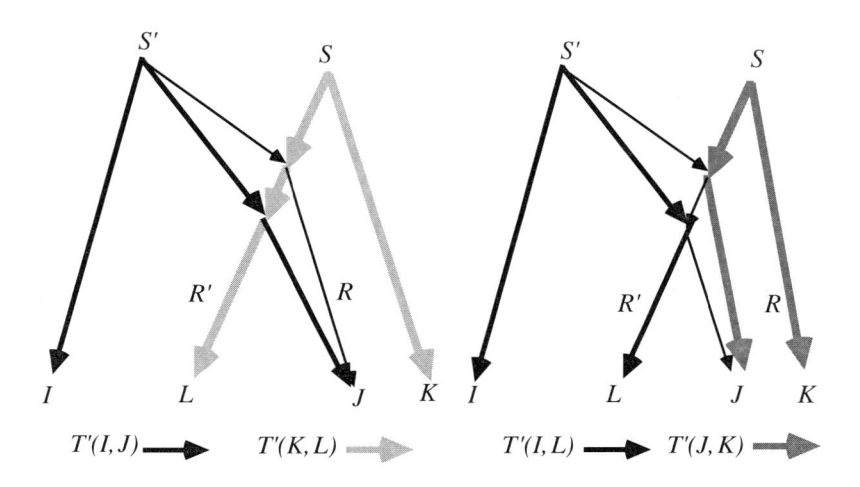

Figure 13.13

LEMMA 6.10.13: In an acyclic LCF G, if there is a trek $T'(I,J)$ such that there is no vertex V' that occurs in the intersection of all $K(T(K,L))$ with $I(T'(I,J))$, then either there are treks $T''(I,J)$ and $T''(K,L)$ such that $I(T''(I,J))$ does not intersect $K(T''(K,L))$ or $\rho_{IJ}\rho_{KL} - \rho_{IL}\rho_{JK} = 0$ is not linearly implied by G.

LEMMA 6.10.14: In an acyclic LCF G, if there is a trek $T'(K,L)$ such that there is no vertex V' that occurs in the intersection of all $I(T(I,J))$ with $K(T'(K,L))$, then either there are treks $T''(I,J)$ and $T''(K,L)$ such that $I(T''(I,J))$ does not intersect $K(T''(K,L))$ or $\rho_{IJ}\rho_{KL} - \rho_{IL}\rho_{JK} = 0$ is not linearly implied by G.

The proofs of lemmas 6.10.12, 6.10.13, and 6.10.14 can all be obtained from the proof of lemma 6.10.11 by permuting I, J, K, and L.

LEMMA 6.10.15: In an acyclic LCF G, if there is no $LJ(T(I,J),T(K,L))$ choke point, and there is no $IK(T(I,J),T(K,L))$ choke vertex, then there exist treks $T'(I,J)$, $T'(K,L)$, $T''(I,J)$, and $T''(K,L)$ such that $I(T'(I,J))$ does not intersect $K(T'(K,L))$ and $J(T''(I,J))$ does not intersect $L(T''(K,L))$, or $\rho_{IJ}\rho_{KL} - \rho_{IL}\rho_{JK} = 0$ is not linearly implied by G.

Proof. This follows directly from lemmas 6.10.9 through 6.10.14. ∴

LEMMA 6.10.16: In an acyclic LCF G, if there is no $LJ(T(I,J),T(K,L))$ choke point, and there is no $IK(T(I,J),T(K,L))$ choke point, then $\rho_{IJ}\rho_{KL} - \rho_{IL}\rho_{JK} = 0$ is not linearly implied by G.

Proof. Assume that there is no $LJ(T(I,J),T(K,L))$ choke point, and there is no $IK(T(I,J),T(K,L))$ choke point. By lemma 6.10.15 either $\rho_{IJ}\rho_{KL} - \rho_{IL}\rho_{JK} = 0$ is not linearly implied by G or there exist treks $T'(I,J)$, $T'(K,L)$, $T''(I,J)$, and $T''(K,L)$ such that $I(T'(I,J))$ does not intersect $K(T'(K,L))$ and $J(T''(I,J))$ does not intersect $L(T''(K,L))$. If $\rho_{IJ}\rho_{KL} - \rho_{IL}\rho_{JK} = 0$ is not linearly implied by G, the proof is done. Assume then that there exist treks $T'(I,J)$, $T'(K,L)$, $T''(I,J)$, and $T''(K,L)$ such that $I(T'(I,J))$ does not intersect $K(T'(K,L))$ and $J(T''(I,J))$ does not intersect $L(T''(K,L))$. There are three cases.

1. Suppose for all $T(I,J)$, $J(T(I,J))$ intersects $L(T'(K,L))$ at each vertex in a non-empty set of vertices **P′**, and all $L(T(K,L))$ intersects $J(T'(I,J))$ at each vertex in a non-empty set of vertices **P**. Hence, all $L(T(K,L))$ contain every vertex in **P** and all $J(T(I,J))$ contain every vertex in **P′**. Since there is no $LJ(T(I,J),T(K,L))$ choke point, there is no vertex Z such that for all $T(I,J)$ and all $T(K,L)$, Z occurs in the intersection of $L(T(I,J))$ and $J(T(I,J))$. Hence **P** and **P′** do not intersect.

Let A be the first vertex in **P**, and B be the first vertex in **P′**. Suppose w.l.g. that A occurs before B. Let $S'(I,J)$ be the source of $T'(I,J)$, $S'(K,L)$ the source of $T'(K,L)$ and $S''(I,J)$ the source of $T''(I,J)$, and $S''(K,L)$ the source of $T''(K,L)$. $L(T''(K,L))$ contains A (since all $L(T(K,L))$ contain A), and $J(T''(I,J))$ contains B (since all $J(T(I,J))$ contain B.) There are two cases.

a. Suppose $K(T''(K,L))$ does not intersect $I(T''(I,J))$. Then, since $K(T''(K,L))$ does not intersect $I(T''(I,J))$ and $J(T''(K,L))$ does not intersect $L(T''(K,L))$, by lemma 6.10.7, $\rho_{IJ}\rho_{KL} - \rho_{IL}\rho_{JK} = 0$ is not linearly implied by G.

b. Suppose $K(T''(K,L))$ does intersect $I(T''(I,J))$ at a vertex X. (See figure 13.14.) Let $R(S''(I,J),X)$ be an initial segment of $I(T''(I,J))$, $R(X,K)$ a final segment of $L(T''(K,L))$. Let $R(S''(I,J),B)$ be an initial segment of $J(T''(I,J))$ and $R(B,L)$ be a final segment of $L(T'(K,L))$. Form the trek $K(T'''(K,L)) = R(S''(I,J),X)\&R(X,K)$, and $L(T'''(K,L)) = R(S''(I,J),B)\&R(B,L)$. $R(S''(I,J),B)$ does not contain A, since it is a subpath of $J(T''(I,J))$ which does not intersect $L(t''(K,L))$, which does contain A. $R(B,L)$ does not contain A, since A occurs before B. Hence $L(T'''(K,L))$ does not contain A; but this is a contradiction.

2. All $L(T(K,L))$ intersect $J(T'(I,J))$, but not at a single vertex, or all $J(T(I,J))$ intersect $L(T'(K,L))$ but not at a single vertex. Assume w.l.g. that the latter is the case. Let S' be the source of $T'(I,J)$ and S be the source of $T'(K,L)$. Let **S** be the set of sources of treks between I and J. By lemma 6.10.8, it is possible to form two paths $R(S'',L)$ and $R(S,J)$ or $R(S'',J)$ and $R(S,L)$ that don't intersect, where S'' is in **S**. Assume that it is possible to form the paths $R(S'',L)$ and $R(S,J)$ that don't intersect. (If the paths that don't intersect are $R(S'',J)$ and $R(S,L)$ the proof is the same except that the indices are permuted.) Let $T''(I,J)$ be a trek with source S'' (See figure 13.15.) Let the first point of intersection of $I(T''(I,J))$ with $I(T'(I,J))$ be M. There are two cases.

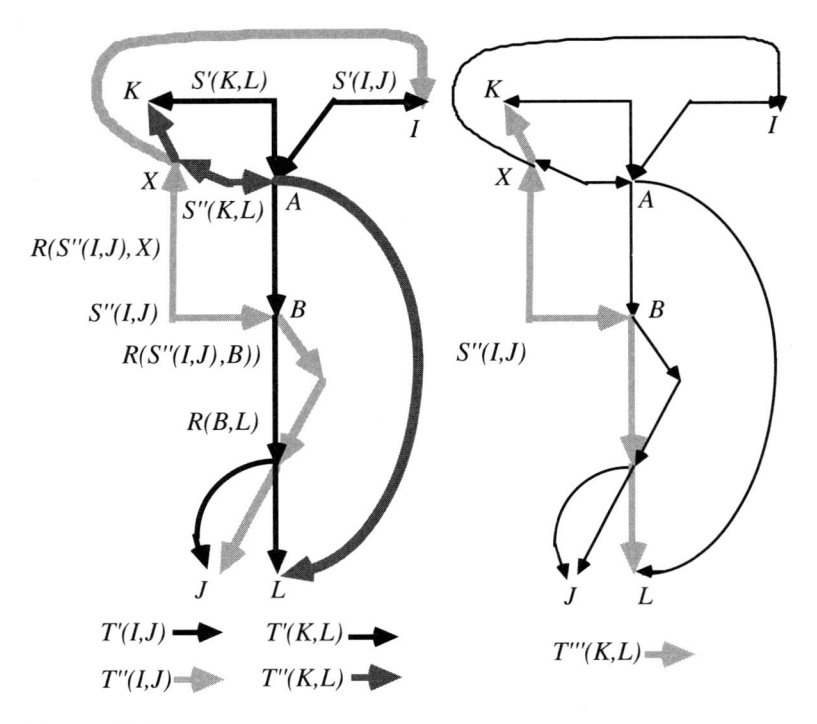

Figure 13.14

a. Assume that $I(T''(I,J))$ does not intersect $K(T'(K,L))$ before it intersects $I(T'(I,J))$ at M. (See figure 13.15.) Let $R(M,I)$ be a final segment of $I(T'(I,J))$ and $R(S'',M)$ be an initial segment of $I(T''(I,J))$. Let $I(T'(I,L)) = R(S'',M)\&R(M,I)$, $L(T'(I,L)) = R(S'',L)$, $J(T'(J,K)) = R(S,J)$ and $K(T'(J,K)) = K(T'(K,L))$. $R(S'',M)$ and $R(M,I)$ do not intersect $K(T'(K,L))$ by hypothesis. By lemma 6.10.7 $\rho_{IJ}\rho_{KL} - \rho_{IL}\rho_{JK} = 0$ is not linearly implied by G.

Proofs of Theorems 445

[Figure 13.15 diagrams]

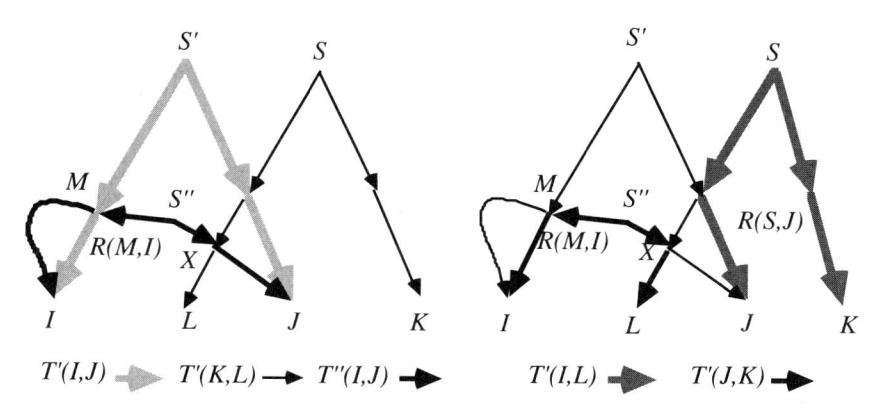

$T'(I,J)$ ⇨ $T'(K,L)$ → $T''(I,J)$ → $T'(I,L)$ ⇨ $T'(J,K)$ →

Figure 13.15

b. Assume that $I(T''(I,J))$ does intersect $K(T'(K,L))$ before it intersects $I(T'(I,J))$, and the first point of intersection is Q. Let $R(Q,K)$ be a final segment of $K(T'(K,L))$ and $R(S'',Q)$ be an initial segment of $I(T''(I,J))$. Let Y be the first point of intersection of $R(S,J)$ and $J(T'(I,J))$, and $R(S',Y)$ be an initial segment of $J(T'(I,J))$. There are two cases.

1. Assume that $R(S'',L)$ intersects $R(S',Y)$ and the first point of intersection is Z. Let $R(S',Z)$ be an initial segment of $J(T'(I,J))$, $R(Z,L)$ be a final segment of $R(S'',L)$, $L(T'(I,L)) = R(S',Z)\&R(Z,L)$, $I(T'(I,L)) = I(T'(I,J))$, $J(T'(J,K)) = R(S,J)$, and $K(T'(J,K)) = K(T'(K,L))$. (See figure 13.16.)

[Figure 13.16 diagrams]

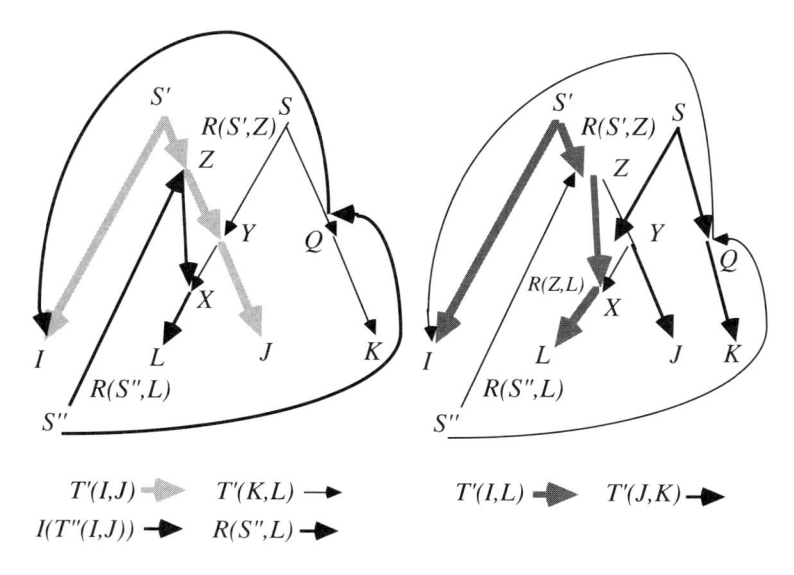

$T'(I,J)$ ⇨ $T'(K,L)$ → $T'(I,L)$ ⇨ $T'(J,K)$ →
$I(T''(I,J))$ → $R(S'',L)$ →

Figure 13.16

$K(T'(J,K))$ does not intersect $I(T'(I,L))$ by hypothesis. $J(T'(J,K))$ does not intersect $L(T'(I,L))$ for the following reasons. $R(S',Z)$ does not intersect $R(S,J)$ because $R(S',Z)$ is a subpath of $J(T'(I,J))$, Z is before Y, and the first point of intersection of $J(T'(I,J))$ and $R(S,J)$ is Y. $R(Z,L)$ does not intersect $R(S,J)$ because it is a subpath of $R(S'',L)$ which does not intersect $R(S,J)$ by construction. By lemma 6.10.7 $\rho_{IJ}\rho_{KL} - \rho_{IL}\rho_{JK} = 0$ is not linearly implied by G.

2. Assume that $R(S'',L)$ does not intersect $R(S',Y)$. Let $L(T''(K,L)) = R(S'',L)$, $K(T''(K,L)) = R(S'',Q)\&R(Q,K)$, $I(T'''(I,J)) = I(T'(I,J))$, and $J(T'''(I,J)) = R(S',Y)\&R(Y,J)$. (See figure 13.17.) $K(T''(K,L))$ does not intersect $I(T'''(I,J))$ for the following reasons. $R(S'',Q)$ does not intersect $I(T'(I,J))$ since $R(S'',Q)$ is an initial segment of $I(T''(I,J))$, and Q occurs before the first point of intersection of $I(T''(I,J))$ and $I(T'(I,J))$. $R(Q,K)$ does not intersect $I(T'(I,J))$ because it is a final segment of $K(T'(K,L))$, which does not intersect $I(T'(I,J))$ by hypothesis. $L(T''(K,L))$ does not intersect $J(T'''(I,J))$ for the following reasons. $R(S',Y)$ does not intersect $R(S'',L)$ by hypothesis, and $R(Y,J)$ is a subpath of $R(S,J)$ which does not intersect $R(S'',L)$ by construction. By lemma 6.10.7 $\rho_{IJ}\rho_{KL} - \rho_{IL}\rho_{JK} = 0$ is not linearly implied by G.

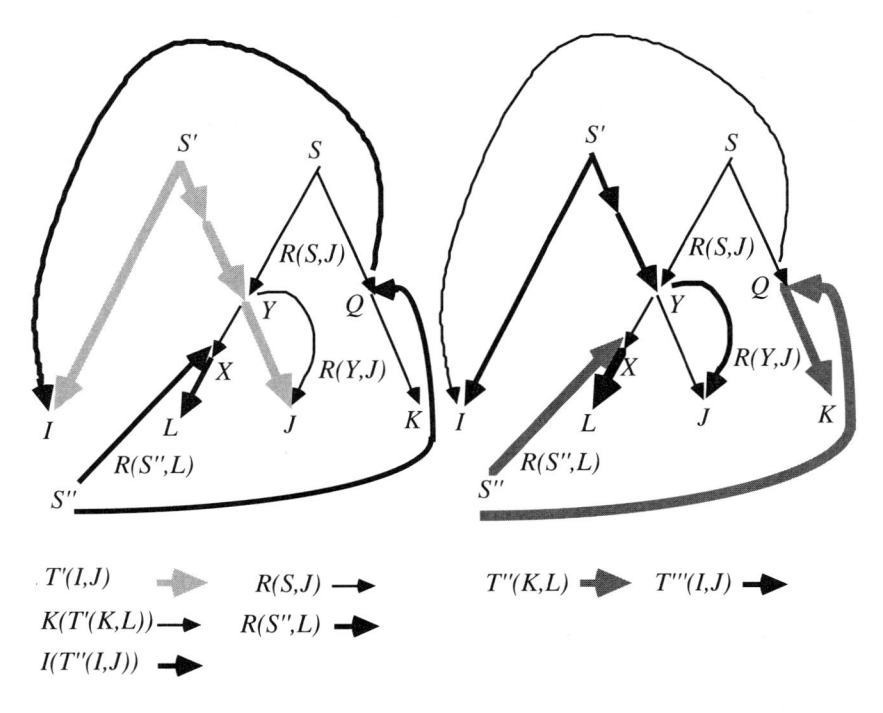

Figure 13.17

Proofs of Theorems   447

3. Either there is an $L(T''(K,L))$ that does not intersect $J(T'(I,J))$ or there is a $J(T''(I,J))$ that does not intersect $L(T'(K,L))$. Assume w.l.g. that $J(T''(I,J))$ with source $S''(I,J)$ does not intersect $L(T'(K,L))$. There are two cases.

a. Suppose that $I(T''(I,J))$ does not intersect $K(T'(K,L))$ before it intersects $I(T'(I,J))$ at vertex X. (See figure 13.18.)

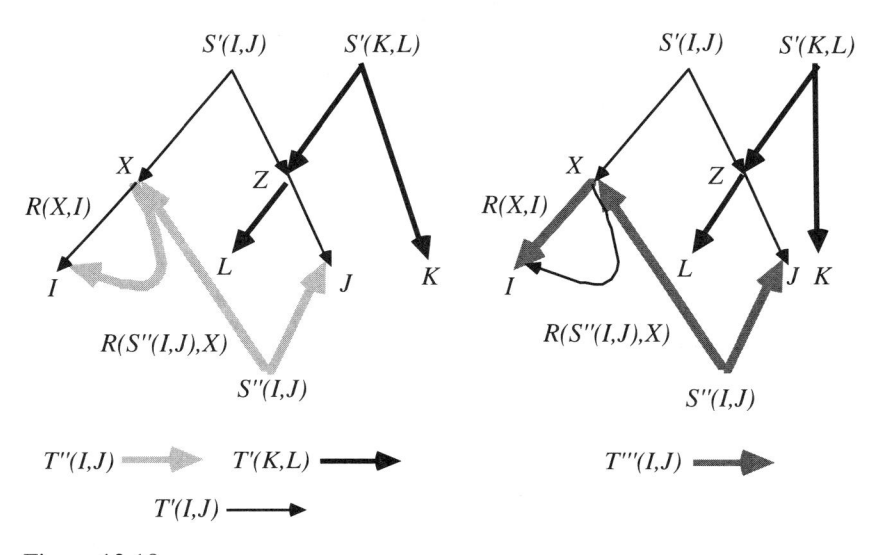

Figure 13.18

Let $R(X,I)$ be a final segment of $I(T'(I,J))$ and $R(S''(I,J),X)$ be an initial segment of $I(T''(I,J))$. The trek $T'''(I,J)$ can be formed as follows. $J(T'''(I,J)) = J(T''(I,J))$ and $I(T'''(I,J)) = R(S''(I,J),X) \& R(X,I)$. $R(S''(I,J),X)$ does not intersect $K(T'(K,L))$ because by hypothesis X occurs on $I(T''(I,J))$ before it intersects $K(T'(K,L))$. $R(X,I)$ does not intersect $K(T'(K,L))$ because it is a subpath of $I(T'(I,J))$ which does not intersect $K(T'(K,L))$ by hypothesis. Hence $I(T'''(I,J))$ does not intersect $K(T'(K,L))$. $J(T'''(I,J)) = J(T''(I,J))$ does not intersect $L(T'(K,L))$ by hypothesis. By lemma 6.10.7, $\rho_{IJ}\rho_{KL} - \rho_{IL}\rho_{JK} = 0$ is not linearly implied by G.

b. Suppose $I(T''(I,J))$ intersects $K(T'(I,J))$ at Y before it intersects $I(T'(I,J))$ at X. Let Z be the first point of intersection of $J(T'(I,J))$ and $L(T'(K,L))$. (If no such vertex exists, then $J(T'(I,J))$ and $L(T'(K,L))$ do not intersect, $I(T'(I,J))$ and $K(T'(K,L))$ do not intersect by hypothesis, and by lemma 6.10.7 $\rho_{IJ}\rho_{KL} - \rho_{IL}\rho_{JK} = 0$ is not linearly implied by G.). Let $R(S'(I,J),Z)$ be an initial segment of $I(T'(I,J))$, and $R(Z,L)$ be a final segment of $L(T'(K,L))$. There are two cases.

1. Suppose that $J(T''(I,J))$ does not intersect $R(S'(I,J),Z)$. (See figure 13.19.)

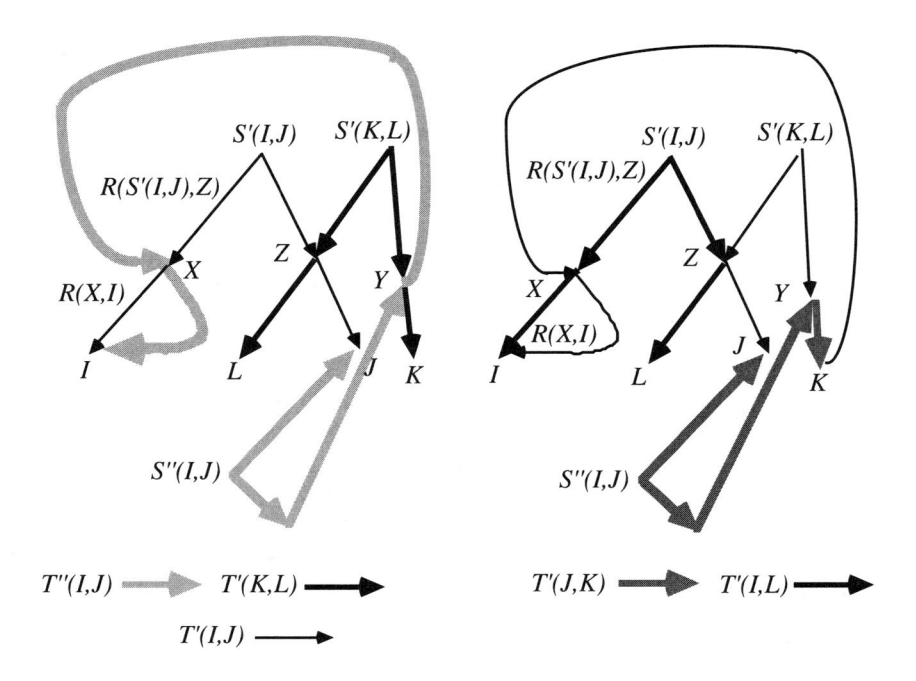

Figure 13.19

Let $R(Y,K)$ be a final segment of $K(T'(K,L))$ and $R(S''(I,J),Y)$ be an initial segment of $I(T''(I,J))$. Let $J(T'(J,K)) = J(T''(I,J))$, $K(T'(J,K)) = R(S''(I,J),Y)\&R(Y,K)$, $I(T'(I,L)) = I(T'(I,J))$, $L(T''(I,L)) = R(S'(I,J),Z)\&R(Z,L)$. $I(T'(I,L))$ and $K(T'(J,K))$ do not intersect for the following reasons. $I(T'(I,L))$ does not intersect $R(S''(I,J),Y)$ because by hypothesis, $I(T''(I,J))$ intersects $K(T'(K,L))$ at Y before it intersects $I(T'(I,J))$. $I(T'(I,L))$ does not intersect $R(Y,K)$ because $I(T'(I,L)) = I(T'(I,J))$ and $R(Y,K)$ is a subpath of $K(T'(K,L))$, which does not intersect $I(T'(I,J))$ by hypothesis. $J(T'(J,K))$ does not intersect $L(T'(I,L))$ for the following reasons. $J(T'(J,K))$ does not intersect $R(S'(I,J),Z)$ because $J(T'(J,K)) = J(T''(I,J))$, which does not intersect $R(S'(I,J),Z)$ by hypothesis. $J(T'(J,K))$ does not intersect $R(Z,L)$ because $J(T'(J,K)) = J(T''(I,J))$ which does not intersect $L(T'(K,L))$ (which contains $R(Z,L)$) by hypothesis. By lemma 6.10.7, $\rho_{IJ}\rho_{KL} - \rho_{IL}\rho_{JK} = 0$ is not linearly implied by G.

2. Suppose that $J(T''(I,J))$ does intersect $R(S'(I,J),Z)$ and the first point of intersection is M. (See figure 13.20.) $M \neq Z$ because $J(T''(I,J))$ does not intersect $L(T'(K,L))$ which contains Z. Let $R(S'(I,J),M)$ be an initial segment of $J(T'(I,J))$ and $R(M,J)$ be a final segment of $J(T''(I,J))$. Let $I(T'''(I,J)) = I(T'(I,J))$ and $J(T'''(I,J)) = R(S'(I,J),M)\&R(M,J)$. $I(T'''(I,J))$ does not intersect $K(T'(K,L))$ by hypothesis. $J(T'''(I,J))$ does not intersect $L(T'(K,L))$ for the following reasons. $R(S'(I,J),M)$ does not intersect $L(T'(K,L))$ since M is

Proofs of Theorems

before Z on J(T'(I,J)), and the first point of intersection of J(T'(I,J)) with L(T'(K,L)) is Z. R(M,J) does not intersect L(T'(K,L)) because it is a subpath of J(T''(I,J)) which does not intersect L(T'(K,L)) by hypothesis. By lemma 6.10.7, $\rho_{IJ}\rho_{KL} - \rho_{IL}\rho_{JK} = 0$ is not linearly implied by G.

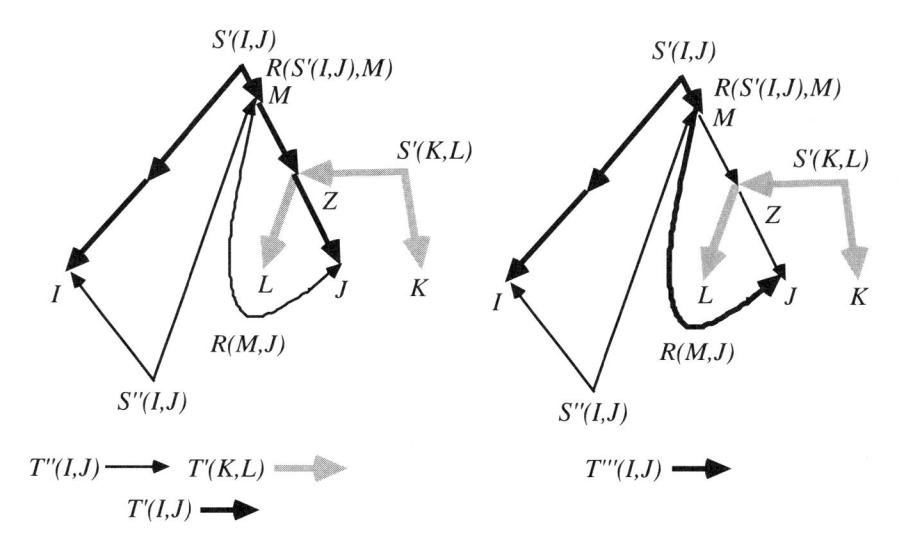

Figure 13.20

∴

LEMMA 6.10.17: In an acyclic LCF G, if there is no LJ(T(I,L),T(J,K)) choke point, and there is no IK(T(I,L),T(J,K)) choke point, then $\rho_{IJ}\rho_{KL} - \rho_{IL}\rho_{JK} = 0$ is not linearly implied by G.

Proof. The proof is the same as that of lemma 6.10.16, with the indices permuted. ∴

LEMMA 6.10.18: In an acyclic LCF G, if G linearly implies $\rho_{IJ}\rho_{KL} - \rho_{IL}\rho_{JK} = 0$, then either there is an LJ(T(I,J),T(K,L)) choke point and an LJ(T(I,L),T(J,K)) choke point, or there is an IK(T(I,J),T(K,L)) choke point and an IK(T(I,L),T(J,K)) choke point.

Proof. Assume that G linearly implies $\rho_{IJ}\rho_{KL} - \rho_{IL}\rho_{JK} = 0$. By lemmas 6.10.16 and 6.10.17, if G linearly implies $\rho_{IJ}\rho_{KL} - \rho_{IL}\rho_{JK} = 0$ then either there is an LJ(T(I,J),T(K,L)) choke point or an IK(T(I,J),T(K,L)) choke point, and there is either an LJ(T(I,L),T(J,K)) choke point or an IK(T(I,L),T(J,K)) choke point. If there is an LJ(T(I,J),T(K,L)) choke point and an LJ(T(I,L),T(J,K)) choke point, or there is an IK(T(I,J),T(K,L)) choke point and an IK(T(I,L),T(J,K)) choke point, the proof is done. Suppose then that there is an LJ(T(I,J),T(K,L)) choke point and an IK(T(I,L),T(J,K)) choke point, but no

$IK(T(I,J),T(K,L))$ choke point and no $LJ(T(I,L),T(J,K))$ choke point. (The case where there is an $LJ(T(I,L),T(J,K))$ choke point and an $IK(T(I,J),T(K,L))$ choke point, but no $LJ(T(I,J),T(K,L))$ choke point and no $IK(T(I,L),T(J,K))$ choke point is essentially the same, with the indices permuted.)

By lemmas 6.10.9 through 6.10.14, if there is no $LJ(T(I,L),T(J,K))$ choke point, then either there is a pair of treks $T'(I,L)$ and $T'(J,K)$ such that $L(T'(I,L))$ does not intersect $J(T'(J,K))$ or $\rho_{IJ}\rho_{KL} - \rho_{IL}\rho_{JK} = 0$ is not linearly implied by G. Since the latter possibility contradicts our hypothesis, assume that there is a pair of treks $T'(I,L)$ and $T'(J,K)$ such that $L(T'(I,L))$ does not intersect $J(T'(J,K))$. There are two cases.

If $I(T'(I,L))$ does not intersect $K(T'(J,K))$ then by lemma 6.10.7, G does not linearly imply $\rho_{IJ}\rho_{KL} - \rho_{IL}\rho_{JK} = 0$, contrary to our hypothesis. Suppose then that $I(T'(I,L))$ does intersect $K(T'(J,K))$ at a vertex Y. (See figure 13.21.)

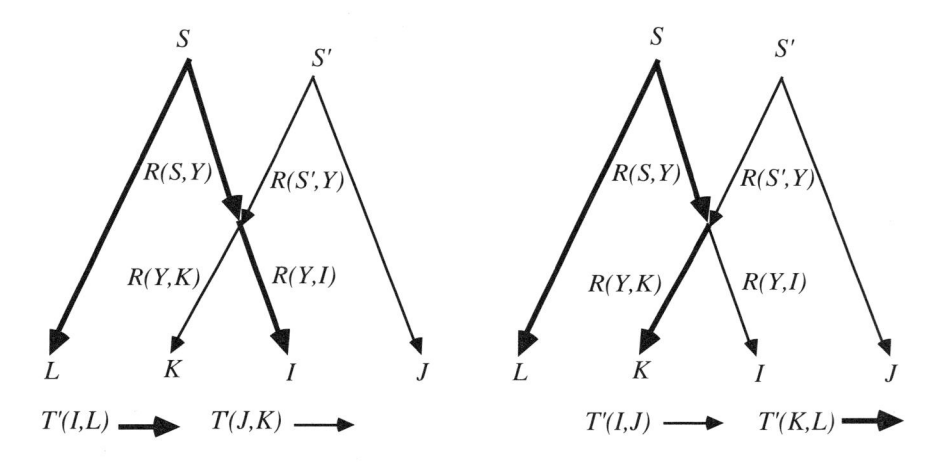

Figure 13.21

Let S be the source of $T'(I,L)$, S' the source of $T'(J,K)$, $R(S,Y)$ an initial segment of $I(T'(I,L))$, $R(Y,K)$ a final segment of $K(T'(J,K))$, $R(S',Y)$ an initial segment of $K(T'(J,K))$, $R(Y,I)$ a final segment of $I(T'(I,L))$, $I(T'(I,J)) = R(S',Y)\&R(Y,I)$, $J(T'(I,J)) = J(T'(J,K))$, $K(T'(K,L)) = R(S,Y)\&R(Y,K)$, and $L(T'(K,L)) = L(T'(I,L))$. But since $J(T'(I,J)) = J(T'(J,K))$ does not intersect $L(T'(K,L)) = L(T'(I,L))$, there is no $LJ(T(I,J),T(K,L))$ choke point, contrary to our hypothesis. ∴

LEMMA 6.10.19: In an acyclic LCF G, if G linearly implies $\rho_{IJ}\rho_{KL} - \rho_{IL}\rho_{JK} = 0$, then either there is an $LJ(T(I,J),T(K,L),T(I,L),T(J,K))$ choke point, or there is an $IK(T(I,J),T(K,L),T(I,L),T(J,K))$ choke point.

Proof. Assume that G linearly implies $\rho_{IJ}\rho_{KL} - \rho_{IL}\rho_{JK} = 0$. By lemma 6.10.18, either there is an $LJ(T(I,J),T(K,L))$ choke point and an $LJ(T(I,L),T(J,K))$ choke point, or there is an

$IK(T(I,J),T(K,L))$ choke point and an $IK(T(I,L),T(J,K))$ choke point. Suppose w.l.g. that the former is the case. If some $LJ(T(I,J),T(K,L))$ choke point is also an $LJ(T(I,L),T(J,K))$ choke point, the proof is done. Suppose then that no $LJ(T(I,J),T(K,L))$ choke point is also an $LJ(T(I,L),T(J,K))$ choke point. Let C be an $LJ(T(I,J),T(K,L))$ choke point. By hypothesis C is not an $LJ(T(I,L),T(J,K))$ choke point, so there exist a pair of treks $T'(I,L)$ and $T'(J,K)$ with sources S and S' respectively, such that $L(T'(I,L))$ and $J(T'(J,K))$ do not intersect at C. (See figure 13.22.)

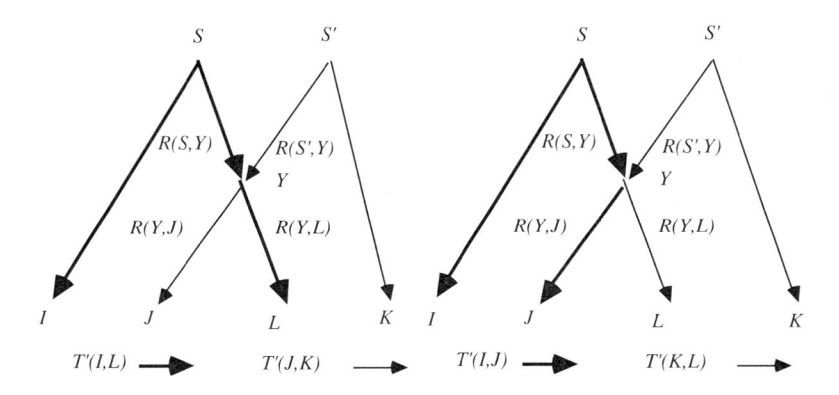

Figure 13.22

Hence there is at most one occurrence of C in the pair of paths $L(T'(I,L))$ and $J(T'(J,K))$. Since there is an $LJ(T(I,L),T(J,K))$ choke point, $L(T'(I,L))$ and $J(T'(J,K))$ intersect at a point Y. Let $R(S,Y)$ be an initial segment of $L(T'(I,L))$, $R(Y,J)$ be a final segment of $J(T'(J,K))$, $R(S',Y)$ an initial segment of $J(T'(J,K))$, $R(Y,L)$ a final segment of $L(T'(I,L))$, $I(T'(I,J)) = I(T'(I,L))$, $J(T'(I,J)) = R(S,Y)\&R(Y,J)$, $K(T'(K,L)) = K(T'(J,K))$ and $L(T'(K,L)) = R(S',Y)\&R(Y,L)$. Since $L(T'(K,L))$ and $J(T'(I,J))$ are rearrangements of the vertices in $J(T'(J,K))$ and $L(T'(I,L))$, the number of occurrences of any vertex in $L(T'(K,L))$ and $J(T'(I,J))$ is less than or equal to the number of occurrences of that vertex in $J(T'(J,K))$ and $L(T'(I,L))$. Since C occurs at most once in $J(T'(J,K))$ and $L(T'(I,L))$, it occurs at most once in $L(T'(K,L))$ and $J(T'(I,J))$. Hence $L(T'(K,L))$ and $J(T'(I,J))$ do not intersect at C, contrary to the hypothesis that C is an $LJ(T(I,J),T(K,L))$ choke point. ∴.

LEMMA 6.10.20: For any probability distribution over a set of random variables **W**, if there exists a subset **P** of **W** such that $\rho_{IJ.\mathbf{P}}\rho_{KL.\mathbf{P}} - \rho_{IL.\mathbf{P}}\rho_{JK.\mathbf{P}} = 0$, and for all variables U in **P** and all subsets **V** of **P** not containing U, either $\rho_{IU.\mathbf{V}} = 0$ and $\rho_{KU.\mathbf{V}} = 0$, or $\rho_{JU.\mathbf{V}} = 0$ and $\rho_{LU.\mathbf{V}} = 0$, then $\rho_{IJ}\rho_{KL} - \rho_{IL}\rho_{JK} = 0$.

Proof. The proof is by induction on the cardinality of **P**.

Base Case: Suppose the cardinality of **P** is zero. Then $\rho_{IJ}\rho_{Kl} - \rho_{IL}\rho_{JK} = 0$ is equivalent to $\rho_{IJ.\mathbf{P}}\rho_{KL.\mathbf{P}} - \rho_{IL.\mathbf{P}}\rho_{JK.\mathbf{P}} = 0$.

Induction Case: Suppose that the lemma is true for all sets of cardinality n or less. Let **P** have cardinality $n+1$. Assume that $\rho_{IJ.\mathbf{P}}\rho_{KL.\mathbf{P}} - \rho_{IL.\mathbf{P}}\rho_{JK.\mathbf{P}} = 0$.

Let Y be a variable in **P**, and $\mathbf{P'} = \mathbf{P} - \{Y\}$. Since $\rho_{IJ.\mathbf{P}}\rho_{KL.\mathbf{P}} - \rho_{IL.\mathbf{P}}\rho_{JK.\mathbf{P}}$, by the recursion formula for partial correlation,

$$\left(\frac{\rho_{IJ.\mathbf{P'}} - \rho_{IY.\mathbf{P'}}\rho_{JY.\mathbf{P'}}}{\left(\sqrt{1-\rho_{IY.\mathbf{P'}}^2}\right)\left(\sqrt{1-\rho_{JY.\mathbf{P'}}^2}\right)}\right)\left(\frac{\rho_{KL.\mathbf{P'}} - \rho_{KY.\mathbf{P'}}\rho_{LY.\mathbf{P'}}}{\left(\sqrt{1-\rho_{KY.\mathbf{P'}}^2}\right)\left(\sqrt{1-\rho_{LY.\mathbf{P'}}^2}\right)}\right) =$$

$$\left(\frac{\rho_{IL.\mathbf{P'}} - \rho_{IY.\mathbf{P'}}\rho_{LY.\mathbf{P'}}}{\left(\sqrt{1-\rho_{IY.\mathbf{P'}}^2}\right)\left(\sqrt{1-\rho_{LY.\mathbf{P'}}^2}\right)}\right)\left(\frac{\rho_{JK.\mathbf{P'}} - \rho_{JY.\mathbf{P'}}\rho_{KY.\mathbf{P'}}}{\left(\sqrt{1-\rho_{JY.\mathbf{P'}}^2}\right)\left(\sqrt{1-\rho_{KY.\mathbf{P'}}^2}\right)}\right)$$

The denominator of the l.h.s. equals the denominator of the r.h.s., so the numerator of the l.h.s. equals the numerator of the r.h.s. Expanding the numerators of each side,

$\rho_{IJ.\mathbf{P'}}\rho_{KL.\mathbf{P'}} - \rho_{IJ.\mathbf{P'}}\rho_{KY.\mathbf{P'}}\rho_{LY.\mathbf{P'}} - \rho_{KL.\mathbf{P'}}\rho_{IY.\mathbf{P'}}\rho_{JY.\mathbf{P'}} - \rho_{IY.\mathbf{P'}}\rho_{JY.\mathbf{P'}}\rho_{KY.\mathbf{P'}}\rho_{LY.\mathbf{P'}} =$
$\rho_{IL.\mathbf{P'}}\rho_{JK.\mathbf{P'}} - \rho_{IL.\mathbf{P'}}\rho_{JY.\mathbf{P'}}\rho_{KY.\mathbf{P'}} - \rho_{JK.\mathbf{P'}}\rho_{IY.\mathbf{P'}}\rho_{LY.\mathbf{P'}} - \rho_{IY.\mathbf{P'}}\rho_{JY.\mathbf{P'}}\rho_{KY.\mathbf{P'}}\rho_{LY.\mathbf{P'}}$

The fourth terms on both sides are equal. By hypothesis, either $\rho_{IY.\mathbf{P'}} = \rho_{KY.\mathbf{P'}} = 0$, or $\rho_{JY.\mathbf{P'}} = \rho_{LY.\mathbf{P'}} = 0$. In either case, the second and third terms on each side are equal to zero. It follows that $\rho_{IJ.\mathbf{P'}}\rho_{KL.\mathbf{P'}} - \rho_{IL.\mathbf{P'}}\rho_{JK.\mathbf{P'}} = 0$. Since **P'** has one less member than **P**, by the induction hypothesis, $\rho_{IJ}\rho_{KL} - \rho_{IL}\rho_{JK} = 0$. ∴

LEMMA 6.10.21: In an acyclic LCF G, if there exists an $LJ(T(I,J),T(K,L),T(I,L),T(J,K))$ choke point or an $IK(T(I,J),T(K,L),T(I,L),T(J,K))$ choke point, then G linearly implies $\rho_{IJ}\rho_{KL} - \rho_{IL}\rho_{JK} = 0$.

Proof. Suppose w.l.g. that X is the last $LJ(T(I,J),T(K,L),T(I,L),T(J,K))$ choke point. There are two cases.

First consider the case where there is no trek between at least one of the pairs I and J, and K and L, and there is no trek between at least one of the pairs I and L, and J and K. It follows that at least one of ρ_{IJ} and ρ_{KL} equals 0, and at least one of ρ_{IL} and ρ_{JK} is equal to zero. Hence $\rho_{IJ}\rho_{KL} - \rho_{IL}\rho_{JK} = 0$.

Next suppose w.l.g. that there are treks $T'(I,J)$ and $T'(K,L)$. We will prove that $\rho_{IJ}\rho_{KL} - \rho_{IL}\rho_{JK} = 0$ by proving that there exists a set $\mathbf{Q'}$ of variables such that $\rho_{IJ.\mathbf{Q'}}\rho_{KL.\mathbf{Q'}} - \rho_{IL.\mathbf{Q'}}\rho_{JK.\mathbf{Q'}}$

= 0, and for all variables U in $\mathbf{Q'}$ and all subsets $\mathbf{V}$ of $\mathbf{Q'}$ not containing U, either $\rho_{IU.\mathbf{V}} = 0$ and $\rho_{KU.\mathbf{V}} = 0$, or $\rho_{JU.\mathbf{V}} = 0$ and $\rho_{LU.\mathbf{V}} = 0$, and applying lemma 6.10.20.

Let $\mathbf{Q}$ = {sources of treks between X and J or X and L}. Since X is on $J(T'(I,J))$ and $L(T'(K,L))$, and by definition the sink of $J(T'(I,J))$ is J, and the sink of $L(T'(K,L))$ is L, there are directed paths $R(X,J)$ and $R(X,L)$; hence X is in $\mathbf{Q}$. We will now demonstrate that $I \perp\!\!\!\perp J|\mathbf{Q}$ by showing that I and J are d-separated given $\mathbf{Q}$. We will show that I and J are d-separated given $\mathbf{Q}$ by showing that every undirected path between I and J either contains a vertex V that is a collider that is not the source of a directed path from V to any vertex in $\mathbf{Q}$, or it contains some vertex in $\mathbf{Q}$ that is not a collider.

Consider first the undirected paths between I and J without colliders. If there is an undirected path with no collider between I and J that does not contain X, there is a trek between I and J that does not contain X. But, every $T(I,J)$ contains X, since X is a choke point. Hence, there does not exist an undirected path between I and J without colliders that does not contain X. Since X is in $\mathbf{Q}$, every undirected path that does not contain a collider contains a vertex in $\mathbf{Q}$.

Consider now undirected paths between I and J that contain colliders. If some vertex W is a collider and is not the source of a directed path from W to some vertex in $\mathbf{Q}$, the proof is done. Suppose then that every vertex W that is a collider is the source of a directed path from W to some vertex in $\mathbf{Q}$. Consider w.l.g. an arbitrary undirected path $R(J,I)$ from J to I. Let Z be the first vertex on $R(J,I)$ that is a collider. By hypothesis, there is a directed path $R(Z,U)$ where U is a vertex in $\mathbf{Q}$. Since the undirected path from J to Z does not contain any colliders, there is a vertex S that is the source of a pair of directed paths $R(S,J)$ and $R(S,Z)$. Since Z has an edge directed into it, $S \neq Z$. There are two cases.

a. $S = J$. (See figure 13.23.) There is a directed path $R(J,Z)$. There is a directed path $R(Z,U)$. Since U is the source of a trek between X and J, there is a directed path $R(U,X)$. We have already shown that there is a directed path $R(X,J)$. Hence there is a cyclic path $R(J,Z)\&R(Z,U)\&R(U,X)\&R(X,J)$.

b. $S \neq J$. (See figure 13.24.) There is a directed path $R(S,J)$, and a directed path $R(S,Z)\&R(Z,U)\&R(U,X)$. By lemma 6.10.2 there is a trek $T'(J,X)$ with source M, where M is the last point of intersection of $R(S,J)$ and $R(S,Z)\&R(Z,U)\&R(U,X)$, and $J(T'(J,X))$ is a subpath of $R(S,J)$. Since M is on $R(S,J)$, and S occurs before Z on $R(J,I)$, M occurs before Z on $R(J,I)$. Hence there is no collision at M in $R(J,I)$. Also, M is in $\mathbf{Q}$, since it is the source of a trek between X and J. The undirected path $R(J,I)$ contains a vertex in $\mathbf{Q}$ that is not a collider.

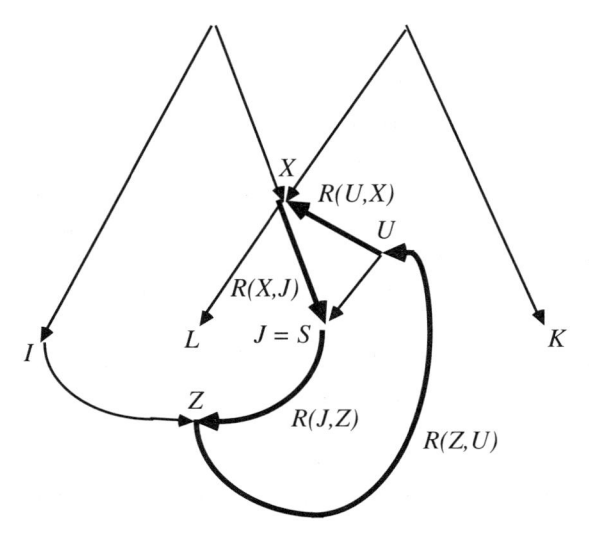

Figure 13.23

In either case **Q** d-separates X and Y, so $I \perp\!\!\!\perp J|\mathbf{Q}$. Similarly, it can be shown that $K \perp\!\!\!\perp L|\mathbf{Q}$, $I \perp\!\!\!\perp L|\mathbf{Q}$, and $J \perp\!\!\!\perp K|\mathbf{Q}$. It follows that $\rho_{IJ.\mathbf{Q}}=0$, $\rho_{KL.\mathbf{Q}}=0$, $\rho_{IL.\mathbf{Q}}=0$, and $\rho_{JK.\mathbf{Q}}=0$. Let $\mathbf{Q}' = \mathbf{Q} \setminus \{X\}$. By the recursion formula for partial correlation, $\rho_{IJ.\mathbf{Q}'} = \rho_{IX.\mathbf{Q}'}\rho_{JX.\mathbf{Q}'}$, $\rho_{KL.\mathbf{Q}'} = \rho_{KX.\mathbf{Q}'}\rho_{LX.\mathbf{Q}'}$, $\rho_{IL.\mathbf{Q}'} = \rho_{IX.\mathbf{Q}'}\rho_{LX.\mathbf{Q}'}$, and $\rho_{JK.\mathbf{Q}'} = \rho_{JX.\mathbf{Q}'}\rho_{KX.\mathbf{Q}'}$. Hence $\rho_{IJ.\mathbf{Q}'} \rho_{KL.\mathbf{Q}'} = \rho_{IX.\mathbf{Q}'} \rho_{JX.\mathbf{Q}'} \rho_{KX.\mathbf{Q}'} \rho_{LX.\mathbf{Q}'} = \rho_{IX.\mathbf{Q}'} \rho_{LX.\mathbf{Q}'} \rho_{JX.\mathbf{Q}'} \rho_{KX.\mathbf{Q}'} = \rho_{IL.\mathbf{Q}'}\rho_{JK.\mathbf{Q}'}$.

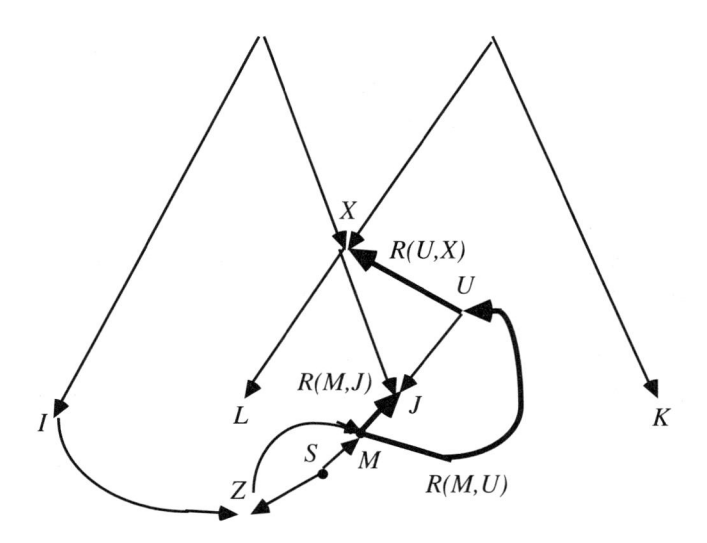

Figure 13.24

We will next demonstrate that for each variable U in $\mathbf{Q}'$, and each subset $\mathbf{V}$ of $\mathbf{Q}'$ not containing U, $I \perp\!\!\!\perp U|\mathbf{V}$, by showing that I and U are d-separated given $\mathbf{V}$. We will show that I and U are d-separated given V by showing that every undirected path between I and U either contains a vertex W that is a collider that is not the source of a directed path from W to any vertex in $\mathbf{V}$, or it contains some vertex in $\mathbf{V}$ that is not a collider.

For U in $\mathbf{Q}'$, consider an arbitrary undirected path $R(I,U)$ that contains colliders. Let Z be the first point of $R(I,U)$ after I that is a collider, and $R(I,Z)$ be an initial segment of $R(I,U)$. If Z is not the source of a path to some vertex M in $\mathbf{V}$, then the path does not d-connect I and U given $\mathbf{V}$, and the proof is done. Suppose then that there is a directed path $R(Z,M)$ to some M in $\mathbf{V}$. Since $R(I,Z)$ contains no colliders, there is a vertex S on $R(I,Z)$ that is the source of directed paths $R(S,I)$ and $R(S,Z)$. Hence S is the source of directed paths to I and M, $R(S,I)$ and $R(S,M) = R(S,Z)\&R(Z,M)$ respectively. (If $R(I,U)$ is an undirected path that contains no colliders, then it still follows that there is a vertex S on $R(I,U)$ that is the source of directed paths $R(S,I)$ and $R(S,U)$.) M is either the source of a trek between X and J or X and L. Suppose w.l.g. that M is the source of a trek between X and J. Then M is the source of a directed path $R(M,J)$ and a directed path $R(M,X)$. M does not equal X by hypothesis. Hence $R(M,J)$ does not contain X, since $R(M,J)$ is a branch of a trek between J and X, and the two branches of the trek intersect only at M. $R(S,M)$ does not contain X, else there is a cycle. Because X is not on the J branch of the trek between I and J just constructed, it is not an $LJ(T(I,J),T(K,L),T(I,L),T(J,K))$ choke point, contrary to the assumption. ∴

TETRAD REPRESENTATION THEOREM 6.10: In an acyclic LCF G, there exists an $LJ(T(I,J),T(K,L),T(I,L),T(J,K))$ choke point or an $IK(T(I,J),T(K,L),T(I,L),T(J,K))$ choke point iff G linearly implies $\rho_{IJ}\rho_{KL} - \rho_{IL}\rho_{JK} = 0$.

Proof. This follows directly from lemma 6.10.19 and lemma 6.10.21. ∴

COROLLARY 6.10.1: If an acyclic LCF G' is a subgraph of an acyclic LCF G, and G linearly implies $\rho_{IJ}\rho_{KL} - \rho_{IL}\rho_{JK} = 0$, then G' linearly implies $\rho_{IJ}\rho_{KL} - \rho_{IL}\rho_{JK} = 0$.

Proof. If G linearly implies $\rho_{IJ}\rho_{KL} - \rho_{IL}\rho_{JK} = 0$, then by lemma 6.10.21 G has either an $LJ(T(I,J),T(K,L),T(I,L),T(J,K))$ choke point or an $IK(T(I,J),T(K,L),T(I,L),T(J,K))$ choke point. If G has either an $LJ(T(I,J),T(K,L),T(I,L),T(J,K))$ choke point or an $IK(T(I,J),T(K,L),T(I,L),T(J,K))$ choke point, then G' has either an $LJ(T(I,J),T(K,L),T(I,L),T(J,K))$ choke point or an $IK(T(I,J),T(K,L),T(I,L),T(J,K))$ choke point. By lemma 6.10.21, G' linearly implies $\rho_{IJ}\rho_{KL} - \rho_{IL}\rho_{JK} = 0$. ∴

13.26 Theorem 6.11

THEOREM 6.11: An acyclic LCF G linearly implies $\rho_{IJ}\rho_{KL} - \rho_{IL}\rho_{JK} = 0$ only if either it linearly implies that ρ_{IJ} or $\rho_{KL} = 0$, and ρ_{IL} or $\rho_{JK} = 0$, or there is a (possibly empty) set $\mathbf{Q}$ of random variables in G that does not contain both I and K or both J and L such that G linearly implies that $\rho_{IJ.\mathbf{Q}} = \rho_{KL.\mathbf{Q}} = \rho_{IL.\mathbf{Q}} = \rho_{JK.\mathbf{Q}} = 0$.

Proof. By theorem 6.10, if G linearly implies $\rho_{IJ}\rho_{KL} - \rho_{IL}\rho_{JK} = 0$, then there is either an $LJ(T(I,J),T(K,L),T(I,L),T(J,K))$ choke point or an $IK(T(I,J),T(K,L),T(I,L),T(J,K))$ choke point in G. In the proof of lemma 6.10.21 we demonstrated that the existence of an $LJ(T(I,J),T(K,L),T(I,L),T(J,K))$ choke point or an $IK(T(I,J),T(K,L),T(I,L),T(J,K))$ choke point then either ρ_{IJ} or $\rho_{KL} = 0$, and ρ_{IL} or $\rho_{JK} = 0$, or there exists a set $\mathbf{Q}$ of random variables such that $\rho_{IJ.\mathbf{Q}} = 0$, $\rho_{KL.\mathbf{Q}} = 0$, $\rho_{IL.\mathbf{Q}} = 0$, and $\rho_{JK.\mathbf{Q}} = 0$.

Suppose without loss of generality that G does not linearly entail that ρ_{IJ} or ρ_{KL} equals 0, does not linearly entail that ρ_{IL} or ρ_{JK} equals 0, there is a $JL(T(I,J),T(K,L),T(I,L),T(J,K))$ choke point C, and $\mathbf{Q}$ is the set of sources of treks between C and J or C and L. Now we will show that $\mathbf{Q}$ does not contain both I and K, and $\mathbf{Q}$ does not contain both J and L.

If $J \neq C$, then J is not the source of a trek between J and C for the following reasons. Because ρ_{IJ} or ρ_{JK} is not linearly entailed to be zero, there is a trek between I and J, or between J and K. Suppose without loss of generality that there is a trek t between I and J. Because C is a $JL(T(I,J),T(K,L),T(I,L),T(J,K))$ choke point, it lies on the J branch of t. If $J \neq C$, then J cannot the source of t. Hence C lies on a directed path from the source of t to J, and there is a directed path from C to J. If J is the source of trek between J and C, then there is a directed path from J to C. It follows then that the directed graph is cyclic, contrary to our assumption. Similarly, if $L \neq C$, then L is not the source of a trek between L and C.

Suppose that $\mathbf{Q}$ contains J and L. Consider first the case where $J = C$. Because $L \neq J$, it follows that $L \neq C$. L is the source of a trek between C and L or C and J. There is no trek between C and J, because $C = J$. Because $L \neq C$, L is not the source of a trek between C and L. This is a contradiction, so $J \neq C$. Similarly, $L \neq C$.

Consider the case where $J \neq C$, and $L \neq C$. It follows that J is the source of a trek between C and L, and L is the source of a trek between C and J. If J is the source of a trek between C and L, there is a directed path from J to L, and if L is the source of a trek between C and J, there is a directed path from L to J. Hence they cannot both be in $\mathbf{Q}$ because the graph is acyclic.

Suppose then that $\mathbf{Q}$ contains both I and K. It follows that I and K are sources of treks between C and J or C and L. If $I \neq C$ then I is the source of a trek between C and J or C and L, and there is a directed path from I to J or I to L that does not contain C. That directed path is a trek that does not contain C, and hence C is not a $JL(T(I,J),T(K,L),T(I,L),T(J,K))$ choke point, contrary to the hypothesis. If $I = C$, then K is the source of a trek between I and J or I and L. It follows that there is directed path from

K to J or K to L that does not contain C, and hence C is not a $JL(T(I,J),T(K,L),T(I,L),T(J,K))$ choke point, contrary to the hypothesis. ∴

13.27 Theorem 7.1

If G is a directed acyclic graph over a set of variables $\mathbf{V} \cup \mathbf{W}$, $\mathbf{W}$ is exogenous with respect to $\mathbf{V}$ in G, $\mathbf{Y}$ and $\mathbf{Z}$ are disjoint subsets of $\mathbf{V}$, $P(\mathbf{V} \cup \mathbf{W})$ is a distribution that satisfies the Markov condition for G, and **Manipulated**($\mathbf{W}$) = $\mathbf{X}$, then $P(\mathbf{Y}|\mathbf{Z})$ is **invariant** under direct manipulation of $\mathbf{X}$ in G by changing $\mathbf{W}$ from $\mathbf{w}_1$ to $\mathbf{w}_2$ if and only if $P(\mathbf{Y}|\mathbf{Z},\mathbf{W} = \mathbf{w}_1) = P(\mathbf{Y}|\mathbf{Z},\mathbf{W} = \mathbf{w}_2)$ wherever they are both defined. .

THEOREM 7.1: If G_{Comb} is a directed acyclic graph over $\mathbf{V} \cup \mathbf{W}$, $\mathbf{W}$ is exogenous with respect to V in G_{Comb}, $\mathbf{Y}$ and $\mathbf{Z}$ are disjoint subsets of $\mathbf{V}$, $P(\mathbf{V} \cup \mathbf{W})$ is a distribution that satisfies the Markov condition for G_{Comb}, no member of $\mathbf{X} \cap \mathbf{Z}$ is a member of **IP(Y,Z)** in G_{Unman}, and no member of $\mathbf{X}\backslash\mathbf{Z}$ is a member of **IV(Y,Z)** in G_{Unman}, then $P(\mathbf{Y}|\mathbf{Z})$ is invariant under a direct manipulation of $\mathbf{X}$ in G_{Comb} by changing $\mathbf{W}$ from $\mathbf{w}_1$ to $\mathbf{w}_2$.

Proof. Suppose that G_{Comb} is a directed acyclic graph over $\mathbf{V} \cup \mathbf{W}$, $\mathbf{W}$ is exogenous with respect to $\mathbf{V}$, G_{Unman} is the subgraph of G_{Comb} over $\mathbf{V}$, $P(\mathbf{V} \cup \mathbf{W})$ is a distribution that satisfies the Markov condition for G_{Comb}, $\mathbf{X}$ = **Manipulated**($\mathbf{W}$), $P(\mathbf{Y}|\mathbf{Z},\mathbf{W}= \mathbf{w}_1) \neq P(\mathbf{Y}|\mathbf{Z},\mathbf{W}= \mathbf{w}_2)$ when G_{Comb} is manipulated by changing the value of $\mathbf{W}$ from $\mathbf{w}_1$ to $\mathbf{w}_2$, $\mathbf{Y}$ and $\mathbf{Z}$ are disjoint subsets of $\mathbf{V}$, no member of $\mathbf{X} \cap \mathbf{Z}$ is a member of **IP(Y,Z)** in G_{Unman}, and no member of $\mathbf{X}\backslash\mathbf{Z}$ is a member of **IV(Y,Z)** in G_{Unman}, but $P(\mathbf{Y}|\mathbf{Z})$ is not invariant when $\mathbf{X}$ is manipulated. Hence there is an undirected path U in G_{Comb} that d-connects some R in $\mathbf{W}$ to some Y in $\mathbf{Y}$ given $\mathbf{Z}$. Let W be the vertex on U closest to Y that is in $\mathbf{W}$. By lemma 3.3.2, $U(W,Y)$ d-connects W and Y given $\mathbf{Z}\backslash\{W,Y\} = \mathbf{Z}$. Because $U(W,Y)$ contains no member of $\mathbf{W}$ except W, every subpath of $U(W,Y)$ that does not contain W is an undirected path in G_{Unman}. Because $U(W,Y)$ is an undirected path between W and Y, it contains some variable X in **Manipulated**($\mathbf{W}$). There are two cases: either X is in $\mathbf{Z}$ or it is not in $\mathbf{Z}$.

If X is in $\mathbf{Z}$ then X is a collider on U in G_{Unman}, and the vertex T adjacent to X on U and between X and Y is a parent of X, and hence not a collider on U. Because T is not a collider on U, T is not in $\mathbf{Z}$, and $\mathbf{Z}\backslash\{T\} = \mathbf{Z}$. If T is in $\mathbf{Y}$, then X is in **IP(Y,Z)**, contrary to our assumption. If T is not in $\mathbf{Y}$, then $U(T,Y)$ d-connects T and Y given $\mathbf{Z}\backslash\{T,Y\} = \mathbf{Z}$ in G_{Unman}. T has a descendant (X) in $\mathbf{Z}$ in G_{Unman}, and hence T is in **IV(Y,Z)** in G_{Unman}. But then X is in **IP(Y,Z)** in G, contrary to our assumption.

If X is not in $\mathbf{Z}$, then $U(X,Y)$ d-connects Y and X given $\mathbf{Z}\backslash\{X\} = \mathbf{Z}$ in G_{Unman}. If X is a collider on U then X has a descendant in $\mathbf{Z}$ in G_{Unman}. If X is not a collider on U then $U(X,Y)$ is out of X because X is a child of W. Either X is an ancestor of a collider on $U(X,Y)$, in which case it is an ancestor of some member of $\mathbf{Z}$ in G_{Comb}, or $U(X,Y)$ is a directed path to Y, in which case it is an ancestor of some member of $\mathbf{Y}$ in G_{Comb}. If X has a descendant in $\mathbf{Z} \cup \mathbf{Y}$ in G_{Comb}, then X has a descendant in $\mathbf{Z} \cup \mathbf{Y}$ in G_{Unman}, because W

is exogenous with respect to **V**. Hence X has a descendant in $\mathbf{Y} \cup \mathbf{Z}$ in G_{Unman}. It follows that X is in **IV(Y,Z)** in G_{Unman}, contrary to our assumption. ∴

13.28 Theorem 7.2

THEOREM 7.2: If $P(\mathbf{O})$ is the marginal of a distribution faithful to G over **V**, π is a partially oriented inducing path graph of G over **O**, and Ord is an ordering of variables in **O** acceptable for some inducing path graph over **O** with partially oriented inducing path graph π, then there is a minimal I-map G_{Min} of $P(\mathbf{O})$ in which **Definite-SP**(Ord,X) in π is included in **Parents**(G_{Min},X) which is included in **Possible-SP**(Ord,X) in π.

Proof. Suppose that G_{IP} is an inducing path graph over **O** with partially oriented inducing path graph π. By lemma 6.2.4 if G_{IP} is an inducing path graph over **O** and Ord an acceptable total ordering of variables for G_{IP}, then **Predecessors**$(Ord,X)\setminus$ **SP**(Ord,G_{IP},X) is d-separated from X given **SP**(Ord,G_{IP},X). Hence, if **Parents**(G_{Min},X) = **SP**(Ord,G_{IP},X) then G_{Min} is an I-map of $P(\mathbf{O})$.

We will now show that no subgraph of G_{Min} is an I-map of $P(\mathbf{O})$. Suppose in G_{Sub} **Parents**(G_{Sub},X) is properly included in **Parents**(G_{Min},X) and hence properly included in **SP**(Ord,G_{IP},X). Let V be some variable in **Parents**$(G_{Min},X)\setminus$**Parents**(G_{Sub},X). Because V is in **SP**(Ord,G_{IP},X) there is an undirected path U in G_{IP} between V and X on which all of the vertices except the endpoints are colliders, and precede X in Ord. Let W be the vertex on U closest to X but not equal to X that is in **Parents**$(G_{Min},X)\setminus$ **Parents**(G_{Sub},X). It follows that $U(W,X)$ is an undirected path in G_{IP} between W and X such that every vertex on $U(W,X)$ except for the endpoints is a collider and in **Parents**(G_{Sub},X). Hence W is in **Predecessors**$(Ord,X)\setminus$**Parents**(G_{Sub},X) and is d-connected to X given **Parents**(G_{Sub},X) in G_{IP}. Hence W is d-connected to X given **Parents**(G_{Sub},X) in G, and because $P(\mathbf{V})$ is faithful to G, W and X are dependent given **Parents**(G_{Sub},X). Hence $P(\mathbf{O})$ does not satisfy the Markov Condition for G_{Sub}.

For a partially oriented inducing path graph π and ordering Ord acceptable for π, V is in **Possible-SP**(Ord,X) if and only if $V \neq X$ and there is an undirected path U in π between V and X such that every vertex on U except for X is a predecessor of X in Ord, and no vertex on U except for the endpoints is a definite-noncollider on U. For a partially oriented inducing path graph π and ordering Ord acceptable for π, V is in **Definite-SP**(Ord,X) if and only if $V \neq X$ and there is an undirected path U in π between V and X such that every vertex on U except for X is a predecessor of X in Ord, and every vertex on U except for the endpoints is a collider on U. From these definitions and the definition of partially oriented inducing path graph it follows that **Definite-SP**(Ord,X) is included in **Parents**(G_{Min},X) which is included in **Possible-SP**(Ord,X). ∴

13.29 Theorem 7.3

THEOREM 7.3: If G is a directed acyclic graph over $\mathbf{V} \cup \mathbf{W}$, $\mathbf{W}$ is exogenous with respect to $\mathbf{V}$ in G, $\mathbf{O}$ is included in $\mathbf{V}$, G_{Unman} is the subgraph of G over $\mathbf{V}$, π is the FCI partially oriented inducing path graph over $\mathbf{O}$ of G_{Unman}, $\mathbf{Y}$ and $\mathbf{Z}$ are included in $\mathbf{O}$, $\mathbf{X}$ is included in $\mathbf{Z}$, $\mathbf{Y}$ and $\mathbf{Z}$ are disjoint, and no X in $\mathbf{X}$ is in **Possibly-IP(Y,Z)** in π, then $P(\mathbf{Y}|\mathbf{Z})$ is invariant under direct manipulation of $\mathbf{X}$ in G by changing the value of $\mathbf{W}$ from $\mathbf{w}_1$ to $\mathbf{w}_2$.

If A and B are not in $\mathbf{Z}$, and $A \neq B$, then an undirected path U between A and B in a partially oriented inducing path graph π over $\mathbf{O}$ is a **possibly d-connecting** path of A and B given $\mathbf{Z}$ if and only if every collider on U is the source of a semidirected path to a member of $\mathbf{Z}$, and every definite noncollider is not in $\mathbf{Z}$.

LEMMA 7.3.1: If G is a directed acyclic graph, U is a path that d-connects V and Y given $\mathbf{Z}$, X is in $\mathbf{Z}$, and X is on U, then there is a path that d-connects X and Y given $\mathbf{Z}\setminus\{X\}$ that is into X and that contains only edges that lie on a directed path to X, and a subpath of $U(X,Y)$.

Proof. Suppose that G is a directed acyclic graph, U is a path that d-connects V and Y given $\mathbf{Z}$, X is in $\mathbf{Z}$, and X is on U. Because X is in $\mathbf{Z}$ and on U, it follows that X is a collider on U, and hence $U(X,Y)$ is into X. No noncollider on $U(X,Y)$ except for the endpoints is in $\mathbf{Z}$, so no noncollider on $U(X,Y)$ except for the endpoints is in $\mathbf{Z}\setminus\{X\}$. Every collider on $U(X,Y)$ has a descendant in $\mathbf{Z}$. If every collider on $U(X,Y)$ has a descendant in $\mathbf{Z}\setminus\{X\}$ then $U(X,Y)$ d-connects X and Y given $\mathbf{Z}\setminus\{X\}$. Suppose then that some collider on $U(X,Y)$ has X as a descendant but no other member of $\mathbf{Z}$ as a descendant, and let C be the closest such collider on U to Y. $U(C,Y)$ d-connects C and Y given $\mathbf{Z}\setminus\{X\}$ because C is not in $\mathbf{Z}\setminus\{X\}$, every collider on $U(C,Y)$ has a descendant in $\mathbf{Z}\setminus\{X\}$, and no noncollider on $U(C,Y)$ is in $\mathbf{Z}\setminus\{X\}$. There is a directed path from C to X that contains no member of $\mathbf{Z}\setminus\{X\}$. Hence by lemma 3.3.3 X is d-connected to Y given $\mathbf{Z}\setminus\{X\}$ by a path that is into X, and that contains only edges that lie on a directed path to X and a subpath of $U(X,Y)$.

LEMMA 7.3.2: If G' is the inducing path graph for G over $\mathbf{O}$, X and Y are in $\mathbf{O}$, $\mathbf{Z}$ is included in $\mathbf{O}$, and there is a path U d-connecting X and Y given $\mathbf{Z}$ in G, then there is a path T d-connecting X and Y given $\mathbf{Z}$ in G' such that if U is into X in G, then T is into X in G' and if U is into Y in G then T is into Y in G'.

Proof. Suppose that in G with inducing path graph G' that U is a path d-connecting X and Y given $\mathbf{Z}$. We will use the following algorithm to construct two sequences of vertices, *Ancestor*, and *D-Path*. (We are actually interested only in the undirected path *D-path*; *Ancestor* is used solely as a device to construct *D-path*.) The vertices in *D-Path* are always observed (i.e., vertices in $\mathbf{O}$), but might not be on U; vertices in *Ancestor* are always on the path U, but might not be observed. For any sequence of vertices R of vertices, $R(n)$ refers to the nth vertex in R. We will say that for any pair of variables V and W on U that W is after V on U if V is between W and X on U or $V = X$.

Algorithm D-Path
$Ancestor(0) = <X>$.
$D\text{-}path(0) = <X>$.
$n = 0$.
repeat
   if $Ancestor(n) = D\text{-}path(n)$ then
      if there is no collider between $Ancestor(n)$ and the next observed variable V on U, $Ancestor(n+1) = D\text{-}path(n+1) = V$;
      else $Ancestor(n+1)$ = first collider on U after $Ancestor(n)$ and $D\text{-}path(n+1)$ = first observed variable on a path from $Ancestor(n+1)$ to a member of **Z**;
   else if $Ancestor(n) \neq D\text{-}path(n)$ then
      if on U there is no collider C after $Ancestor(n)$ that has $D\text{-}path(n)$ as the first observed variable on a directed path from C to a member of **Z**, then $Ancestor(n+1) = D\text{-}path(n+1)$ = first observed variable on U after $Ancestor(n)$
      else
         let C_2 be the collider closest to Y that has $D\text{-}path(n)$ as the first observed variable on a directed path from C_2 to a member of **Z**;
         if there is no collider between C_2 and the first observed variable after C_2 on U then $Ancestor(n+1) = D\text{-}path(n+1)$ = first observed variable after C_2 on U;
         else let C_1 be the first collider after C_2, let $Ancestor(n+1) = C_1$ and $D\text{-}path(n+1)$ = the first observed variable on a directed path from C_1 to a member of **Z**;
   $n = n + 1$.
until Y is in $D\text{-}path$.

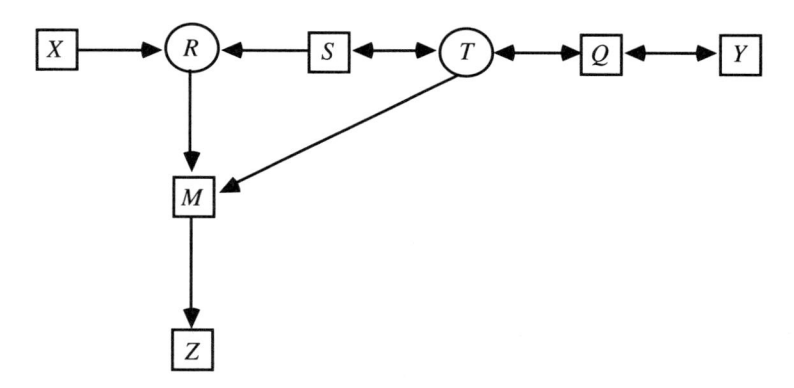

Figure 13.25

For example, when the algorithm is applied to the graph in figure 13.25 (where the circled vertices are not observed, and **Z** = {Z,Q}), for U = <X,R,S,T,Q,Y>, and the result is *Ancestor* = <X,R,Q,Y> and *D-path* = <X,M,Q,Y>.

We will now show that either *D-path* d-connects X and Y given **Z** in G', or some other path in G' d-connects X and Y given **Z**.

All of the vertices in *D-path* are observed variables, and hence in G'. By the way that *D-path* is constructed, each adjacent pair of vertices A and B in *D-path* is connected in G by a trek T(A,B) that contains no observed variables, except for the endpoints. If A and B are both on U then T(A,B) contains the edges in U(A,B); if A is on U and B is not then T(A,B) contains the edges in U(A,Ancestor(B)) and a directed path from *Ancestor*(B) to B; if A is not on U and B is, then T(A,B) consists of a directed path from *Ancestor*(A) to A and U(Ancestor(A),B); and if neither is on U, then T(A,B) contains the edges in a directed path from *Ancestor*(A) to A, U(Ancestor(A),Ancestor(B)), and a directed path from *Ancestor*(B) to B. T(A,B) is constructed out of subpaths of U, and subpaths of directed paths from colliders on U to vertices in **Z**. T(A,B) is an inducing path in G, and hence each adjacent pair of vertices in *D-path* is adjacent in G'. The method of construction of *D-path* makes *D-path* acyclic. It follows that *D-path* is an acyclic undirected path from X to Y in G'.

If W is on *D-path*, but is not a collider on *D-path*, then W is on U in G, and is not a collider on U. It follows that W is not in **Z**.

We will now show that we can transform *D-path* into a path *D-path'* in G' such that every collider B on *D-path'* has a descendant in **Z** in G. Let B be the vertex on *D-Path* closest to X that is a collider on *D-path* but that in G does not have a descendant in **Z**, and A be the predecessor of B on *D-path*, and C be the successor of B on *D-path*. If in G T(A,B) and T(B,C) are both into B, then by the construction of *D-path*, B has a descendant in **Z** in G. Hence at least one of T(A,B) and T(B,C) is out of B in G. Suppose without loss of generality that T(B,C) is out of B in G, and B is between X and C on *D-path*. It follows that B is an ancestor of C in G. In addition since there is an arrowhead at B in G', there is an inducing path between B and C that is into B and C. By lemma 6.6.2, there is a vertex V on *D-path*(X,C) such that there is an edge between V and C in G' that is substitutable for *D-path*(V,C). Let *D-path'* be the concatenation of *D-path*(X,V) with the edge between V and C. By lemma 6.6.2, *D-path'* is into X if *D-path* is. Every collider on *D-path'* is a collider on *D-path*, and every noncollider on *D-path'* is a noncollider on *D-path*. Furthermore, *D-path'* does not contain the vertex B which in G does not have a descendant in **Z**. Repeat this process until every vertex on the modified *D-path* that in G does not have a descendant in **Z** has been removed from the path. Call the result *D-path'*.

Suppose now that some collider B on *D-path'* has a descendant in **Z** in G but not in G'. We will show how to transform *D-path'* into a path in G' in which every collider has a descendant in **Z** in G'. Let P be a directed path in G from B to some Z that is a member of **Z**. In G', let P' be the undirected path from B to Z that consists of the observed variables on P in the order in which they occur. P' is an undirected path in G' because in G the directed path between any two observed variables on *P* is an inducing path. Let *S* be the vertex on *P'* closest to *B* such that there is no directed path from *B* to *S* in *G'*. Let *R* be the predecessor of *S* on *P'*. If *P'*(*B*,*R*) is not a directed path from *B* to *R* then form *P''* by substituting some directed path from *B* to *R* in *G'* for *P'*(*B*,*R*) in *P'*. There is an inducing path between *R* and *S* in *G* that is into *S*, so in *G'* the edge between *R* and *S* is into *S*. Because *P''*(*B*,*S*) is not a directed path from *B* to *S*, but *P''*(*B*,*R*) is a directed path from *B* to *R*, it follows that *R* ↔ *S* in *G'*.

We will now demonstrate that there is an edge *B* ↔ *S* in *G'*. If *B* = *R*, it follows from what we have just shown. Suppose then that *R* ≠ *B*. In that case let *Q* be the predecessor of *R* on *P''*. Because *P''*(*B*,*R*) is a directed path from *B* to *R*, *Q* → *R* in *G'*. By lemma 6.6.2, there is a vertex *E* on *P''*(*B*,*R*) such that there is an edge between *E* and *S* that is into *S* and is substitutable for *P''*(*E*,*S*) in *P''*(*B*,*S*). If the edge between *E* and *S* is out of *E*, then there is a directed path from *B* to *S* in *G'*, contrary to our assumption. It follows that the edge between *E* and *S* is into *E*. But because *P''*(*B*,*R*) is a directed path from *B* to *R*, if the edge between *E* and *S* is into *E*, the edge between *E* and *S* is not substitutable for *P''*(*E*,*S*) in *P''*(*B*,*S*) unless *E* = *B*. It follows then that *B* ↔ *S* in *G'*.

We will now form a path *D-path''* between *X* and *Y* by the following iteration, where at each stage of the iteration the vertices *B* and *S* are defined as above. Let the 0th stage *D-path''* equal *D-path'*. If *S* is on the $n-1^{th}$ stage *D-path''*(*X*,*B*) let the nth stage *D-path''*(*X*,*S*) equal the $n-1^{th}$ stage *D-path''*(*X*,*S*). If *S* is not on the $n-1^{th}$ stage *D-path''*(*X*,*B*) let *V* equal the concatenation of the $n-1^{th}$ stage *D-path''*(*X*,*B*) and *B* ↔ *S*. By lemma 6.6.2 there is a vertex *E* on *V* that is not equal to *B* and not equal to *S* such that there is an edge from *E* to *S* that is into *S*, and is a collider on *V* if and only if it is a collider on the concatenation of *V*(*X*,*E*) with the edge between *E* and *S*. Let the nth stage *D-path''*(*X*,*S*) equal the concatenation of *V*(*X*,*E*) and the edge between *E* and *S*. Similarly, form the nth stage *D-path''*(*Y*,*S*). The nth stage *D-path''*(*X*,*S*) does not intersect the nth stage *D-path''*(*Y*,*S*) except at *S* because except for the edges containing *S*, they are subpaths of paths that do not intersect except possibly at *S*. Let the nth stage *D-path''* be the concatenation of *D-path''*(*X*,*S*) and *D-path''*(*Y*,*S*). If *S* does not have a descendant in *Z* in *G'*, repeat this process until some vertex *M* on *P'* that does have a descendant in *Z* in *G'* is on *D-path''*. (See figure 13.26, where *D-path'* is <*X*,*E*,*B*,*F*,*Y*> and *D-path''* consists of the edges in boldface.)

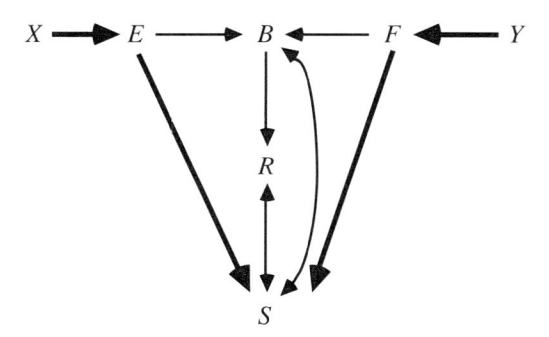

Figure 13.26

The nth stage D-$path''$ is into X if the n–1th stage D-$path''$ is, and into Y if the n–1th stage D-$path''$ is. Moreover, the 0th stage D-$path''$ (D-$path'$) is into X if U is, and into Y if U is. Every noncollider on the nth stage D-$path''$ is a noncollider on the n–1th stage D-$path''$. Because every noncollider on D-$path'$ is not in **Z**, every noncollider on the nth stage D-$path''$ is not in **Z**. Every collider on the nth stage D-$path''$ with the possible exception of M is a collider on the n–1th stage D-$path''$, and hence a collider on D-$path'$. M is a collider on the nth stage D-$path''$, but it has a descendant in **Z**. There is at least one fewer collider on n^{th} stage D-$path''$ that does not have a descendant in **Z** than there is on D-$path'$ (because D-$path'$ contains B, and the nth stage D-$path''$ does not.) This process can be repeated until every collider on D-$path''$ has a descendant in **Z**. The resulting path d-connects X and Y given **Z** in G', is into X if U is, and into Y if U is. ∴

LEMMA 7.3.3: If G is a directed acyclic graph over **V**, π is the FCI partially oriented inducing path graph of G over **O**, and some path U in G d-connects X and Y given **Z**, then there is a path U'' in π that possibly d-connects X and Y given **Z**. Furthermore if U is into X, then U'' is not out of X.

Proof. Suppose that some path U in G d-connects X and Y given **Z**. Let G' be the inducing path graph of G. By lemma 7.3.2, there is a path U' in G' that d-connects X and Y given **Z**, and if U is into X then U' is into X. Let U'' be the path in π that corresponds to U' in G'. If R is a collider on U'', then by the definition of partially oriented inducing path graph R is a collider on U'. Because R is a collider on U', and U' d-connects X and Y given **Z**, R has a descendant in **Z** in G'. By theorem 6.6, there is a semidirected path from R to a member of **Z** in π. If R is a definite noncollider on U'', then by definition of partially oriented inducing path graph R is a noncollider on U'. Because R is a noncollider on U', and U' d-connects X and Y given **Z**, R is not in **Z**. Hence U'' is a possibly d-connecting path between X and Y given **Z**. Furthermore, if U' is into X, then by definition of partially oriented inducing path graph U'' is not out of X. ∴

If π is a partially oriented inducing path graph of G over $\mathbf{O}$, then X is in **Possibly-IV(Y,Z)** if and only if X is not in $\mathbf{Z}$, there is a possibly d-connecting path between X and some Y in $\mathbf{Y}$ given $\mathbf{Z}$, and there is a semidirected path from X to a member of $\mathbf{Y} \cup \mathbf{Z}$. If π is a partially oriented inducing path graph of G over $\mathbf{O}$, then X is in **Possibly-IP(Y,Z)** if and only if $\mathbf{Y}$ and $\mathbf{Z}$ are disjoint, X is in $\mathbf{Z}$, and there is a possibly d-connecting path between X and some Y in $\mathbf{Y}$ given $\mathbf{Z}\backslash\{X\}$ that is not out of X. If π is the FCI partially oriented inducing path graph of G over $\mathbf{O}$, then X is in **Definite-Nondescendants(Y)** if and only if there is no semidirected path from any member of $\mathbf{Y}$ to X in π.

LEMMA 7.3.4: If X is in **IV(Y,Z)** in directed acyclic graph G, $\mathbf{Y}$ and $\mathbf{Z}$ are disjoint subsets of $\mathbf{O}$, X is in $\mathbf{O}$, and π is the FCI partially oriented inducing path graph of G over $\mathbf{O}$, then X is in **Possibly-IV(Y,Z)** in π.

Proof. Suppose that X is in **IV(Y,Z)** in G, $\mathbf{Y}$ and $\mathbf{Z}$ are disjoint subsets of $\mathbf{O}$, X is in $\mathbf{O}$, and π is the FCI partially oriented inducing path graph of G over $\mathbf{O}$. Because X is in **IV(Y,Z)** in G, X has a descendant in $\mathbf{Y} \cup \mathbf{Z}$ in G. Hence, by theorem 6.6, there is a semidirected path from X to a member of $\mathbf{Y} \cup \mathbf{Z}$ in π. Also, there is a path that d-connects X and some member Y of $\mathbf{Y}$ given $\mathbf{Z}$ in G. Hence, by lemma 7.3.3 there is a path that possibly d-connects X and some member Y of $\mathbf{Y}$ given $\mathbf{Z}$ in π. By definition X is in **Possibly-IV(Y,Z)** in π. ∴

LEMMA 7.3.5: If X is in **IP(Y,Z)** in directed acyclic graph G, $\mathbf{Y}$ and $\mathbf{Z}$ are disjoint subsets of $\mathbf{O}$, and π is the FCI partially oriented inducing path graph of G over $\mathbf{O}$, then X is in **Possibly-IP(Y,Z)** in π.

Proof. Suppose that X is in **IP(Y,Z)** in G, $\mathbf{Y}$ and $\mathbf{Z}$ are disjoint subsets of $\mathbf{O}$, and π is the FCI partially oriented inducing path graph of G over $\mathbf{O}$. Because X is in **IP(Y,Z)** in G, some variable T in G is a parent of X and in **IV(Y,Z)** or $\mathbf{Y}$. If T is in $\mathbf{Y}$ then there is a directed path from a member T of $\mathbf{Y}$ to X that d-connects T and X given $\mathbf{Z}\backslash\{X\}$. If T is in **IV(Y,Z)** then T is d-connected to some Y in $\mathbf{Y}$ given $\mathbf{Z}$ by some path U. If X is on U then X is a collider on U and $U(X,Y)$ is into X; furthermore, by lemma 7.3.1 there is an undirected path that d-connects X and $\mathbf{Y}$ given $\mathbf{Z}\backslash\{X\}$ that is into X. If X is not on U then the concatenation of the edge from T to X and U is a path that d-connects X and $\mathbf{Y}$ given $\mathbf{Z}\backslash\{X\}$ and is into X. Hence, by lemma 7.3.3 there is a path that possibly d-connects X and $\mathbf{Y}$ given $\mathbf{Z}\backslash\{X\}$ in π that is not out of X. By definition X is in **Possibly-IP(Y,Z)** in π. ∴

THEOREM 7.3: If G is a directed acyclic graph over $\mathbf{V} \cup \mathbf{W}$, $\mathbf{W}$ is exogenous with respect to $\mathbf{V}$ in G, $\mathbf{O}$ is included in $\mathbf{V}$, G_{Unman} is the subgraph of G over $\mathbf{V}$, π is the FCI partially oriented inducing path graph over $\mathbf{O}$ of G_{Unman}, $\mathbf{Y}$ and $\mathbf{Z}$ are included in $\mathbf{O}$, $\mathbf{X}$ is included in $\mathbf{Z}$, $\mathbf{Y}$ and $\mathbf{Z}$ are disjoint, and no X in $\mathbf{X}$ is in **Possibly-IP(Y,Z)** in π, then $P(\mathbf{Y}|\mathbf{Z})$ is invariant under direct manipulation of $\mathbf{X}$ in G by changing the value of $\mathbf{W}$ from $\mathbf{w_1}$ to $\mathbf{w_2}$.

Proofs of Theorems   465

Proof. Suppose that G is a directed acyclic graph over $\mathbf{V} \cup \mathbf{W}$, $\mathbf{O}$ is included in $\mathbf{V}$, $\mathbf{W}$ is exogenous with respect to $\mathbf{V}$ in G, G_{Unman} is the subgraph of G over $\mathbf{V}$, π is the FCI partially oriented inducing path over $\mathbf{O}$ of G_{Unman}, $\mathbf{Y}$ and $\mathbf{Z}$ are included in $\mathbf{O}$, $\mathbf{X}$ is included in $\mathbf{Z}$, $\mathbf{Y}$ and $\mathbf{Z}$ are disjoint, and no X in $\mathbf{X}$ is in **Possibly-IP(Y,Z)** in π. If $P(\mathbf{Y}|\mathbf{Z})$ is not invariant when $\mathbf{X}$ is manipulated by changing the value of $\mathbf{W}$ from $\mathbf{w_1}$ to $\mathbf{w_2}$ then $\mathbf{W}$ is d-connected to $\mathbf{Y}$ given $\mathbf{Z}$ in G. Suppose that $\mathbf{W}$ is d-connected to $\mathbf{Y}$ given $\mathbf{Z}$ in G. Let W be a member of $\mathbf{W}$ that is d-connected to some Y in $\mathbf{Y}$ by an undirected path U in G that contains no other member of $\mathbf{W}$. No noncollider on U is in $\mathbf{Z}$, and every collider on U has a descendant in $\mathbf{Z}$.

Note that if R and N are in $\mathbf{V}$ and R is a descendant of N in G, then R is a descendant of N in G_{Unman}, because there is no edge from any member of $\mathbf{V}$ into a member of $\mathbf{W}$. In G, U contains some X in $\mathbf{X}$. Because X is in $\mathbf{Z}$, X is a collider on U, and $U(X,Y)$ is into X. By lemma 7.3.1 in G there is an undirected path M that d-connects X and Y given $\mathbf{Z}\setminus\{X\}$, is into X, and contains only edges that lie on a directed path to X and a subpath of $U(X,Y)$. Hence M is an undirected path in G_{Unman}, no noncollider on M is in $\mathbf{Z}\setminus\{X\}$, and every collider on M has a descendant in $\mathbf{Z}\setminus\{X\}$ in G, and hence in G_{Unman}. It follows that M d-connects X and Y given $\mathbf{Z}\setminus\{X\}$ in G_{Unman}. Let T be the vertex adjacent to X on M. If $T = Y$ then X is in **IP(Y,Z)** in G_{Unman}. If $T \neq Y$ then T has a descendant in $\mathbf{Z}$ (namely X) in G_{Unman}. Also T is not a collider on $U(X,Y)$, and hence not in $\mathbf{Z}$. By lemma 3.3.2 T is d-connected to Y given $\mathbf{Z}\setminus\{T\} = \mathbf{Z}$ in G_{Unman}. It follows that T is in **IV(Y,Z)** in G_{Unman}, and hence X is in **IP(Y,Z)** in G_{Unman}. In either case X is in **IP(Y,Z)** in G_{Unman} and by lemma 7.3.5, X is in **Possibly-IP(Y,Z)** in π, contrary to our assumption. ∴

13.30 Theorem 7.4

THEOREM 7.4: If G is a directed acyclic graph over $\mathbf{V} \cup \mathbf{W}$, $\mathbf{W}$ is exogenous with respect to $\mathbf{V}$ in G, $\mathbf{O}$ is included in $\mathbf{V}$, G_{Unman} is the subgraph of G over $\mathbf{V}$, π is the FCI partially oriented inducing path graph over $\mathbf{O}$ of G_{Unman}, $\mathbf{X}$, $\mathbf{Y}$ and $\mathbf{Z}$ are included in $\mathbf{O}$, $\mathbf{X}$, $\mathbf{Y}$ and $\mathbf{Z}$ are pairwise disjoint, and no X in $\mathbf{X}$ is in **Possibly-IV(Y,Z)** in π, then $P(\mathbf{Y}|\mathbf{Z})$ is invariant under direct manipulation of $\mathbf{X}$ in G by changing the value of $\mathbf{W}$ from $\mathbf{w_1}$ to $\mathbf{w_2}$.

Proof. Suppose G is a directed acyclic graph over $\mathbf{V} \cup \mathbf{W}$, $\mathbf{W}$ is exogenous with respect to $\mathbf{V}$ in G, $\mathbf{O}$ is included in $\mathbf{V}$, G_{Unman} is the subgraph of G over $\mathbf{V}$, π is the FCI partially oriented inducing path over $\mathbf{O}$ of G_{Unman}, $\mathbf{Y}$ and $\mathbf{Z}$ are included in $\mathbf{O}$, $\mathbf{X}$, $\mathbf{Y}$ and $\mathbf{Z}$ are pairwise disjoint, and no X in $\mathbf{X}$ is in **Possibly-IV(Y,Z)**. If $P(\mathbf{Y}|\mathbf{Z})$ is not invariant when $\mathbf{X}$ is manipulated by changing the value of $\mathbf{W}$ from $\mathbf{w_1}$ to $\mathbf{w_2}$ then $\mathbf{W}$ is d-connected to $\mathbf{Y}$ given $\mathbf{Z}$ in G. Let W be a member of $\mathbf{W}$ that is d-connected to some Y in $\mathbf{Y}$ given $\mathbf{Z}$ by an undirected path U in G that contains no other member of W.

Because U d-connects W and Y given $\mathbf{Z}$, no noncollider on U is in $\mathbf{Z}$, and every collider on U has a descendant in $\mathbf{Z}$. U contains some X in $\mathbf{X}$. By lemma 3.3.2 $U(X,Y)$ is an undirected path that d-connects X and Y given $\mathbf{Z}$ in G. There is a path $U'(X,Y)$ in

G_{Unman} with the same edges as $U(X,Y)$ in G, because $U(X,Y)$ contains no member of **W**. No noncollider on $U'(X,Y)$ is in **Z**. In G, every collider on $U(X,Y)$ has a descendant in **Z**; hence every collider on $U'(X,Y)$ has a descendant in **Z** in G_{Unman}. Hence $U(X,Y)$ d-connects X and Y given **Z** in G_{Unman}. By lemma 7.3.3 there is a possibly d-connecting path between X and some Y in **Y** given **Z** in π.

Now we will show that X has a descendant in **Y** $\cup$ **Z** in G_{Unman}. If X is a collider on U, then X has a descendant in **Z** in G, and hence in G_{Unman}. Suppose then that X is not a collider on U. The edge from W to X on U is into X, so the edge containing X on $U(X,Y)$ is out of X. If $U(X,Y)$ contains no colliders then Y is a descendant of **X**. If $U(X,Y)$ contains a collider, then the collider on $U(X,Y)$ closest to X is a descendant of X, and an ancestor of a member of **Z**. Hence X is an ancestor of a member of **Z**. In either case, X has a descendant in **Y** $\cup$ **Z** in G, and hence in G_{Unman}.

It follows that X is in **IV(Y,Z)** in G_{Unman}, and hence by lemma 7.3.4 X is in in **Possibly-IV(Y,Z)**, contrary to our assumption. ∴

13.31 Theorem 7.5

THEOREM 7.5: If G is a directed acyclic graph over **V** $\cup$ **W**, **W** is exogenous with respect to **V** in G, G_{Unman} is the subgraph of G over **V**, $P_{Unman(\mathbf{W})}(\mathbf{V}) = P(\mathbf{V}|\mathbf{W} = \mathbf{w_1})$ is faithful to G_{Unman}, and changing the value of **W** from $\mathbf{w_1}$ to $\mathbf{w_2}$ is a direct manipulation of **X** in G, then the Prediction Algorithm is correct.

Proof. Let G_{Man} be the manipulated graph, and F the minimal I-map of $P_{Unman\ (\mathbf{W})}(\mathbf{V})$ constructed by the algorithm for the given ordering of variables *Ord*. Step A) is trivial. Step B) is correct by theorem 6.4. Step C1) is correct by theorem 7.2. In step C2, by lemma 3.3.5, for all values of **V** for which the conditional distributions in the factorization are defined

$$P_{Unman(\mathbf{W})}(\mathbf{Y}|\mathbf{Z}) = \frac{\sum_{\overrightarrow{\mathbf{IV}(\mathbf{Y},\mathbf{Z})}} \prod_{V \in \mathbf{IV}(\mathbf{Y},\mathbf{Z}) \cup \mathbf{IP}(\mathbf{Y},\mathbf{Z}) \cup \mathbf{Y}} P_{Unman(\mathbf{W})}(V|\mathbf{Parents}(F,V))}{\sum_{\overrightarrow{\mathbf{IV}(\mathbf{Y},\mathbf{Z}) \cup \mathbf{Y}}} \prod_{V \in \mathbf{IV}(\mathbf{Y},\mathbf{Z}) \cup \mathbf{IP}(\mathbf{Y},\mathbf{Z}) \cup \mathbf{Y}} P_{Unman(\mathbf{W})}(V|\mathbf{Parents}(F,V))}$$

for all values **z** of **Z** such that $P_{Man}(\mathbf{z}) \neq 0$.

Because G_{Man} is a subgraph of G_{Unman}, if F is an I-map of $P_{Unman\ (\mathbf{W})}(\mathbf{V})$ then F is an I-map of $P_{Man\ (\mathbf{W})}(\mathbf{V})$. Hence $P_{Man\ (\mathbf{W})}(\mathbf{V})$ satisfies the Markov condition for F, and by lemma 3.3.5

(1)

$$P_{Man(\mathbf{W})}(\mathbf{Y}|\mathbf{Z}) = \frac{\sum_{\mathbf{IV}(\mathbf{Y},\mathbf{Z})} \prod_{V \in \mathbf{IV}(\mathbf{Y},\mathbf{Z}) \cup \mathbf{IP}(\mathbf{Y},\mathbf{Z}) \cup \mathbf{Y}} P_{Man(\mathbf{W})}(V|\mathbf{Parents}(F,V))}{\sum_{\mathbf{IV}(\mathbf{Y},\mathbf{Z}) \cup \mathbf{Y}} \prod_{V \in \mathbf{IV}(\mathbf{Y},\mathbf{Z}) \cup \mathbf{IP}(\mathbf{Y},\mathbf{Z}) \cup \mathbf{Y}} P_{Man(\mathbf{W})}(V|\mathbf{Parents}(F,V))}$$

for all values **z** of **Z** such that $P_{Man}(\mathbf{z}) \neq 0$, and for all values for which the conditional distributions in the factorization exist.

$P_{Man(\mathbf{W})}(\mathbf{V})$ satisfies the Markov condition for G_{Man} by hypothesis. Hence in $P_{Man(\mathbf{W})}(\mathbf{V})$ X is independent of its nonparental nondescendants in G_{Man} given **Parents**(G_{Man},X). The predecessors of X in Ord by hypothesis are either in **Definite-Nondescendants**(π,X), in which case they are in **Nondescendants**(G_{Unman},X) or they are in **Parents**(G_{Man},X). G_{Man} is a subgraph of G_{Unman}, so any vertex that is a nondescendant of X in G_{Unman} is a nondescendant of X in G_{Man}. Hence each predecessor of X in Ord is a nondescendant of X in G_{Man}. The algorithm guarantees that **Parents**(G_{Man},X) is included in **Predecessors**(Ord,X). It follows that **Parents**(G_{Man},X) is a subset of **Predecessors**(Ord,X) such that **Predecessors**(Ord,X)**Parents**(G_{Man},X) is independent of X given **Parents**(G_{Man},X) in $P_{Man(\mathbf{W})}(\mathbf{V})$. Hence, if **Parents**$(G_{Man},X)$ is substituted for **Parents**(F,X) in F, the resulting graph is still an I-map of $P_{Man(\mathbf{W})}(\mathbf{V})$, by lemma 3.7.1. So in (1) we can substitute $P(X|\mathbf{Parents}(G_{Man},X))$ for $P(X|\mathbf{Parents}(F,X))$ By assumption the algorithm returns a value only if $P_{Man(\mathbf{W})}(V|\mathbf{Parents}(F,V)) = P_{Unman(\mathbf{W})}(V|\mathbf{Parents}(F,V))$ for each $V \neq X$, so we can substitute $P_{Unman(\mathbf{W})}(V|\mathbf{Parents}(F,V))$ for $P_{Man(\mathbf{W})}(V|\mathbf{Parents}(F,V))$ in (1). ∴

13.32 Theorem 9.1

THEOREM 9.1: If $P(S)$ is faithful to $G(S)$, and **X** and **Y** are sets of variables in $G(S)$ not containing S, then $P(\mathbf{Y}|\mathbf{X}) = P(\mathbf{Y}|\mathbf{X},S)$ if and only if **X** d-separates **Y** and S in $G(S)$.

Proof. This follows from theorem 3.3. ∴

13.33 Theorem 9.2

THEOREM 9.2: For a joint distribution, P, faithful to graph G, exactly one of $<Y \perp\!\!\!\perp X|\mathbf{Z}$; $Y \perp\!\!\!\perp X|\mathbf{Z} \cup \{S\}>$ is true in P if and only if the corresponding member and only that member of $<\mathbf{Z}$ d-separates X, Y; $\mathbf{Z} \cup \{S\}$ d-separates $X, Y>$ is true in G.

Proof. This follows from theorem 3.3. ∴

13.34 Theorem 10.1

THEOREM 10.1: If G is an almost pure latent variable graph over $\mathbf{V} \cup \mathbf{T} \cup \mathbf{C}$, $\mathbf{T}$ is causally sufficient, and each latent variable in $\mathbf{T}$ has at least two measured indicators, then latent variables T_1 and T_3, whose measured indicators include J and L respectively, are d-separated given latent variable T_2, whose measured indicators include I and K, if and only if G linearly implies $\rho_{JI}\rho_{LK} = \rho_{JL}\rho_{KI} = \rho_{JK}\rho_{IL}$.

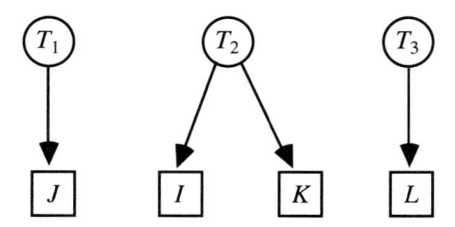

Figure 13.27

We say that a measurement model is **almost pure** if the only kind of impurities among the measured variables are common cause impurities. An **almost pure latent variable graph** is one in which the measurement model is almost pure.

LEMMA 10.1.1: If G' is an almost pure latent variable graph over $\mathbf{V} \cup \mathbf{T} \cup \mathbf{C}$, $\mathbf{T}$ is causally sufficient, and each latent variable in $\mathbf{T}$ has at least two measured indicators, and latent variables T_1 and T_3, whose measured indicators include J and L respectively, are d-separated given latent variable T_2, whose measured indicators include I and K, then G' linearly implies $\rho_{JI}\rho_{LK} = \rho_{JL}\rho_{KI} = \rho_{JK}\rho_{IL}$.

Proof. Let G be a pure latent variable subgraph of G', formed by removing the sources of all treks creating common cause impurities. If T_1 and T_2 are d-separated given T_2 in G' then they are d-separated given T_2 in G. Because I and K are pure indicators of T_2 and G' and thus children only of T_2, T_2 is a noncollider on all undirected paths between I and any other indicator or K and any other indicator. Therefore J and I are d-separated given T_2, K and L are d-separated given T_2, and K and I are d-separated given T_2.

Since T_1 and T_3 are d-separated given T_2, and again J and L are children only of T_1 and T_3 respectively, then J and L are d-separated given T_2. X and Z are d-separated given Y if and and only if G linearly implies $PXZ.T = 0$. Hence G linearly implies $\rho_{IJ.T_2} = 0$, and $\rho_{IJ} = \rho_{IT_2} \times \rho_{JT_2}$. Similarly, G linearly implies $\rho_{KL} = \rho_{KT_2} \times \rho_{LT_2}$, $\rho_{JL} = \rho_{JT_2} \times \rho_{LT_2}$ and $\rho_{IK} = \rho_{IT_2} \times \rho_{KT_2}$. Hence G linearly implies $\rho_{JI}\rho_{LK} = \rho_{JT_2} \times \rho_{IT_2} \times \rho_{LT_2} \times \rho_{KT_2} = \rho_{JT_2} \times \rho_{LT_2} \times \rho_{KT_2} \times \rho_{IT_2} = \rho_{JL}\rho_{KI}$. G linearly implies the same vanishing tetrad differences as G', so G' linearly implies $\rho_{JI}\rho_{LK} = \rho_{JL}\rho_{KI}$. The proof that $\rho_{JL}\rho_{KI} = \rho_{JK}\rho_{IL}$ is linearly implied by G' is essentially the same. ∴

Proofs of Theorems

LEMMA 10.1.2: If G is an almost pure latent variable graph over $\mathbf{V} \cup \mathbf{T} \cup \mathbf{C}$, $\mathbf{T}$ is causally sufficient, and each latent variable in $\mathbf{T}$ has at least two measured indicators, then latent variables T_1 and T_3, whose measure indicators respectively include J and L, are d-separated given latent variable T_2, whose measured indicators include I and K, if G linearly implies $\rho_{JI}\rho_{LK} = \rho_{JL}\rho_{KI}$.

Proof. Suppose that G linearly implies $\rho_{JI}\rho_{LK} = \rho_{JL}\rho_{KI}$ but T_1 and T_3 are not d-separated given T_2.

By the Tetrad Representation Theorem, if G linearly implies $\rho_{JI}\rho_{LK} = \rho_{JL}\rho_{KI}$ then either there is an $IL(T(I,J),T(L,K),T(L,J),T(I,K))$ choke point, or there is a $JK(T(I,J),T(L,K),T(L,J),T(I,K))$ choke point.

Let $T(I,K)$ be the trek consisting of the edges from T_2 to I and T_2 to K. Suppose first that there is an $IL(T(I,J),T(L,K),T(L,J),T(I,K))$ choke point. The choke point is either I or T_2 because those are the only vertices in $I(T(I,K))$. I is not the choke point because it does not lie on any trek between L and K. Hence T_2 is the choke point. Similarly, if there is a $JK(T(I,J),T(L,K),T(L,J),T(I,K))$ choke point it is T_2. Hence, in either case T_2 is a choke point.

There are two ways that T_1 and T_3 might fail to be d-separated given T_2. Either there is a trek between T_1 and T_3 that does not contain T_2, or there is some undirected path U between T_1 and T_3 such that T_2 is a descendent of every collider on U, and T_2 is not a noncollider on U.

First assume that there is some trek between T_1 and T_3 that does not contain T_2. Then there is a trek between J and L that does not contain T_2. But then T_2 is not a choke point, contrary to what we have just proved.

Now assume that there is some undirected path U between T_1 and T_3 such that T_2 is a descendent of every collider on U, and T_2 is not a noncollider on U. In that case U d-connects T_1 and T_3 given T_2. Again there are two cases.

Suppose first that T_2 is an $IL(T(I,J), T(L,K), T(L,J), T(I,K))$ choke point. Let C be the collider on the undirected path U that is closest to T_3. (See figure 13.28.)

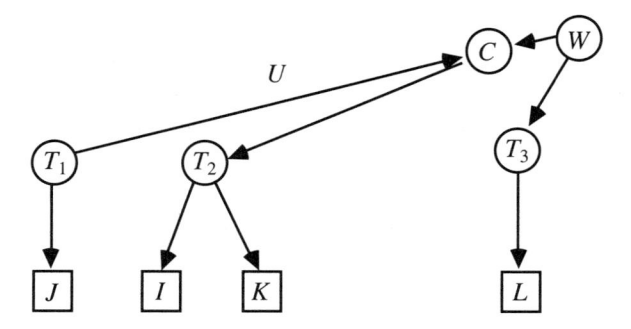

[Figure 13.28]

Figure 13.28

$U(T_3,C)$ does not contain any colliders on U except C because C is the closest collider to T_3 on U; hence $U(T_3,C)$ is a trek between T_3 and C. There is a vertex W on $U(T_3,C)$ that is the source of a trek between T_3 and C. $W \neq C$ because W is not a collider on U, but C is. Hence $U(W, T_3)$ contains no colliders on U. It follows that $U(W, T_3)$ does not contain T_2, because T_2 is not a noncollider on U. Hence there is a trek $T(K,L)$ between K and L whose K branch consists of the concatenation of $U(W,C)$, a directed path from C to T_2, and the edge from T_2 to K, and whose L branch consists of the concatenation of $U(W, T_3)$ and the edge from T_3 to L. Because neither $U(W, T_3)$ nor the edge from T_3 to L contains T_2, T_2 is not in $L(T(K,L))$, and hence is not an $IL(T(I,J),T(L,K), T(L,J),T(I,K))$ choke point, not in $L(T(K,L))$, and hence is not an $IL(T(I,J),T(L,K),T(L,J),T(I,K))$ choke point, contrary to our hypothesis.

such that T_2 is a descendent of every collider on U and T_2 is not a noncollider on U, then there is no $JK(T(I,J), T(L,K), T(L,J), T(I,K))$ choke point.

Therefore T_1 and T_3 are d-separated given T_2. ∴

THEOREM 10.1: If G is an almost pure latent variable graph over $\mathbf{V} \cup \mathbf{T} \cup \mathbf{C}$, $\mathbf{T}$ is causally sufficient, each latent variable in $\mathbf{T}$ has at least two measured indicators, then latent variables T_1 and T_3, whose measured indicators include J and L respectively, are d-separated given latent variable T_2, whose measured indicators include I and K, if and only if G linearly implies $\rho_{JI}\rho_{LK} = \rho_{JL}\rho_{KI} = \rho_{JK}\rho_{IL}$.

Proof. The theorem follows from lemmas 10.1.1 and 10.1.2.

13.35 Theorem 10.2

THEOREM 10.2: If G is an almost pure latent variable graph over $\mathbf{V} \cup \mathbf{T} \cup \mathbf{C}$, $\mathbf{T}$ is causally sufficient, each variable in $\mathbf{T}$ has at least two measured indicators, the input to MIMBuild is a list of all vanishing zero and first order correlations among the latent variables linearly implied by G, and Π is the output of MIMBuild then

Proofs of Theorems

A–1) If X and Y are not adjacent in Π, then they are not adjacent in G.
A–2) If X and Y are adjacent in Π and the edge is not labeled with a "?," then X and Y are adjacent in G.
O–1) If $X \rightarrow Y$ is in Π, then every trek in G between X and Y is into Y.
O–2) If $X \rightarrow Y$ is in Π and the edge between X and Y is not labeled with a "?," then $X \rightarrow Y$ is in G.

LEMMA 10.2.1: If G is an almost pure latent variable graph over $\mathbf{V} \cup \mathbf{T} \cup \mathbf{C}$, $\mathbf{T}$ is causally sufficient, each variable in $\mathbf{T}$ has at least two measured indicators, the input to MIMBuild is a list of all vanishing zero and first order correlations among the latent variables linearly implied by G, Π is the output of MIMBuild, and X and Y are not adjacent in Π, then they are not adjacent in G.

Proof. This follows directly from theorem 3.4.

LEMMA 10.2.2: If G is an almost pure latent variable graph over $\mathbf{V} \cup \mathbf{T} \cup \mathbf{C}$, $\mathbf{T}$ is causally sufficient, each variable in $\mathbf{T}$ has at least two measured indicators, the input to MIMBuild is a list of all vanishing zero and first order correlations among the latent variables linearly implied by G, Π is the output of MIMBuild, and $X \rightarrow Y$ is in Π, then every trek in G between X and Y is into Y.

Proof. Suppose $X \rightarrow Y$ is in Π. The proof is by induction on the number of iterations of the repeat loop in step D) in the PC Algorithm.

Base Case: There is a trek between X and Y in G, because otherwise X and Y are d-separated given the empty set and therefore not adjacent in Π. Suppose that $X \rightarrow Y$ is oriented as $X \rightarrow Y \leftarrow Z$ by step C) of the PC Algorithm (i.e., X and Z are d-separated by some set not containing Y.) If in G, there is a trek between X and Y, and a trek between Y and Z that are not both into Y, then there is a trek between X and Z and hence X and Z are not d-separated given the empty set. Suppose then that X and Z are d-separated by some $W \neq Y$ in G. Because X and Y are adjacent in Π, W does not d-separate X and Y in G. Similarly, W does not d-separate Y and Z. If there is a trek in G between X and Y that is out of Y then there is a directed path U from Y to X in G. If U does not contain W then U d-connects X and Y given W in G. There is also a path V in G that d-connects Y and Z given W. Because U is out of Y, U and V do not collide at Y in G. Hence by lemma 3.3.1 X and Z are d-connected given W in G, contrary to our assumption. If U does contain W, then W is a descendant of Y, and by lemma 3.3.1 X and Z are d-connected given W, contrary to our assumption. Hence no trek in G between X and Y is out of Y.

Induction Case: Suppose after $n-1$ iterations of the repeat loop in step D) of the PC Algorithm, if $Z \rightarrow X$ in Π, then every trek between Z and X in G is into X. Suppose that the $X \to Y$ edge is oriented because there is some vertex Z such that $Z \to X$ - Y in Π and Z is not adjacent to Y in Π. Because the edge between X and Y in Π was not oriented into Y, X and Z are d-separated given Y. There are treks between X and Y, and between Y and Z in G, because they are adjacent in Π. If there is a trek between Y and X that is into X, then by lemma 3.3.1, X and Z are d-connected given Y, contrary to our assumption. ∴

Y is a **definite noncollider** on an undirected path U in pattern Π if and only if either X *-* $Y \to Z$, or $X \leftarrow Y$ *-* Z are subpaths of U, or X and Z are not adjacent and not $X \to Y \leftarrow Z$ on U.

LEMMA 10.2.3: If G is an almost pure latent variable graph over $\mathbf{V} \cup \mathbf{T} \cup \mathbf{C}$, $\mathbf{T}$ is causally sufficient, each variable in $\mathbf{T}$ has at least two measured indicators, the input to MIMBuild is a list of all vanishing zero and first order correlations among the latent variables linearly implied by G, Π is the output of MIMBuild, and Y is a definite noncollider on undirected path U in P, and the corresponding path U' exists in G, then Y is a noncollider on U'.

Proof. If U contains X *-* $Y \to Z$ in Π, then by lemma 10.2.2, if the corresponding path U' exists in G, then the edge between Y and Z in G is out of Y; hence Y is not a collider on U'. Similarly, if $X \leftarrow Y$ *-* Z in Π, then Y is not a collider on U'. Suppose then that X and Z are not adjacent and not $X \to Y \leftarrow Z$ on U in Π. It follows that X and Z are d-separated given Y in G. Hence if the edges between X and Y and between Y and Z exist in G, they do not collide at Y.

LEMMA 10.2.4: If G is an almost pure latent variable graph over $\mathbf{V} \cup \mathbf{T} \cup \mathbf{C}$, $\mathbf{T}$ is causally sufficient, each variable in $\mathbf{T}$ has at least two measured indicators, the input to MIMBuild is a list of all vanishing zero and first order correlations among the latent variables linearly implied by G, Π is the output of MIMBuild, and X - Y or $X \to Y$ is in Π, and the edge is not labeled by a "?," then X and Y are adjacent in G.

Proof. Suppose that X - Y or $X \to Y$ is in Π, the edge is not labeled by a "?," but that X and Y are not adjacent in G. Then there is some set $\mathbf{S}$ that d-separates X and Y in G. Let $\mathbf{P}$ be the set of undirected paths in P between X and Y of length ≥ 2. Any such $\mathbf{S}$ has cardinality ≥ 2, because otherwise MIMBuild would have found it with some test of vanishing zero or first order partial correlations. X - Y or $X \to Y$ was not labeled with a "?" so either (i) $\mathbf{P}$ is empty, or (ii) every path in $\mathbf{P}$ contains a collider, or (iii) there is some vertex Z that is a definite noncollider on every path in $\mathbf{P}$, or (iv) every path in P contains some subpath $<A,B,C>$.

Suppose $\mathbf{P}$ is empty. Because by lemma 10.2.1 nonadjacencies in Π are nonadjacencies in G, the adjacencies in Π are a superset of those in G, and thus the set of undirected paths in Π is a superset of the undirected paths in G. It follows that there is no

Proofs of Theorems

undirected path of length ≥ 2 in G. If in G there is also no edge between X and Y, then X and Y are d-separated given the empty set in G. But since there is an edge between X and Y in Π, X and Y are not d-separated given the empty set in G. Hence there is an edge between X and Y in G.

Suppose every path in **P** contains a collider and there is no edge between X and Y in G. By lemmas 10.2.1 and 10.2.2 every path in G between X and Y contains a collider. Hence there is no trek between X and Y in G. But then there is no edge between X and Y in Π, contrary to our assumption.

Suppose there is some vertex Z that is a definite noncollider on every path in **P**. It follows from lemma 10.2.1, 10.2.2, and 10.2.3 that if there is no edge between X and Y in G, then Z is a noncollider on every undirected path between X and Y in G. Hence X and Y are d-separated by Z. It follows that there is no edge between X and Y in Π, contrary to our assumption.

Suppose every path in **P** contains some subpath <A,B,C>. If there is no edge between X and Y in G, then every undirected path in G between X and Y contains <A,B,C>. It follows that B is either a collider on every path between X and Y in G, in which case X and Y are d-separated given the empty set, or B is a noncollider on every path between X and Y in G, in which case, X and Y are d-separated given B in G. In either case, there is no edge between X and Y in Π, contrary to our assumption. ∴

LEMMA 10.2.5: If G is an almost pure latent variable graph over $\mathbf{V} \cup \mathbf{T} \cup \mathbf{C}$, $\mathbf{T}$ is causally sufficient, each variable in **T** has at least two measured indicators, the input to MIMBuild is a list of all vanishing zero and first order correlations among the latent variables linearly implied by G, Π is the output of MIMBuild, and $X \rightarrow Y$ is in Π, and the edge is not labeled by a "?," then $X \rightarrow Y$ is in G.

Proof. This follows from lemmas 10.2.2 and 10.2.4. ∴

THEOREM 10.2: If G is an almost pure latent variable graph over $\mathbf{V} \cup \mathbf{T} \cup \mathbf{C}$, $\mathbf{T}$ is causally sufficient, each variable in **T** has at least two measured indicators, the input to MIMBuild is a list of all vanishing zero and first order correlations among the latent variables linearly implied by G, and Π is the output of MIMBuild then

A–1) If X and Y are not adjacent in Π, then they are not adjacent in G.
A–2) If X and Y are adjacent in Π and the edge is not labeled with a "?," then X and Y are adjacent in G.
O–1) If $X \rightarrow Y$ is in Π, then every trek in G between X and Y is into Y.
O–2) If $X \rightarrow Y$ is in Π and the edge between X and Y is not labeled with a "?," then $X \rightarrow Y$ is in G.

Proof. This follows from lemmas 10.2.1 through 10.2.5. ∴

13.36 Theorem 11.1

THEOREM 11.1: If G is a subgraph of directed acyclic graph G', than the set of tetrad equations among variables of G that are linearly implied by G' is a subset of those linearly implied by G.

Proof. If G is a subgraph of directed acyclic graph G', then the treks in G are a subset of the treks in G'. Hence if there is a choke point in G', there is a choke point in G. By the Tetrad Representation Theorem, if G' linearly implies that a tetrad difference t vanishes, then G linearly implies t vanishes. ∴

Notes

Chapter 2
1. It is customary to represent the ordered pair A, B with angle brackets as <A, B>, but for endpoints of an edge we use square brackets so that the angle brackets will not be misread as arrowheads.
2. Some writers, especially in statistics, understand "clique" as we have defined maximal clique.
3. We do not include trivial independence relations, for example, $C \perp\!\!\!\perp \varnothing \mid \varnothing$ which are true by definition.

Chapter 3
1. Strictly, we require for causal sufficiency of **V** for a population that if X is not in **V** and is a common cause of two or more variables in **V**, that the joint probability of all variables in **V** be the same on each value of X that occurs in the population.
2. Using the notion of identifiability, Simon (1953) proposed a general means to derive causal structure from a set of equations describing a system; later in the same paper Simon also proposed an account of causation using invariances under perturbations of linear coefficients.
3. Since causation for variables is assumed to be transitive and irreflexive, the directed graph representing a causal structure must be acyclic. Introducing cyclic directed graphs requires a systematic reinterpretation.
4. A better practical arrangement might be a query system that, besides inferring the causal graph or graphs, responds to the user's questions about the effects of the manipulation of variables.
5. P. 319. Q is Yule's $Q = (ad - bc)/(ad+bc)$ when the first row is a,b and the second c,d in a 2×2 table.
6. (Sic.) Kendall means, of course, that the symbols denote the respective treatment and recovery states, not vice-versa.
7. Fienberg (1977), citing Darroch, attributes the issue to Yule "since Yule discussed it in the final section of his 1903 paper on the theory of association of attributes" (p. 51). But save for the first sentence of that section, Yule actually discusses the reverse issue of mixtures, namely circumstances in which variables are statistically dependent in a population but independent in sub-populations.
8. The subsequent literature has confused it with a number of other questions about how independence and dependence relations in a population may be related to independence and dependence relations in sub-populations, and the causal significance of such facts. The unfortunate aspect of collapsing these questions is that they have distinct answers. A circumstance attributed to Simpson and now often called "Simpson's paradox," but nonetheless distinct from the question Simpson actually posed, was described by Colin Blyth (1972):

It is possible to have simultaneously

(1) $P(A|B) < P(A|B')$

and

(2) $P(A|BC) \geq P(A|B'C)$
(3) $P(A|BC') \geq P(A|B'C')$

In fact, Simpson has equality in (1) and > in (2) and (3).
9. The point is implicit in Blalock 1961 and no doubt other sources as well.
10. Treks were defined in section 2.3.1. A trek between X an Y is either i) a directed path from X to Y, ii) a directed path from Y to X, or iii) a pair of directed paths from Z to X and Z to Y that have only Z in common.
11. We thank Marek Druzdel for suggesting this example, and pointing out the problem of reversible mechanisms to us.

Chapter 5

1. In particular, when the method is idealized to give up the greedy algorithm. Because of the greedy algorithm, we would expect the specific search procedure to be asymptotically unreliable when there are two or more treks between a pair of nonadjacent variables, say X and Y, that result in a close statistical association between those variables. This is the circumstance in the case of the one edge the procedure erroneously introduces in the ALARM network. In practice, such structures may be sufficiently uncommon for the error to be tolerable and Cooper and his colleagues are investigating techniques to ameliorate the problem.
2. Indeed, any statistical constraint can be used as input for the algorithms for any pairing of distributions with graphs such that the constraint is satisfied in the distribution if and only if the corresponding d-separation relation holds in the graph.
3. In the following heuristics, "high probabilistic dependence" means high partial correlation in the linear case, and high G^2 statistic in the discrete case.
4. For causally sufficient structures, if a distribution P, obtained by imposing a linear distribution compatible with a graph G, implies some vanishing partial correlation not linearly implied by G, is then P not faithful to G? If P is not faithful to G, does P necessarily imply some vanishing partial correlation not linearly implied by G? We don't know the answer to either question.
5. An exact general rule for calculating the reduction of degrees of freedom given cells with zero entries seems not to be known. See Bishop, Fienberg, and Holland 1975.
6. It is not clear from the article how the correlations of the latent variables, *GPQ* and *ABILITY,* with other variables such as publishing productivity and *QFJ* were obtained. They can be obtained, for example, by using the factor structure as a regression model to

calculate estimated factor scores for each subject, or by including the covariances of the latents among the free parameters in a set of structural equations and letting a program such as LISREL estimate their values. In general the results of these procedures will be different.

7. The small differences are presumably attributable to round-off errors.

8. We do not know whether this method of graph generation produces "realistic" graphs. One feature of some of the graphs generated in this fashion that may not be desirable is the existence of isolated variables. An informal examination showed topologies not unlike the Alarm network.

Chapter 6

1. We thank Thomas Verma (personal communication) for pointing out an error in the original formulation of the CI algorithm.

2. N.B. "P_1," "M," "R," in this line do **not** refer to vertices on the definite discriminating path U.

Chapter 7

1. This section is based on Spirtes, Glymour, Scheines, Meek, Fienberg, and Slate 1992.

Chapter 8

1. In linear regression, we understand the "direct influence" of X_i on Y to mean (i) the change in value of a variable Y that would be produced in each member of a population by a unit change in X_i, with all other X variables forced to be unchanged. Other meanings might be given, for example: (ii) the population average change in Y for unit change in X_i, with all other X variables forced to be unchanged; (iii) the change in Y in each member of the population for unit change in X_i; (iv) the population average change in Y for unit change in X_i; etc. Under interpretations (iii) and (iv) the regression coefficient is an unreliable estimate whenever X_i also influences other regressors that influence Y. Interpretation (ii) is equivalent to (i) if the units are homogeneous and the stochastic properties are due to sampling; otherwise, regression will be unreliable under interpretation (i) except in special cases, for example, when the linear coefficients, as random variables, are independently distributed (in which case the analysis given here still applies [Glymour, Spirtes, and Scheines 1991a]).

2. In fact, we were inadvertently misinformed that all seven tests are components of AFQT and we first discovered otherwise with the SGS algorithm.

3. The correlation matrix given in Rawlings 1988 incorrectly gives the correlation between *CU* and *NH4* as 0.93.

4. The "maximum R-square" and "stepwise" options in PROC REG in the SAS program.

5. Although the definition of the population in this case is unclear, and must in any case be drawn quite narrowly.

6. More exactly, at .05, with the exception of *MG* the partial correlation of every regressor with *BIO* vanishes when some set containing *PH* is controlled for; the correlation of *MG* with *BIO* vanishes when *CA* is controlled for.

7. Searches at lower significance levels remove the adjacency between *FI* and *EN*.

Chapter 9

1. Personal communication.

2. We thank Jay Kadane for pointing out that the causal relationship between *Preference* and other variables might be different in the experimental and non-experimental populations, even if *Preference* is not directly manipulated.

Chapter 11

1. This chapter is an abbreviated version of Spirtes, Scheines, and Glymour 1990, and is reprinted with the permission of Sage Publications.

2. The original TETRAD program (Glymour, Scheines, Spirtes, and Kelly 1987) had no such scoring function. It was left to the user to balance the Explanatory and Falsification principles.

3. We have also implemented heuristic search procedures that are theoretically less reliable than that described here but are much faster and in practice about equally reliable.

4. LISREL VII retains the same architecture but with an altered modification index.

5. LISREL VI outputs a number of other measures that could be used to suggest modifications to a starting model, but these are not used in the automatic search. See Costner and Herting 1985.

6. As long as they are not in the list of parameters not to be freed.

7. Since the Lagrange Multiplier statistic, like the modification indices of LISREL VI, estimates the effect on the χ^2 of freeing a parameter, in subsequent sections we will use the term "modification index" to refer to either of these statistics.

8. EQS allows the user to specify several different types of searches. We have only described the one used in our Monte Carlo simulation tests.

9. We are indebted to Peter Bentler for suggesting this transformation.

10. We did not provide LISREL or EQS with the values of the parameters in the original models that generated our covariance matrices because the input to LISREL and EQS was a pseudocorrelation matrix, not the original covariance matrix. We therefore provided the programs with the population parameters of transformed models that would generate the pseudo correlation matrices. The detailed transformations are given in Spirtes (1990).

11. For LISREL IV, the details of this procedure are described in Glymour et al. 1987. The same procedure works for LISREL VI with the exception of the Beta matrix. See Joreskog and Sorbom (1984).

12. To simplify the calculations, we assumed that the length of the lists output by TETRAD II for all of the covariance matrices generated by a single model was in each case equal to the average length of the lists. This is a fairly good approximation in most cases.

13. The expression "X C Y" means that the error terms for X and Y are correlated, or, equivalently, that there is an additional, common cause of X and Y.

14. TETRAD II will, on request, automatically generate EQS input files for all models that it suggests.

Chapter 12

1. Lauritzen's proposal was given at a lecture at the Santa Fe Institute in 1997. At this writing, Lauritzen and Richardson are working on the details of the required parameterization. We thank Thomas Richardson for very helpful discussions.

2. We wish to thank Larry Wasserman, Teddy Seidenfeld, and Jamie Robins for many valuable conversations on the issue of consistency, although this does not imply that they endorse any of our conclusions.

3. We are grateful to David Heckerman, Greg Cooper, and Christopher Meek for permission to use their article.

4. In sections 12.5.1 through 12.5.6, "we" refers to Heckerman, Meek, and Cooper.

5. Bernardo and Smith (1994) provide a summary of likelihoods from the exponential family and their conjugate priors.

6. Discussions of equivalent sample size can be found in Winkler 1967 and Heckerman et al. 1995.

7. The algorithm assumes that there are no hidden variables. See section 12.5.5 for a discussion of hidden-variable models and methods for learning them. A modification of the PC algorithm has been implemented in Pronel, which can be used in conjunction with Hugin, a package for updating Bayesian networks and helping users construct Bayesian networks. BIFROST (Hojsgaard and Thieson 1995) constructs block recursive models which can also be used in conjunction with Hugin. See *http://www.hugin.dk*.

8. One of the technical assumptions used to derive this approximation is that the prior is bounded and bounded away from zero around $\hat{\theta}_m$.

9. The MAP configuration $\bar{\theta}_m$ depends on the coordinate system in which the parameter variables are expressed. The MAP given here corresponds to the canonical coordinate system for the multinomial distribution (see, for example, Bernardo and Smith 1994, pp. 199–202.)

10. In particular, Heckerman (1995) showed that strong likelihood equivalence is not consistent with parameter independence and parameter modularity.

11. We say as "typically employed" because the FCI and PC algorithms take a significance level as a parameter. We will assume that for samples of size between 100 and 10000 the significance levels are in the range 0.001 to 0.1.

12. We wish to thank Larry Wasserman for valuable discussions regarding the priors and the resulting posteriors, although the conclusions about the plausibility of various priors are our own.

13. For the purposes of comparing the prior suggested by Robins and Wasserman with some common alternatives, we have changed some minor details about how DAGs are counted; the results are essentially the same, however.

14. Needleman's regression had 6 independent variables and an R^2 of .271. Ours has 3 independent variables with an R^2 of .243.

15. Personal communication.

Glossary

A: In a graph G, Let $\mathbf{A}(A,B)$ be the union of the ancestors of A or B.

Acceptable: Let a total order Ord of variables in a graph G' be **acceptable** for G if and only if whenever $A \neq B$ and there is a directed path from A to B in G', A precedes B in Ord.

After: In a graph G, vertex X is **after** vertex Y if and only if there is a directed path from Y to X in G.

Almost Pure: We say that a measurement model is **almost pure** if the only kind of impurities among the measured variables are common cause impurities. An **almost pure latent variable graph** is one in which the measurement model is almost pure.

Before: In a graph G, vertex X is **before** vertex Y if and only if there is a directed path from X to Y in G.

C.F: See constant factor.

Choke point: In a directed acyclic graph G, if for all $T(K,L)$ in $\mathbf{T}(K,L)$ and all $T(I,J)$ in $\mathbf{T}(I,J)$, $L(T(K,L))$ and $J(T(I,J))$ intersect at a vertex Q, then Q is an $LJ(T(I,J),T(K,L))$ **choke point**. Similarly, if for all $T(K,L)$ in $\mathbf{T}(K,L)$ and all $T(I,J)$ in $\mathbf{T}(I,J)$, $L(T(K,L))$ and all $J(T(I,J))$ intersect at a vertex Q, and for all $T(I,L)$ in $\mathbf{T}(I,L)$ and all $T(J,K)$ in $\mathbf{T}(J,K)$, $L(T(I,L))$ and $J(T(J,K))$ also intersect at Q, then Q is an $LJ(T(I,J),T(K,L),T(I,L),T(J,K))$ **choke point**. Also see the definition of trek.

Combined graph: See manipulation.

Constant factor: In an LCF or LCT T, if an expression is equal to ce, where c is a nonzero constant, and e is a product of equation coefficients raised to positive integral powers, then c is the **constant factor** (c.f.) of ce.

Contains: In a directed acyclic graph, directed paths $R(U,I)$ and $R(U,J)$ **contain trek** T iff $I(T(I,J))$ is a final segment of $R(U,I)$ and $J(T(I,J))$ is a final segment of $R(U,J)$.

D: Given a directed acyclic graph G, $\mathbf{D}(X_i,X_j)$ is the set of all directed paths from X_i to X_j.

D-connection: See D-separation.

Definite discriminating path: In a partially oriented inducing path graph π, U is a **definite discriminating path** for B if and only if U is an undirected path between X and Y containing B, $B \neq X$, $B \neq Y$, every vertex on U except for B and the endpoints is a collider or a definite noncollider on U, and

(i) if V and V' are adjacent on U, and V' is between V and B on U, then $V *\!\!\rightarrow V'$ on U,
(ii) if V is between X and B on U and V is a collider on U then $V \rightarrow Y$ in π, else $V \leftarrow\!* Y$ in π,
(iii) if V is between Y and B on U and V is a collider on U then $V \rightarrow X$ in π, else $V \leftarrow\!* X$ in π,
(iv) X and Y are not adjacent in π.

Definite noncollider: A vertex B is a **definite noncollider** on undirected path U if and only if either B is an endpoint of U, or there exist vertices A and C such that U contains one of the subpaths $A \leftarrow B *\!\!-\!* C$, $A *\!\!-\!* B \rightarrow C$, or $A *\!\!-\!* \underline{B *\!\!-\!*} C$.

Definite nondescendant: If π is the FCI partially oriented inducing path graph of G over **O**, then X is in **Definite-Nondescendants(Y)** if and only if there is no semidirected path from any member of **Y** to X in π.

Definite-SP: For a partially oriented inducing path graph π over **O** and ordering Ord acceptable for π, V is in **Definite-SP**(Ord,X) if and only if $V \neq X$ and there is an undirected path U in π between V and X such that every vertex on U except for X is a predecessor of X in Ord, and every vertex on U except for the endpoints is a collider on U.

Dependent: In an LCT or LCF S, a variable X_i is **dependent** iff X_i does not have zero indegree.

Det: **Det(Z)** is the set of variables determined by any subset of **Z**.

Determines: A set of variables **Z** **determines** the set of variables **A**, when every variable in **A** is a deterministic function of the variables in **Z**, and not every variable in **A** is a deterministic function of any proper subset of **Z**.

Det-connected: See Det-separation.

Det-separated: If G is a directed acyclic graph over **V**, **Z** is a subset of **V** that does not contain X or Y, and $X \neq Y$, then X and Y are **det-separated** given **Z** and **Deterministic(V)** if and only if either X and Y are d-separated given $\mathbf{Z} \cup \mathbf{Det(Z)}$ in some **Mod**(G) relative to **Deterministic(V)** and **Z**, or X or Y is in **Det(Z)**; otherwise if $X \neq Y$ and X and Y are not in **Z**, then *X* and *Y* are **det-connected** given **Z** and **Deterministic**(**V**). If **X**, **Y** and **Z** are disjoint sets of variables in **V**, and **X** and **Y** are non-empty, then **X** and **Y** are **det-separated** given **Z** if and only if every member *X* of **X** and every member *Y* of **Y** are det-separated given **Z**; otherise if **X**, **Y** and **Z** are disjoint sets of variables in **V**, and **X** and **Y** are non-empty, then **X** and **Y** are **det-connected** given **Z** and **Deterministic**(**V**).

Discriminating path: In an inducing path graph G', U is a **discriminating path** for B if and only if U is an undirected path between X and Y containing B, $B \neq X$, $B \neq Y$, and

(i) if V and V' are adjacent on U, and V' is between V and B on U, then $V \mathbin{*}{\rightarrow} V'$ on U,
(ii) if V is between X and B on U and V is a collider on U then $V \rightarrow Y$ in G', else $V \leftarrow^* Y$ in G',
(iii) if V is between Y and B on U and V is a collider on U then $V \rightarrow X$ in G', else $V \leftarrow^* X$ in G',
(iv) X and Y are not adjacent in G'.

Distributed form: The **distributed form** of an expression or equation E is the result of carrying out every multiplication, but no additions, subtractions, or divisions in E. If there are no divisions in an equation then its distributed form is a sum of terms. For example, the distributed form of the equation $u = (a + b)(c + d)v$ is $u = acv + adv + bcv + bdv$.

D-map: An acyclic graph G over **V** is a **D-map** of probability distribution $P(\mathbf{V})$ iff for every **X**, **Y**, and **Z** that are disjoint sets of random variables in **V**, if **X** is not d-separated from **Y** given **Z** in G then **X** is not independent of **Y** given **Z** in $P(\mathbf{V})$. However, when D-map is applied to the graph in an LCT, the quantifiers in the definitions apply only to sets of *non-error* variables.

D-Sep: If G' is an inducing path graph over **O** and $A \neq B$, let $V \in$ **D-SEP**(*A,B*) if and only if $A \neq V$ and there is an undirected path U between A and V such that every vertex on U is an ancestor of A or B, and (except for the endpoints) is a collider on U.

D-separated: If G is a directed acyclic graph with vertex set **V**, **Z** is a set of vertices not containing X or Y, $X \neq Y$, and X and Y are not in **Z**, then X and Y are **D-separated** given **Z** and **Deterministic**(**V**) if and only if there is no undirected path U in G between X and Y such that each collider on U has a descendant in **Z**, and no other vertex on U is in **Det**(Z); otherwise if $X \neq Y$ and X and Y are not in **Z**, then X and Y are **D-connected** given **Z** and **Deterministic**(**V**). Similarly, if **X**, **Y**, and **Z** are disjoint sets of variables, and **X** and **Y** are non-empty, then **X** and **Y** are D-separated given **Z** and **Deterministic**(**V**) if and only if each pair <*X,Y*> in the Cartesian product of **X** and **Y** are **D-separated** given **Z** and **Deterministic**(**V**); otherwise if **X**, **Y**, and **Z** are disjoint, and **X** and **Y** are non-empty, then **X** and **Y** are **D-connected** given **Z** and **Deterministic**(**V**).(Note that this is different from d-separation, which begins with a lowercase "d," and d-connection, which also begins with a lowercase "d.")

e: In an LCF F, **e(S)** is equal to S if S is an independent variable, and it is equal to the error variable into S if S is not an independent variable.

E: If X is a random variable, $E(X)$ is the expected value of X.

Equiv(G'): If G' is an inducing path graph over **O**, **Equiv(G')** is the set of inducing path graphs over the same vertices with the same d-connections as G.

E.C.F: See equation coefficient factor.

Equation coefficient: See linear causal theory, linear causal form.

Equation coefficient factor: In an LCF or LCT T, if an expression is equal to ce, where c is a nonzero constant, and e is a product of equation coefficients raised to positive integral powers, then e is the **equation coefficient factor**(e.c.f.) of ce.

Equivalent to a polynomial: In an LCF, a quantity (e.g., a covariance) X **is equivalent to a polynomial in the coefficients and variances of exogenous variables** if and only if for each LCF F = <<R,M,E>, **C**, **V**, EQ,L,Err> and in every LCT S = <<R',M',E'>, (Ω,f,P), EQ',L',Err'> that is an instance of F, there is a polynomial in the variables in **C** and **V** such that X is equal to the result of substituting the linear coefficients of S in as values for the corresponding variables in **C**, and the variances of the exogenous variables in S as values for the corresponding variables in **V**.

Error variable: See linear causal theory, linear causal form.

Exogenous: If G is a directed acyclic graph over a set of variables $\mathbf{V} \cup \mathbf{W}$, and $\mathbf{V} \cap \mathbf{W} = \varnothing$, then **W is exogenous with respect to V** in G if and only if there is no directed edge from any member of **V** to any member of **W**.

Faithfully indistinguishable: We will say that two directed acyclic graphs, G, G' are **faithfully indistinguishable** (f.i.) if and only if every distribution faithful to G is faithful to G' and vice-versa.

F.I.: See faithfully indistinguishable.

Final segment: In a graph G, a path U of length n is a **final segment** of path V of length m iff $m \geq n$, and for $1 \leq i \leq n+1$, the ith vertex of V equals the $(m\text{-}n\text{+}i)^{th}$ vertex of U.

I-Map: An acyclic directed graph G over **V** is an **I-map** of probability distribution $P(\mathbf{V})$ iff for every **X**, **Y**, and **Z** that are disjoint sets of random variables in **V**, if **X** is d-separated from **Y** given **Z** in G then **X** is independent of **Y** given **Z** in $P(\mathbf{V})$. However, when I-map is applied to the graph in an LCT, the quantifiers in the definitions apply only to sets of *non-error* variables.

Ind: For a directed acyclic graph G, **Ind** is the set of independent variables in G.

$^{Ind}a_{IJ}$: $^{Ind}a_{IJ}$ is the coefficient of J in the independent equational for I. See also independent equational.

Independent: In an LCT or LCF S, a variable X_i is **independent** iff X_i has zero indegree (i.e., there are no edges directed into it). Note that the *property* of independence is completely distinct from the *relation* of statistical independence. The context will make clear in which of these senses the term is used.

Independent equational: In an LCF <<**R,M,E**>, **C**, **V**, **EQ,L,S**> an equation is an **independent equational for a dependent variable** X_j if and only if it is implied by **EQ** and the variables in **R** which appear on the r.h.s. are independent and occur at most once.

Inducing path: If G is a directed acyclic graph over a set of variables **V**, **O** is a subset of **V** containing A and B, and $A \neq B$, then an undirected path U between A and B is an **inducing path relative to O** if and only if every member of **O** on U except for the endpoints is a collider on U, and every collider on U is an ancestor of either A or B. We will sometimes refer to members of **O** as **observed** variables.

Inducing path graph: G' is an **inducing path graph over O for directed acyclic graph** G if and only if **O** is a subset of the vertices in G, there is an edge between variables A and B with an arrowhead at A if and only if A and B are in **O**, and there is an inducing path in G between A and B relative to **O** that is into A. (Using the notation of chapter 2, the set of marks in an inducing path graph is {>, EM}.)

Initial segment: In a graph G, a path U of length n is an **initial segment** of path V of length m iff $m \geq n$, and for $1 \leq i \leq n+1$, the ith vertex of V equals the ith vertex of U.

Into: In a graph G, an edge between A and B is into A if and only if the mark at the A end of the edge is an ">." If an undirected path U between A and B contains an edge into A we will say that U is **into** A.

Invariant: If G is a directed acyclic graph over a set of variables $\mathbf{V} \cup \mathbf{W}$, **W** is exogenous with respect to **V** in G, **Y** and **Z** are disjoint subsets of **V**, $P(\mathbf{V} \cup \mathbf{W})$ is a distribution that satisfies the Markov condition for G, and **Manipulated(W) = X**, then $P(Y|Z)$ is **invariant** under direct manipulation of **X** in G by changing **W** from w_1 to w_2 if and only if $P(Y|Z, W = w_1) = P(Y|Z, W = w_2)$ wherever they are both defined.

Instance: An LCT S is an **instance** of an LCF F if and only if the graph of S is isomorphic to the graph of F.

IP: In a directed acyclic graph G, if $Y \cap Z = \emptyset$, W is in **IP(Y,Z)** (W has a parent that is an informative variable for **Y** given **Z**) if and only if W is a member of **Z**, and W has a parent in **IV(Y,Z)** $\cup$ **Y**.

IV: In a directed acyclic graph G, if $Y \cap Z = \emptyset$, then V is in **IV(Y,Z)** (informative variables for **Y** given **Z**) if and only if V is d-connected to **Y** given **Z**, and V is not in **ND(YZ)**. (This entails that V is not in $Y \cup Z$.)

Label: See linear causal theory, linear causal form.

Length: In a graph G, the **length** of a path equals the number of vertices in the path minus one.

Last point of intersection: In a directed acyclic graph G, the **last point of intersection** of directed path $R(U,I)$ with directed path $R(V,J)$ is the last vertex on $R(U,I)$ that is also on $R(V,J)$. Note that if G is a directed acyclic graph, the last point of intersection of directed path $R(U,I)$ with directed path $R(V,J)$ equals the last point of intersection of $R(V,J)$ with $R(U,I)$; this is not true of directed cyclic paths.

LCF: See linear causal form.

LCT: See linear causal theory.

Linear causal form: A **linear causal form** is an unestimated LCT in which the linear coefficients and the variances of the exogenous variables are real variables instead of constants. This entails that an edge label in an LCF is a real variable instead of a constant (except that the label of an edge from an error variable is fixed at one.) More formally, let a linear causal form (**LCF**) be <<**R,M,E**>, **C**, **V**, **EQ,L,Err**> where

(i) <**R,M,E**> is a directed acyclic graph. **Err** is a subset of **R** called the **error variables**. Each error variable is of indegree 0 and outdegree 1. For every X_i in **R** of indegree $\neq 0$ there is exactly one error variable with an edge into X_i.

Glossary 487

(ii) c_{ij} is a unique real variable associated with an edge from X_j to X_i, and **C** is the set of c_{ij}. **V** is the set of variables σ_i^2, where X_i is an exogenous variable in <**R,M,E**> and σ_i^2 is a variable that ranges over the positive real numbers.

(iii) L is a function with domain E such that for each e in E, $L(e) = c_{ij}$ iff $head(e) = X_i$ and $tail(e) = X_j$. $L(e)$ will be called the **label** of e. By extension, the product of labels of edges in any acyclic undirected path U will be denoted by $L(U)$, and $L(U)$ will be called the **label** of U. The label of an empty path is fixed at 1.

(iv) **EQ** is a consistent set of independent homogeneous linear equationals in variables in **R**. For each X_i in **R** of positive indegree there is an equation in **EQ** of the form

$$X_i = \sum_{X_j \in \mathbf{Parents}(X_i)} c_{ij} \, X_j$$

where each c_{ij} is a real variable in **C** and each X_i is in **R**. There are no other equations in **EQ**. c_{ij} is the **equation coefficient** of X_j in the equation for X_i.

Linear causal theory: Let a **linear causal theory** be (**LCT**) be <<**R,M,E**>, (Ω, f, P), **EQ,L,Err**> where

(i) (Ω, f, P) is a probability space, where Ω is the sample space, f is a sigma-field over Ω, and P is a probability distribution over f.

(ii) <**R,M,E**> is a directed acyclic graph. **R** is a set of random variables over (Ω, f, P).

(iii) The variables in **R** have a joint distribution. Every variable in **R** has a nonzero variance. **E** is a set of directed edges between variables in **R**. (**M** is the set of marks that occur in a directed graph, that is, {EM, >}.

(iv) **EQ** is a consistent set of independent homogeneous linear equations in random variables in **R**. For each X_i in **R** of positive indegree there is an equation in **EQ** of the form

$$X_i = \sum_{X_j \in \mathbf{Parents}(X_i)} a_{ij} \, X_j$$

where each a_{ij} is a nonzero real number and each X_i is in **R**. This implies that each vertex X_i in **R** of positive indegree can be expressed as a linear function of all and only its parents. There are no other equations in **EQ**. A nonzero value of a_{ij} is the **equation coefficient** of X_j in the equation for X_i.

(v) If vertices (random variables) X_i and X_j are exogenous, then X_i and X_j are pairwise statistically independent.

(vi) L is a function with domain E such that for each e in E, $L(e) = a_{ij}$ iff $head(e) = X_i$ and $tail(e) = X_j$. $L(e)$ will be called the **label** of e. By extension, the product of labels of edges in any acyclic undirected path U will be denoted by $L(U)$, and $L(U)$ will be called the **label** of U. The label of an empty path is fixed at 1.

(vii) There is a subset of **S** of **R** called the **error variables**, each of indegree 0 and outdegree 1. Note that the variance of any endogenous variable I conditional on any set of variables that does not contain the error variable of I is not equal to zero.

Linear Representation: A directed acyclic graph G over **V linearly represents** a distribution $P(\mathbf{V})$ if and only if there exists a a directed acyclic graph G' over $\mathbf{V}'$ and a distribution $P''(\mathbf{V}')$ such that

(i) **V** is included in $\mathbf{V}'$;

(ii) for each endogenous (that is, with positive indegree) variable X in **V**, there is a unique variable ε_X in $\mathbf{V}'\backslash\mathbf{V}$ with zero indegree, positive variance, outdegree equal to one, and a directed edge from ε_X to X;

(iii) G is the subgraph of G' over **V**;

(iv) each endogenous variable in G is a linear function of its parents in G';

(v) in $P''(\mathbf{V}')$ the correlation between any two exogenous variables in G' is zero;

(vi) $P(\mathbf{V})$ is the marginal of $P''(\mathbf{V}')$ over **V**.

The members of $\mathbf{V}'\backslash\mathbf{V}$ are called **error variables** and we call G' the **expanded graph**.

Linearly implies: A directed acyclic graph G **linearly implies** $\rho_{AB.\mathbf{H}} = 0$ if and only if $\rho_{AB.\mathbf{H}} = 0$ in all distributions linearly represented by G. (We assume all partial correlations are defined for the distribution.)

Manipulate: See manipulation.

Manipulated graph: See manipulation.

Manipulation: If G is a directed acyclic graph over a set of variables $\mathbf{V} \cup \mathbf{W}$, and $\mathbf{V} \cap \mathbf{W} = \emptyset$, then **W is exogenous with respect to V** in G if and only if there is no directed edge from any member of **V** to any member of **W**. If G_{Comb} is a directed acyclic graph over a set of variables $\mathbf{V} \cup \mathbf{W}$, and $P(\mathbf{V} \cup \mathbf{W})$ satisfies the Markov condition for G_{Comb}, then changing the value of **W** from $\mathbf{w}_1$ to $\mathbf{w}_2$ is a **manipulation** of G_{Comb} with respect to **V** if and only if **W** is exogenous with respect to **V**, and $P(\mathbf{V}|\mathbf{W} = \mathbf{w}_1) \neq P(\mathbf{V}|\mathbf{W} = \mathbf{w}_2)$. We define $P_{Unman(\mathbf{W})}(\mathbf{V}) = P(\mathbf{V}|\mathbf{W} = \mathbf{w}_1)$, and $P_{Man(\mathbf{W})}(\mathbf{V}) = P(\mathbf{V}|\mathbf{W} = \mathbf{w}_2)$, and similarly for various marginal and conditional distributions formed from $P(\mathbf{V})$. We refer to G_{Comb} as the **combined graph**, and the subgraph of G_{Comb} over **V** as the **unmanipulated graph** G_{Unman}. V is in **Manipulated(W)** (that is, V is a variable directly influenced by one of the manipulation variables) if and only if V is in **Children(W)** $\cap$ **V**; we will also say that the variables in **Manipulated(W)** have been **directly manipulated**. We will refer to the variables in **W** as **policy variables**. The **manipulated graph**, G_{Man} is a subgraph of

G_{Unman} for which $P_{Man(W)}(\mathbf{V})$ satisfies the Markov Condition and which differs from G_{Unman} in at most the parents of members of **Manipulated(W)**.

Minimal I-map: An acyclic graph G **is a minimal I-map** of probability distribution P iff G is an I-map of P, and no subgraph of G is an I-map of P. However, when minimal I-map is applied to the graph in an LCT, the quantifiers in the definitions apply only to sets of *non-error* variables.

Mod: If G is a directed acyclic graph over $\mathbf{V}$, and $\mathbf{Z}$ is included in $\mathbf{V}$, then G' is in **Mod**(G) relative to **Deterministic(V)** and $\mathbf{Z}$ if and only if for each V in $\mathbf{V}$

(i) if there exists a set of vertices included in $\mathbf{Z}$ that are nondescendants of V in G and that determine V, then **Parents**$(G',V) = \mathbf{X}$, where $\mathbf{X}$ is some set of vertices included in $\mathbf{Z}$ that are nondescendants of V in G and that determine V;
(ii) if there is no set $\mathbf{X}$ of vertices included in $\mathbf{Z}$ that are nondescendants of V in G and that determine V, then **Parents**$(G',V) = $ **Parents**(G,V).

ND: In a directed acyclic graph G, **ND(Y)** is the set of all vertices that do not have a descendant in $\mathbf{Y}$.

Nondescendants: In a directed acyclic graph G, X is in **Nondescendants(Y)** if and only if there is no directed path from any member of $\mathbf{Y}$ to X in G.

Observed: See inducing path graph, inducing path.

Out of: In a graph G, an edge between A and B is out of A if and only if the mark at the A endpoint is the empty mark. If an undirected path U between A and B contains an edge out of A we will say that U is **out of** A.

Parallel embedding: Directed acyclic graphs G_1 and G_2 with common vertex set $\mathbf{O}$ have a **parallel embedding** in directed acyclic graphs H_1 and H_2 having a common set $\mathbf{U}$ of vertices that includes $\mathbf{O}$ if and only if

(i) G_1 is the subgraph of H_1 over $\mathbf{O}$ and G_2 is the subgraph of H_2 over $\mathbf{O}$;.
(ii) every directed edge in H_1 but not in G_1 is in H_2 and every directed edge in H_2 but not in G_2 is in H_1.

Path form: If G is a directed acyclic graph, let Let $\mathbf{P}_{XY}$ be the set of all directed paths in G from X to Y. In an LCF S, the **path form of a product of covariances** $\gamma_{IJ}\gamma_{KL}$ is the distributed form of

$$\left(\sum_{U \in \mathbf{U}_{IJ}} \left(\sum_{R \in \mathbf{P}_{UI}} \sum_{R' \in \mathbf{P}_{UJ}} L(R)L(R')\sigma_U^2\right)\right) \left(\sum_{V \in \mathbf{U}_{KL}} \left(\sum_{R'' \in \mathbf{P}_{VK}} \sum_{R''' \in \mathbf{P}_{VL}} L(R'')L(R''')\sigma_V^2\right)\right)$$

$\gamma_{IJ}\gamma_{KL} - \gamma_{IL}\gamma_{JK}$ is in **path form** iff both terms are in path form. $\gamma_{IJ}\gamma_{KL} - \gamma_{IL}\gamma_{JK}$ is in **path form** iff both terms are in path form.

Policy variables: See manipulate.

Possible-D-SEP(A,B): If $A \neq B$ in partially oriented inducing path graph π, V is in **Possible-D-Sep**(A,B) in π if and only if $V \neq A$, and there is an undirected path U between A and V in π such that for every subpath $<X,Y,Z>$ of U either Y is a collider on the subpath, or Y is not a definite noncollider on U, and X, Y, and Z form a triangle in π.

Possibly d-connecting: If A and B are not in $\mathbf{Z}$, and $A \neq B$, then an undirected path U between A and B in a partially oriented inducing path graph π over $\mathbf{O}$ is a **possibly d-connecting** path of A and B given $\mathbf{Z}$ if and only if every collider on U is the source of a semidirected path to a member of $\mathbf{Z}$, and every definite noncollider is not in $\mathbf{Z}$.

Possibly-IP: If π is a partially oriented inducing path graph of G over $\mathbf{O}$, then X is in **Possibly-IP(Y,Z)** if and only if $\mathbf{Y}$ and $\mathbf{Z}$ are disjoint, X is in $\mathbf{Z}$, and there is a possibly d-connecting path between X and some Y in $\mathbf{Y}$ given $\mathbf{Z}\backslash\{X\}$ that is not out of X.

Possibly-IV: If π is a partially oriented inducing path graph of G over $\mathbf{O}$, then X is in **Possibly-IV(Y,Z)** if and only if X is not in $\mathbf{Z}$, there is a possibly d-connecting path between X and some Y in $\mathbf{Y}$ given $\mathbf{Z}$, and there is a semidirected path from X to a member of $\mathbf{Y} \cup \mathbf{Z}$.

Possible-SP: For a partially oriented inducing path graph π and ordering *Ord* acceptable for π, let V be in **Possible-SP**(*Ord*,X) if and only if $V \neq X$ and there is an undirected path U in π between V and X such that every vertex on U except for X is a predecessor of X in *Ord*, and no vertex on U except for the endpoints is a definite-noncollider on U.

Predecessors: For inducing path graph G' and acceptable total ordering *Ord*, let **Predecessors**(*Ord*,V) equal the set of all variables that precede V (not including V) according to *Ord*.

Proper final segment: A path U of length n is a **proper final segment** of path V of length m iff U is a final segment of V and $U \neq V$.

Proper initial segment: A path U of length n is a **proper initial segment** of path V of length m iff U is an initial segment of V and $U \neq V$.

$P_{Man(W)}(\mathbf{V})$: See manipulate.

$P_{Unman(W)}(\mathbf{V})$: See manipulate.

Pure Latent Variable Graph: A **pure latent variable graph** is a directed acyclic graph in which each measured variable is a child of exactly one latent variable, and a parent of no other variable.

Random coefficient linear causal theory: The definition of a **random coefficient linear causal theory** is the same as that of a linear causal theory except that each linear coefficient is a random variable independent of the set of all other random variables in the model.

Rigidly statistically indistinguishable: If directed acyclic graphs G and G' are strongly statistically indistinguishable and every parallel embedding of G and G' is strongly statistically indistinguishable then structures G and G' are **rigidly statistically indistinguishable (r.s.i.).**

R.S.I.: See rigidly statistically indistinguishable.

Semi-directed: A **semidirected path from** A to B in partially oriented inducing path graph π is an undirected path U from A to B in which no edge contains an arrowhead pointing toward A, that is, there is no arrowhead at A on U, and if X and Y are adjacent on the path, and X is between A and Y on the path, then there is no arrowhead at the X end of the edge between X and Y.

Source: See trek.

SP: For inducing path graph G' and acceptable total ordering Ord, W is in $\mathbf{SP}(Ord,G',V)$ (separating predecessors of V in G' for ordering Ord) if and only if $W \neq V$ and there is an undirected path U between W and V such that each vertex on U except for V precedes V in Ord and every vertex on U except for the endpoints is a collider on U.

S.S.I.: See strongly statistically indistinguisable.

Strongly statistically indistinguishable: Two directed acyclic graphs G, G' are **strongly statistically indistinguishable** if and only if they have the same vertex set $\mathbf{V}$ and every distribution P on **V** satisfying the Minimality and Markov Conditions for *G* satisfies those conditions for *G'*, and vice-versa.

Substituable: In an inducing path or directed acyclic graph *G* that contains an undirected path *U* between *X* and *Y*, the edge between *V* and *W* is **substitutable** for *U*(*V*,*W*) in *U* if and only if *V* and *W* are on *U*, *V* is between *X* and *W* on *U*, *G* contains an edge between *V* and *W*, *V* is a collider on the concatenation of *U*(*X*,*V*) and the edge between *V* and *W* if and only if it is a collider on *U*, and *W* is a collider on the concatenation of *U*(*Y*,*W*) and the edge between *V* and *W* if and only if it is a collider on *U*.

T: See trek.

Termini: See trek.

Trek: A **trek** *T*(*I*,*J*) between two distinct vertices *I* and *J* is an unordered pair of acyclic directed paths from some vertex *K* to *I* and *J* respectively that intersect only at *K*. The source of the paths in the trek is called the **source** of the trek. *I* and *J* are called the **termini of the trek**. Given a trek *T*(*I*,*J*) between *I* and *J*, ***I***(*T*(*I*,*J*)) will denote the path in *T*(*I*,*J*) from the source of *T*(*I*,*J*) to *I* and ***J***(*T*(*I*,*J*)) will denote the path in *T*(*I*,*J*) from the source of *T*(*I*,*J*) to *J*. One of the paths in a trek may be an empty path. However, since the termini of a trek are distinct, only one path in a trek can be empty. **T**(*I*,*J*) is the set of all treks between *I* and *J*. *T*(*I*,*J*) will represent a trek in **T**(*I*,*J*). *S*(*T*(*I*,*J*)) represents the source of the trek *T*(*I*,*J*).

Undirected: In a graph *G*, Let *V* be in **Undirected**(*X*,*Y*) if and only if *V* lies on some undirected path between *X* and *Y*.

Unmanipulated graph: See manipulation.

U_X: In an LCF *S*, U_X is the set of all independent variables that are the source of a directed path to *X*. (Note that if *X* is independent then $X \in U_X$ since there is an empty path from every vertex to itself.)

U_{XY}: In an LCF *S*, U_{XY} is $U_X \cap U_Y$.

Weakly faithfully indistinguishable: Two directed acyclic graphs are **weakly faithfully indistinguishable** (w.f.i.) if and only if there exists a probability distribution faithful to both of them.

Weakly statistically indistinguishable: Two directed acyclic graphs are **weakly statistically indistinguishable** (w.s.i.) if and only if there exists a probability distribution meeting the Minimality and Markov Conditions for both of them.

W.F.I.: See weakly faithfully indistinguishable.

W.S.I.: See weakly statistically indistinguishable.

References

Ahn, W., Kalish, C., Medin, D., and Gelman, S. (1995). The role of covariation versus mechanism information in causal attribution. *Cognition* 54: 299–352.

Ahn, W., and Bailenson, J. (1996). Causal attribution as a search for underlying mechanisms: An explanation of the conjunction fallacy and the discounting principle. *Cognitive Psychology*.

Aigner, D., and Goldberger, S. (1977). *Latent Variables in Socio-economic Models*. Amsterdam: North-Holland.

Aitkin, M. (1979). A simultaneous test procedure for contingency table models. *Applied Statistics* 28: 233–242.

Akleman, D., Bessler, D., and Burton, D. (1999). Modeling corn exports and exchange rates with directed graphs and statistical loss functions. In *Computation, Causation and Discovery*, edited by C. Glymour and G. Cooper. Cambridge, Mass.: MIT Press.

Alchourrón, C., Gärdenfors, P., and Mackinson, C. (1985). On the logic of theory change: Partial meet contraction and revision functions. *Journal of Symbolic Logic* 50: 510–530.

Aliferis, C., and Cooper, G. (1994). An evaluation of an algorithm for inductive learning of Bayesian belief networks using simulated data sets. *Proceedings of the Tenth Conference on Uncertainty in Artificial Intelligence*, Seattle: Morgan Kaufmann, 8–14.

Allen, D. (1974). The relationship between variable selection and data augmentation and a method for prediction. *Technometrics* 17: 125–127.

Anderson, J., and Gerbing, D. (1982). Some methods for respecifying measurement models to obtain unidimensional construct measurement. *Journal of Marketing Research* 19: 453–460.

Anderson, T. W. (1984). *An Introduction to Multivariate Statistical Analysis*. New York: Wiley.

Andersson, S., Madigan, D., and Perlman, M. (1996). An alternative Markov property for chain graphs. *Proceedings of the 12th Conference on Uncertainty in AI*, Portland, Ore.: Morgan Kaufmann, 40–48.

Artzenius, F. (1992). The Common Cause Principle. *Philosophy of Science Association*: 227–237.

Asher, Herbert B. (1976). *Causal Modeling*. Beverly Hills, Calif.: Sage Publications.

Asmussen, S., and Edwards, D. (1983). Collapsibility and response variables in contingency tables. *Biometrika* 70: 567–578.

Bagozzi, R. (1980). *Causal Models in Marketing*. New York: Wiley.

Baldwin, B. (1986). The effects of structural model misspecification and sample size on the robustness of LISREL maximum likelihood parameter estimates. Department of Administrative and Foundational Services, Louisiana State University.

Balke, A., and Pearl, J. (1994). Probabilistic Evaluation of Counterfactual Queries. *Proceedings of the Twelfth National Conference on Artificial Intelligence*. MIT Press, volume 1, 230–237.

Bartlett, M. (1935). Contingency table interaction. *J. Roy. Statist. Soc. Suppl.* 2: 248–252.

Bartlett, M. (1954). A note on the multiplying factors for various chi-squared approximations. *J. Roy. Statist. Soc. Ser. B* 16: 196–198.

Basmann, R. (1965). A note on the statistical testability of 'explicit causal chains' against the class of 'interdependent' models. *JASA* 60: 1080–1093.

Beale, E., Kendall, M., and Mann, D. (1967). The discarding of variables in multivariate analysis. *Biometrika* 54: 357–366.

Bearden, W. O., Sharma, S., and Teel, J. E. (1982). Sample size effects on chi-square and other statistics used in evaluating causal models. *Journal of Marketing Research* 19: 425–430.

Becker, G. (1964). Human capital; a theoretical and empirical analysis, with special reference to education. New York: National Bureau of Economic Research; distributed by Columbia University Press.

Becker, S., and LeCun, Y. (1989). Improving the convergence of back-propagation learning with second order methods. *Proceedings of the 1988 Connectionist Models Summer School.* Morgan Kaufmann.

Beinlich, I., Suermondt, H., Chavez, R., and Cooper, G. (1989). The ALARM monitoring system: a case study with two probabilistic inference techniques for belief networks. *Proceedings of the Second European Conference on Artificial Intelligence in Medicine*. London: England.

Bentler, P. (1980). Multivariate analysis with latent variables: causal modeling. *Annual Review of Psychology* 31: 419–456.

Bentler, P. (1985). Theory and Implementation of EQS: A Structural Equations Program. Los Angeles, BMDP Statistical Software.

Bentler, P. (1995). *EQS: Structural equations program manual (Version 5.0).* Encino, Calif.: Multivariate Software.

Bentler, P., and Bonett, D. (1980). Significance tests and goodness of fit in the analysis of covariance structures. *Psychological Bulletin* 88: 588–606.

Bentler, P., and Peeler, W. (1979). Models of female orgasm. *Archives of Sexual Behavior* 8: 405–423.

Bernardo, J., and Smith, A. (1994). *Bayesian Theory.* New York: Wiley.

Bessler, D., and Akleman, D. (1998). Farm prices, retail prices and directed graphs: results for pork and beef. *American Journal of Agricultural Economics* 80: 1144–1149.

Bessler, D., and Güven, D. (1997). A note on directed graphs and time series data. submitted to *Econometric Reviews*.

Birch, M. (1963). Maximum likelihood in three-way contingency tables. *J. Roy. Statist. Soc.* 25: 220–223.

Bishop, Y, Fienberg, S., and Holland, P. (1975). *Discrete Multivariate Analysis: Theory and Practice.* Cambridge, Mass.: MIT Press.

Blalock, H. (1969). *Theory Construction; from Verbal to Mathematical Formulations.* Englewood Cliffs, N.J.: Prentice-Hall.

Blalock, H. (1971). *Causal Models in the Social Sciences.* Chicago: Aldine·Atherton.

Blalock, H. (1972). *Causal Inferences in Nonexperimental Research.* New York: Norton.

Blau, P., and Duncan, O. (1967). *The American Occupational Structure.* New York: Wiley.

Blum, R. (1984). Discovery, confirmation and incorporation of causal relationship from a time-oriented clinical data base: The RX project. *Readings in Medical Artificial Intelligence,* edited by W. Clancey and E. Shortliffe. Reading, Mass.: Addison-Wesley.

Blyth, C. (1972). On Simpson's paradox and the sure-thing principal. *JASA* 67: 364–366.

Bollen, K. (1990). Outlier screening and a distribution-free test for vanishing tetrads. *Sociological Methods and Research* 19: 80–92.

Bollen, K. (1989). *Structural Equations with Latent Variables*. New York: Wiley.

Boomsma, A. (1982). The robustness of LISREL against small sample sizes in factor analysis models. In *Systems Under Indirect Observation: Causality, Structure, Prediction (Part I)*, edited by K. Jöreskog and H. Wold. Amsterdam: North-Holland, 149–173.

Boomsma, A. (1983). On the robustness of LISREL (maximum likelihood estimation) against small sample size and nonnormality. Amsterdam, Sociometric Research Foundation.

Bowden, R., and Turkington, D. (1984). *Instrumental Variables*. Cambridge: Cambridge University Press.

Boyen, X., Friedman, N., and Koller, D. (1999) Discovering the structure of complex dynamic systems. *Proceedings of the Fifteenth Conference on Uncertainty in Artificial Intelligence*. Morgan Kaufman, 91–100.

Breslow, N., and Day, N. (1980). *Statistical Methods in Cancer research, Vol. 1: The analysis of Case-Control Studies*. Lyon, IARC.

Brownlee, K. (1965). A review of "Smoking and Health." *JASA*, 722–739.

Bunker, J., Forrest, W., Mosteller, F., and Vandam, L. (1969). The National Halothane Study: Report of the Subcommittee on the National Halothane Study of the Committee on Anesthesia. Washington, D.C., Division of Medical Sciences, National Academy of Sciences, National Research Council.

Buntine, W. (1991). Theory refinement on Bayesian networks. *Proceedings of the Seventh Conference on Uncertainty in AI,* Los Angeles, Calif.: Morgan Kaufmann, 52–61.

Buntine, W. (1994). Operations for learning with graphical models. *Journal for Theoretical and Experimental Artificial Intelligence* 2: 159–225.

Buntine, W. (1996). A guide to the literature on learning graphical models. *IEEE Transactions on Knowledge and Data Engineering* 8: 195–210.

Burch, P. (1978). Smoking and lung cancer: The problem of inferring cause (with discussion). *J. Roy. Statist. Soc.* 141(Series A): 437–477.

Burch, P. (1983). The Surgeon General's "Epidemiologic Criteria for Causality." A critique. *Journal of Chronic Diseases* 36: 821–836.

Burch, P. (1984). The Surgeon General's "Epidemiologic Criteria for Causality." Reply to Lilenfeld. *Journal of Chronic Diseases* 37: 148–157.

Byron, R. (1972). Testing for misspecification in econometric systems using full information. *International Economic Review* 28: 138–151.

Callahan, J., and Sorenson, S. (1992). Using TETRAD II as an automated exploratory tool. *Social Science Computer Review* 10: 329–336.

Campbell, D., Schwartz, R., Sechrest, L., and Webb, Eugene J. (1966). *Unobtrusive Measures; Nonreactive Research in the Social Sciences*. Chicago: Rand McNally.

Campbell, D., Stanley, J., and Gage, N. (1966). *Experimental and quasi-experimental designs for research*. Chicago: R. McNally.

Caramazza, A. (1986). On drawing inferences about the structure of normal cognitive processes from patterns of impaired performance: the case for single patient studies. *Brain and Cognition* 5: 41–66.

Cartwright, N. (1983). *How the Laws of Physics Lie*. Oxford, New York: Clarendon Press; Oxford University Press.

Cartwright, N. (1989). *Nature's Capacities and their Measurement*. Oxford, New York: Clarendon Press; Oxford University Press.

Cartwright, N. (1993). Marks and probabilities: two ways to find a causal structure. In *Scientific Philosophy: Origins and Development,* edited by F. Stadler. Kluwer, Dordrecht.

Casella, G., and George, E. (1992). Explaining the Gibbs sampler. *The American Statistician* 46: 167–174.

Cavallo, R., and Klir, G. (1979). Reconstructability analysis of multi-dimensional relations: A theoretical basis for computer-aided determination of acceptable systems models. *International Journal of General Systems* 5: 143–171.

Cederlof, R., Friberg, L., and Lundman, T. (1972). The interactions of smoking, environment and heredity and their implications for disease etiology. *Acta Med Scand.* 612 (Suppl.).

Cheeseman, P., and Stutz, J. (1995). Bayesian classification (AutoClass): Theory and results. *Advances in Knowledge Discovery and Data Mining*, edited by U. Fayyad, G. Piatesky-Shapiro, P. Smyth, and R. Uthurusamy, Menlo Park, Calif.: AAAI Press, 153–180.

Chib, S. (1995). Marginal likelihood from the Gibbs output. *JASA* 90: 1313–1321.

Chib, S., and Greenberg, E. (1995). Understanding the Metropolis-Hastings algorithm. *The American Statistician* 49: 327–335.

Chickering, D. (1996). Learning Bayesian networks is NP-complete. *Learning from Data*, edited by D. Lenz and H. Fisher. Springer-Verlag, 121–130.

Chickering, D. (1996). Learning equivalence classes of Bayesian-network structures. *Proceedings of the Twelfth Conference on Uncertainty in AI*, Portland, Ore.: Morgan Kaufmann, 150–157.

Chickering, D., and Heckerman, D. (1997). Efficient approximations for the marginal likelihood of Bayesian networks with hidden variables. *Machine Learning* 29: 181–212.

Chou, C., Bentler, P., and Satorra, A. (1991). Scaled test statistics and robust standard errors for non-normal data in covariance structure analysis: A Monte Carlo study. *British Journal of Mathematical and Statistical Psychology* 85: 398–409.

Chow, C., and Liu, C. (1968). Approximating discrete probability distributions with dependence trees. *IEEE Trans. on Info. Theory* IT–14: 462–467.

Chow, C., and Wagner, T. (1973). Consistency of an estimate of tree-dependent probability distributions. *IEEE Trans.on Info. Theory* IT–19: 369–371.

Christensen, R. (1990). *Log-linear Models.* New York: Springer-Verlag.

Coleman, J. (1964). *Introduction to Mathematical Sociology.* New York: Free Press of Glencoe.

Cooper, G. (1989). Current research in the development of expert systems based on belief networks. *Applied Stochastic Models and Data Analysis* 5: 39–52.

Cooper, G. (1995). Causal discovery from data in the presence of selection bias. *Proceedings of the Fifth International Workshop on Artificial Intelligence and Statistics*, Fort Lauderdale, FL.

Cooper, G. (1997). A simple constraint-based algorithm for efficiently mining causal observational databases for causal relationships. *Data Mining and Knowledge Discovery* 1: 201–224.

Cooper, G., and Herskovits, E. (1991). A Bayesian method for constructing Bayesian belief networks from databases. *Proceedings of the Seventh Annual Conference on Uncertainty in AI*, Los Angeles, Calif.: Morgan Kaufmann, 86–94.

Cooper, G., and Herskovits, E. (1992). A Bayesian method for the induction of probabilistic networks from data. *Machine Learning* 9: 309–347.

Cooper, G., and Yoo, C. (2000) Causal discovery from a mixture of experimental and observational data, in *Proceedings of the 15th Annul Conference on Uncertainty in Artificial Intelligence*. Morgan Kaufman, San Francisco, CA 116-125.

Cornfield, J., Haenszel, W., Hammond, E., Lilienfeld, A., Shimkin, M., and Wynder, E. (1959). Smoking and lung cancer: Recent evidence and a discussion of some questions. *Journal of the National Cancer Institute* 22: 173–203.

Costner, H. (1971). Theory, deduction and rules of correspondence. In *Causal Models in the Social Sciences*, edited by H. Blalock. Chicago: Aldine.

Costner, H., and Herting, J. (1985). Respecification in multiple indicator models. In *Causal Models in the Social Sciences*, edited by H. Blalock. New York: Aldine, 321–393.

Costner, H., and Schoenberg, R. (1973). Diagnosing indicator ills in multiple indicator models. In *Structural Equation Models in the Social Sciences*, edited by A. Duncan and O. Goldberger. New York: Seminar Press.

Cox, D. (1958). *Planning of Experiments*. New York: Wiley.

Cox, D., and Wermuth, N. (1996). *Multivariate Dependencies: Models, Analysis and Interpretation.*,Boca Raton, Fla.: CRC Press.

Cox, D., and Wermuth, N. (1999). On the generation of the chordless four-cycle, ZUMA-Arbeitsbericht 99/04. *Submitted to Biometrika.*

Crawford, S. (1994). An application of the Laplace method to finite mixture distributions. *JASA* 89: 259–267.

Crawford, S., and Fung, R. (1990). An analysis of two probabilistic model induction techniques. *Proceedings of the Third Annual Workshop on Artificial Intelligence and Statistics*, Fort Lauderdale, Fla.

Dai, H., Korb, K., Wallace C., and Wu, X. (1997). A study of causal discovery with weak links and small samples. *15th International Joint Conference on Artificial Intelligence*, Nagoya, Japan, 23–29 August, 1997.

Darroch, J., Lauritzen, S., and Speed, T. (1980). Markov fields and log linear interaction models for contingency tables. *Ann. Stat.* 8: 522–539.

Davis, W. (1988). Probabilistic theories of causation. In *Probability and Causality*, edited by J. Fetzer. Dordrecht: D. Reidel.

Dawid, A. (1979). Conditional independence in statistical theory (with discussion). *J. Roy. Statist. Soc.* Ser. B 41: 1–31.

Dawid, A. (1997). Causal inference without counterfactuals. Technical Report, Department of Statistical Science, University College, London.

de Campos, L., and Huete, J. Approximating causal orderings for Bayesian networks using genetic algorithms and simulated annealing. Technical Report, #DECSAI–990212, Department of Computer Science, University of Grenada, Spain.

de Fazio, J., Ramsey, J., Roush, T., Gazis, P., and Glymour, C. (2000) Automated Mineral Identification from Reflectance Spectra, Carnegie Mellon University Department of Philosophy Technical Report.

Dempster, A. (1972). Covariance selection. *Biometrics* 28: 157–175.

Dempster, A., Laird, N., and Rubin, D. (1977). Maximum likelihood from incomplete data via the EM algorithm. *J. Roy. Stat. Soc.* 39(Ser. B): 1–38.

Desjardins, B. (1999). The Limits of Causal Discovery. Department of History and Philosophy of Science. Pittsburgh: University of Pittsburgh.

DiCiccio, T., Kass, R., Raftery, A., and Wasserman, L. (1995). Computing Bayes factors by combining simulation and asymptotic approximations. Pittsburgh: Department of Statistics, Carnegie Mellon University.

Doll, R., and Hill, A. (1952). A study of the aetiology of carcinoma of the lung. *Brit. Med. Journal* 2: 1271–1286.

Druzdzel, M., and Glymour, C. (1999). Causal inferences from databases: why universities lose students. *Computation, Causation and Discovery,* edited by C. Glymour and G. Cooper. Cambridge, Mass.: MIT Press.

Duncan, O. (1975). *Introduction to Structural Equation Models.* New York: Academic Press.

Duncan, O., Featherman, D., and Duncan, B. (1972). *Socioeconomic Background and Achievement.* New York: Seminar Press.

Edwards, A. (1976). *An Introduction to Linear Regression and Correlation.* San Francisco: W. H. Freeman.

Edwards, D. (1995). *Introduction to Graphical Modeling.* New York: Springer-Verlag.

Edwards, D., and Havranek, T. (1985). A fast procedure for model search in multi-dimensional contingency tables. *Biometrika* 72: 339–351.

Edwards, D., and Havranek, T. (1987). A fast model selection procedure for large families of models. *J. Amer. Statist. Assoc.* 82: 205–211.

Edwards, D., and Kreiner, S. (1983). The analysis of contingency tables by graphical models. *Biometrika* 70: 553–565.

Elby, A. (1992). Should we explain the EPR causally? *Philosophy of Science* 59: 16–25.

Fienberg, S. (1977). *The Analysis of Cross-classified Categorical Data.* Cambridge, Mass.: MIT Press.

Fine, A. (1982). Hidden variables, joint probability, and the Bell inequalities. *Physical Review Letters* 48: 291–295.

Fine, M., Auble, T., Yealy, D., Hanusa, B., Weissfeld, L., Singer, D., Coley, C., Marrie, T., Kapoor, W. (1997). A prediction rule to identify low-risk patients with community-acquired pneumonia. *New England Journal of Medicine* 336: 243–250.

Fine, T. (1973). *Theories of Probability; an Examination of Foundations.* New York: Academic Press.

Fisher, F. (1966). *The Identification Problem in Economics*. New York: McGraw-Hill.

Fisher, F. (1970). A correspondence principle for simultaneous equation models. *Econometrica* 38: 73–92.

Fisher, R. (1951). *The Design of Experiments*. Edinburgh: Oliver and Boyd.

Fisher, R. (1959). *Smoking. The Cancer Controversy*. Edinburgh: Oliver and Boyd.

Flack, V., and Chang, P. (1987). Frequency of selecting noise variables in subset regression analysis: A simulation study. *American Statistician* 41: 84–86.

Forbes, H., and Tufte, E. (1968). A note of caution in causal modeling. *American Political Science Review* 62: 1258–1264.

Fox, J. (1984). *Linear Statistical Models and Related Methods, with Applications to Social Research*. New York: Wiley.

Freedman, D. (1983a). A note on screening regression equations. *American Statistician* 37: 152–155.

Freedman, D. (1983b). Structural-equation models: A case study. Berkeley: University of California.

Freedman, D., Navidi, W., and Peters, S. (1986). On the impact of variable selection in fitting regression equations. *On Model Uncertainty and its Statistical Implications. Lecture Notes in Economics and Mathematical Systems,* edited by T. Dijkstra. Berlin: Springer-Verlag, 307.

Friedman, M. (1957). *A Theory of Consumption Function*. Princeton, N.J.: Princeton University Press.

Friedman, N. (1997). Learning belief networks in the presence of missing values and hidden variables. *Proceedings of the 14th International Conference on Machine Learning.*

Friedman, N., and Goldszmidt, M. (1996). Discretization of continuous attributes while learning Bayesian networks. *Proceedings of the 13th International Conference on Machine Learning*. L. Saitta, 157–165.

Friedman, N., and Goldszmidt, M. (1999a). Data analysis with Bayesian networks: a bootstrap approach. *Proceedings of the Fifteenth Conference on Uncertainty in Artificial Intelligence*. San Francisco: Morgan Kaufman, 196–205.

Friedman, N., Goldszmidt, M., and Wyner, A. (1999b). On the application of the bootstrap for computing confidence measures on features of induced Bayesian networks. *Artificial Intelligence and Statistics 99*. San Francisco: Morgan Kaufman, 197–202.

Friedman, N., Nachman, I., and Pe'er, D. (1999c). Learning Bayesian networks from massive datasets: The "Sparse Candidate Algorithm." *Proceedings of the Fifteenth Conference on Uncertainty in Artificial Intelligence*. San Francisco: Morgan Kaufman, 206–215.

Frydenberg, M. (1990). The chain graph Markov property. *Scandanavian Journal of Statistics* 17: 333–353.

Fung, R., and Crawford, S. (1990). Constructor: A system for the induction of probabilistic models. *Proceedings of the Eighth National Conference on AI*, Boston: AAAI.

Furnival, G., and Wilson, R. (1974). Regression by leaps and bounds. *Technometrics* 16: 4990–5111.

Galles, D., and Pearl, J. (1995). Testing Identifiability of Causal Effects. *Proceedings of the Eleventh Annual Conference on Uncertainty in Artificial Intelligence*. San Francisco: Morgan Kaufmann, 185–195.

Galles, D., and Pearl, J. (1998a). Axioms of causal relevance. *Artificial Intelligence* 97: 9–43.

Galles, D., and Pearl, J. (1998b). An axiomatic characterization of causal counterfactuals. *Foundations of Science* 3: 151–182.

Geiger, D. (1990). Graphoids: A Qualitative Framework for Probabilistic Inference. Los Angeles: University of California.

Geiger, D., Heckerman, D., and Meek, C. (1996). Asymptotic model selection for directed networks with hidden variables. *Proceedings of the Twelfth Conference on Uncertainty in AI*. San Francisco: Morgan Kaufmann, 283–290.

Geiger, D., and Heckerman, D. (1991). Advances in probabilistic reasoning. *Proceedings of the Seventh Annual Conference on Uncertainty in AI.* San Francisco: Morgan Kaufmann, 118–126.

Geiger, D., and Heckerman, D. (1994). Learning Gaussian networks. *Proceedings of the Tenth Conference on Uncertainty in AI.* San Francisco: Morgan Kaufmann, 235–243.

Geiger, D., and Heckerman, D. (1995). A characterization of the Dirichlet distribution applicable to learning Bayesian networks. Redmond, Wash.: Microsoft Research.

Geiger, D., and Meek, C. (1999). Quantifier elimination for statistical problems, *Proceedings of the 15th Annual Conference on Artificial Intelligence and Statistics*, San Francisco, Morgan Kaufman, 226–235.

Geiger, D., and Pearl, J. (1989a). Axioms and Algorithms for Inferences Involving Conditional Independence. Los Angeles: Cognitive Systems Laboratory, University of California.

Geiger, D., and Pearl, J. (1989a). Logical and Algorithmic Properties of Conditional Independence and Qualitative Independence. Los Angeles, Cognitive Systems Laboratory, University of California.

Geiger, D., Heckerman, D., King, H., and Meek, C. (1999). On the geometry of DAG models with hidden variables. *Artificial Intelligence and Statistics 99.* San Francisco: Morgan Kauffman.

Geiger, D., Verma, T., and Pearl, J. (1990). Identifying independence in Bayesian Networks. *Networks* 20: 507–533.

Geisser, S., and Eddy, W. (1979). A predictive approach to model selection. *JASA* 74: 153–160.

Gelfand, A., and Smith, A. (1990). Sampling based approaches to calculating marginal densities. *Journal of the American Statistical Association* 85: 398–409.

Gelman, A., Carlin, J., Stern, H., and Rubin, D. (1995). *Bayesian Data Analysis*. London: Chapman and Hall.

Gelman, A., Meng, X., and Stern, H.S. (1996). Posterior predictive assessment of model fitness via realized discrepancies (with discussion). *Statistica Sinica* 6: 733–807.

Gelman, A., and Rubin, D. (1992). Inference from iterative simulation using multiple sequences. *Statistical Science* 7: 457–511.

Geman, S., and Geman, D. (1984). Stochastic relaxation, Gibbs distributions and the Bayesian restoration of images. *IEEE Transactions on Pattern Analysis and Machine Intelligence* 6: 721–742.

Geweke, J., Meese, R., and Dent, W. (1983). Comparing alternative tests of causality in temporal systems. *Journal of Econometrics* 21: 161–194.

Glymour, C. (1983). Social science and social physics. *Behavioral Science* 28: 126–133.

Glymour, C., Scheines, R., and Spirtes, P. (1989). Why Aviators Leave the Navy: Applications of Artificial Intelligence Procedures in Manpower Research, Naval Personnel Research Development Center.

Glymour, C., Spirtes P., and Scheines, R. (1991). From probability to causality. *Philosophical Studies* 64: 1–36.

Glymour, C., Spirtes, P., and Scheines, R. (1991). Independence relations produced by parameter values. *Philosophical Topics* 18: 55–70.

Glymour, C., Scheines, R., Spirtes, P., and Kelly, K. (1987). *Discovering Causal Structure*. San Diego, Calif.: Academic Press.

Gold, E. (1965). Limiting recursion. *Journal of Symbolic Logic* 30: 27–48.

Gold, E. (1967). Language identification in the limit. *Information and Control* 10: 447–474.

Goldszmidt, M., and Pearl, J. (1992). Rank-based systems: A simple approach to belief revision, belief update, and reasoning about evidence and actions. *Proceedings of the Third International Conference on Knowledge Representation and Reasoning*, edited by B. Nebel, C. Rich, W. Swartout. San Francisco: Morgan Kaufman, 661–672.

Goldberg, A., and Duncan, O., eds. (1973). *Structural Equation Models in the Social Sciences*. New York: Seminar Press.

Goodman, L. (1973a). The analysis of multidimensional contingency tables when some variables are posterior to others: A modified path analysis approach. *Biometrika* 60: 179–192.

Goodman, L. (1973b). Causal analysis of data from panel studies and other kinds of surveys. *American Journal of Sociology* 78: 1135–1191.

Granger, C. (1969). Investigating causal relations by econometric models and cross-spectral methods. *Econometrica* 37: 424–438.

Greene, T., and Ernhart, C. (1993). Dentine lead and intelligence prior to school entry: A statistical sensitivity analysis. *Journal of Clinical Epidemiology* 46: 323–329.

Greenland, S. (1989). Modelling variable selection in epidemiologic analysis. *American Journal of Public Health* 79: 340–349.

Griffiths, W., Hill, R., and Pope, P. (1987). Small sample properties of probit model estimators. *JASA* 82: 929–937.

Güven D., and Bessler, D. (1997). A note on directed graphs and time series data. Submitted to *Econometric Reviews*.

Güven, D., and Tolun, M. (1991). Analyzing the inter and intra vocabulary performances in isolated speech recognition. *Journal of American Voice Input/Output Society* 10: 19–37.

Haberman, S. (1979). *Analysis of Qualitative Data.* San Diego, Calif.: Academic Press.

Harary, F., Norman R., and Cartwright, D. (1965). *Structural Models: An Introduction to the Theory of Directed Graphs.* New York: Wiley.

Harary, F., and Palmer, E. (1973). *Graphical Enumeration.* New York: Academic Press.

Haughton, D. (1988). On the choice of a model to fit data from an exponential family. *Annals of Statistics* 16: 342–355.

Hausman, D. (1984). Causal priority. *Nous* 18: 261–279.

Hausman, D. (1998). Causal Asymmetries. Cambridge: Cambridge University Press.

Hausman, D., and Woodward, J. (in press). The causal Markov condition. *British Journal for Philosophy of Science.*

Havranek, T. (1984). A procedure for model search in multi-dimensional contingency tables. *Biometrics* 40: 95–100.

Heckerman, D. (1995). A Bayesian approach for learning causal networks. *Proceedings of the Eleventh Conference on Uncertainty in AI.* San Francisco: Morgan Kaufmann, 274–284.

Heckerman, D. (1998). A tutorial on learning with Bayesian networks. In *Learning in Graphical Models,* edited by M. Jordan. Cambridge, Mass.: MIT Press.

Heckerman, D., and Shacter, R. (1995). Decision-theoretic foundations for causal reasoning. *Journal of Artificial Intelligence Research* 3: 405–430.

Heckerman, D., and Geiger, D. (1996). Likelihoods and Priors for Bayesian Networks. Redmond, Wash.: Microsoft Research.

Heckerman, D., Geiger, D, and Chickering, D. (1994). Learning Bayesian networks: The combination of knowledge and statistical data, *Proceedings of the Tenth Conference on Uncertainty in Artificial Intelligence.* San Francisco: Morgan Kaufmann, 293–302.

Heckerman, D., Geiger, D., and Chickering, D. (1995). Learning Bayesian networks: The combination of knowledge and statistical data. *Machine Learning* 20: 197–243.

Heckerman, D., Meek, C., and Cooper, G. (1999) A Bayesian approach to causal discovery. *Computation, Causation and Discovery*, edited by C. Glymour and G. Cooper. Cambridge, Mass.: MIT Press.

Heise, D. (1975). *Causal Analysis*. New York: Wiley.

Herskovits, E. (1991). Computer-based probabilistic network construction. Medical Information Sciences. Stanford, Calif.: Stanford University.

Herskovits, E. (1992). Computer Based Probabilistic-Network Construction. Computer Science and Medicine, Stanford University.

Herskovits, E., and Cooper, G. (1990). Kutato: An entropy-driven system for construction of probabilistic expert systems from databases, *Proceedings of the Sixth Conference on Uncertainty in AI.* Mountain View, Calif.: Association for Uncertainty in AI.

Herting, J., and Costner, J. (1985). Respecification in multiple indicator models. *Causal Models in the Social Sciences*, edited by H. Blalock. New York: Aldine.

Hocking, R., and Leslie, R. (1967). Selection of the best subset in regression analysis. *Technometrics* 9: 531–540.

Hojsgaard, S., and Thiessan, B. (1995). Block recursive models induced from observations and statistical techniques, *Computational Statistics and Data Analysis*. 19: 155-175.

Holland, P., and Rosenbaum, P. (1986). Conditional association and unidimensionality in monotone latent variable models. *Annals of Statistics* 14: 1523–1543.

Hoogland, J., and Boomsma, A. (1998). Robustness studies in covariance structure modeling: An overview and a meta-analysis. *Sociological Methods and Research* 26: 329–368.

Hosmer, D., and Lemeshow, S. (1989). *Applied Logistic Regression*. New York: Wiley.

Hu, L., Bentler, P., and Kano, Y. (1992). Can test statistics in covariance structure analysis be trusted? *Psychological Bulletin* 112: 351–362.

Humphreys, P., and Freedman, D. (1996). The grand leap. *British Journal for the Philosophy of Science* 47: 113–118.

Iwasaki, Y., and Simon, H. (1994). Causality and model abstraction. *Artificial Intelligence*, 67, 143–194.

James, L., Mulaik, S., and Brett, J. (1982). *Causal Analysis: Assumptions, Models and Data*. Beverly Hills, Calif.: Sage Publications.

Jeffreys, H. (1957). *Scientific Inference*. New York: Cambridge University Press.

Jensen, F., Lauritzen, S., and Oleson, K. (1990). Bayesian updating in recursive graphical models by local computations. *Computational Statistics Quarterly* 4: 269–282.

Jordan, M. (1998). *Learning in Graphical Models*. NATO Science Series D, Behavioral Social Sciences, vol. 89. Kluwer. Also published in 1999 by MIT Press, Cambridge, Mass.

Jöreskog, K. (1973). A general method for estimating a linear structural equation. *Structural Equation Models in the Social Sciences*. A. Goldberger and O. Duncan. New York: Seminar Press.

Jöreskog, K. (1978). Structural analysis of covariance and correlation matrices. *Psychometrika* 43: 443–447.

Jöreskog, K. (1981). Analysis of covariance structures. *Scandinavian Journal of Statistics* 8: 65–92.

Jöreskog, K., and Sörbom, D. (1984). *LISREL VI User's Guide*. Mooresville, Ind.: Scientific Software.

Jöreskog, K., and Sörbom, D. (1990). Model search with TETRAD II and LISREL. *Sociological Methods and Research* 19: 93–106.

Jöreskog, K. G., and Sörbom, D. (1993). *LISREL 8: User's reference guide.* Chicago: Scientific Software International.

Kadane, J., and Sedransk, N. (1996). *Bayesian Ethics in a Clinical Trial Design,* edited by J. Kadane. New York: Wiley.

Kadane, J., and Seidenfeld, T. (1990). Randomization in a Bayesian Perspective. *Journal of Statistical Planning and Inference* 25: 329–345.

Kano, Y., and Harada, A. (in press). Stepwise variable selection in factor analysis. *Psychometrika.*

Kass, R., and Raftery, A. (1995). Bayes factors. *JASA* 90: 773–795.

Kass, R., and Wasserman, L. (1995). A reference Bayesian test for nested hypotheses and its relationship to the Schwarz criterion. *JASA* 90: 928–934.

Kass, R., Tierney, L., and Kadane, J. (1988). Asymptotics in Bayesian computation. *Bayesian Statistics 3,* edited by J. Bernardo, M. DeGroot, D. Lindley, and A. Smith. Oxford: Oxford University Press.

Katsumo, H., and Mendelson, A. (1991). On the difference between updating a knowledge base and revising it. *Principles of Knowledge Representation and Reasoning: Proceedings of the Second International Conference,* Boston, Mass.

Kelley, T. (1928). *Crossroads in the Mind of Man.* Stanford: Stanford University Press.

Kelly, K. (1996). *The Logic of Reliable Inquiry.* New York: Oxford University Press.

Kendall, M. (1948). *The Advanced Theory of Statistics.* London: Charles Griffin and Co.

Kenny, D. (1979). *Correlation and Causality.* New York: Wiley.

Kiiveri, H. (1982). A Unified Approach to Causal Models, University of Western Australia.

Kiiveri, H., Speed, T., and Carlin, J. (1984). Recursive causal models. *Journal of the Australian Mathematical Society* 36: 30–52.

Kiiveri, H., and Speed, T. (1982). Structural analysis of multivariate data: A review. *Sociological Methodology*, edited by S. Leinhardt. San Francisco: Jossey-Bass.

Klein, L. (1961). *An Econometric Model of the United Kingdom*. Oxford: Oxford University, Institute of Statistics.

Kleinbaum, D, Kupper, L., and Morgenstern, H. (1982). *Epidemiologic Research*. Belmont, Calif.: Lifetime Learning Publications.

Klepper, S., and Leamer, E. (1984). Consistent sets of estimates for regressions with errors in all variables. *Econometrica* 52: 163–183.

Klepper, S. (1988). Regressor diagnostics for the classical errors-in-variables model. *Journal of Econometrics* 37: 225–250.

Klir, G., and Parviz, B. (1986). General reconstruction characteristics of probabilistic and possibilistic systems. *International Journal of Man-Machine Studies* 25: 367–397.

Kohn, M. (1969). *Class and Conformity*. Homewood, Ill.: Dorsey Press.

Korb, K., and Wallace, C. (1997). In search of the philosopher's stone: remarks on Humphreys and Freedman's critique of causal discovery. *British Journal for the Philosophy of Science* 48: 543–553.

Koster, J. (1995). Gibbs-factorization and the Markov property. Unpublished manuscript.

Koster, J. (1996). Markov properties of nonrecursive causal models. *Annals of Statistics* 24: 2148–2178.

Koster, J. (forthcoming). On the validity of the Markov interpretation of path diagrams of Gaussian structural equations systems with correlated errors. *Scandinavian Journal of Statistics*.

Kullback, S. (1959, 1968). *Information Theory and Statistics*. New York: Wiley.

Kullback, S. (1967). A lower bound for discrimination information in terms of variation. *IEEE Trans. Information Theory* 13: 126–127.

Kullback, S. (1968). Probability densities with given marginal. *Annals of Mathematical Statistics* 39: 79–86.

Lauritzen, S., Speed, T., and Vijayan, K. (1978). Decomposable Graphs and Hypergraphs. University of Copenhagen, Institute of Mathematical Studies. Preprint 9.

Lauritzen, S. (1996). *Graphical Models*. Oxford: Oxford University Press.

Lauritzen, S. (2000). Causal Inference from Graphical Models, in *Complex Stochastic Systems,* edted by O. Barnsdorf-Nielsen, O. Cox, and C. Klippelberg. London: Chapman and Hall.

Lauritzen, S., and Wermuth, N. (1989). Graphical models for association between variables, some of which are qualitative and some quantitative. *Annals of Statistics* 17: 31–57.

Lauritzen, S., Dawid, A., Larsen, B., and Leimer, H. (1990). Independence properties of directed Markov fields. *Networks* 20: 491–505.

Lawley, D., and Maxwell, A. (1971). *Factor Analysis as a Statistical Method*. London: Butterworth.

Lazarsfeld, P., and Henry, N. (1968). *Latent Structure Analysis*. Boston: Houghton Mifflin.

Leamer, E. (1978). *Specification Searches: Ad Hoc Inference with Non-experimental Data*. New York: Wiley.

Lee, S. (1985). Analysis of covariance and correlation structures. *Computational Statistics and Data Analysis* 2: 279–295.

Lee, S. (1987). Model Equivalence in Covariance Structure Modeling. Department of Psychology, Ohio State University.

Lemmer, J. 1996. The Causal Markov Condition, Fact or Artifact. *SIGART Bulletin* 7: 3–16.

Lewis, D. (1973a). Causation. *Journal of Philosophy* 70: 556–572.

Lewis, D. (1973b). *Counterfactuals*. Cambridge, Mass.: Harvard University Press.

Lilienfeld, A. (1983). The Surgeon General's "Epidemiologic Criteria for Causality." A Criticism of Burch's Critique 36: 837–845.

Linthurst, R. A. (1979). Aeration, nitrogen, pH and salinity as factors affecting Spartina Alterniflora growth and dieback. North Carolina State University.

Lohmoller, J. (1989). *Latent Variable Path Modeling with Partial Least Squares*. Heidelberg, Physica-Verlag.

Long, J. (1983a). *Confirmatory Factory Analysis*. Beverly Hills, Calif.: Sage Publications.

Long, J. (1983b). *Covariance Structure Models*. Beverly Hills, Calif.: Sage Publications.

Luijben, T., Boomsma, A., and Molenaar, I. (1986). Modification of factor analysis models in covariance structure analysis. A Monte Carlo Study. *On Model Uncertainty and its Statistical Implications*. Berlin: Springer-Verlag. 307.

MacCallum, R. (1986). Specification searches in covariance structure modeling. *Psychological Bulletin* 100: 107–120.

Mackie, J. (1974). *The Cement of the Universe*. New York: Oxford University Press.

Madigan, D. (1999). Bayesian graphical models, intention to treat, and the Rubin causal model. *Artifical Intelligence and Statistics 99*. San Francisco: Morgan Kaufman, 123–132.

Madigan, D., Garvin, J., and Raftery, A. (1995). Eliciting prior information to enhance the predictive performance of Bayesian graphical models. *Communications in Statistics: Theory and Methods* 24: 2271–2292.

Madigan, D., Raftery, A., Volinsky, C., and Hoeting, J. (1996). Bayesian model averaging. *AAAI Workshop on Integrating Multiple Learned Models*, Portland, Ore.

Madigan, D., and York, J. (1995). Bayesian graphical models for discrete data. *International Statistical Review* 63: 215–232.

Mallows, C. (1973). Some comments on C_p. *Technometrics* 15: 661–676.

Mani, S., and Cooper, G. (forthcoming) A study in causal discovery from population-based infant birth and death records. *Proceedings of the Annual Fall Symposium of the American Medical Informatics Association*. Hanley and Belfus, Philadelphia.

Manksi, C. (1995). *Identification Problems in the Social Sciences*. Harvard University Press, Cambridge, Mass.

Mardia, K., Kent, J., and Bibby, J. (1979). *Multivariate Analysis*. New York: Academic Press.

Maruyama, G., and McGarvey, B. (1980). Evaluating causal models: An application of maximum likelihood analysis of structural equations. *Psychological Bulletin* 87: 502–512.

MathSoft (1993). *Mathcad 4.0. User's guide. Windows version*. Cambridge, Mass.

Matuš, F. (1995). Conditional independence properties among four random variables II. *Combinatorics, Probability, and Computing* 4, 407–417.

Matuš, F., and Studený, M. (1995). Conditional independence properties among four random variables I. *Combinatorics, Probability, and Computing* 4: 269–278.

Maudlin, T. (1994). Quantum Non-Locality and Relativity. Blackwell, Cambridge, UK.

McKenna, M., and Shipley, W. 1999. Interacting determinants of interspecific relative growth: empirical patterns and a theoretical explanation. *Ecoscience* (in press).

McLachlan, G., and Krishnan, T. (1997). *The EM Algorithm and Extensions*. New York: Wiley.

McPherson, J., Welch, S., and Clark, C. (1977). The stability and reliability of political efficacy: Using path analysis to test alternative models. *American Political Science Review* 71: 509–521.

Meek, C. (1995). Strong completeness and faithfulness in Bayesian networks. *Proceedings of the Eleventh Conference on Uncertainty in AI*. San Francisco: Morgan Kaufmann, 411–418.

Meek, C., and Glymour, C. (1994). Conditioning and intervening. *British Journal for Philosophy of Science* 45: 1001–1021.

Meng, X., and Rubin, D. (1991). Using EM to obtain asymptotic variance-covariance matrices: The SEM algorithm. *JASA* 86: 899–909.

Miller, J., Slomczynski, K., and Schoenberg, R. (1981). Assessing comparability of measurement in cross-national research: Authoritarian-conservatism in different sociocultural settings. *Social Psychology Quarterly* 44: 178–191.

Miller, R., Jr. (1981). *Simultaneous Statistical Inference*. New York: McGraw-Hill.

Miller, W., and Stokes, D. (1963). Constituency influence in Congress. *American Political Science Review* 1963: 45–456.

Mitchell, T. (1977). Version spaces: A candidate elimination approach to rule learning. *Proceedings of the Fifth International Joint Conference on AI*, Pittsburgh.

Mitchell, T., and Beauchamp, J. (1988). Bayesian variable selection in linear regression. *JASA* 83: 1023–1032.

Monti, S., and Cooper, G. (1998). A multivariate discretization method for learning bayesian networks from mixed data. *Proceedings of the Fourteenth Annual Conference on Uncertainty in Artificial Intelligence*. San Francisco: Morgan Kaufmann, 404–413.

Mosteller, F., and Tukey, J. (1977). *Data Analysis and Regression, A Second Coure in Regression*. Reading, Mass.: Addison-Wesley.

MRFIT Research Group (1982). Multiple risk factor intervention trial; risk factor changes and mortality results. *JAMA* 248: 1465–1477.

Muthen, B. (1984). A general structural equation model with dichotomous, ordered categorical and continuous latent variable indicators. *Psychometrika* 49: 115–132.

Neal, R. (1993). Probabilistic inference using Markov chain Monte Carlo methods. Toronto: Department of Computer Science, University of Toronto.

Neal, R. (2000). On deducing conditional independence from d-separation in causal graphs with feedback: The Uniqueness Condition is not sufficient. *Journal of Artificial intelligence Research*, 12:87-91.

Neapolitan, R. (1990). *Probabilistic Reasoning in Expert Systems*. New York: Wiley.

Needleman, H., Geiger, S., and Frank, R. (1985). Lead and IQ Scores: A Reanalysis. *Science* 227: 701–704.

Neyman, J. (1935). Statistical problems with agricultural experimentation. *J. Roy. Stat. Soc. Suppl.* 2: 107–180.

Osherson, D., Stob, T., and Weinstein, S. (1986). *Systems That Learn*. Cambridge, Mass.: MIT Press.

Pearl, J. (1988). *Probabilistic Reasoning in Intelligent Systems*. San Mateo, Calif.: Morgan Kaufmann.

Pearl, J. (1994). A probablistic calculus of actions. *Proceedings of the Tenth Conference on Uncertainty in Artificial Intelligence.* Seattle: Morgan Kaufmann, 454–462.

Pearl, J. (1995). Causal diagrams for empirical research. *Biometrika* 82: 669–709.

Pearl, J. (1999a). Graphs, structural models and causality. *Computation, Causation and Discovery.* C. Glymour and G. Cooper. Cambridge, Mass.: MIT Press.

Pearl, J. (1999b). Reasoning with cause and effect. UCLA Cognitive Science Laboratory Technical Report, R–265.

Pearl, J. (forthcoming). Probabilities of causation: Three counterfactual interpretations and their identification. *Synthese.*

Pearl, J., and Dechter, R. (1989). *Learning structure from data: A survey.* proceedings COLT '89, Santa Cruz, Calif., 230–423.

Pearl, J., and Dechter, R. (1996). Identifying independencies in causal graphs with feedback. *Proceedings of the Twelfth Annual Conference on Uncertainty in Artificial Intelligence.* San Francisco: Morgan Kaufman, 420–426.

Pearl, J., and Robins, J. (1995). Probabilistic evaluation of sequential plans from causal models with hidden variables. *Proceedings of the Eleventh Annual Conference on Uncertainty in Artificial Intelligence.* San Francisco: Morgan Kaufmann, 444–453.

Pearl, J., and Tarsi, M. (1986). Structuring causal trees. *Journal of Complexity* 2: 60–77.

Pearl, J., and Verma, T. (1987). The Logic of Representing Dependencies by Directed Graphs. Los Angeles: UCLA Cognitive Systems Laboratory.

Pearl, J., and Verma, T. (1990). A Formal Theory of Inductive Causation. Los Angeles: UCLA Cognitive Systems Laboratory, Computer Science Department.

Pearl, J., and Verma, T. (1991). A theory of inferred causation. *Proceedings of the Second International Conference of Representation and Reasoning,* San Francisco: Morgan Kaufmann.

Pearl, J., Geiger, D., and Verma, T. (1990). The logic of influence diagrams. *Influence Diagrams, Belief Nets and Decision Analysis,* edited by R. Oliver and J. Smith. New York: Wiley.

Pratt, J., and Schlaifer, R. (1988). On the interpretation and observation of laws. *Journal of Econometrics* 39: 23–52.

Putnam, H. (1965). Trial and error predicates and a solution to a problem of Mostowski. *Journal of Symbolic Logic* 30: 49–57.

Pyankov, V., Kondratchuk, A., and Shipley, W. (1999). Leaf structure and specific leaf mass: the alpine desert plants of the Eastern Pamirs (Tadjikistan). *The New Phytologist* (in press).

Raftery, A. (1993). Bayesian model selection in structural equation models. *Testing Structural Equation Models*, edited by K. A. Bollen and J. S. Long. Newbury Park, Calif.: Sage, 163–180.

Raftery, A. (1994). Bayesian model selection in social research. Working paper no. 94-12, University of Washington, Center for Studies in Demography and Ecology.

Raftery, A. (1995). Bayesian model selection in social research. In *Sociological Methodology,* edited by P. Marsden. Cambridge, Mass.: Blackwell.

Raftery, A. (1996). Hypothesis testing and model selection. *Markov Chain Monte Carlo in Practice*, edited by W. R. Gilks, S. Richardson, and D. Spiegelhalter. London: Chapman and Hall, 163–187.

Ramoni, M (1996) Learning Bayesian networks from incomplete databases. *Proceedings of the Thirteenth Conference on Uncertainty in Artificial Intelligence*. Morgan Kaufman, 401–408.

Rawlings, J. (1988). *Applied Regression Analysis*. Belmont, Calif.: Wadsworth.

Reichenbach, H. (1956). *The Direction of Time*. Berkeley: University of California Press.

Reiss, I., Banwart, A., and Forman, H. (1975). Premarital contraceptive usage: A study and some theoretical explorations. *Journal of Marriage and the Family* 37: 619–630.

Richardson, T. (1996a). Models of Feedback: Interpretation and Discovery. Department of Philosophy. Pittsburgh: Carnegie Mellon University.

Richardson, T. (1996b) A discovery algorithm for directed cyclic graphs. *Proceedings of the 12th Conference of Uncertainty in AI*, Portland, Ore.: Morgan Kaufmann, 454–461.

Richardson, T. (1996c). A polynomial-time algorithm for deciding Markov equivalence of directed cyclic graphical models. *Proceedings of the 12th Conference of Uncertainty in AI*, Portland, Ore.: Morgan Kaufmann, 462–469.

Richardson, T. (1998). Chain graphs and symmetric associations. *Learning in Graphical Models*. M. Jordan. Cambridge, Mass.: MIT Press.

Richardson, T. (1999b). A local Markov property for acyclic directed mixed graphs. *Proceedings of 52nd Session of the International Statistical Institute*, Aug. 10–18, Helsinki, Finland.

Richardson, T., Bailer, H., and Banerjee, M. (1999). Specification searches using MAG models. *Proceedings of 52nd Session of the International Statistical Institute*, Aug. 10–18, Helsinki, Finland.

Richardson, T., and Spirtes, P. (1999). Parameterizing and Scoring Mixed Ancestral Graphs. Carnegie Mellon Department of Philosophy Technical Report Phil 102.

Rindfuss, R., Bumpass, L., and St. John, C. (1980). Education and fertility: Implications for the roles women occupy. *American Sociological Review* 45: 431–447.

Rissanen, J. (1987). Stochastic complexity (with discussion). *J. Roy. Stat. Soc.* 49(Ser. B): 223–239 and 253–265.

Robins, J. (1986). A new approach to causal inference in mortality studies with sustained exposure period—applications to control of healthy workers survivor effect. *Mathematical Modelling* 7: 1393–1512.

Robins, J. (1987). Addendum to "A new approach to causal inference in mortality studies with sustained exposure period – applications to control of healthy workers survivor effect." *Computers and Mathematics with Applications* 14: 923–945

Robins, J. (1993). Analytic methods for estimating HIV-treatment and cofactor effects. Methodological Issues in AIDS Mental Health Research, edited by D. Ostrow and R. Kessler. New York: Plenum, 213–290.

Robins, J. (1994). Correcting for noncompliance in randomized trials using structural nested mean models. *Communications in Statistics* 23: 2379–2412.

Robins, J. (1995). Discussion of "Causal Diagrams for Empirical Research" by J. Pearl. *Biometrika* 82: 695–698.

Robins, J. (1997). Causal inference from complex longitudinal data. *Latent Variable Modeling and Applications to Causality*, M. Berkane. Lecture Notes in Statistics 120. New York: Springer Verlag.

Robins, J. (1998). Marginal Structural models. *Proceedings of the American Statistical Association*, Chicago, 1–10.

Robins, J. (1999). Testing and estimation of direct effects by reparameterizing directed acyclic graphs with structural nested models. *Computation, Causation and Discovery*. C. Glymour and G. Cooper. Cambridge, Mass.: MIT Press.

Robins, J., and Greenland, S. (1989). The probability of causation under a stochastic model for individual risk. *Biometrics* 45: 1125–1138.

Robins, J., Scheines, R., Spirtes, P., and Wasserman, L. (1999). The Limits of Causal Knowledge. Carnegie Mellon University Philosophy Department Technical Report Phil 97.

Robins, J., and Wasserman, L. (1999). On the impossibility of inferring causation from association without background knowledge. In *Computation, Cuasation, and Discovery*, edited by C. Glymour and G. Cooper. Cambridge, Mass.: MIT Press.

Rodgers, R., and Maranto, C. (1989). Causal models of publishing productivity in psychology. *Journal of Applied Psychology* 74: 636–649.

Rose, G. Hamilton, P. Colwell, L., and Shipley, J. (1982). A randomised controlled trial of anti-smoking advice: 10-year results. *Journal of Epidemiology and Community Health* 36: 102–108.

Rosenbaum, P. (1984). From association to causation in observational studies. *JASA* 79: 41–48.

Rosenbaum, P. (1995). *Observational Studies*. New York: Springer-Verlag.

Rubin, D. (1974). Estimating causal effects of treatments in randomized and nonrandomized studies. *Journal of Educational Psychology* 66: 688–701.

Rubin, D. (1977). Assignment to treatment group on the basis of a covariate. *Journal of Educational Statistics* 2: 1–26.

Rubin, D. (1978). Bayesian inference for causal effects: The role of randomizations. *Annals of Statistics* 6: 34–58.

Rubin, D. (1986). Comment: Which ifs have causal answers. *Journal of the American Statistical Association* 81: 396.

Russell, S., Binder, J., Koller, D., and Kanazawa, K. (1995). Local learning in probabilistic networks with hidden variables. *Proceedings of the International Joint Conference on AI.* San Francisco: Morgan Kaufmann.

Salmon, W. (1980). Probabilistic causality. *Pacific Philosophical Quarterly* 61: 50–74.

Salmon, W. (1984). *Scientific Explanation and the Causal Structure of the World.* Princeton, N.J.: Princeton University Press.

Saris, W., and Stronkhorst, H. (1984). *Causal Modeling in Nonexperimental Research.* Amsterdam, Sociometric Research Foundation.

Scheines, R. (1988). Automating creativity. *Aspects of Artificial Intelligence.* J. Fetzer. Boston: Kluwer.

Scheines, R., Spirtes, P., Glymour, G., and Sorensen, S. (1990). Causes of Success and Satisfaction Among Naval Recruiters. San Diego, Calif.: Navy Personnel Research Development Center.

Scheines, R., Spirtes, P., and Glymour, C. (1990). A qualitative approach to causal modeling. *Qualitative Simulation Modeling and Analysis.* P. Luker and P. Fishwick. New York: Springer-Verlag, 72–97.

Scheines, R. (1993). Unidimensional linear latent variable models. *Technical Report CMU-PHIL–39.* Carnegie Mellon University, Pittsburgh.

Scheines, R., Spirtes, P., Glymour, C., and Meek, C. (1994). *Tetrad II: User's Manual.* HIllsdale, N.J.: Lawrence Erlbaum.

Scheines, R. (1997). Estimating latent causal influence: TETRAD II model selection and Bayesian parameter estimation. *Proceedings of the 6th International Workshop on Artificial Intelligence and Statistics.* Fort Lauderdale, Fla., 445–456.

Scheines, R., Boomsma, A., and Hoijtink, H. (1997). The mulitmodality of the likelihood function in structural equation models. Pittsburgh.: Department of Philosophy, Carnegie Mellon University.

Scheines, R., Spirtes, P., Glymour, C., Meek, C., and Richardson, T. (1998). The TETRAD Project: constraint based aids to causal model specification. *Multivariate Behavioral Research* 33: 65–118.

Scheines, R., Boomsma, A., and Hoijtink, H. (1999). Bayesian Estimation and Testing of Structural Equation Models. *Psychometrika* 64: 37–52.

Scheines, R., and Spirtes, P. (1992). Finding latent variable models in large data bases. *International Journal of Intelligent Systems* 7: 609–622.

Schwarz, G. (1978). Estimating the dimension of a model. *Annals of Statistics* 6: 461–464.

Sclove, S. (undated). On Criteria for Choosing a Regression Equation for Prediction. Pittsburgh: Department of Statistics, Carnegie Mellon University.

Settimi, R., and Smith, J. Geometry, moments, and Bayesian networks with hidden variables. *Artificial Intelligence and Statistics 99*. San Francisco: Morgan Kaufman, 293–298.

Sewell, W., and Shah, V. (1968). Social class, parental encouragement, and educational aspirations. *American Journal of Sociology* 73: 559–572.

Shafer, S. (1996). *The Art of Causal Conjecture*. Cambridge, Mass.: MIT Press.

Shigemasu, K. (1999). Computing Bayes factors for structural equation models. *Proceedings of the 1999 Meetings of the European Psychometric Society*, Luneburg, Germany.

Shipley, W. (1995). Structured interspecific determinants of specific leaf area in 34 species of herbaceous angeosperms. *Functional Ecology* 9: 312–319.

Shipley, W. (1997). Exploratory path analysis with applications in ecology and evolution. *American Naturalist* 149: 1113–1138.

Shipley, W. (1999). Exploring hypothesis space: examples from organismal biology. *Computation, Causation and Discovery,* edited by C. Glymour and G. Cooper. Cambridge, Mass.: MIT Press.

Shipley, W. (1999). Testing causal explanations in organismal biology: causation, correlation and structural equation modelling. Oikos (in press).

Simon, H. (1953). Causal ordering and identifiability. *Studies in Econometric Methods*. Hood and Koopmans. New York: Wiley, 49–74.

Simon, H. (1954). Spurious correlation: a causal interpretation. *JASA* 49: 467–479.

Simon, H. (1977). *Models of Discovery*. Dordrecht: D. Reidel.

Simpson, C. (1951). The interpretation of interaction in contingency tables. *J. Roy. Statist. Soc.* 13(Ser. B): 238–241.

Sims, C. (1972). Money, income, and causality. *American Economic Review* 62: 540–552.

Singh, M., and Valtorta, M. (1993). An algorithm for the construction of Bayesian network structure from data. *Proceedings of the Ninth Conference on Uncertainty in AI*. San Francisco: Morgan Kaufmann, 259–265.

Skyrms, B. (1980). *Causal Necessity: A Pragmatic Investigation of the Necessity of Laws*. New Haven: Yale University Press.

Smith, A., and Roberts, G. (1993). Bayesian computation via the Gibbs sampler and related Markov chain Monte Carlo methods. *Journal of the Royal Statistical Society* 55(Ser. B): 3–23.

Sober, E. (1987). The principle of the common cause. *Probability and Causality*. J. Fetzer. Dordrecht: D. Reidel.

Sorbom, D. (1975). Detection of correlated errors in longitudinal data. *British Journal of Mathematical and Statistical Psychology* 28: 138–151.

Sosa, E., and Tooley, M., eds. (1993). *Causation*. New York: Oxford University Press.

Spearman, C. (1904). General intelligence objectively determined and measured. *American Journal of Psychology* 15: 201–293.

Spence, M. (1973). Job market signalling. *Quarterly Journal of Economics* 87: 355–379.

Spiegelhalter, D. (1986). Probabilistic reasoning in predictive expert systems. *Uncertainty in Artificial Intelligence,* edited by K. Lemmer and J. Kanal. Amsterdam: North-Holland.

Spiegelhalter, D., and Knell-Jones, R. (1984). Statistical and knowledge-based approaches to clinical decision-support systems. *J. Royal Statist. Soc.* 147(Ser. A): 35–77.

Spiegelhalter, D., and Lauritzen, S. (1990). Sequential updating of conditional probabilities on directed graphical structures. *Networks* 20: 579–605.

Spiegelhalter, D., and Lauritzen, S. (1995). Sequential updating of conditional probabilities on directed graphical structures. *Networks* 20: 579–605.

Spirtes, P. (1989). Fast Geometrical Calculations of Overidentifying Constraints. Pittsburgh: Laboratory for Computational Linguistics, Carnegie Mellon University.

Spirtes, P. (1989). A Necessary and Sufficient Condition for Conditional Independencies to Imply a Vanishing Tetrad Difference. Pittsburgh: Laboratory for Computational Linguistics, Carnegie Mellon University.

Spirtes, P. (1994a). Building causal graphs from statistical data in the presence of latent variables. *Logic, Methodology, and Philsophy of Science IX,* edited by D. Prawitz, B. Skyrms, and D. Westerståhl. Amsterdam: north Holland, 813–832.

Spirtes, P. (1994b). Conditional Independence in Directed Cyclic Graphical Models for Feedback. Pittsburgh: Dept. of Philosophy, Carnegie Mellon University.

Spirtes, P. (1995). Directed cyclic graphical representation of feedback models. *Proceedings of the Eleventh Conference on Uncertainty in Artificial Intelligence*, San Francisco: Morgan Kaufmann, 491–498.

Spirtes, P. (1996). Discovering Causal Relations Among Latent Variables in Directed Acyclic Graphical Models. Pittsburgh: Department of Philosophy, Carnegie Mellon University.

Spirtes, P., and Cooper, G. (1998). An experiment in causal discovery. *Artificial Intelligence and Statistics '99,* edited by D. Heckerman and J. Whittaker. San Francisco: Morgan Kaufman, 162–168.

Spirtes, P., and Glymour, C. (1988). Latent variables, causal models and overidentifying constraints. *Journal of Econometrics* 39: 175–198.

Spirtes, P., and Glymour, C. (1990). Causal Structure Among Measured Variables Preserved with Unmeasured Variables. Pittsburgh: Laboratory for Computational Linguistics, Carnegie Mellon University.

Spirtes, P., and Meek, C. (1995). Learning Bayesian networks with discrete variables from data. *Proceedings of the First International Conference on Knowledge, Discovery and Data Mining.* San Francisco: Morgan Kaufmann.

Spirtes, P., and Richardson, T. (1996). A polynomial time algorithm for determining DAG equivalence in the presence of latent variables and selection bias. *Proceedings of the 6th International Workshop on Artificial Intelligence and Statistics,* Fort Lauderdale, Fla., 489–500.

Spirtes, P., and Verma, T. (1992). Equivalence of causal models with latent variables. Pittsburgh: Carnegie Mellon University.

Spirtes, P., Glymour, C., and Scheines, R. (1990a). Causality from probability. *Evolving Knowledge in Natural Science and Artificial Intelligence,* edited by J. Tiles et al. London: Pitman, 181–199.

Spirtes, P., Glymour, C., and Scheines, R. (1990b). Causality from probability. *Proceedings of the Conference on Advanced Computing for the Social Sciences*, Williamsburg, Va.

Spirtes P., Glymour C., and Scheines, R. (1991a). An algorithm for fast recovery of sparse causal graphs. *Social Science Computer Review* 9: 62–72.

Spirtes P., Glymour C., and Scheines, R. (1991b). From probability to causality. *Philosophical Studies* 64: 1–36.

Spirtes, P., Glymour, C., and Scheines, R.(1997). Reply to Humphreys' and Freedman's review of Causation, Prediction, and Search. *British Journal for the Philosophy of Science*, 48, 555–568.

Spirtes, P., Glymour, C., Scheines, R., and Sorensen, S. (1990). TETRAD Studies of Data for Naval Air Traffic Controller Trainees. San Diego, Calif.: Navy Personnel Research Development Center.

Spirtes, P., Meek, C., and Richardson, T. (1995). Causal inference in the presence of latent variables and selection bias. *Proceedings of the Eleventh Conference on Uncertainty in AI*, San Francisco: Morgan Kaufmann, 499–506.

Spirtes, P., Meek, C., and Richardson, T. (1999). An algorithm for causal inference in the presence of latent variables and selection bias. *Computation, Causation and Discovery*, edited by C. Glymour and G. Cooper. Cambridge, Mass.: MIT Press.

Spirtes, P., Richardson, T., and Meek, C. (1996). Heuristic greedy search algorithms for latent variable models. *Proceedings of the 6th International Workshop on Artificial Intelligence and Statistics,* Fort Lauderdale, Fla., 481–488.

Spirtes, P, Richardson, T., and Meek, C. (1997). The Dimensionality of Mixed Ancestral Graphs. Technical Report CMU-PHIL-83, Department of Philosophy, Carnegie Mellon University, November, 1997.

Spirtes, P., Richardson, T., Meek, C., Scheines, R., and Glymour, C. (1996). Using D-separation to calculate zero partial correlations in linear models with correlated errors. Tehcnical Report Phil-72, Deparatment of Philosophy, Carnegie Mellon University.

Spirtes, P., Richardson, T., Meek, C., Scheines, R., and Glymour, C. (1998). Using path diagrams as a structural equation modeling tool. *Sociological Methods and Research*, 27, 148–181.

Spirtes, P., Scheines, R., and Glymour, C. (1990). Simulation studies of the reliability of computer aided specification using the TETRAD II, EQS, and LISREL Programs. *Sociological Methods and Research* 19: 3–66.

Spohn, W. (1983). Deterministic and probabilistic reasons and causes. *Methodology, Epistemology, and Philosophy of Science: Essays in Honour of Wolfgang Stegmuller on the Occasion of his 60th Birthday*, edited by H. Putnam, C. Hempel, and W. Essler. Dordrecht: D. Reidel, 371–396.

Spohn, W. (1990). Direct and indirect causes. *Topoi* 9: 125–145.

Spohn, W. (1991). On Reichenbach's principle of the common cause. *Proceedings of the First Pittsburgh-Konstanz Colloquium.*

Spohn, W. (1992). Causal laws are objectifications of inductive schemes. *Theory of Probability.* J. Dubucs. Dordrecht: Kluwer.

Stein, C. (1960). Multiple regression. *Contributions to Probability and Statistics. Essays in Honor of Harold Hotelling.* I. Olkin. Stanford, Stanford University Press.

Stetzl, I. (1986). Changing causal relationships without changing the fit: Some rules for generating equivalent LISREL models. *Multivariate Behavior Research* 21: 309–331.

Strotz, R., and Wold, H. (1960). Recursive versus nonrecursive systems: an attempt at synthesis. *Econometrica* 28: 417–427.

Studený, M. (1992). Conditional independence relations have no finite complete characterization. *Information Theory, Statistical Decision Functions and Random Processes: Proceedings of the 11th Prague Conference*, ed. by S. Kubík and J. Višek., Dordrecht: Kluwer, 377–396.

Studený, M., and Bouckaert, R. (1998). On chain graph models for descriptions of conditional independence structures. *Annals of Statistics* 26: 1434–1495.

Suppes, P. (1970). *A Probabilistic Theory of Causality*. Amsterdam, North-Holland.

Suppes, P., and Zanotti, M. (1981). When are probabilistic explanations possible. *Synthese* 48: 191–199.

Surgeon General of the United States (1964). *Smoking and Health*. Washington, D.C., U.S. Government Printing Office.

Surgeon General of the United States (1979). *Smoking and Health*. Washington, D.C., U.S. Government Printing Office.

Swamy, P. (1971). *Statistical Inference in Random Coefficient Regression Models*. Berlin, Springer-Verlag.

Tanner, M.A. (1993). *Tools for Statistical Inference: Methods for the Exploration of Posterior Distributions and Likelihood Functions*. New York: Springer.

Thiesson, B. (1995). Score and information for recursive exponential models with incomplete data. Aalborg, Denmark, Institute of Electronic Systems, Aalborg University.

Thomson, G. (1916). A hierarchy without a general factor. *British Journal of Psychology* 8: 271–281.

Thomson, G. (1935). On complete families of correlation coefficients and their tendency to zero tetrad-differences: Including a statement of the sampling theory of abilities. *British Journal of Psychology* 26: 63–92.

Thurstone, L. (1935). *The Vectors of Mind*. Chicago: University of Chicago Press.

Tierney, L. (1994). Markov chains for exploring posterior distributions (with discussion). *Annals of Statistics* 22: 1701–1762.

Timberlake, M., and Williams, K. (1984). Dependence, political exclusion, and government repression: Some cross-national evidence. *American Sociological Review* 49: 141–146.

Tolun, M., and Guven, D. (1991). Analyzing the inter and intra vocabulary performances in isolated speech recognition. *Journal of American Voice Input/Output Society* 10: 19–37.

Verma, T. (1987). Causal networks: semantics and expressiveness. Los Angeles, Cognitive Systems Laboratory, University of California.

Verma, T., and Pearl, J. (1990). Equivalence and synthesis of causal models. *Proceedings of the Sixth Conference on Uncertainty in AI*, Mountain View, Calif.: Association for Uncertainty in AI.

Verma, T., and Pearl, J. (1990). On equivalence of causal models. Los Angeles, Cognitive Systems Laboratory, University of California.

Verma, T., and Pearl, J. (1991). Equivalence and synthesis of causal models. Los Angeles, Cognitive Systems Laboratory, University of California.

Verma, T., and Pearl J. (1990). Causal networks: semantics and expressiveness. *Uncertainty in Artificial Intelligence 4*. T. Levitt R. Shacter, L. Kanal, J. Lemer. North-Holland, Elsevier Science.

Waldemark, J., and Norqvist, P. In-flight calibration of satellite ion compositon data using artificial intelligence methods. *Computation, Causation and Discovery,* edited by C. Glymour and G. Cooper. Cambridge, Mass.: MIT Press.

Wallace, C., and Freeman, P. (1987). Estimation and inference by compact coding. *J. Roy. Statist. Soc.* 49(Ser. B): 240–265.

Wedelin D. (1996). Efficient estimation and model selection in large graphical models. *Statistics and Computing* **6**: 319–323.

Weisberg, S. (1985). *Applied Linear Regression*. New York: Wiley.

Wermuth, N. (1976). Model search among multiplicative models. *Biometrika* 32: 253–363.

Wermuth, N. (1980). Linear recursive equations, covariance selection and path analysis. *JASA* 75: 963–972.

Wermuth, N., and Lauritzen, S. (1983). Graphical and recursive models for contingency tables. *Biometrika* 72: 537–552.

Wermuth, N., and Lauritzen, S. (1990). On substantive research hypotheses, conditional independence graphs and graphical chain models. *J. Roy. Statist. Soc.* 52(Ser. B): 21–50.

Wermuth, N., Cox, D., and Pearl, J. (1994, revised 1998) Explanations for multivariate structures derived from univariate recursive regressions. Center of Survey Research and Methodology, ZUMA, Mannheim, FRG.

Wermuth, N., Cox, D., Richardson, T., and Glonek, G. (1999). On transforming and generating cyclic graph models. ZUMA Tech report, July 1999.

Wheaton, B., Muthen, B., Alwin, D., and Summers, G. (1977). Assessing Reliability and Stability in Panel Models. *Sociological Methodology*. D. Heise. San Francisco, Jossey-Bass.

Whitney, H. (1957). Elementary structures of real algebraic varieties. *Annals of Mathematics* 66: 545–556.

Whittaker, J. (1990). *Graphical Models in Applied Multivariate Statistics*. New York: Wiley.

Winkler, R. (1967). The assessment of prior distribution in Bayesian analysis. *American Statistical Association Journal* 62: 776–800.

Wise, D. (1975). Academic achievement and job performance. *American Economic Review* 65: 350–366.

Wishart, J. (1928). Sampling errors in the theory of two factors. *British Journal of Psychology,* 19: 180–187.

Wright, S. (1934). The method of path coefficients. *Annals of Mathematical Statistics* 5: 161–215.

Younger, M. (1978). *Handbook for Linear Regression*. North Scituate, Mass.: Duxbury Press.

Yule, G. (1903). Notes on the theory of association of attributes in statistics. *Biometrika* 2: 121–134.

Index

2x2 Foursomes, 257–259
3x1 Foursomes, 259–260
Acceptable Ordering, 176
Active, 44
Acyclic Path, 8
Adjacency, 7–8
Adjacent, 7–8
Aggregation, 296
Aitkin, M. 174
Akleman, D. 369
Alchourron, C. 376
Aliferis, C., 333
Almost Pure Latent Variable Graph, 255
Almost Pure Measurement Model, 254, 468
Ancestor, 10, 302, 304
Anderson, J., 362
Anderson, T., 94
Andersson, S., 303, 304
Anterior, 303
Artzenius, F., 296
Augmented Bi-Flag, 304
Axioms
  Causal Markov Condition, 29–30, 32–38
    Statement, 29
  Causal Minimality Condition, 30–31
    Statement, 31
  Faithfulness, 31
  Faithfulness Condition
    Statement, 31
  Markov, 11

Baldwin, B., 364
Balke, A., 308, 309
Banwart, A., 109
Basmann, R., 71
Bayes Consistency, 317–320
Bayesian Information Criterion, 343
Bayesianism, 41–43, 226–234, 269
  Search Strategies, 78–79, 329–346
Bearden, W., 364
Becker, S., 343
Belief Revision, 376
Bell, J., 37
Bentler, P., 74, 103–105, 269, 347, 364, 478
Bernardo, J., 479
Bessler, D., 369
Best, N., 364
Bibby, J., 152
BIC. *See* Bayesian Information Criterion
Bi-flag, 304
  Augmented, 304
BIFROST, 479
Birch, M., 74, 75
Bishop, E., 17
Bishop, Y., 74, 476
Blalock, H., 47, 57
Blau, P., 105–108
Blyth, C., 217, 475
Bollen, K., 76, 151, 271, 297
Boomsma, A., 362, 364, 365
Bootstrapping, 352
  Non-Parameteric, 352
  Parameteric, 352
Bounds, 328, 366
Boyen, X., 310, 349
Breslow, N., 124
Brownlee, K., 239, 244–246, 249
Bumpass, L., 102–103
Burch, P., 239, 246–249
Byron, R., 74
BUGS, 364
Buntine, W., 329, 336, 344, 346

Califano, J., 239

Calculability, 313–314
Cambell, D., 282
Carlin, J., 17, 122
Caramazza, A., 34
Cartwright, N., 67, 296
Causal Chain, 22
Causal Graph, 24
Causal Inference Algorithm, 138–140
   Statement, 138–139
Causal Markov Condition, 29–30, 32–38, 295–297, 330
   Statement, 29
Causal Mediary, 20
Causal Minimality Condition, 30–31
   Statement, 31
Causal Representation Convention, 24
Causal Structure, 22
   Deterministic, 25–27
   Generates Probability Distribution, 27, 30
   Generation, 22
   Genuinely Indeterministic, 28–29
   Indeterministic, 27
   Isomorphism, 22
   Linear Deterministic, 25
   Linear Pseudo-indeterministic, 28
Casella, G., 364
Causal System
   Causally Sufficient, 22
Causally Connected, 22
Causally Sufficient, 22, 296
Cause
   Boolean, 21
   Common, 22, 24
   Direct, 20
   Indirect, 22
   Representation, 24
   Scaled Variable, 21
Cederlof, R., 245
Chain Graph, 303–305
Cheeseman, P., 343
Chib, S., 341, 364
Chickering, D.. 333, 343
Child. 8
Children. 8
Choke Point. 151
Chou, C.. 364
Christensen, R.. 74, 112–113
CI. *See* Causal Inference Algorithm
CI Partially Oriented Inducing Path Graph, 146
Clique, 10
   Maximal, 10
Coleman, J., 110–111
Collider, 10, 138, 302
Combined Graph, 50
Common Cause, 22, 24
Common Cause Impure, 255
Complete Graph, 9
Complex, 303
Composition, 312
Concatenation, 8
Conditional, 19–20
Conditional Independence, 10–11
   Observed, 299
Connected, 304
Connected Graph, 9
Consistency
   Bayes, 317–320
   Pointwise, 321–322
   Uniform, 322–326
Constraints
   Non-independence, 147–148
Cooper, G., 78, 109, 269, 329, 330, 332, 333, 339, 350, 369, 370, 373, 476, 479
Cornfield, J., 239, 241–244
Costner, H., 57, 282, 478
Covariance Structures, 74
Cox, D,. 303, 305
Crawford, S., 77, 91, 346
Cross-Construct Foursomes, 256–260
Cross-Construct Impure, 255
Cyclic, 8, 297, 301

Cyclic Discovery Algorithm, 301, 348

d-connection, 14, 44, 55
D-SEP, 134
d-separation, 14, 43–47, 54–55, 81
  Cyclic, 298
  Faithful, 81
  Order, 261
DAG-Isomorph, 13
Darroch, J., 17, 475
Data Faithful, 81
Data Sets
  Abortion Opinions, 112–113
  AFQT, 195–196, 476, 477
  Alarm Network, 108–109, 476
  American Occupational Structure, 105–108
  Biological, 374
  College Dropouts, 367–368
  College Plans, 111–112
  Education and Fertility, 102–103, 351
  Economic Forecasting, 369
  Female Orgasm, 103–105
  Infant Mortality, 373–374
  Lead and IQ, 365–367
  Leading Crowd, 110–111
  Mathematical Marks, 152–154
  Mineral Identification, 375–376
  Pneumonia, 369–371
  Political Exclusion, 200–201
  Publishing Productivity, 97–102, 126
  Rat Liver 203–207, 352, 374
  Recalibration, 368–369
  Simulation Studie,s 121, 201–202, 265–266, 276–277, 293
    Results, 115–121
    Sample Generation, 114–115
  Spartina grass, 196–200, 351, 368, 477
  Virginity, 109–110
Dai, H., 349
Daughter, 8

Davis, W., 35
Dawid, P., 311
Day, N., 124
de Campos, L., 348
Dechter, R., 298
DeFazio, J., 375
Definite d-connection, 145–146
Definite Discriminating Path, 138
Definite Non-Collider, 136, 178, 264
Definite-Non-Descendant, 176
Definite-SP, 177–178
Definition, 296
Degree, 8
Dempster, A., 74, 344
Descendant, 10
Descendants, 11
Desjardins, B., 302
Det, 54
Deterministic, 54–55
Deterministic Causal Structure, 25–27
Deterministic Graph, 15
Deterministic System, 15–16
Deterministic Variable, 54
DiCiccio, T., 341
Direct Cause, 20
Direct Manipulation, 50
Directed Acyclic Graph, 9, 11–13
  DAG-Isomorph, 13
  I-map, 11
  Linear Implication, 151
  Minimal I-map, 12
  Perfect Map, 13
  Represents Probability Distribution, 12
Directed Edge, 8
  Head, 9
  Tail, 9
Directed Graph, 5, 6, 8
  Acyclic, 9, 11–13
Directed Independence Graph, 13, 79
Dirichlet Distribution, 332, 340
Directed Path, 5, 8–9

Discovery, 73
Discovery Problems, 73
Discretization, 350
Distribution Equivalence, 305, 334
Doll, R., 239
Druzdzel, M,. 367, 476
Duncan, O., 105–108

Edge, 7, 8
  Directed, 8, 300, 302
  Double-headed, 300, 302
  Edge-end, 7
  Endpoint, 7
  Into, 8
  Out of, 8
  Semidirected, 300, 302
  Undirected, 300, 302
Edge-end, 7
Effectiveness, 312
Edwards, A., 94
Edwards, D., 78, 295
Elby, A., 37
Empty Path, 8
EM Algorithm, 344–345
Endogenous Variable, 25
Endpoint, 7
EQS, 74, 77, 102, 106, 108, 269, 347, 478
  Automatic Model Modification, 274–276
  Boldness, 289
  Lagrange Multiplier Statistic, 276
  Reliability, 289
  Search, 276
Equiv, 135
Equivalence
  Distribution, 305, 334
  Markov, 300, 334
  O-distribution, 305
  O-Markov, 300
  Likelihood, 334
  Strong Likelihood, 345
Ernhart, C., 366

Error
  Edge Direction of Commision, 109
  Edge Direction of Omission, 109
  Edge Existence of Commision, 109
  Edge Existence of Omission, 109
Error Variable, 15
Estimation, 97
  Multinomial Distributions, 97
Ethics
  Experimental Design, 234–238
Exogenous Variable, 25, 50, 312
Expanded Graph, 15
Experimental design, 226–234
  Ethical, 234–238
  Prospective, 223–224
  Randomized, 210
  Retrospective, 223–224
Exogenous Vertex, 12

F.I. *See* Faithful Indistinguishability
Factor Analysis, 76, 153
Faithful Indistinguishability, 61. *See also*
  Markov equivalence
Faithfulness, 13–14, 38–42
Faithfulness Condition, 13, 31, 38–42
  Statement, 31
Fast Causal Inference Algorithm, 140–146, 480
  Statement, 144–145
FCI. *See* Fast Causal Inference Algorithm
FCI Partially Oriented Inducing Path
  Graph, 146
Featherman, D., 107
Fienberg, S., 17, 57, 74, 75, 110, 190, 475, 476, 477
Fine, M., 371
Fisher, F., 71, 299
Fisher, R., 209–210, 239–241, 248
Foreman, H., 109
Foursomes
  Cross-Construct, 256–260

Intra-Construct, 256–257
Fox, J., 191
Frank, R., 365
Freedman, D., 105, 297
Friberg, L., 245
Friedman, M., 92
Friedman, N., 344, 349, 350, 352, 374
Frydenberg, M., 303, 304
Functional Determination, 54
Fung, R., 77, 91

Galles, D., 308, 312, 313
Gazis, P.,375
Gaussian Approximation, 341–343
Geiger, D., 25, 44, 57–58, 178, 302, 306, 307, 334, 343, 346
Geiger, S., 365
Gelfand, A., 364
Geman, D., 340, 364
Geman, S., 340, 364
Genetic, 348
Generation, 22, 27, 29
Genuinely Indeterministic Causal Structure, 28–29
George, E., 364
Gerbing, D., 362
Gibbs Sampling, 340–341, 364
Gilks, W., 365
Glonek, G. 305
Glymour, C., 57–58, 76, 122, 155, 367, 477, 478
Gold, M., 121
Goldszmidt, M., 297, 350, 376
Gradient Ascent, 344
Graph, 5, 7
  Adjacent, 7
  Causal, 24
  Chain, 303–305
    Andersson-Madigan-Perlman, 304
    Lauritzen-Wermuth-Frydenberg, 303
  Clique, 10

Combined, 50
Complete, 9
Connected, 9
d-connection, 44, 55
d-separation, 44, 55
Deterministic, 15
Directed, 5, 6
Directed Acyclic, 9, 11–13
Directed Cyclic, 297
Directed Independence, 13, 79
Edge, 7, 8
Expanded, 15
Faithful Representation of D-separations, 81
Inducing Path, 7, 132–135
Manipulated, 50–51
Mixed Ancestral, 301–303
Moral, 303
Over, 7
Parallel Embedding, 65
Partial Ancestral, 299–301
Partially Oriented Inducing Path, 5, 135–138
Pattern, 6, 61, 81
Subgraph, 10
Summary, 303
Trek, 67, 151
Trek Sum, 67
Triangle, 10
Underlying, 15
Undirected, 5
Undirected Independence, 15, 77
Unmanipulated, 50
Greenberg, E., 364
Greene, T., 366
Greenland, S., 312
Guven, D., 369

Haenszel, W., 239
Halpern, J., 312
Hammond, E., 239

Haughton, D., 343
Hausman, D., 297
Havranek, T., 78
Head, 9
Heckerman, D., 25, 329, 333, 334, 335, 336, 343, 345, 346, 371, 376, 479
Heise, D., 47
Henry, N., 57
Herskovits, E., 78, 109, 269, 330, 332, 333
Herting, J., 478
Hidden Variables. *See* Missing Data
Hierarchical Models, 75
Hill, A., 239
Hoijtink, H., 362, 365
Holland, P., 17, 74, 124, 159–167, 476
Hojsgaard, S., 479
Hoogland, J., 364
Howson, C., 209
Hu, L., 364
HUGIN, 479
Huete, J., 348
Humphreys, P., 297

I-map, 11
IG Algorithm, 90–91
Identification, 307
Identifiability, 313–314
Implication
  Linear, 15
Impure Indicators
  Common Cause, 255
  Cross-Construct, 255
  Intra-Construct, 255–257
  Latent-Measured, 255
Indegree, 8
Independence, 10–11
Indeterministic Causal Structure, 27
Indeterministic System, 16
Indirect Cause, 22
Inducing Path Graph, 5–6, 132–135
  D-separability, 134

SP, 177
Informative Parents, 164
Informative Variables, 164
Instrumental Variable, 171, 366
Instrumental Variable Algorithm, 370–371
Intersect, 8
Into, 8, 138
Intra-Construct Foursomes, 256–257
Intra-Construct Impure, 255–257
Invariance, 163–164, 457–458
IP, 164
Isomophic Causal Structures, 22
IV, 164
Iwasaki, Y. 376

Jeffreys, H., 122
Jensen, F., 342
Joint Independence, 10
Jordan, M., 329
Joreskog, K., 74, 76, 97, 269, 347, 364

Kadane, J., 210, 226–234, 478
Kano, Y., 364
Kass, R., 341, 343
Katsumo, H., 376
Kelly, K., 57, 155, 324, 376, 478
Kelly, T., 123
Kendall, M., 38–40, 475
Kent, J., 152
Kiiveri, H., 17, 57, 97, 122
Klein, L., 123
Kleinbaum, D., 124
Klepper, S., 92, 365, 366
Klir, G., 74
Kohn, M., 254, 277
Korb, K., 297
Koster, J., 298
Krishnan, T., 345
Kullback, S., 75
Kullback-Liebler Distance, 234
Kupper, L., 124

Lagrange Multiplier Statistic, 276
Laplace, P., 343
Laplace Approximation, 343
Latent Variable Graph
  Almost Pure, 254
Latent Variable Models
  Cross-Construct Foursome, 256
  Intra-Construct Foursome, 256
Latent Variables, 22, 253
Latent-Measured Impure, 255
Lauritzen, S., 15, 17, 79, 295, 303, 304, 340, 478
Lazarsfeld, P., 57
LCD2 Algorithm, 373–374
Leamer, E., 78, 366
Lecun, Y., 343
Lee, S. 71
Lemmer, J., 296,
Lewis, D., 376
Lilienfeld, A., 239, 241, 247–249
Linear Deterministic Causal Structure, 25
Linear Faithfulness, 47
Linear Implication, 15, 151
Linear Model, 151
Linear Pseudo-Indeterministic, 16
Linear Pseudo-indeterministic Causal Structure, 28
Linear Pseudo-Indeterministic Models, 253
Linear Regression, 71, 77
Linear Representation, 14
Linear Statistical Indistinguishability, 65–69
Linthurst, R., 196–200
LISREL ,74, 76, 77, 106, 269, 347, 477, 478
  Automatic Model Modification, 274–276
  Boldness, 289
  Modification Indices, 275
  Reliability, 289
  Search, 275–276

Log-linear models, 74–77
  Hierarchical Models, 75–77
  Representing Colliders, 76
Lohmoller, J., 97
Loper, T., 369
Lundman, T., 245
Lung Cancer, 239–249

Madigan, D., 303, 304, 309, 333, 336, 341
MAG. *See* Mixed Ancestral Graph
Mani, S., 373
Manipulated, 50
Manipulated Graph, 50–51
Manipulation, 50
  Direct, 50
  Invariance, 163–164, 457–458
Manipulation Theorem, 47–53
  Statement, 51
Manski, C., 328
Marginal Likelihood 330
MAP Approximation, 344–345
Maranto, C., 97–102, 126
Mardia, K., 152
Marginal, 11
Markov Condition, 11–12
  Andersson-Madigan-Perlman, 304
  Lauritzen-Wermuth-Frydenberg, 303
Markov Equivalence, 300, 334
Marks, 7
Maruyama, G., 282
Mathematica, 480
Matuš, F., 376
Maudlin, T., 295
Maximal Clique, 10
Maximally Informative Partially Oriented Inducing Path Graph, 136
M-connection, 302
M-separation, 302
Maximum Likelihood Estimation, 344
  Fitting Function, 275
McKenna, M., 374

McLachlan, G., 345
McGarvey, B., 282
McPherson J., 277
Measurement Model
  Almost Pure, 254, 468
Meek, C., 57, 301, 306, 307, 329, 330, 333, 479
Mendelson, A., 376
Meng, X., 342
MIMBuild Algorithm
  Complexity, 265
  Generalization, 362–364
  Reliability, 264–265
  Simulation Study Results, 265
Minimal I-map, 12
Minimality Condition, 12
Minimum Message Length, 349
MINITAB, 192
Missing Data, 339–335
  Gaussian Approximation, 341–343
  EM Algorithm, 344–345
  Laplace Approximation, 343
  MAG, 346–347
  MAP Approximation, 344–345
  ML Approximation, 344–345
  Monte-Carlo, 340–341
  Open Problems, 345–346
  PAG, 347
Mitchell, T., 78
Mixed Ancestral Graph, 301–303
  Search, 346–347
Mixture, 33
ML Approximation, 344–345
Model Selection, 329
Modification Indices, 275
Modified PC Algorithm, 125–127
  Statement, 125–126
Monotonic, 312
  Stochastically, 312
Monte-Carlo Simulation, 340–341
Monti, S., 350

Moral Graph, 303
Morgenstern, H., 124
Mosteller, F., 157, 191
Multiple Risk Factor Intervention Tial
  Research Group, 248

ND, 164
Neal, R., 298, 341
Needleman, H., 365, 367, 480
Neyman, J., 239
No Descendants, 164
Node
  Visit, 272
Noncollider, 10
Norqvist, P., 368

Observed, 130
O-distribution Equivalence, 305
O-Markov Equivalence, 300
O-oracle, 306
Order
  d-separability, 261
Osherson, D., 121
Out of, 8, 138
Outdegree, 8
Over, 7

Parallel Embedding, 65
Parameter Independence, 335
Parameter Modularity, 335
Parent, 5, 8, 138
Parents, 8
Partially Oriented Inducing Path Graph, 5, 6, 135–138
  Acceptable Ordering, 176
  CI, 146–147
  Collider, 138
  Definite d-connection, 145–146
  Definite Discriminating Path, 138
  Definite Non-Collider, 136, 178
  Definite-SP, 177–178

FCI, 147
Into, 138
Maximally Oriented, 18, 37
Out of, 138
PAG. *See* Partial Ancestral Graph
Parent, 138
Possible-D-SEP, 143
Possible-SP, 177–178
Possibly d-connecting, 178
Possibly-IP, 178–179
Possibly-IV, 178–179
Semi-Directed Path, 9, 146
Partial Ancestral Graph, 299–301
  Search, 347
Parviz, B., 74
Path
  Acyclic, 8
  Adjacency, 7–8
  Collider on, 10
  Concatenation of, 8
  Cyclic, 8
  Definite Discriminating, 138
  Directed, 5, 8–9, 302
  Empty, 8
  Intersect, 8
  Into, 8
  Noncollider on, 10
  Out of, 8
  Point of Intersection, 8
  Undirected, 5, 8, 302
  Unshielded Collider on, 10
Pattern
  Definite Non-Collider, 264
  Output, 109
  True, 109
Pattern Graph, 6, 61, 81
PC Algorithm, 84–88, 338, 375, 479
  Applications, 97
  Discrete Distributions, 109
  Linear Structural Equation Models, 98
  Applied to Latent Variables, 261
  Complexity, 85–87
  Stability, 87–88
  Statement, 84–85
PC* Algorithm, 89–90
  Heuristics, 90
  Statement, 89
Pearl, J., 14, 17, 41, 44, 57–58, 72, 90–92,
    122, 147, 154, 178, 190, 207, 210, 297,
    298, 307, 308, 309, 311, 312, 313, 376
Peeler, W., 103–105
Perfect Map, 13
Perlman, M., 303, 304
PMan, 48
PN (Probability of Necessity), 312
PNS (Probability of Necessity and
    Sufficiency), 312
Point of Intersection, 8
Pointwise Consistency, 320–322
Policy Variable, 50
Political Exclusion, 200–201
Population
  Manipulated, 48
  Unmanipulated, 48
Possible-D-SEP, 143–144
Possible-SP, 177–178
Possibly D-connecting, 178
Possibly-IP, 178–179
Possibly-IV, 278–179
Power, 205
Pratt, J., 52, 158–167, 191
Predictable, 168
Prediction Algorithm, 176–190
  Examples, 181–190
  Statement, 176
Prior Model, 335
Prior Probabilities, 334–336, 353–361
  Model, 335–336, 359–361
  Parameter, 334–335, 353–359
Probability, 10–11
  Conditional Independence, 10–11
  Independence, 10–11

Probability Distribution
  Generated by Causal Structure, 27, 29
PRONEL, 479
Prospective Design, 223–224
PS (Probability of Sufficiency), 312
Pseudo-Indeterministic Causal Structure, 27–29
Pseudo-Indeterministic System, 15–16
Pseudocorrelation Matrix, 283
P_{Unman}, 50
Putnam, H., 121
Pyankov, V., 374

R.S.I. *See* Rigid Statistical
  Indistinguishability)
Raftery, A., 341, 342, 343, 346
Ramoni, M., 349
Rawlings, J., 196–200, 350, 477
Reconstructability Analysis, 74
Recursive Diagram, 79
Regression, 191–194
  Logistic, 74
  Stepwise, 74
Reiss, I., 109
Representation
  Linear, 14
Represents, 12
Retrospective Design, 223–224
Reversibility, 296
Richardson, T., 299, 300, 301, 303, 304, 305, 306, 314, 346, 347, 348
Rigid Statistical Indistinguishability, 64–65
Rindfuss, R., 102–103, 350
Rissanen, J., 343
Roberts, G., 364
Robins, J., 312, 313, 315, 320, 323, 339, 359, 479, 480
Rodgers, R., 97–102, 126
Rose, G., 247
Rosenbaum, P., 124, 328

Rubin, D., 52, 158–167, 190, 307, 309, 312, 339, 342
Russell, S., 344

S.S.I. *See* Strong Statistical
  Indistinguishability
Salmon, W., 35–37
Sampling, 222–226
Sampling Error, 296
SAS, 197, 477
Satorra, A., 364
Scheines, R., 57–58, 122, 155, 300, 315, 323, 338, 362, 365, 477, 478
Schlaifer, R., 52, 158–167, 191
Schoenberg, R., 272
Schwartz, G., 343
Selective Model Averaging, 329
Search Algorithms
  Attitude Towards Output, 350–352
  Bayesian, 329–346
    Model Priors, 335–336
    Parameter Priors, 334–335
  Cyclic Discovery Algorithm, 301, 348
  Discretization, 350
  Error Probabilitites, 203–207
  FCI Algorithm, 140–146
    Statement, 144–145
  Genetic, 348
  Gradient Ascent, 344
  Hidden Variables. *See* Missing Data
  Incorporating Background Knowledge, 93
  LCD2 Algorithm, 373–374
  Minimum Message Length, 349
  Missing Data, 339–345
    EM Algorithm, 344–345
    Gaussian Approximation, 341–343
    Laplace Approximation, 343
    MAG, 346–347
    MAP Approximation, 344–345, 479
    ML Approximation, 344–345
    Monte-Carlo, 340–341

Open Problems, 345–346
PAG, 347
Model Selection, 329
PC Algorithm, 84–88, 338, 375
  Applications, 97
  Discrete Distributions, 109
  Linear Structural Equation Models, 98
  Applied to Latent Variables, 261
  Complexity, 85–87
  Stability, 87–88
  Statement, 84–85
PC* Algorithm, 89–90
  Heuristics, 90
  Statement, 89
Probabilities of Error, 95–96
Selective Model Averaging, 329
SGS Algorithm, 82–83
  Complexity, 82
  Stability, 82–83
  Statement, 82
Simulated Annealing, 348
Sparse Candidate Algorithm, 349
Strategies, 73–77
Statistical Decisions, 93–95
Structural EM Algorithm, 349–350
Variable Selections, 91–93
Wedelin, 349
Sedransk, N., 226–234
Seidenfeld, T., 210, 226–234, 479
Selection Bias, 299
Semi-Directed Path, 9, 146
Sensitivity Analysis, 327–328, 366
Settimi, R., 302
Sewell, W., 111–112
SGS Algorithm, 82–83
  Complexity, 82
  Stability, 82–83
  Statement, 82
Shacter, R., 376
Shafer, G,. 295, 314, 376
Shah, V., 111–112

Sharma, S., 364
Shimkin, M., 239
Shipley, W., 352, 374
Simon, H., 45, 47, 376, 475
Simpson's Paradox, 38–42, 63, 217, 223
Simpson, C., 38, 40–41, 217, 475
Simulated Annealing, 348
Simultaneous Equation Model ,71
Singh, M. ,333
Sink, 8–9
Slate, E., 57, 477
Smith, A., 364, 479
Smith, J., 301
Smoking, 239–249
  Surgeon General's Report, 239, 243–249
Sober, E., 37, 296
Sorbom, D., 74, 76, 269, 347, 364
Sosa, E., 376
Source, 8–9, 12–13
Sparse Candidate Algorithm, 349
SP, 177–178
Spearman, C., 57, 121, 123, 155
Speed, T., 17, 97, 122
Spiegelhalter, D., 340, 365
Spirtes, P., 57–58, 82, 122, 154, 155, 297, 298, 300, 301, 306, 315, 323, 333, 339, 346, 347, 361, 362, 369, 370, 477, 478
St. John, C., 102–103
Statistical Indistinguishability, 59
  Faithful, 61
  Linear, 65–69
  Rigid, 64–65
  Strong, 60
  Weak, 62–64
Statistical Monotonicity, 312
Statistical Tests
  Degrees of Freedom, 95
  G^2, 95
  Vanishing Partial Correlation, 93–94
  Vanishing Tetrad Difference, 270
  X^2, 95

Stepwise Regression, 74
Stetzl, I., 71
Stob, T., 121
Strong Likelihood Equivalence, 345
Strong Statistical Indistinguishability, 60–61
Strotz, R., 299, 309, 312
Structural Class, 360
Structural EM Algorithm, 349–350
Structural Equation Models, 253, 361–367
  Bayesian Estimation, 364–367
  Measurement Model, 254
  Structural Model, 253
Studený, M., 376
Stutz, J., 343
Subgraph, 9–10
Summary Graph, 303
Subjunctives, 307–312
Suppes, P., 72, 124
Surgeon General's Report on Smoking and Health, 124, 239–249

Tail, 9
Tarski, A., 306, 307
Teel, J., 364
Tetrad Differences, 270
  Vanishing, 150, 256–260
  Variance of Sampling Distribution, 271
TETRAD II, 93, 96, 102, 108–109, 269, 368, 478, 479
  Search Procedure, 269–270
    Boldness, 289
    Implied-H, 271
    ImpliedH, 271
    Limitations, 293
    Reliability, 289
    Scoring Principles, 270–272
  T-maxscore, 272
  Tetrad-Score, 270–272
  Weight, 271
TETRAD III, 365

Tetrad Representation Theorem, 151–152, 260
  Statement, 151
Thiesson, B., 342, 344, 346, 479
Thomas, A., 365
Thomson, G., 155
Thurstone, L., 76, 122, 153
Tierney, L., 364
Timberlake, M., 200–201
Tooley, M., 376
Trek, 12–13, 67, 151
  Source, 12–13
Trek Sum, 67
Triangle, 10
Triplex, 304
Tukey, J., 157, 191
Type 1 Error, 204
Type 2 Error, 204

Underdetermination, 62
Underlying Graph, 15
Undirected Graph, 5, 8
Undirected Independence Graph, 15, 77
  Algorithm, 191
Undirected Path, 5, 8
Uniform Consistency, 322–325
Unmanipulated Graph, 50
Unshielded Collider, 10
Urbach, P., 209

Valtorta, M., 333
Vanishing Tetrad Differences, 150, 256–260
  Linear Implication, 151
Variable
  Determined, 53
  Deterministic, 54
  Endogenous, 25
  Error, 15
  Exogenous, 25, 50
  Functionally Determined, 54

Hidden, 299
Instrumental, 71
Latent, 399
Observed, 130
Policy, 50
Redefining, 69–71
Selection, 299, 329
Variable Aggregation, 91
Variable Redefinition, 69–71
Variable Selection, 221–222
Verma, T., 44, 58, 72, 90–91, 122, 147, 154, 178, 207, 210, 301, 306, 334, 477
Version Spaces, 78
Vertex
  Ancestor, 10
  Child, 8
  Children, 8
  Daughter, 8
  Degree, 8
  Descendant, 10
  Descendants, 11
  Indegree, 8
  Outdegree, 8
  Parent, 8
  Parents, 8
Vijayan, K. 17
Visit, 272

Waldemark, J. 368
Wallace, C., 297, 349
Wasserman, L., 315, 320, 323, 359, 479, 480
W.F.I. *See* Weak Faithful Indistinguishability
W.S.I. *See* Weak Statistical Indistinguishability
Weak Faithful Indistinguishability, 62
Weak Statistical Indistinguishability, 62–64
Wedelin, D., 349
Weinstein, S., 121
Weisberg, S., 203–207, 350, 374

Wermuth, N., 15, 17, 74, 79, 299, 303, 304, 305
Wermuth-Lauritzen Algorithm, 79–80
Wheaton, B., 277
Whittaker, J., 74, 79, 91, 152
Williams, K., 200–201
Wishart, J., 271
Winkler, R., 479
Wold, H., 299, 309, 312
Woodward, J., 297
Wright, S., 47, 57
Wynder, E., 239

Yoo, C., 332
York, J., 341
Yule, G., 32, 475
Yung, Y., 364

Zanotti, M., 72, 124

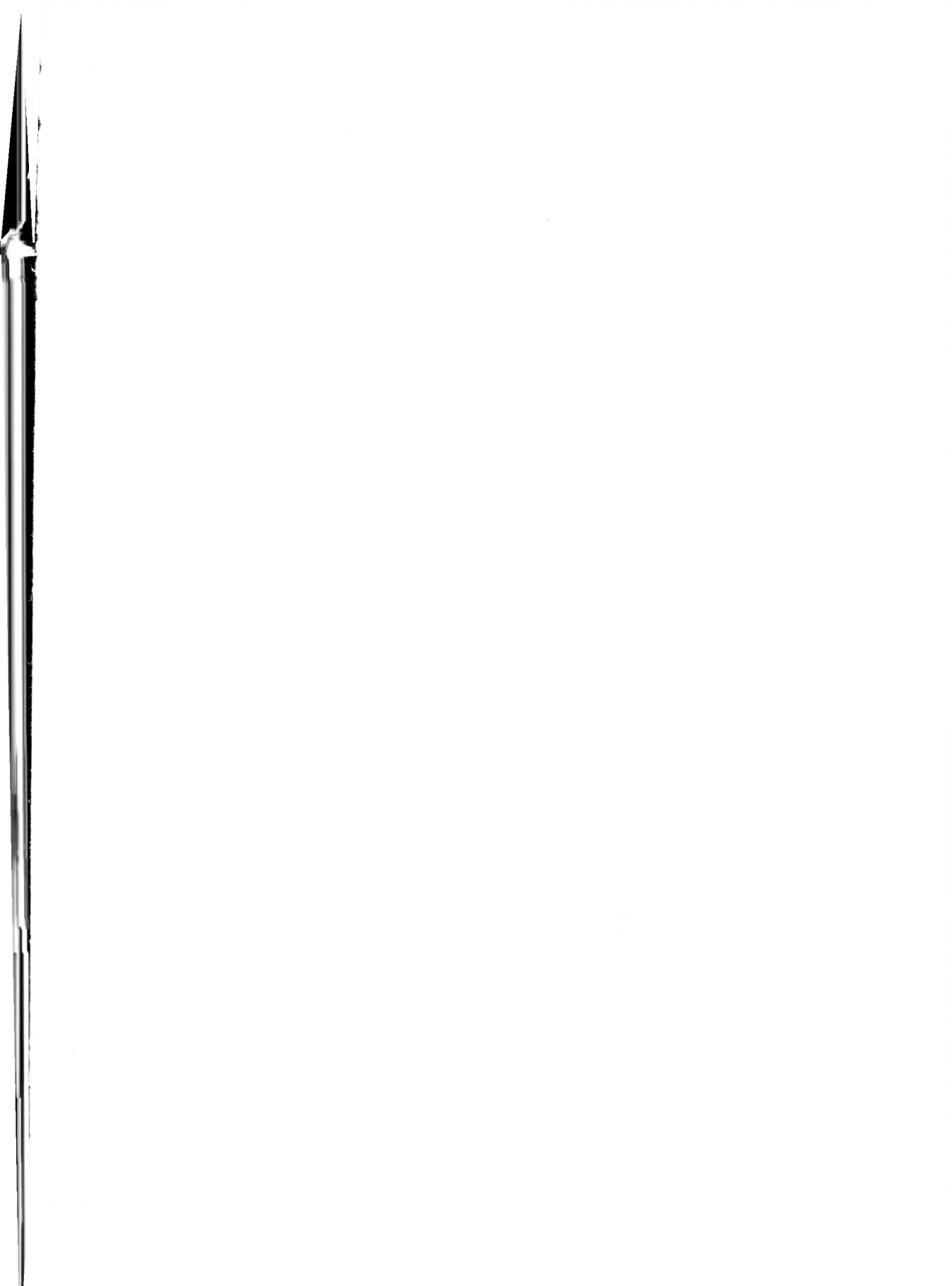